AF688305

# Vegetation of Australian Riverine Landscapes

## BIOLOGY, ECOLOGY AND MANAGEMENT

Editors: Samantha Capon, Cassandra James and Michael Reid

*For Christy*

# Vegetation of Australian Riverine Landscapes

## BIOLOGY, ECOLOGY AND MANAGEMENT

Editors: Samantha Capon, Cassandra James and Michael Reid

CSIRO
PUBLISHING

© CSIRO 2016

All rights reserved. Except under the conditions described in the *Australian Copyright Act 1968* and subsequent amendments, no part of this publication may be reproduced, stored in a retrieval system or transmitted in any form or by any means, electronic, mechanical, photocopying, recording, duplicating or otherwise, without the prior permission of the copyright owner. Contact CSIRO Publishing for all permission requests.

National Library of Australia Cataloguing-in-Publication entry

Vegetation of Australian riverine landscapes : biology,
ecology and management / Samantha Capon, Cassandra
James and Michael Reid (editors).

9780643096318 (paperback)
9780643104525 (epdf)
9780643104532 (epub)

Includes bibliographical references and index.

Vegetation management – Australia.
Ecosystem management – Australia.
Wetland management – Australia.
Wetland conservation – Australia.

Capon, Samantha, editor.
James, Cassandra, editor.
Reid, Michael Alister, 1966– editor.

333.9100994

Published by

CSIRO Publishing
36 Gardiner Road, Clayton VIC 3168
Private Bag 10, Clayton South VIC 3169
Australia

Telephone: [+613] 9545 8555
Email: csiropublishing@csiro.au
Website: www.publishing.csiro.au

Front cover: (top) Coolibah (*Eucalyptus coolabah*) growing on a floodplain of the Georgina River; (bottom) Kanuka Box (*Tristaniopsis exiliflora*) growing along Behana creek in the Wet Tropics. Source: Cassandra James.
Back cover: caption and acknowledgement to come

Set in 10.5/14 Palatino & Optima
Edited by Anna Cutler
Cover design by Andrew Weatherill
Typeset by Thomson Digital
Index by Bruce Gillespie
Printed by Ingram Lightning Source

CSIRO Publishing publishes and distributes scientific, technical and health science books and journals from Australia to a worldwide audience and conducts these activities autonomously from the research activities of the Commonwealth Scientific and Industrial Research Organisation (CSIRO). The views expressed in this publication are those of the author(s) and do not necessarily represent those of, and should not be attributed to, the publisher or CSIRO. The copyright owner shall not be liable for technical or other errors or omissions contained herein. The reader/user accepts all risks and responsibility for losses, damages, costs and other consequences resulting directly or indirectly from using this information.

Feb26_RP_ILS

# FOREWORD

Our appreciation of the vegetation of riverine landscapes starts from 40 000 years of indigenous knowledge passed on aurally and in pictures. Written records began less than 250 years ago with the 1804 diary of botanist Robert Brown, which alludes to the spatial and temporal dynamism of rivers and vegetation in the Hunter system through descriptions of floodplains, dry lagoons, river blockages and flood debris stranded 35 feet above the river level. Human changes, both intentional and unintentional, to catchments, basins and channels altered riparian landscapes and vegetation long before patterns and processes were described and understood. Over the past 50 years our scientific understanding of the structure and functioning of plant communities in our water landscapes has increased exponentially, building from individual studies on selected plants, rivers and catchments to a comprehensive coverage of the diversity of plant groups and riverine landscapes across the continent.

In 2016 we at last see a volume that synthesises our knowledge and management of riparian landscapes from the perspective of the vegetation. The functional importance and intrinsic values of these plants and ecosystems are captured by the three editors who have drawn together the perspectives of their 39 authors in 21 chapters to set the scene for research and management into the future.

*Labitur et labetur in omne volubilis aevum*

*Still on the river glides and on it will glide,*
*rolling its flood forever*

Horace, *Epistles*, 65–8BC

Dr Margaret Brock
Wetland botanist and ecologist,
Honorary Associate, University of Tasmania

# CONTENTS

# LIST OF CONTRIBUTORS

**Sophie Bickford**
Central Victorian Biolinks, Kyneton, Vic.

**Paul I. Boon**
Institute for Sustainability and Innovation, Victoria University, Melbourne, Vic.

**Samantha J. Capon**
Australian Rivers Institute, Griffith University, Nathan, Qld.

**Michelle T. Casanova**
Water Research Network, Faculty of Science & Technology, Federation University Australia, Mt Helen, Vic.

**Jane A. Catford**
School of BioSciences, The University of Melbourne, Parkville, Vic.; Fenner School of Environment and Society, The Australian National University, Canberra, ACT.

**Tim Cavagnaro**
School of Food, Wine and Agriculture, University of Adelaide, Adelaide, SA.

**Matthew J. Colloff**
CSIRO Ecosystem Sciences, Canberra, ACT.

**Ian H. Dixon**
Charles Darwin University, Casuarina, NT.

**Tanya M. Doody**
CSIRO Ecosystem Sciences, Adelaide, SA.

**Michael M. Douglas**
School of Earth and Environment, University of Western Australia, Perth, WA.

**John L. Dowe**
Australian Tropical Herbarium, James Cook University, Smithfield, Qld.

**George Ganf**
School of Biological Sciences, The University of Adelaide, Adelaide, SA.

**Peter A. Gell**
Water Research Network, Federation University Australia, Mt Helen, Vic.

**Amy K. George**
Department of the Environment, Environmental Research Institute of the Supervising Scientist, Darwin, NT.

**Cassandra S. James**
Centre for Tropical Water & Aquatic Ecosystem Research, James Cook University, Townsville, Qld.

**Kasper Johansen**
Remote Sensing Research Centre, School of Geography, Planning and Environmental Management, The University of Queensland, St Lucia, Qld.

**Christopher Jones**
School of BioSciences, The University of Melbourne, Parkville, Vic.

**David A. Keith**
Centre for Ecosystem Science, School of Biological, Earth and Environmental Sciences, UNSW Australia, Sydney, NSW.

**Christine E. Kenyon**
School of Geography, University of Melbourne, Parkville, Vic.

**Richard T. Kingsford**
Centre for Ecosystem Science, School of Biological, Earth and Environmental Sciences, UNSW Australia, Sydney, NSW.

**Jamie B. Kirkpatrick**
Discipline of Geography and Spatial Sciences, School of Land and Food, University of Tasmania, Hobart, Tas.

**Garreth Kyle**
Commonwealth Environment Research Facility (Landscape Logic), Arthur Rylah Institute for Environmental Research, Heidelberg, Vic.

**P. Sam Lake**
School of Biological Sciences, Monash University, Clayton, Vic.

**Stephen J. Mackay**
Water Monitoring, Department of Natural Resources and Mines; and Australian Rivers Institute, Townsville, Qld.

**Kaylene Morris**
Arthur Rylah Institute for Environmental Research, Department of Land, Water Environment and Planning, Heidelberg, Vic.

**Lucy C. Nairn**
Centre for Ecosystem Science, School of Biological, Earth and Environmental Sciences, UNSW Australia, Sydney, NSW.

**Melissa Parsons**
Riverine Landscapes Research Laboratory, Geography and Planning, University of New England, Armidale, NSW.

**Neil E. Pettit**
Centre of Excellence in Natural Resource Management, The University of Western Australia, Albany, WA; and School of Natural Sciences, Edith Cowan University, Joondalup, WA.

**John Porter**
Centre for Ecosystem Science, School of Biological, Earth and Environmental Sciences, UNSW Australia, Sydney, NSW; and Ecosystem Management Science, Science Division, Office of Environment and Heritage, Hurstville, NSW.

**Elisa Raulings**
Greening Australia, LaTrobe University, Bundoora, Vic.

**Paul Reich**
Arthur Rylah Institute for Environmental Research, Department of Environment, Land, Water and Planning, Heidelberg, Vic.

**Michael A. Reid**
Riverine Landscapes Research Laboratory, Geography and Planning, University of New England, Armidale, NSW.

**Jane Roberts**
Institute for Applied Ecology, University of Canberra, O'Connor, ACT.

**Samantha A. Setterfield**
School of Earth and Environment, University of Western Australia, Perth, WA.

**Nicholas J. Souter**
Fauna & Flora International, Cambodia Programme, Phnom Penh, Cambodia.

**Mark Southwell**
Riverine Landscapes Research Laboratory, Geography and Planning, University of New England, Armidale, NSW.

**Martin Thoms**
Riverine Landscapes Research Laboratory, Geography and Planning, University of New England, Armidale, NSW.

**Peter Vesk**
School of BioSciences, The University of Melbourne, Parkville, Vic.

**Laura Williams**
Department of Ecology, Evolution and Behavior, University of Minnesota, Saint Paul, MN, USA.

# 1

# Introduction

*Samantha Capon, Michael Reid and Cassandra James*

Of all the iconic landscapes in Australia, those associated with rivers, floodplains and their wetlands perhaps best capture both the highly dynamic nature and the enormous regional diversity of the continent. From the lush gallery rainforests of the wet tropics to the seasonally flooded savannas of the monsoonal north, the vast ephemeral herbfields of the desert channel country to the majestic river red gum forests along the Murray River and the extensive mangrove swamps of virtually all but the southernmost of mainland estuaries, vegetation is integral to the character of Australia's riverine ecosystems.

As elsewhere, rivers in Australia have played a central role in shaping the physical landscapes through which they flow, strongly influencing the development, over millennia, of the biological and human communities that they sustain. Vegetation in riverine systems both reflects these processes, through its composition and structure, and contributes significantly to a plethora of critical ecological functions which operate from local to catchment scales and beyond. Indeed, vegetation can be seen as the living architecture of many riverine ecosystems, which themselves can be perceived as keystones of the landscapes in which they are situated due to the disproportionate influence they exert. Riverine vegetation is particularly crucial in regulating many fundamental processes (e.g. water and nutrient cycling), providing essential habitat for terrestrial and aquatic fauna and supplying a wide range of highly valued economic and cultural goods and services, the benefits of which extend far beyond the area it occupies (Capon *et al.* 2013).

The functional significance of riverine vegetation has long been recognised in Australia. For thousands of years, indigenous Australians have relied on a wide diversity of riparian and wetland plants for food, shelter, wood, fibre, ornaments and medicinal ingredients. With the arrival of Europeans, riverine plants have also been exploited for timber, pastures and fodder as well as inspiring new cultural connections associated with their aesthetic and scientific values. Unfortunately, many activities associated with European agriculture, industry and human settlement have also led to considerable modification and degradation of riverine ecosystems and their vegetation as a result of clearing, grazing, altered water and fire regimes, salinisation, invasive species and so on. In recent decades, the destruction and visible deterioration of riparian vegetation (e.g. dying river red gums) has come to symbolise many of the environmental issues surrounding these threats, sparking major public and political concern. Among the scientific community, the declining condition and transformation of riparian and wetland vegetation has triggered alarm about the implications of these changes for aquatic ecosystem health, water quality and patterns of erosion and sedimentation in catchments as well as in coastal nearshore ecosystems such as the Great Barrier Reef. These concerns have led to substantial investment in protecting and restoring riparian, floodplain and wetland vegetation through the improved management of clearing, grazing and flows, as well as revegetation actions, to restore the functions of riverine ecosystems (Brooks and Lake 2007). Climate change makes such efforts even more important (Seavy *et al.* 2009; Capon *et al.* 2013).

The development in Australia of policy and management aimed at restoring the function of riparian, floodplain and wetland ecosystems has been accompanied by growing research interest in the ecology of riverine plants and vegetation. While much of this work has focused on the role of riparian vegetation in broader ecosystem function (e.g. effects on bank stability or aquatic food webs; Lovett and Price 2007), significant research has also sought to understand the patterns and processes that characterise riverine plants and vegetation communities themselves. This book aims to collect and consolidate the information gleaned from these latter studies in order to survey the current state of knowledge, inform those with interests in riverine systems, and guide management and future research. In the face of its tremendous functional significance, the intrinsic value of riverine vegetation is often overlooked. Consequently, this book also aims to celebrate the diversity of riverine plants and vegetation as well as increase appreciation of the mechanisms via which these are sustained rather than focusing on how riverine vegetation sustains other ecosystem components. The current volume builds on previous syntheses including those provided by Brock (1994), Specht (1990), Sainty and Jacobs (2003) and Capon and Dowe (2007), and complements recent related works on iconic taxa (e.g. Colloff 2014) and significant regions (e.g. Roberts and Marston 2011).

The book is divided into four sections: (1) riverine landscapes, (2) riverine plants, (3) riverine vegetation, and (4) conservation and management. Within each section, chapters have been selected that capture the diversity of landscapes, plants, vegetation communities and the issues facing them, as well as a range of different scientific approaches to understanding these. The first section provides context for the remainder of the book with respect to the spatial and temporal dimensions of riverine landscapes in Australia. In Chapter 2, Thoms, Parsons and Southwell provide a description of physical patterns and processes in Australian river habitats across a range of scales, from mesohabitats to catchments, with an emphasis on the interdependence of fluvial geomorphology and riverine vegetation ecology. In Chapter 3, Thoms and Parsons extend this synthesis to floodplains and their wetlands, highlighting the physical components that influence vegetation and presenting a framework for conceptualising the role of the physical floodplain template in structuring floodplain ecosystems. The long-term environmental history of Australian riverine landscapes and their vegetation is presented in Chapter 4. In particular, Reid, Bickford, Gell and Kenyon focus on the past 30 000 years to track these systems from the height of the most recent glacial period, around 22 000 years ago, through the transition to warmer conditions and sea level rise during the Holocene as well as responses to human activities, especially over the past 200 years.

Chapters in the second section examine key groups of riverine plants. Casanova and Nairn present a survey of riverine bryophytes and aquatic algae, including charophytes, in Chapter 5. These groups are often overlooked in ecological studies but are likely to be highly significant functionally and can also contribute greatly to plant species diversity, with a high degree of endemism evident compared with other parts of the riverine flora. In Chapter 6, Mackay and James discuss vascular aquatic macrophytes of riverine systems, lotic habitats in particular. The diversity, adaptations, distributions and dynamics of these ecosystem engineers are examined. In Chapter 7, Capon considers a more amorphous group, the riparian herbs, which are presented as a broad group encompassing all non-woody amphibious and terrestrial plants inhabiting mainly lentic riverine environments.

This chapter covers the diversity, both taxonomic and in terms of traits, and dynamics of this ubiquitous plant group which is present in virtually all riverine habitats. In Chapter 8, Capon, James and George survey the woody plants of Australia's riverine landscapes. The biogeography and traits of trees and shrubs at different life history stages are discussed as are woody vegetation dynamics in riverine forests, woodlands and shrublands.

The third section of the book provides an overview of riverine vegetation in five major regions of the Australian continent that broadly differ in their location and climate. Each chapter offers a description of major vegetation patterns and the processes sustaining these as well as a discussion of significant threats to vegetation and their management in relation to each region. In Chapter 9, Boon, Keith and Raulings present a comprehensive description of vegetation communities of coastal floodplains and wetlands from northern NSW to western Victoria including mangroves, coastal saltmarsh, freshwater and brackish meadows, swamps and other forested wetlands. Chapter 10 concerns the riverine vegetation in Australia that is probably both the most well studied and of the greatest management concern, that of inland south-eastern Australia, the Murray–Darling Basin. Roberts, Colloff and Doody focus on the woody component of the vegetation in this region and its adaptation to habitats characterised largely by differences in flood frequency and flow energy. In Chapter 11, Kirkpatrick provides a thorough overview of riverine vegetation in the treeless high country of New South Wales, Victoria and Tasmania, emphasising the role of vegetation in the functioning of these vulnerable ecosystems. In Chapter 12, Pettit, Dowe and Dixon consider riparian and floodplain vegetation across the tropical northern third of Australia including that of the wet tropics, the wet/dry tropical savannah and tropical floodplains. Finally, in Chapter 13, Capon, Porter and James discuss the plants and vegetation communities of riverine landscapes in Australia's arid and semi-arid inland.

The fourth and final section of this volume provides chapters on major issues associated with the conservation and management of plants and vegetation in Australian riverine habitats. In Chapter 14, Kingsford presents a brief synthesis of water management issues including the history of flow alteration and its effects on vegetation and approaches to environmental flow management. Ganf and Morris consider the closely related issue of salinisation in Chapter 15, which provides a thorough synthesis of the effects of salt on riverine plants along with an overview of salinisation trends in Australia and approaches to its management, including a case study on the Coorong. In Chapter 16, Douglas, Pettit and Setterfield discuss the effects on vegetation and management of fire in riverine landscapes with an emphasis on northern Australia. Jones and Vesk consider grazing in Chapter 17, presenting a history of grazing in Australian riverine habitats and its effects on vegetation along with an overview of management options. Alien plant invasions in Australian riparian zones are the topic of Chapter 18, in which Catford and Kyle examine the characteristics of riparian invaders and invasions as well as management strategies. In Chapter 19, Reich, Williams, Cavagnaro and Lake discuss restoration including its context in riparian ecosystems, approaches to and challenges associated with restoring riparian vegetation and recommendations for the future. Souter, Johansen and Reid review monitoring techniques in relation to riverine vegetation in Chapter 20, considering both on-ground and remotely sensed methods. Finally, in Chapter 21, James, Reid and Capon ponder the impacts of climate change, both past and present, on Australia's riverine vegetation and its management.

## NOTE FROM THE EDITORS

As editors, we hope this book inspires interest in the ecology of riverine plants and vegetation in Australia. In particular, we hope this volume fosters awareness of the incredible diversity and dynamic nature of riverine vegetation across Australia both for its own sake and for its vital functional role. We wish to thank all of the contributors to this book including the authors, reviewers and CSIRO Publishing for their effort and dedication to providing thorough and up to date chapters. We look forward to further collaborations in pursuit of greater knowledge with which to better understand and manage these essential components of our Australian landscapes.

## REFERENCES

Brock MA (1994) Aquatic vegetation of inland wetlands. In *Australian Vegetation*. 2nd edn. (Ed. RH Groves) pp. 437–466. Cambridge University Press, Cambridge, UK.

Brooks SS, Lake PS (2007) River restoration in Victoria, Australia: change is in the wind, and none too soon. *Restoration Ecology* **15**, 584–591. doi:10.1111/j.1526-100X.2007.00253.x

Capon SJ, Dowe JL (2007) Diversity and dynamics of riparian vegetation. In *Principles for Riparian Lands Management*. (Eds S Lovett and P Price) pp. 3–33. Land and Water, Canberra.

Capon SJ, Chambers LE, Mac Nally R, Naiman RJ, Davies P, Marshall N, Pittock J, Reid M, Capon T, Douglas M, Catford J, Baldwin DS, Stewardson M, Roberts J, Parsons M, Williams SE (2013) Riparian ecosystems in the 21st century: hotspots for climate change adaptation? *Ecosystems* **16**, 359–381. doi:10.1007/s10021-013-9656-1

Colloff M (2014) *Flooded Forest and Desert Creek: Ecology and History of the River Red Gum*. CSIRO Publishing, Melbourne.

Lovett S, Price P (Eds) (2007) *Principles for Riparian Lands Management*. Land and Water, Canberra.

Roberts J, Marston F (2011) *Water Regime for Wetland and Floodplain Plants: a Source Book for the Murray–Darling Basin*. National Water Commission, Canberra.

Sainty GR, Jacobs SW (2003) *Waterplants in Australia*. 4th edn. Sainty and Associates, Potts Point, NSW.

Seavy NE, Gardali T, Golet GH, Griggs FT, Howell CA, Kelsey R, Small SL, Viers JH, Weigana JF (2009) Why climate change makes riparian restoration more important than ever: recommendations for practice and research. *Ecological Research* **27**, 330–338. doi:10.3368/er.27.3.330

Specht RL (1990) Forested wetlands in Australia. In *Ecosystems of the World*. (Eds AE Lugo, MM Brinson and S Brown) pp. 387–406. Elsevier, Amsterdam.

# Section 1

Australia's riverine vegscapes in space and time

# 2

# The physical template of Australia's rivers

*Martin Thoms, Melissa Parsons and Mark Southwell*

## INTRODUCTION

The physical character of Australian rivers is as diverse as the country's people and their cultural origins. Many rivers meander slowly through the lowland interior of the country, never making their way to the sea, while others do so and often rush down steep rocky gorges or flow hidden beneath the ground in limestone caves. Some rivers only flow after prolonged rainfall has saturated desert surfaces and some flow all year round with little variation in water levels.

Rivers convey water and sediment downstream and, combined with the nature of the river valley, this governs the physical character of river systems. Physical differences between and along rivers demonstrate how catchment geology, climate and topography interact to govern the amount and rate of water and sediment supplied to a river, and how water and sediment supply control river form and process. As geology, climate and topography change with time and in space, the physical character of river systems changes in response. Rivers are therefore process-response systems. Given Australia's long geological history and its inherent climatic variability, the physical character of many Australian river systems is complex and does not always conform to accepted popular models, many of which are based largely on research conducted elsewhere in more predictable and less complex environments (Tooth and Nanson 1995; Thoms *et al*. 2004).

Plants usually exhibit adaptations to the conditions of their physical environment. Southwood (1977) proposed that habitat provides the template on which evolution acts to forge characteristic life history strategies of organisms. In rivers, hydrogeomorphic processes generate a template within which organisms (individuals, populations or communities) respond. The presence and movement of water and sediment, the drivers of the river template, constrained by catchment morphology, valley setting, climate and flows of different magnitude, frequency and duration, erode and deposit sediment within the river channel (Charlton 2008). This erosion and deposition generates the readily recognisable physical features of river channels such as banks, bars, pools, riffles and floodplains. Physical features are often associated with unique ecological communities because of the way that sediment and water creates suitable habitat for organisms (Parsons and Norris 1996; van Coller *et al*. 1997). The presence and action of organisms within the template also feeds back to influence the physical features of river channels, for example, through the formation of bars around instream vegetation (e.g. Edwards *et al*. 1999; Gurnell 2014). Thus, the interplay between hydrogeomorphic processes and organisms generates a dynamic and heterogeneous mosaic of physical features that constitute the river template.

There have been several recent calls for greater interdisciplinary effort in river science (Petts *et al*. 2006;

Dollar *et al.* 2007), driven largely by the deteriorating state of the world's riverine ecosystems. Information about the condition of rivers, the predicted consequences of anthropogenic activities and the location of areas of high conservation significance or severe degradation inputs into policy and legislative instruments that, in most cases, are mandated to achieve some notion of sustainability (Dovers and Hussey 2013). The physical template supporting biological assemblages is often poorly considered in assessments of river condition, which often focus largely on biological elements (Parsons *et al.* 2004). Understanding the distribution, dynamics and condition of organisms within the river template, however, requires an understanding of the form and function of the river template itself.

In this chapter we examine the factors that influence the form and function of the physical template of Australian rivers at different spatial and temporal scales and explore the diversity of Australia's riverine habitats. We argue that an understanding of scale is key to deciphering the physical river template because fluvial processes operating at different spatial and temporal scales influence the formation of different physical features. Thus, understanding the distribution, dynamics and condition of riverine vegetation requires a focus at the correct scale of physical template influence for the problem at hand.

## THE GEOLOGICAL AND CLIMATIC SETTING OF AUSTRALIA'S RIVER TEMPLATES

The diversity of river forms present in Australia reflects the patterns of geology and climate present across the continent. The influence of geology and geomorphology on this diversity is illustrated by the spatial organisation of distinct physiographic regions across Australia (Fig. 2.1). A physiographic region is a morphological unit that has a high degree of internal landform coherence (Jennings and Mabbutt 1986). They are discrete areas containing landforms of similar character in terms of underlying geology, regolith and soils. Physiographic regions can be subdivided into divisions, provinces and sections. There are three physiographic divisions of Australia (Fig. 2.1a): the western craton or the broad plateau in the western regions of the continent; the intra-continental basin or the interior plains; and the eastern orogen, represented by the south-eastern highlands and the escarpment of eastern Australia (Jennings and Mabbutt 1986). These divisions comprise 22 physiographic provinces (Fig. 2.1b), which are smaller areas distinguished on the basis of tectonic influences and/or longer-term exogenic processes such as fluvial, aeolian or littoral processes. Physiographic sections are the smallest landform area, averaging 39 000 km$^2$, and represent the finest level of landform delineation based on detailed knowledge of landform evolution and their modifications. There are 197 physiographic sections within Australia with between three and 29 sections within the various physiographic provinces. This classification of landforms, highlighting the action of underlying controls of geology and climate, provides a basis for an understanding of the physical character of riverine landscapes at different scales.

Australia is a dry continent because of its size, latitudinal position and lack of topographic relief. Mean annual precipitation for the continent is 420 mm, of which 50 mm is converted to runoff and the remaining 370 mm is lost through evapotranspiration processes. These statistics are low compared with world averages of 660 mm for precipitation, 250 mm of runoff and 410 mm for evapotranspiration (Thoms 1994). Rainfall is the main form of precipitation in Australia and most of this occurs along a relatively narrow coastal perimeter in the north, east and south-east of the country. Most of the continent (~75%) has a semi-arid to arid climate with an annual rainfall of less than 406 mm and evaporation in excess of 2000 mm. Overall, Australian rainfall is influenced by three factors. First, the majority of Australia's landmass is located between 15 and 35 degrees south of the equator. This is a latitudinal zone dominated by sub-tropical anticyclones that are associated with stable climatic conditions and low moisture. Second, Australia is relatively flat. Only 7% of the landmass is more than 600 m above

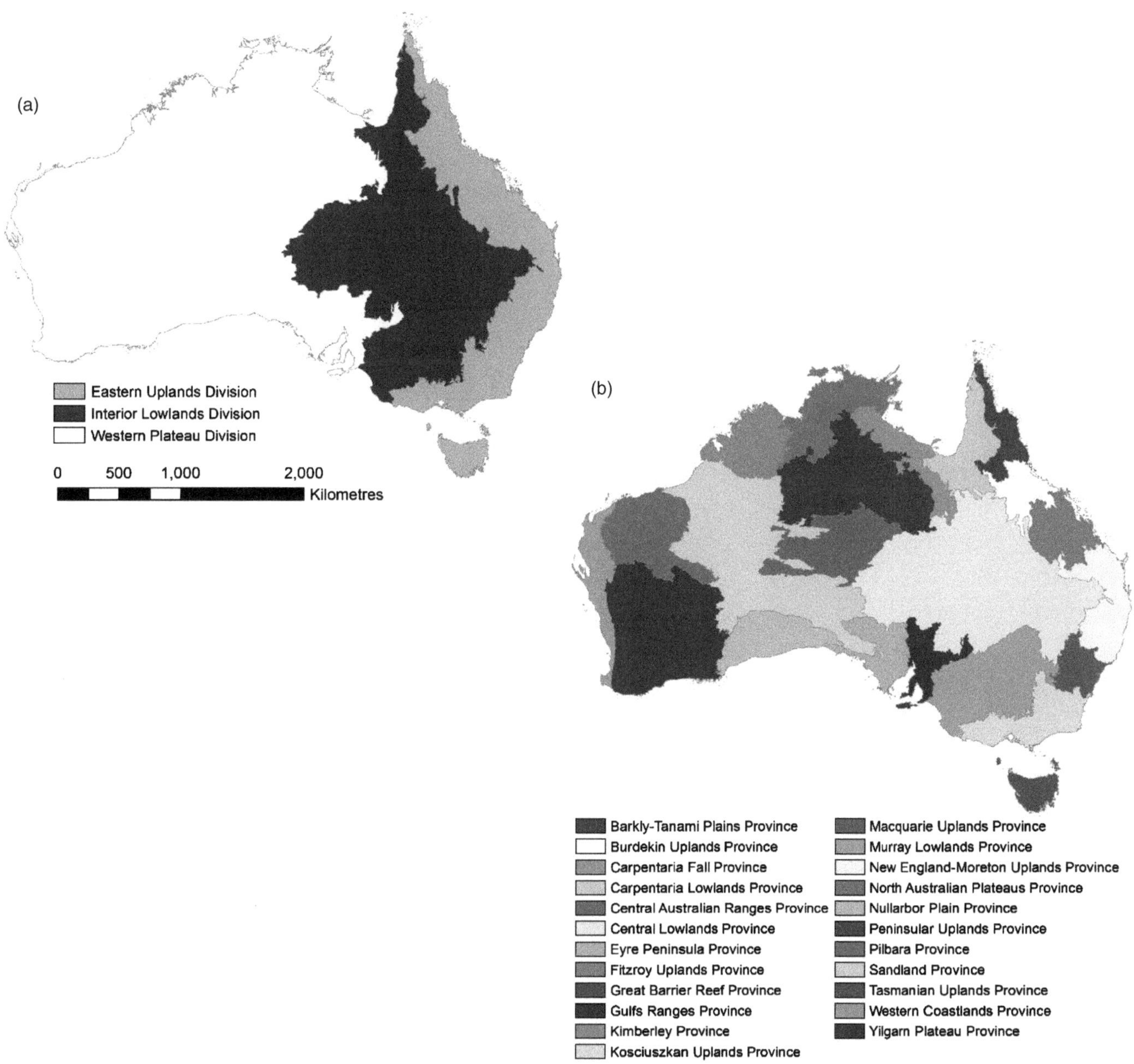


**Figure 2.1:** The physiographic regions of Australia showing (a) physiographic divisions and (b) physiographic provinces. After Jennings and Mabbutt (1986).

sea level and there are a limited number of mountain ranges to serve as climatic barriers to induce orographic rainfall. Monsoonal activity does produce well-defined summer maxima in northern Australia, whereas the passage of low-pressure systems in the southern regions produce marked winter maxima in the south-east and south-west of the country. Third, the occurrence of rainfall is highly variable.

Australia has experienced marked climatic variations over the past 130 000 years. The legacy of these variations to more contemporary river systems is well documented (e.g. Ollier *et al.* 1988; Clarke 1994). For example, three phases of channel development have been identified in the Goulburn and Murray river systems (Bowler 1986). The Tallygaroopna and Kotupna phases both preserve evidence of river channels much greater than those

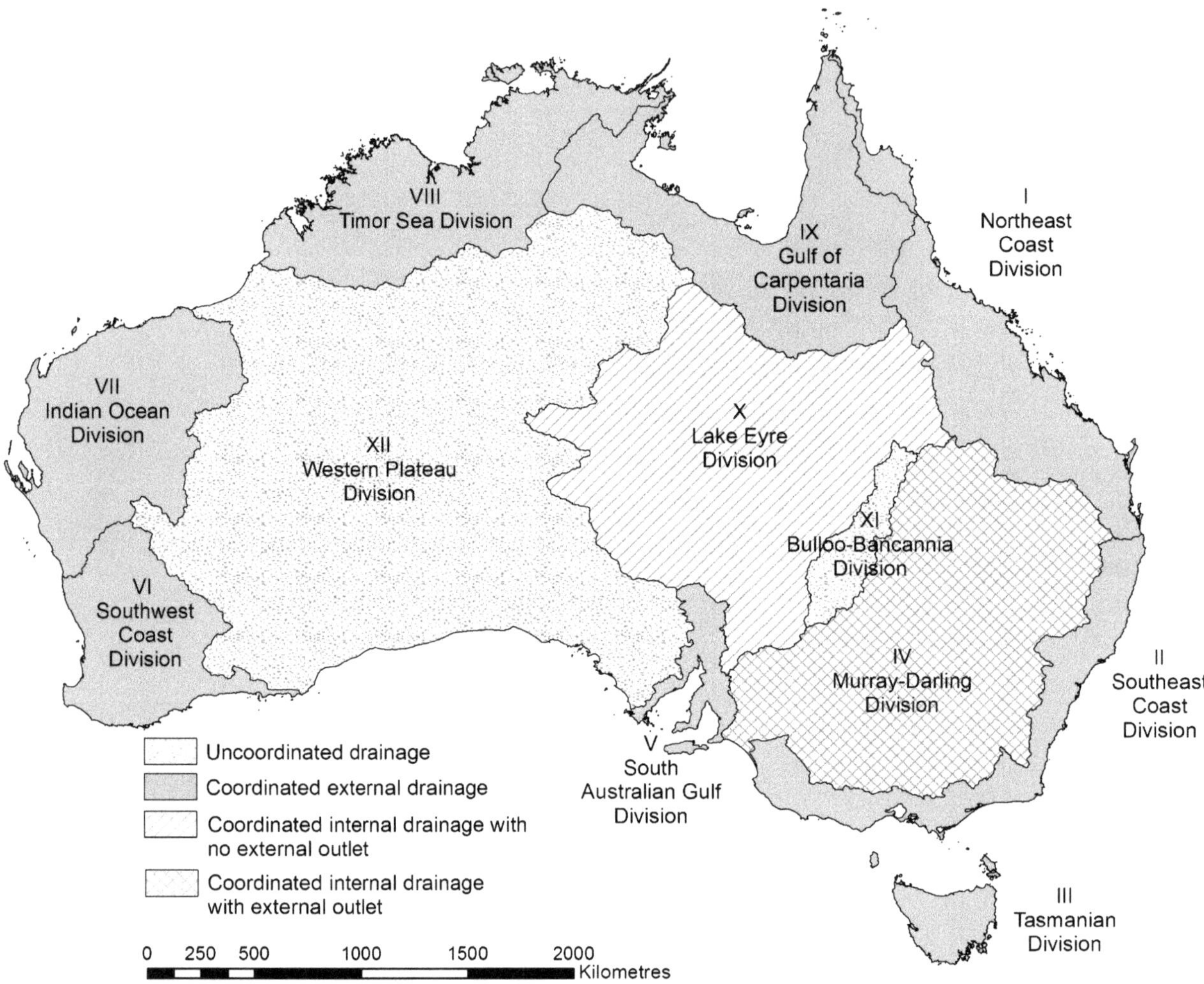


**Figure 2.2:** Drainage divisions of Australia.

of the present climatic regime. The modern expression of the Murray River, in the riverine plains, is controlled to some degree by the presence of these prior systems. Studies of palaeo-drainage patterns in various regions throughout Australia all document past climates that are significantly different to contemporary regimes and thus the high degree of temporal variability in rainfall and runoff. In addition, analyses of long-term records show distinct changes in rainfall patterns since records began in the late 1700s. Distinct alternating periods, of various lengths, of above or below average annual rainfall, have been documented for many parts of the country. For example, Pickup (1976), using data for Meangle, west of Sydney, showed a sequence of 'wet' and 'dry' periods. Severe droughts occurred in the early 1900s and the late 1930s, while the 1880s and 1890s were significantly wetter periods. The early 1950s and 1960s were periods of extreme wet for the region. The implications of these alternating wet and dry periods on river channel morphology have been well documented by Erskine and Warner (1988), among others.

Australia has 12 drainage divisions (Fig. 2.2), each with a distinct hydrology in terms of rainfall and runoff. Seven of these divisions discharge water directly to the sea. Tasmania comprises a single division. The Murray–Darling division drains most of the western side of the south-east highlands and reaches the sea after crossing an arid interior. It is the only division with a single

**Table 2.1.** Surface runoff within Australia's 12 drainage divisions
See Fig. 2.2 for the locations of the Australian drainage divisions.

| Drainage division | | Area (km$^2$) | Mean annual runoff | |
|---|---|---|---|---|
| | | | Mega litres^10 | mm |
| I | North-east coast | 451 000 | 91.5 | 203 |
| II | South-east coast | 274 000 | 45.4 | 166 |
| III | Tasmanian | 68 00 | 53.4 | 785 |
| IV | Murray–Darling | 1 063 000 | 22.7 | 21 |
| V | South Australian gulf | 82 000 | 1.0 | 12 |
| VI | South-west coast | 314 000 | 6.7 | 21 |
| VII | Indian Ocean | 519 000 | 4.0 | 7.5 |
| VIII | Timor Sea | 547 000 | 81.2 | 148 |
| IX | Gulf of Carpentaria | 648 000 | 130.5 | 205 |
| X | Lake Eyre | 1 170 000 | 3.3 | 2.8 |
| XI | Bulloo-Bancannia | 101 000 | 0.6 | 5.7 |
| XII | Western Plateau | 2 455 000 | 0 | 0 |
| Australia | | 7 682 000 | 440.3 | 57 |

outlet to the sea. There are two internal divisions, Lake Eyre and Bulloo-Bancannia, both of which discharge into terminal lake systems. The Western Plateau is the largest division, covering an area about one-third of the Australian landmass. Runoff within this division is so small that coordinated river networks do not generally occur. Additionally, surface runoff only occurs in this region for short periods of time during high intensity rainfall events and this runoff does not travel far because of high rates of infiltration and evaporation.

Surface runoff is highly variable between and within each of the divisions, reflecting inherent climatic variations (Thoms 1994). Mean annual runoff across the continent varies from zero runoff for most of inland Australia to over 700 mm in Tasmania. Overall, runoff from the Australian continent is dominated by three divisions: the Timor Sea, Gulf of Carpentaria and north-east coast divisions, which produce 75% of the total runoff (Table 2.1). The greatest depth of runoff per unit area occurs in Tasmania, with 785 mm/yr, and this is the only division that has a relatively uniform spatial distribution of runoff. Average runoff patterns for each division are variable. Divisions I, II, III, VIII and IX produce 92% of all external runoff, i.e. one-third of the continent yields 41 300 million m$^3$/yr of the total water (Table 2.1). These divisions are dominated by either summer rainfall or uniform annual rain patterns. The remaining two-thirds of the continent yields only 5.3% of the external drainage and most of this is through the mouth of the Murray River. Annual variations in discharge are also high for Australian river systems and the index of flow variability (the ratio of the maximum to minimum flow) for Australian rivers ranges from 300 to >3000, which is extremely high compared with Europe (3–10) and the USA (3–15).

The variability of climate and geology are important in moulding and sustaining Australia's rivers and give rise to a hydro-geomorphic template that is not uniform in space and time. As a result, a scaled approach is needed to encompass the variation in the physical template of Australia's rivers.

## THE HIERARCHICAL NATURE OF RIVER TEMPLATES

### Hierarchical organisation of river networks

Hierarchy theory was derived from general systems theory and has been applied in ecology as a conceptual framework for the examination of ecological phenomena at multiple spatial and temporal scales (Allen and Starr 1982; O'Neill *et al.* 1986). In essence, a hierarchy can be viewed as a series of organisational levels, or holons, that are constrained within a nested vertical structure (O'Neill *et al.* 1986). The boundary of a level is generally identified on the basis of functional process rates (O'Neill and King 1998) or structural spatial criteria (O'Neill *et al.* 1986; Bergkamp 1995). There are three main properties that govern the exchange of information between levels within a hierarchy (Bergkamp 1995). First, specific levels of organisation are linked to specific spatial and temporal scales, where higher levels correspond to larger spatial scales and longer temporal scales, and lower levels correspond to smaller spatial scales and shorter temporal scales. Second, rate differences of at least one order of magnitude exist between different levels, so that higher levels have lower frequencies of behaviour than lower levels. Third, higher levels constrain lower levels through their larger time constraints.

Hierarchical principles are the foundation of many aspects of fluvial geomorphology. There are many interrelated hydro-geomorphic factors that operate within a river system, and these factors sit within a cascade of hierarchical influence (Schumm 1988; de Boer 1992). Geology and climate are considered to be independent catchment factors because they directly or indirectly control the formation of all other factors in the hierarchy (Schumm and Lichty 1965; Knighton 1984; Montgomery 1999). At the top of the hierarchy, climate and geology act to control physiography of the catchment, the types of vegetation and soils that are present in a catchment and the ways humans use the land. In turn, catchment physiography, vegetation, soils and land use control a set of independent channel factors such as stream discharge, sediment load, sediment transport, sediment calibre, bank characteristics and valley slope (Knighton 1984). These channel-forming factors subsequently control a set of dependent channel geometry and flow factors that occur at the bottom of the hierarchy, such as channel slope, water velocity, channel width and depth, sinuosity, meander wavelength and the arrangement of bedforms in the channel (Knighton 1984). Thus, in a fluvial system, geomorphological factors and processes operating at one level of the hierarchy constrain the formation of factors at successively lower levels (Schumm 1991).

Overlaid across this hierarchy of interrelated hydro-geomorphic factors is a temporal context that determines the strength of constraint between successive levels, as outlined in the classic time-space causality paper of Schumm and Lichty (1965). The relationship between factors at different hierarchical levels is a function of the persistence of each factor over time. In general, the strength of influence between factors sitting at the top and bottom of the geomorphological hierarchy decreases with an increase in time span. The independent catchment factors, climate and geology, sit at the top of the hierarchy because over a geological time span of millions of years they constrain the formation of factors such as catchment physiography, vegetation, soils and valley dimensions. However, at this time span, factors that sit at the intermediate and lower levels within the cascade, such as sediment discharge, water discharge, channel morphology and flow character are not influenced by any of the higher level factors. Over a time span of thousands of years, channel morphology becomes the only factor that is influenced by factors operating at higher levels of the cascade, and lower level factors such as sediment discharge, water discharge and flow character are unresponsive. This pattern of constraint continues until, at a time span of one year, sediment discharge, water discharge and flow character are the only factors that are influenced by higher level factors on a yearly basis. Thus, interrelationships between any of the geomorphological factors in a river system are directly linked to the evolution and behaviour of the system at different temporal scales.

## Hierarchical river characterisation

River characterisation involves the ordering of sets of observations or characteristics into meaningful groups based on their similarities or differences.

Implicit in this exercise is the assumption that relatively distinct boundaries exist and that these may be identified by a discrete set of variables. Although river systems are continuously evolving and often display complexity, the grouping of a set of elements with a definable structure can aid in relating broader physical factors. River characterisation can also contribute to understanding why rivers have certain biological characteristics.

In a review of the geomorphic classification of river systems, Kondolf *et al.* (2003) recognised five different broad characterisation groups. The first group is simply termed 'early characterisation schemes'. These are mainly based on the generic relation of rivers to geological structure and the evolution of landscapes. The geographic cycle of Davis (1899) is a good example, as it was derived from evolutionary theory in which rivers were divided into three stages of an evolutionary cycle: young, youthful and mature. Process-based characterisations form the second group, and there are numerous examples of these. The work of Leopold and Wolman (1957) provides a quantitative basis for differentiating braided, meandering and straight channels based on relationships between slope and discharge. The third group is based on stream power approaches, with stream power being a key variable in shaping river channels. In these approaches there is a focus on identifying discontinuities in specific stream power as a means to delineate between reaches. Many river characterisations have been devised explicitly for river management, the fourth group of river characterisation approaches, and are focused on the specific needs of the management context.

The fifth group of characterisation approaches is composed of those employing hierarchical methods and examples include Frissell *et al.* (1986), van Niekerk *et al.* (1995), Montgomery and Buffington (1998) and Thoms *et al.* (2004). This group recognises that broadscale parameters determine the range of behaviour of physical processes in smaller scale units. There is an interaction between physical units of a river system at different scales, and this determines the character and behaviour of river systems (Naiman *et al.* 1992). Using a hierarchical approach to characterisation, river systems can be examined across a range of scales and each level of the river system hierarchy will have a different degree of sensitivity and recovery time to disturbances such as floods or droughts. Thus, a hierarchical approach not only defines the structural components of a river but also recognises the relative importance of factors controlling the long- and short-term behaviour of river (habitat) changes within a spatial scale. The capacity of hierarchical characterisations to reflect the spatial and temporal nature of fluvial geomorphic processes makes them ideal for understanding highly variable systems like Australia's rivers.

## THE HIERARCHICAL RIVER TEMPLATE

In any study of riverine vegetation in relation to its physical template, consideration must be given to the hierarchical situation of relevant geomorphic features and processes, so that appropriate vegetation attributes can be examined. Regardless of the levels included in any hierarchical river characterisation, each level is associated with characteristic geomorphic features and processes. These fluvial geomorphic features and processes form the physical template for the structure and function of riverine vegetation. Therefore the physical template forms as a result of geomorphic processes and so conveniently sits between the physical forces that structure river systems and the biological communities that inhabit them (Harper and Everard 1998).

### Hierarchical organisation of river systems

A seven-tier hierarchical framework for characterising river systems has been developed by Thoms *et al.* (2004). This framework considers the catchment to be the primary unit for investigating river systems and nested within this are the riverine landscape, functional processes zones, river reaches, functional sets, functional units and mesohabitats. A description of each and their spatial scale are provided below.

### The catchment

This is the primary unit for investigating river systems. Its structure is a result of geological processes and climatic influences operating at times scales over $10^5$ to $10^6$ years. There are 245 catchments draining the Australian continental landmass and these are contained within 77 hydrological regions, which are located in the 12 major Australian drainage divisions (BOM 2012). Although the majority of these catchments (90%) flow directly to the sea they only drain ~22% of the continental landmass.

### The riverine landscape

Nested within catchments are riverine landscapes. Located in catchment valleys, they are composed of two components: the riverscape and floodscape (*sensu* Thorp *et al.* 2008). The riverscape contains the active river channel, its riparian zone, and those marginal areas of the active channel that interact frequently with it and may include features such as side channels or anabranches. The floodscape is that area of the riverine landscape formed as a result of alluvial processes and contains active and inactive floodplain surfaces and can be defined in terms of its character of inundation. It does contain an array of predominantly terrestrial to aquatic ecosystems as well as flora and fauna associated with it. It has long been recognised that there is a strong relationship between a river and its valley (Hynes 1975). Valleys exert lateral and vertical control over the channel and, thus, riverine landscapes may be confined by narrow valleys, or flow unconstrained across broad floodplains (Church 2002). Within confined valleys the river channel is relatively constrained and at high flows, river depth and velocity increase rapidly with increasing discharge. In unconfined valleys the river channel has a floodplain and at high flows the river spreads across the broad valley floor, dissipating much of the energy of the current (Gregory 2006). These differences in stream energy potential between valley types have important implications for sediment transport (Bisson and Montgomery 1996) and the hydrologic regime.

There are marked differences in the riverine landscape of Australian rivers, especially those that drain the coastal perimeter of the country compared with those that are endorheic (i.e. have no outlet to the sea) or drain inland then to the sea. The majority of Australia's coastal catchments have limited floodscapes and are characterised by constrained floodplains extending up to 5 km wide. In comparison, Australia's inland rivers are dominated by vast unconstrained floodplain landscapes that can extend up to 80 km wide (see Chapter 3).

### Functional process zone

Functional process zones are lengths of the river system that have similar discharge and sediment regimes contained within similar river valley settings. Channel pattern is an indicator of differences between functional process zones; however, functional process zones also have contrasting flow and sediment regimes. The scale of functional process zones reflects the size of the river system being studied and they can vary from several kilometres to over 500 km (cf. Thoms *et al.* 2004).

### River reach

River reaches are repeatable lengths of river channel within a functional process zone and therefore have a similar river channel style. Typically, they are based on river channel planform or bedform character, or the character of the riverbed sediment. Commonly, a reach is delineated as several meander bends or riffle-pool sequences. In many biomonitoring approaches, for example, a reach is defined as a riffle-pool sequence, while fisheries ecology many delineate a river reach as several meander bends. The reach scale generally describes river planform, which in turn is set by fluvial characteristics such as discharge character (Cullum *et al.* 2008), depositional or erosional sediment transport mode (Montgomery and Buffington 1998) and morphological channel dimensions (Cullum *et al.* 2008).

### Functional set

Each river reach may be divided into functional sets of morphological units associated with specific landforms within the riverscape or floodscape. Typical landforms include side channels or anabranches, wetlands located within the floodscape, cut-offs and the main active channel. Riffle-pool systems are typical functional sets within the active channel. The character of each functional set within the riverine landscape is determined by the magnitude, frequency and duration of longitudinal, lateral and vertical fluxes, which are related to the geomorphology of the each landform. Petts and Amoros (1996) suggest the appropriate time scale for the formation of functional sets is between 10 and 1000 years.

### Functional unit

A functional unit is indicative of the physical conditions at a smaller scale. They commonly represent habitats associated with animal and plant communities present at a site and commonly range in size from 1 to 10 $m^2$. Essentially all functional units within a functional set evolve from a single origin by progressive changes over different time periods (Petts and Amoros 1996). Functional units typically include individual pools, riffles, bars and riverbanks.

### Mesohabitat

Mesohabitat subsystems are patches that have relatively homogeneous substrate type, water depth and velocity (Frissell *et al.* 1986). These structures are sensitive to variations in flow, sediment and nutrient fluxes and often change in character from year to year as a result. Common mesohabitats include sand and gravel bars, in-channel benches, scour holes, gravel patches, undercut riverbanks, and other smaller features such as emergent and submerged vegetation, submerged wood and other substrates.

## Example of the river template in Australia: the Murray–Darling Basin

The Murray–Darling Basin comprises 21 subcatchments. These catchments have distinct geologies and climates. From east to west, catchment geology changes from the bedrock-dominated eastern uplands, which border the basin and provide areas of maximum relief in the basin, to the vast sedimentary basin of the mid to western interior. This broad regional pattern is interrupted by a regional outcrop of old folded synclinal rock that extends from the eastern uplands into the interior sedimentary basin near Cobar. The north-west region of the basin has a complex geological pattern consisting of various geologies that form a series of low elevation ranges. There are three main climatic zones within the Murray–Darling Basin, which correspond with the broad physiographic provinces of the basin. Climatic zone A corresponds closely with the eastern uplands, climatic zone B with the broad interior lowlands of the Murray Basin, and climatic zone C with the Darling Basin (Thoms *et al.* 2004). In general, there is an east–west gradient (latitudinal) that reflect gradients of increasing temperature, lower rainfall and increased evaporation. There is also a north–south gradient of decreasing solar radiation and summer rainfall. There are some exceptions to these broad generalisations because of orographic influences associated with the mountain ranges on the south-western, southern and northern rims of the basin.

There are two main riverine landscape types in the Murray–Darling Basin associated with the degree of valley confinement and catchment elevation. In the upland regions of the basin, valley widths relative to the size of the river channels are low, resulting in highly confined river floodplain systems. In some areas, there is no floodplain development, with the river channel occupying the entire valley floor. However, with a widening of the valley floor, sediment deposits occur with the formation of terrace and contemporary floodplains. Through most of the lowland regions in the basin, rivers flow through unconfined valleys with floodplain widths often exceeding 50 km.

Eight functional process zones have been identified in the Murray–Darling Basin (Thoms *et al.* 2004). A description of each functional process zone and the nested river reach, functional set, functional unit and main mesohabitat within each is provided below. The eight functional process zones have distinct valley–river channel associations in terms of the degree of valley confinement, slope, stream energy, boundary sediment and flow and sediment transport regimes. Moreover, the individual zones have different physical river channel character as reflected in the presence and structure of instream habitat and associated biological communities.

## Pool zone

Pool zones are characterised by long pools separated by short channel constrictions. The pools form upstream of channel constrictions and are the dominant morphological feature in the zones. Channel constrictions are generally associated with major bedrock bars that extend across the channel or substantial localised gravel deposits that act as riffle areas. Local riverbed slopes increase significantly at these constrictions; they therefore are small areas of relatively high energy in contrast to the relatively low bed slopes and energies of the pool environment.

The planform or channel configuration of pool reaches is controlled by valley morphology. Generally, the river channel may have a small flanking floodplain because of the narrow valley floor configuration. Hence, valley conditions limit floodplain development. Bankfull channels can have dimensions of up to 30 m wide and 3–4 m deep with width-to-depth ratios of 10. The nature of channel sediment or substratum in these reaches consists of fine silt or clay material overlying a bedrock or cobble base in the pools; however, gravel, cobble or bedrock dominates the short constricted riffle areas. Bankfull flows can mobilise the finer bed substratum, but significantly higher discharges are required to set coarser material in motion.

The functional set habitat is the pool section itself. The main functional units in the pool zones are the riffle or chute areas and the large pool areas. Riffles or chutes provide relatively fast-flowing, shallow, turbulent water. Water in pool regions is deeper and slow-flowing. The major mesohabitats are associated with the various substratum types, namely, cobble, gravel and sand. Mesohabitats appear to be somewhat more diverse within the pool areas and have areas of varying substratum combined with emergent and submerged aquatic vegetation.

## Constrained zone

Constrained zones are commonly represented by gorges, those sections of river that have relatively high energy, are dominated by steep bed slopes often greater than 0.010 and where bankfull channel stream powers may exceed 400 $\omega m^2$. Bedrock chutes, large accumulations of boulders or cobbles, and scour pools often dominate the in-channel environment. The boulder materials are relatively immobile; however, extreme flood flows can move the cobble material, producing well-sorted deposits. There are no or very few floodplains in the constrained zones of the Murray–Darling Basin, therefore sediments are added directly to the channel from adjacent valley slopes. Moreover, because these are high-energy environments, there is a lack of major sedimentary deposits. Channel planforms in the constrained zones are controlled by the structure of the valley.

In this zone, the functional set is the constrained section itself. There are few functional units in the constrained zone of the river system. The main channel is the dominant unit with perhaps some differentiation of riffle areas and pool areas within the channel, but the areas are not as distinct as in the pool zone. Large boulder and cobble accumulations dominate the in-channel environment, and some provide habitat for in-channel vegetation. An array of cobble and gravel accumulations generally provides a complex suite of mesohabitats or substratum habitats. However, stands of riparian vegetation also provide complex habitat, both in themselves and because of their associated fallen timber.

## Armoured zone

Armoured zones are also high energy zones with bankfull stream powers often between 250 and 400 $\omega m^2$ with corresponding bed slopes between 0.01 and 0.002. The riverbed sediments are typically 'armoured', having a surface layer of sediment that is significantly coarser than the sub-surface sediment. This coarse surface layer protects the finer materials underneath, and these sediments are not mobilised by the flow until the armoured layer is disturbed. Armoured zones can have a series of floodplains of different ages inset into valley flats and often resemble a series of broad steps up and away from the river channel. These zones are active sediment transfer zones with little net accumulation of sediment. The river channel is relatively constrained, being controlled by the configuration of the valley but the presence and extent of floodplains determine the degree of control of the valley on its configuration. The river channel often has a meandering pattern that is superimposed on a larger valley pattern. This channel pattern is characteristic of a bed load or mixed load channel and has large bed slopes, lower channel sinuosities, and larger meander arcs and wavelengths. The in-channel environment is dominated by cobble and gravel-sized sediments that are extensively armoured and relatively stable.

The armoured zone of a river marks the point in the downstream array of functional process zones where the functional sets and functional units can begin to be divided into those occurring in the low-flow channel and those in a high-flow channel. In the low-flow section of the armoured zone, common functional sets are riffle-pool sequences, some of which can support well-established stands of in-channel vegetation. Pools are generally large and deep and provide a substantial refuge area for fish and other aquatic organisms during floods. In this zone the high-flow channel is present but often not well developed. Functional units in this region of the channel include flat surfaces or benches within the incised channel and small flood runners. Mesohabitats within the low-flow channel consist of the various accumulations of cobble, gravel and sand-sized particles within the riffles. These often provide an array of complex substratum habitats. In-channel vegetation and snags also provide complex habitat. Large debris dams are often associated with the snags and these are rich in organic matter. Within the pool unit of the low-flow channel, mesohabitats include emergent and submerged vegetation along with some woody debris.

## Mobile zone

The mobile zone is an area characterised by a relatively active river channel with mobile sediment in the riverbed and large sediment storage areas. The presence of well-developed inset floodplain features such as benches, point bar systems, cut-offs and levees testify to the relatively active and unrestricted nature of this river–floodplain environment. The valley floor is generally wider in this zone, therefore allowing floodplain development to occur. Often the river channel freely meanders with irregular planform. Characteristics of the mobile zone are the increased meander wavelengths (up to 2 km) and meander arcs in comparison to an armoured reach. Stream powers may range from 8 to 20 $\omega m^2$.

The morphology of the in-channel environment is extremely variable; bars (point and lateral), benches (at various elevations) and riffle-pool sequences all may be present. These in-channel storage features reflect high rates of sediment transport. Riverbed sediments typically have a bimodal distribution (median grain size of 64–100 mm) with a range of sediment sorting that is highly mobile (relative bed stability less than 1).

The mobile zone is probably the most complex in terms of functional unit development, with distinct and diverse low- and high-flow channels. Within the low-flow section, functional units are again the riffle and pool areas within the main channel. Generally, the riffle sections are large and the pools large and deep, which are often associated with well-established stands of riparian vegetation, as result of relatively stable sediments, relatively high nutrient status and good moisture conditions. The low-flow channel is also characterised by large sandy point bars. In the mobile zone the high-flow channel is normally well-developed with in-channel benches, diverse flood runners and an extensive floodplain. Functional units in the high-flow channel include flat bench surfaces, small flood runners and complex features of the floodplain itself. Mesohabitats within the low-flow channel again depend on substratum composition with accumulations of cobble, gravel and sand within the riffles providing complex substratum habitats. Stands of riparian vegetation also provide complex habitat, as do their associated fallen timber, which, as in the armoured zone, creates large debris dams. Within the pool functional unit of the low-flow channel, microhabitats include emergent vegetation with submerged vegetation not as abundant in these reaches owing to the overall depth of the channel and relative instability of the sediments in this unit. Woody debris is also present within pools and frequently there is limited riparian vegetation present. Mesohabitats within the high-flow channel also depend on the terrestrial environment inundated during floods; however, snags and fallen woody debris form a major microhabitat in this region of the channel.

### Meander zone

A distinguishing feature of the meander zone is the significant increase in the width of the valley floor and the corresponding increase in the area of floodplain surfaces. The river channel in the meandering zone is relatively active and displays a typically meandering style, with sinuosities ranging from 1.8 to 2.35 and meander wavelengths between 200 and 700 m. The presence of well-developed floodplain features, such as flood channels, former channels, avulsions, cut-offs and minor anabranching, attribute to the relatively active and unrestricted nature of the river–floodplain environment in this zone. The river in this zone is typical of a mixed to 'wash load' channel.

The morphology of the in-channel environment is complex with a relatively diverse set of physical features, such as bars, benches and riffle-pool sequences. These in-channel features are important sediment storage areas and reflect the relatively high rates of sediment transport within this zone. The character of the riverbed sediments is typically bimodal in the meander zone, with a coarse mode of gravels and a significant fine sediment component. There is appreciable fining of the bed sediment in this zone compared with the upstream zones. Bank sediments are also finer, being mostly composed of fine sands, silts and clays. The cohesive nature of the bank sediments contributes to relatively steep banks in this zone compared with upstream zones.

There is a complex array of functional units within both the low- and high-flow channel areas. Riffle-pool sequences dominate the main channel. Riffles can be especially large and are often associated with well-established stands of riparian vegetation because of the stability of the sediments and their enhanced nutrient and moisture status. Pool areas are large and deep. The low-flow channel is also characterised by large sandy point bars. In this zone, high-flow functional units are well developed, and in-channel benches, diverse flood runners and an extensive floodplain are all present. Functional units include the flat bench surfaces, which are important and distinct habitats for riverine vegetation, the small flood runners and complex features of the floodplain itself. Mesohabitats include the complex array of substratum types, stands of in-channel vegetation, snags and debris dams. Within the pool functional unit of the low-flow channel, mesohabitats also include emergent and submerged vegetation, although this is generally not abundant owing to the overall depth of the channel.

### Anabranching zone

The channel in this river zone is associated with very broad low angle fan complexes and is accompanied with a series of effluent channels or anabranches. The main river channel is characteristic of a 'wash load' system with low bed slopes, high sinuosities, low bankfull stream power and highly cohesive bank materials. River channels in this zone can be described as being 'underfitted' within an older large palaeochannel system. Usually the contemporary, younger channel has sinuosities up to 2.06, which is contained within the older channel system that has a much larger meander wavelength and channel dimensions. The active channel can have bankfull characteristics of low width to depth ratios, widths of between 30 and 150 m and depths of 10 to 15 m. Anabranch channels begin to flow at approximately one-third to one-half of bankfull discharges; therefore, bankfull capacities in this zone are lower in comparison with other zones.

These anabranch channels are frequently associated with abundant and diverse riparian vegetation. They are, as a result of being relatively frequently inundated and supplied with nutrients, floodplain hotspots in terms of their riparian vegetation productivity (Parsons and Thoms 2013). The anabranch zone is typical of many lowland rivers in the interior of the basin. In this zone the low-flow channel is relatively simple, with most of the habitat diversity occurring at higher-flow levels. Within the low-flow section of the anabranching zone, riffle functional units do not exist and the main functional units are the large pools within the main channel. Sections of the low-flow channel may also be characterised by large sandy point bars. By comparison, the high-flow channel is also well developed with in-channel benches occurring at various levels within the channel, diverse flood runners and large anabranches leaving the main channel at various flow heights, and an extensive floodplain. In-channel benches support abundant stands of *Eucalyptus camaldulensis* (river red gum), *Acacia stenophylla* (river cooba) and the introduced *Xanthium occidentale* (Noogoora burr). The major mesohabitats depend on substratum composition; however, diversity is limited to sandy bars and regions of silt and clay. Woody debris from fallen riparian vegetation is the other major mesohabitat of the low-flow channel. Mesohabitats within the high-flow channel are similar to those within the low-flow channel, with woody debris dominating.

### Distributary zone

A series of bifurcating channels (channels that take off from each other) are the main distinguishing feature of distributary zones. These secondary channels persist relatively independently of the main channel, for some length far in excess of their width. These distributary channels may re-join the main channel or each other. In all channels there is a downstream decrease in bankfull cross-sectional area or channel size. This is attributed to loss of water by evaporation and flood. Sediments within all the channels are composed of very fine sands, silts and clays. Indeed, the percentage weight of silts and clays can be up to 50%. Most of the channels are relatively narrow and featureless with occasional deep holes scattered along their length.

The low-flow channel is relatively simple with most of the functional units occurring within the higher flow channels. Within the low-flow section of the distributary zone, large deep pools are common. Sections of the low-flow channel may have the occasional point bar. In the high-flow section of the main channel, in-channel benches occur at various levels and these are considered to be important for the in-stream ecological processes. Secondary channels and the extensive floodplain surface through which they flow are the dominant functional unit in this zone. The high-flow channels also contain flat bench surfaces, flood runners and anabranches, and a complex array of floodplain features. Associated with this array of physical features are the main floodplain vegetation communities of *Eucalyptus camaldulensis*, *E. coolabah* (coolibah), *E. largiflorens* (black box) and *Acacia stenophylla*. These occur in higher floodplain elevation areas along flood runners. In lower elevation areas that are relatively wetter, areas of dense *Duma florulenta* (tangled lignum) shrubs and small patches of *Phragmites australis* (common reed) can occur along with a diverse mix of sedges and ephemeral herbfields (Thoms and Parsons 2011). Webb *et al.* (2006) recorded distinct zones of vegetation communities occurring within floodplain channels along an elevation gradient of the Narran River. The first zone, which extends from the edge of the channel to halfway down the bank is dominated by *Cuscuta campestris*, and the second zone, extending from mid-bank to the edge of the channel bed, by *Polygonum plebieum*. The final zone, which is dominated by *Helitropium supinum* (creeping heliotrope), covers the lowest elevations of channel bed (Webb *et al.* 2006). Woody debris from fallen riparian vegetation is probably the major mesohabitat of the low-flow channel. Mesohabitats within the high-flow channel are similar to those within the low-flow channel, with woody debris dominating. At high flows mesohabitats relate to the secondary channels and inundated terrestrial environments; however, snags and fallen woody debris again form a major mesohabitat in this region of the channel.

## ASSOCIATIONS BETWEEN PHYSICAL FEATURES AND RIVERINE VEGETATION

As with the hierarchical organisation of physical elements of rivers, the biological elements of rivers can also be divided into levels of organisation. At a large scale, biological elements can be viewed as an ecosphere or biome, and at a much smaller scale at the individual level (Barrett *et al.* 1997). Thus, as with the expression of physical habitat at different hierarchical levels, the expression of riverine vegetation attributes will also differ between levels of the biological hierarchy. Dollar *et al.* (2007) proposed a framework for interdisciplinary understanding of river ecosystems containing three parallel hierarchies of geomorphological, hydrological and ecological organisation. Compatibility between the levels of the parallel hierarchies is achieved through the assignment of scales based on the grain and extent of processes in both space and time (Dollar *et al.* 2007). Organisms perceive and respond to the physical features expressed at different scales of the hierarchy. Different life history stages of an organism may also perceive physical features and respond at different levels of the hierarchy. Despite the complications of physical heterogeneity, biological diversity and biological processes, integrating the ecological hierarchy with the physical hierarchy can facilitate ordering of investigation and pursuit of understanding (Dollar *et al.* 2007).

Applying these ideas of parallel hierarchy to riverine vegetation, different levels of vegetation organisation may correspond to different levels of physical organisation (Table 2.2). At the catchment level, physical template features of rivers are expressed at a large spatial scale ($10^3$–$10^6$ km$^2$) and a long temporal scale ($10^5$–$10^6$ years). Corresponding to this level of physical template expression, the matching foci for riverine vegetation are the biome and ecosystem levels (Table 2.2). Biomes correspond to different broad vegetation types and their associated physiographic settings such as grasslands, forests, deserts or savannas. Within a biome, vegetation elements at the ecosystem level include primary production, the role of vegetation primary production in food webs and the flow and transformation of energy and nutrients in the riparian ecosystem (Leuschner 2005). Evolutionary processes and the generation of diversity are also important at this level of biological organisation.

At the riverine landscape level of physical organisation, the matching focus for riverine vegetation is the ecosystem and community biological levels of organisation (Table 2.2). In addition to the ecosystem elements described above, altitudinal gradients in vegetation communities, the spatial composition and configuration of vegetation communities, particularly in relation to the hydrological regime, and the implications of this composition and configuration on ecosystem structure and function drive vegetation patterns at the riverine landscape level (Turner 1990). Thus, if concerned with the river template at the riverine landscape level these are the elements of riverine vegetation to consider.

Physical features expressed at the functional process zone level also occur at relatively large spatial and temporal scales, and the matching focus for riverine vegetation would also be at the ecosystem and community level (Table 2.2). Here, riverine ecosystem processes such as productivity and energy and nutrient transfer might be particularly associated with different floodplain and river morphologies (McGinness and Arthur 2011). Similarly, vegetation community structure and function may also be associated with different floodplain and river morphologies (Thoms and Parsons 2011; Parsons and Thoms 2013). Thus, if concerned with the river template at the functional process zone level, ecosystem and community level attributes of riverine vegetation should be highlighted.

The physical features expressed at the reach level correspond to the community and population level of ecological organisation (Table 2.2). Communities are an assemblage of interacting

**Table 2.2.** Typical features associated with different hierarchical levels of river system organisation and their association with riparian vegetation structure and function

| Hierarchical level | Spatial scale (km$^2$) | Temporal scale (yr) | Geomorphic processes influencing the formation of physical features | Physical features characteristic of the hierarchical level | Associations between physical features and riparian vegetation |
|---|---|---|---|---|---|
| Catchment | $10^3$–$10^6$ | $10^5$–$10^6$ | Erosion<br>Weathering<br>Planation and denudation<br>Tectonic activity<br>Climate<br>Sedimentary basin infilling | Catchment boundary (size and shape)<br>Catchment topography<br>River network geometry<br>Valley configuration | Biome<br>Ecosystem |
| Riverine landscape | $10^1$–$10^4$ | $10^5$–$10^6$ | Boundary geologies<br>Geological history | Valley confinement<br>Riverscape character<br>Floodscape character | Ecosystem<br>Community |
| Functional process zone | $10^1$–$10^3$ | $10^3$–$10^5$ | Flow regime<br>Sediment regime<br>Valley confinement | Floodplain morphology<br>River planform<br>River cross sectional morphology<br>River slope | Ecosystem<br>Community |
| River reach | $10^0$–$10^2$ | $10^2$–$10^3$ | Sediment supply<br>Floodplain inundation<br>Conveyance of flood and in-channel flows<br>Aggradation/degradation | Floodplain stability<br>Riverbed morphology<br>Riverbank morphology<br>Loading of woody debris | Community<br>Population |
| Functional set | $10^0$–$10^1$ | $10^1$–$10^2$ | Floodplain shedding and accumulation of water and sediment<br>Stream power<br>Sediment mobility<br>Channel stability | Composition and configuration of floodplain geomorphic units<br>Composition and configuration of in-channel geomorphic units<br>Residence time of geomorphic units<br>Composition and configuration of surface flow types | Community<br>Population |
| Functional unit | $10^0$–$10^1$ | $10^0$–$10^1$ | Flow velocity<br>Initiation of motion of riverbed and bank sediment | Size, shape and residence time of erosional and depositional geomorphic units<br>Size, shape and residence time of surface flow types<br>Sediment types and stratigraphy<br>Armouring | Population<br>Individual |
| Mesohabitat | $10^1$–$10^2$ | $10^0$–$10^1$ | Flow hydraulics<br>Particle movement | Sediment texture<br>Sediment geochemistry<br>Biogeochemical processing<br>Soil moistureLarge wood pieces | Individual |

species occurring in the same place. Community level attributes of riverine vegetation include succession, disturbance, inter-specific competition, functional groups, niche partitioning, species richness, species diversity, guilds, gradients and patch dynamics. A prominent example of riverine vegetation responses to physical features at the reach scale is the commonly observed gradients of riverine vegetation guilds in the lateral dimension of rivers associated with flooding frequency (Yang *et al.* 2011). Populations are the members of the same species occurring in the same place. Population level attributes of riverine vegetation largely centre on population dynamics, including the fertility, mortality, survivorship and dispersal (migration) of a species. As part of population dynamics, riverine species recruitment is a population level process. Predator–prey interactions such as herbivory and intra-specific competition for resources such as moisture and nutrients are also population level attributes of riverine vegetation. Thus, if concerned with the river template at the reach level, community and population level attributes of riverine vegetation should be highlighted. The same is required at the functional set level (Table 2.2).

The physical features of rivers expressed at the functional unit level correspond to the population and individual levels of ecological organisation (Table 2.2). In addition to the population level, described above, the individual level of organisation is concerned with the life history, morphological and physiological attributes of individual plants that facilitate their survival in the face of a flooded environment. Life history attributes include the timing of seed production to coincide with seasonal flooding, and mixed sexual and asexual modes of reproduction (Roberts and Marston 2011). Morphological adaptations include the ability to re-sprout from plant fragments following damage by floodwaters (Bond and Midgley 2001) or stem hypertrophy, development of aerenchyma tissue or the formation of adventitious roots to avoid anoxia (Kozlowski 2002). Physiological adaptations of riverine plants to flooding include the capacity to survive anaerobic environments by maintaining a glucose supply and avoiding accumulation of toxic compounds (Hook 1984; Kozlowski 2002). Thus, if concerned with the river template at the functional unit level, population and individual attributes of riverine plants should be highlighted. The same is required at the mesohabitat level (Table 2.2).

## CONCLUSIONS

The Australian continent has a complex geological and climatic setting that directly influences the physical form and function of its rivers. Three broad physiographic regions – based on landforms with similar underlying geology – divide the continent and influence the physical form and function of the rivers that drain it. While Australia is a dry continent overall, rainfall and runoff also differs markedly among 12 drainage divisions, influencing the delivery of water and sediment to river channels. Given the spatial heterogeneity and temporal variability in Australia's river form and function, a hierarchical perspective on the river template is warranted. The hierarchical organisation of rivers is based on principles of hierarchy theory, where hydro-geomorphic processes at larger levels constrain the expression of physical features at smaller nested levels. Australian rivers can be hierarchically divided into catchments, riverine landscapes, functional process zones, river reaches, functional sets, functional units and mesohabitats, each of which is associated with a characteristic spatial and temporal scale of operation, and a characteristic set of physical template features. We argue this organisation will also be matched by a parallel level of ecological organisation (Dollar *et al.* 2007) corresponding to key riverine vegetation structure, function and processes. Thus, it is not sensible to uncritically adopt paradigms developed in less-variable northern hemisphere systems to explain Australia's riparian ecosystems. The approach taken here makes no pretence to be a panacea for understanding fluvial ecosystem functioning, nor the full solution to the challenges of interdisciplinarity. Rather, it presents an initial consideration and re-ordering of the phenomena and material on the relationship between riverine vegetation and the physical template within an Australian context.

The disciplines of vegetation ecology and fluvial geomorphology are intricately related because the physical dimension of a river provides the template within which ecological processes occur. What is sometimes forgotten is that the expression of physical features in rivers occurs in the context of a nested geomorphic hierarchy, so understanding the relationships between riverine vegetation and the physical river template also requires an understanding of the physical and ecological processes associated with each level of the physical template. We provide a hierarchical template that links the description of river form to the matching ecological level. This facilitates appropriate causal explanations for the relationship between vegetation and the physical template at the correct spatial and temporal scales. Such an approach provides a basis for scale-appropriate scientific description and explanation, and the prediction of the response of vegetation pattern to physical processes, and the impact of these vegetation patterns on physical processes. Explanation and prediction are pre-eminent, so that river science and management can solve real-world problems and break down barriers between different disciplines.

## REFERENCES

Allen TFH, Starr TB (1982) *Hierarchy: Perspectives for Ecological Complexity*. The University of Chicago Press, Chicago, IL.

Barrett GW, Peles JD, Odum EP (1997) Transcending processes and the level-of-organization concept. *Bioscience* **47**, 531–535. doi:10.2307/1313121

Bergkamp G (1995) A hierarchical approach for desertification assessment. *Environmental Monitoring and Assessment* **37**, 59–78. doi:10.1007/BF00546880

Bisson PA, Montgomery DR (1996) Valley segments, stream reaches and channel units. In *Methods in Stream Ecology*. (Eds FR Hauer and GA Lamberti) pp. 23–49. Academic Press, San Diego, CA.

BOM (2012) *Australian Hydrological Geospatial Fabric Topographic Drainage Divisions and River Regions. Australian Commonwealth*. Bureau of Meteorology, Melbourne.

Bond WJ, Midgley JJ (2001) Ecology of sprouting in woody plants: the persistence niche. *Trends in Ecology & Evolution* **16**(1), 45–51. doi:10.1016/S0169-5347(00)02033-4

Bowler JM (1986) Quaternary landform evolution. In *Australia – a Geography; the Natural Environment*. (Ed. DN Jeans) pp. 117–147. Sydney University Press, Camperdown, NSW.

Charlton R (2008) *Fundamentals of Fluvial Geomophology*. Routledge, Abingdon, UK.

Church M (2002) Geomorphic thresholds in riverine landscapes. *Freshwater Biology* **47**, 541–557. doi:10.1046/j.1365-2427.2002.00919.x

Clarke JDA (1994) Geomorphology of the Kambalda region, Western Australia. *Australian Journal of Earth Sciences* **41**, 229–239. doi:10.1080/08120099408728132

Cullum C, Brierley G, Thoms MC (2008) The spatial organisation of river systems. In *River Futures*. (Eds GJ Brierley and KA Fryirs) pp. 41–64. Island Press, Washington, DC.

Davis WM (1899) The geographic cycle. *The Geographical Journal* **14**, 481–504. doi:10.2307/1774538

de Boer DH (1992) Hierarchies and spatial scale in process geomorphology: a review. *Geomorphology* **4**, 303–318. doi:10.1016/0169-555X(92)90026-K

Dollar ESJ, James CS, Rogers KH, Thoms MC (2007) A framework for interdisciplinary understanding of rivers as ecosystems. *Geomorphology* **89**, 147–162. doi:10.1016/j.geomorph.2006.07.022

Dovers S, Hussey K (2013) *Environment and Sustainability*. 2nd edn. Federation Press, Annandale, NSW.

Edwards PJ, Kollmann J, Gurnell AM, Petts GE, Tockner K, Ward JV (1999) A conceptual model of vegetation dynamics on gravel bars of a large alpine river. *Wetlands Ecology and Management* **7**, 141–153. doi:10.1023/A:1008411311774

Erskine WD, Warner RF (1988) Geomorphic effects of alternating flood- and drought-dominated regimes on NSW coastal rivers. In *Fluvial Geomorphology of Australia*. (Ed. RF Warner) pp. 223–244. Academic Press, Sydney.

Frissell CA, Liss WJ, Warren CE, Hurley MD (1986) A hierarchical framework for stream habitat classification: viewing streams in a watershed context. *Environmental Management* **10**, 199–214. doi:10.1007/BF01867358

Gregory KJ (2006) The human role in changing river channels. *Geomorphology* **79**, 172–191. doi:10.1016/j.geomorph.2006.06.018

Gurnell A (2014) Plants as river system engineers. *Earth Surface Processes and Landforms* **39**, 4–25. doi:10.1002/esp.3397

Harper D, Everard M (1998) Why should the habitat-level approach underpin holistic river survey and management? *Aquatic Conservation: Marine and Freshwater Ecosystems* **8**, 395–413. doi:10.1002/(SICI)1099-0755(199807/08)8:4<395::AID-AQC297>3.0.CO;2-X

Hook DD (1984) Adaptations to flooding with fresh water. In *Flooding and Plant Growth* (Ed. TT Kozlowski) pp. 265–294. Academic Press, Gainesville, FL.

Hynes HBN (1975) Edgardo Baldi memorial lecture. The stream and its valley. *Verhandlungen der Internationalen Vereinigung für Theoretische und Angewandte Limnologie* **19**, 1–15.

Jennings JN, Mabbutt JA (1986) Physiographic outlines and regions. In *Australia – a Geography; the Natural Environment* (Ed. DN Jeans) pp. 50–80. Sydney University Press, Camperdown, NSW.

Knighton D (1984) *Fluvial Forms and Processes.* Edward Arnold, London.

Kondolf GM, Montgomery DR, Piegay H, Schmitt L (2003) Geomorphic classification of rivers and streams. In *Tools in Fluvial Geomorphology.* (Eds GM Kondolf and H Piegay) pp. 171–204. Wiley, Chichester, UK.

Kozlowski TT (2002) Physiological-ecological impacts of flooding on riparian forest ecosystems. *Wetlands* **22**, 550–561. doi:10.1672/0277-5212(2002)022[0550:PEIOFO]2.0.CO;2

Leopold LB, Wolman MG (1957) *River Channel Patterns: Braided, Meandering and Straight.* United States Geological Survey professional paper, 282B. United States Geological Survey, Boulder, CO.

Leuschner C (2005) Vegetation and ecosystems. In *Vegetation Ecology.* (Ed. E van der Maarel) pp. 85–105. Blackwell Science, Malden, MA.

McGinness HM, Arthur AD (2011) Carbon dynamics during flood events in a lowland river: the importance of anabranches. *Freshwater Biology* **56**, 1593–1605. doi:10.1111/j.1365-2427.2011.02598.x

Montgomery DR (1999) Process domains and the river continuum. *Journal of the American Water Resources Association* **35**, 397–410. doi:10.1111/j.1752-1688.1999.tb03598.x

Montgomery DR, Buffington JM (1998) Channel processes, classification, and response. In *River Ecology and Management: Lessons from the Pacific Coastal Ecoregion.* (Eds RJ Naiman and RE Bilby) pp. 13–42. Springer Verlag, New York, NY.

Naiman RJ, Lonzarich DG, Beechie TJ, Ralph SC (1992) General principles of classification and the assessment of conservation in rivers. In *River Conservation and Management.* (Eds PJ Boon, P Calow and GE Petts) pp. 93–123. American Geophysical Union, Washington, DC.

O'Neill RV, King AW (1998) Homage to St Michael; or, why are there so many books on scale? In *Ecological Scale: Theory and Applications.* (Eds DL Peterson and VT Parker) pp. 3–15. Columbia University Press, New York, NY.

O'Neill RV, DeAngelis DL, Waide JB, Allen TFH (1986) *A Hierarchical Concept of Ecosystems.* Princeton University Press, Princeton, NJ.

Ollier CD, Chan RA, Craig MA, Gibson DL (1988) Aspects of landscape history and regolith in the Kalgoorlie region, Western Australia. *Journal of Australian Geology and Geophysics.* **10**, 309–321.

Parsons M, Norris RH (1996) The effect of habitat-specific sampling on biological assessment of water quality using a predictive model. *Freshwater Biology* **36**, 419–434. doi:10.1046/j.1365-2427.1996.00088.x

Parsons ME, Thoms MC (2013) Patterns of vegetation greenness during flood, rain and dry resource states in a large, unconfined floodplain landscape. *Journal of Arid Environments* **88**, 24–38. doi:10.1016/j.jaridenv.2012.07.023

Parsons M, Thoms MC, Norris RH (2004) Development of a standardized approach to river habitat assessment in Australia. *Environmental Monitoring and Assessment* **98**, 109–130. doi:10.1023/B:EMAS.0000038182.03176.97

Petts GE, Amoros C (1996) *Fluvial Hydrosystems*. Chapman & Hall, London.

Petts GE, Nestler J, Kennedy R (2006) Advancing science for water management. In *Living Rivers: Trends and Challenges in Science and Management; Part Three*. (Eds RSEW Leuven, AMJ Ragas, AJM Smits and G van der Velde) pp. 277–288. Springer, Netherlands.

Pickup G (1976) Geomorphic effects of changes in river runoff, Cumberland Basin, NSW. *The Australian Geographer* **13**, 188–193. doi:10.1080/00049187608702692

Roberts J, Marston F (2011) *Water Regime for Wetland and Floodplain Plants: a Source Book for the Murray–Darling Basin*. National Water Commission, Canberra.

Schumm SA (1988) Variability of the fluvial system in space and time. In *Scales and Global Change*. (Eds T Rosswall, RG Woodmansee and PG Risser) pp. 225–250. John Wiley & Sons, Chichester, UK.

Schumm SA (1991) *To Interpret the Earth: Ten Ways To Be Wrong*. Cambridge University Press, Cambridge, UK.

Schumm SA, Lichty RW (1965) Time, space and causality in geomorphology. *American Journal of Science* **263**, 110–119. doi:10.2475/ajs.263.2.110

Southwood TRE (1977) Habitat, the templet for ecological strategies? *Journal of Animal Ecology* **46**, 337–365. doi:10.2307/3817

Thoms MC (1994) Water resources and its development in Australia. In *Australian Geography for the 1990s*. (Eds P Hirsch and MC Thoms) pp. 12–20. Sydney University Press, Camperdown, NSW.

Thoms MC, Parsons ME (2011) Patterns of vegetation community distribution in a large, semi-arid floodplain landscape. *River Systems* **19**, 271–282. doi:10.1127/1868-5749/2011/019-0050

Thoms MC, Hill SM, Spry MJ, Chen XJ, Mount TJ, Sheldon F (2004) The geomorphology of the Darling River. In *The Darling*. (Eds R Breckwodt, R Boden and J Andrews) The Murray–Darling Basin Commission, Canberra.

Thorp JH, Thoms MC, Delong MD (2008) *The Riverine Ecosystem Synthesis: Toward Conceptual Cohesiveness in River Science*. Academic Press, London.

Tooth S, Nanson GC (1995) The geomorphology of Australia's fluvial systems: retrospect, perspect and prospect. *Progress in Physical Geography* **19**, 35–60. doi:10.1177/030913339501900103

Turner MG (1990) Spatial and temporal analysis of landscape patterns. *Landscape Ecology* **4**, 21–30. doi:10.1007/BF02573948

van Coller AL, Rogers KH, Heritage GL (1997) Riparian vegetation-environment relationships: complimentarity of gradients versus patch hierarchy approaches. *Journal of Vegetation Science* **11**, 337–350. doi:10.2307/3236626

van Niekerk AW, Heritage GL, Moon BP (1995) River classification for management: the geomorphology of the Sabie River in the Eastern Transvaal. *The South African Geographical Journal* **77**, 68–76. doi:10.1080/03736245.1995.9713594

Webb MC, Reid MA, Capon S, Thoms MC, Rayburg S (2006) Are floodplain-wetland plant communities determined by seed bank composition or inundation periods? *International Association of Hydrological Sciences* **306**, 241–250.

Yang J, Dilts TE, Condon LA, Turner PL, Weisberg PJ (2011) Longitudinal- and transverse-scale environmental influences across multiple levels of ecological organization. *Landscape Ecology* **26**, 381–395. doi:10.1007/s10980-010-9565-z

3

# The physical template of Australia's floodplain landscapes

*Martin Thoms and Melissa Parsons*

## INTRODUCTION

The phrase the 'story of our land can be revealed through plants' (Wardle 2010) highlights the strong and reciprocal influence of climate, earth surface processes and human activities on vegetation distribution. Such influences strongly apply in riverine landscapes, from waterways and wetlands to the floodplains with which they are associated. Floodplains occupy the zone between terrestrial uplands and river channels and their inundation, along with associated fluxes of water, sediment and nutrients, provides strong selection on their vegetation so that floodplain plant communities are often distinctly different from surrounding upland areas (Naiman *et al.* 2005; Sabo *et al.* 2005; Thoms and Parsons 2011). Inundation also drives the productivity of floodplain vegetation, so that floodplains are areas of elevated but often intermittent vegetation productivity (Tockner and Stanford 2002; Leigh *et al.* 2010; Reid *et al.* 2011; Parsons and Thoms 2013). Associated with this productivity, humans utilise floodplains for grazing and cropping, water supply, recreation, settlement and flood control (Tockner and Stanford 2002). It is estimated that of the US$33 trillion value provided by the world's ecosystem services per year, 10% is contributed by floodplains and wetlands (MEA 2003). Thus, the distribution and productivity of vegetation within floodplains is integral to sustaining human well-being.

It has been proposed that habitat provides the 'templet' on which evolution acts to forge characteristic life history strategies (Southwood 1977, 1988). Accordingly, the character of the floodplain physical template will influence the types of vegetation communities found there. Floodplains are three-dimensional temporary landforms formed by a complex set of fluvial processes. Their character and evolution are essentially the product of flow and sediment regimes and the character of the valley floor trough because this is the site for alluvial sediment storage, or where floodplains form (Anderson *et al.* 1996). Water and sediment interact hierarchically at different scales to form physical habitat.

At larger scales, flow and sediment regimes interacting over decades to millennia determine the character of floodplains found within a valley and the physical features associated with floodplain landforms (Nanson and Croke 1992). Three floodplain valley floor trough associations have been identified along coastal valleys of NSW, Australia, each with different floodplain morphologies, sediment compositions, quantities of sediment stored and evolutionary sequences (Warner 1992). In the upper, narrower regions of the coastal valleys, type 1 incised floodplains are prominent. These regions, dominated by high-energy river channel systems, represent poor sites for alluvial storage and therefore floodplain development. Most of the floodplain sediment is disturbed frequently and transported downstream to lower-energy environments, so that boulders dominate the material that remains. The lower–mid regions have wider valleys, containing larger volumes of sediment in storage and the presence of type 2 inset floodplains. These floodplains are not as frequently disturbed and are therefore older than type 1 floodplains. Inset floodplains have a wide range of sediment compositions and a higher diversity of morphologies present on their surface. In the lowermost regions, the very wide valleys and the higher rates of sediment accumulation result in type 3 overlapped floodplains. These lower-energy floodplains contain finer sediments and an array of surface morphologies such as levees, back swamps and flood basins.

The character of physical features found on floodplain surfaces is also the product of processes occurring at smaller scales, especially the character of flow and sediment pulses occurring over weeks to years (Anderson *et al.* 1996). Using detailed stratigraphic information obtained from a series of floodplain surfaces, Thoms and Olley (2004) recorded different sediment depositional sequences for flood events along the Barwon-Darling River, Australia. The presence of complex sequences of sediment lamina within the various floodplain deposits indicates highly variable sediment supply during and between flood events. As a result, significant changes in floodplain surface sediment character do occur between flood events (Southwell and Thoms 2011). These smaller-scale changes have the potential to influence not only the amount of sediment supplied to adjacent floodplain environments, but also the physical and chemical characteristics of deposited sediments in time and space.

Many studies have found relationships between water, sediment and vegetation distribution. The spatial organisation of floodplain vegetation, in terms of age and community structure, is significantly associated with factors including floodplain morphology (Latterell *et al.* 2006), flooding condition (Parsons and Thoms 2013) and soil character (Thoms and Parsons 2011). At smaller scales, the presence of vegetation on floodplain surfaces can influence the hydraulic character of overbank flows and the ability of these surfaces to trap sediment and associated nutrients. Increases in floodplain sedimentation by two orders of magnitude in the Barmah Forest, Australia, occurred, in part, because of changes to the understorey vegetation (Thoms 1995). A reduction in the area of the native *Pseudoraphis spinescens* (Moira grass) grasslands and an increase in the presence of *Juncus ingens* (giant rush) increased the roughness coefficient of the floodplain surface, thereby reducing the transmission velocity of overbank flows and increasing the quantity of sediment deposition. The roughness character, or resistance to flow, of various floodplain vegetation assemblages does differ. For example, the Mannings ‘n’ for common rush and sedge species is 75% greater than that of grasses (Klassen and Van Urk 1985). Such empirical evidence typifies the strong influence of water and sediment as drivers of vegetation pattern and process in floodplains and vice versa.

## Genesis of our current understanding of floodplain ecosystems

Interpretations of the relationships between water, sediment and vegetation distribution are influenced by the locations in which floodplain research takes place, the types of floodplains studied, and the disciplinary approaches to floodplain research. We undertook a meta-analysis of 252 scientific journal papers published between 1995 and 2013 in five international journals in the fields of geomorphology, ecology and river science. Each paper was interrogated in terms of the location of the floodplain (temperate, tropical, dryland, other), size of the floodplain (width of <1 km, 1–10 km, >10 km), the enquiring discipline (hydrology, geomorphology, ecology) and whether the study was single- or multi-disciplined. The results show that floodplain research is concentrated in well-watered temperate areas, with dryland and tropical floodplains receiving less attention (Fig. 3.1). The majority of research concerns floodplains up to 10 km wide, with floodplains greater than 10 km wide receiving less attention (Fig. 3.1). Many studies are conducted from a geomorphological perspective, but floodplains are also frequently studied from hydrological and ecological perspectives, either as single discipline studies or in concert with other disciplinary approaches (Fig. 3.1). In addition, much research also takes place in younger, active geological settings, where erosional and depositional processes are actively reshaping the landscape. Arising from these communities of research practice, four ideas about floodplain structure and function dominate floodplain studies:

(1) *Inundation pattern* – floodplains are aquatic terrestrial transition zones, forming a moving littoral zone related to the flood pulse (Junk *et al.* 1989). As a result of the imprint of the flood pulse, vegetation communities are organised along a lateral gradient, often associated with flood frequency (Naiman *et al.* 2005).

(2) *Valley setting* – the valley rules the stream, and the physical dimensions and characteristics of the floodplain (Hynes 1975; Gupta 2007). It is estimated that 78% of the more than 2 million floodplains occurring in the USA are associated with first order streams and are less than 3 m wide (Tockner and Stanford 2002). Thus, much of our knowledge of the structure and function of floodplains comes from those in constrained valley settings.

(3) *Floodplain complexity* – floodplains are often considered a homogeneous unit situated between upland and channel areas. Floodplains can be mistaken as flat, featureless and uniform, with wetlands forming the main element of interest (Alvarez 2011). Geomorphic processes generating hydro-geomorphic complexity and variability in floodplains are important determinants of ecological processes (Hughes 1997; van Coller *et al.* 2000; Gurnell and Petts 2002; Thoms 2003).

(4) *Scaled processes* – awareness of scale is important in floodplain studies. The geomorphic factors shaping the physical character of floodplains operate within a process-based hierarchy. Larger-scale factors constrain the conditions within which smaller-scale factors operate, so that each hierarchical level has factors associated with it that operate at a particular spatial and temporal scale (Dollar *et al.* 2007). Data collected at spatial and temporal scales that do not match the focal scale of the problem can result in misinterpretations of floodplain pattern and process. Floodplains should be studied at scales appropriate to the problem at hand (e.g. Parsons and Thoms 2007).

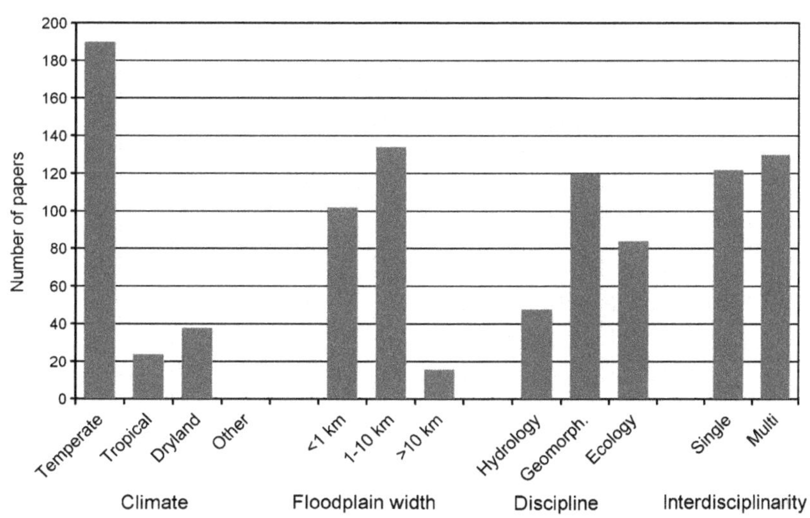


**Figure 3.1:** Location, floodplain character and disciplinarity of contemporary floodplain research. The analysis is based on 252 studies published between 1995 and 2013 in the journals *Geomorphology* (129 papers), *River Research and Applications* (92), *Ecology* (11), *Ecological Applications* (17) and *Ecological Monographs* (3). Papers were gathered in each journal using the search terms 'floodplain', 'floodplain + ecosystem' and 'floodplain landscapes' in the ISI Web of Science database. Each paper was examined further for relevance and only studies specifically of floodplains were included. Each of the 252 floodplain papers was tallied by climatic location, floodplain width, the major discipline of study, and whether it was single- or multi-disciplinary in approach.

There is no doubt that research related to these factors has enhanced our understanding of floodplain structure and function, and informed efforts to conserve and sustainably manage this valuable resource. However, knowledge of the relationships between water, sediment and vegetation gained from well-watered, relatively young geological settings may not fully apply to Australian floodplains. Many of Australia's floodplains are set in unconstrained valleys, or operate as terminal lake systems. Such floodplains are large and often have a surface area greater than 10 000 km$^2$, especially those associated with inland river systems (Thoms and Sheldon 2000). This is in contrast to the estimated largest floodplain surface of 4449 km$^2$ associated with a tenth order stream in the USA (Tockner and Stanford 2002). Australian floodplain surfaces are geomorphically diverse and often contain a suite of physical features, including wetlands, levees, scrolls, swales, distributary channels, in-channel benches, palaeochannels, cut-offs and flat floodplain surfaces. Australia's flow and sediment regimes are highly variable. Puckridge *et al.* (1998) analysed the flow regimes of 52 global rivers with similar catchment character and showed Australia's rivers to be among the most hydrologically variable in the world. Sediment regimes are also highly variable (Thoms and Walker 1993). Moreover, despite many Australian catchments having high rates of erosion, with rates up to 20 900 t/km$^2$, the delivery of this sediment to the oceans is low (Wasson 1996). Australian river systems have an average sediment delivery ratio of 3%, thus most sediment eroded from catchments is stored within the Australian landscape, much in the form of floodplains. We currently lack a framework that incorporates the unique valley setting and variability of water and sediment inputs to Australia's floodplains. Such a framework would enhance floodplain science by providing an interdisciplinary platform for understanding floodplains as ecosystems, including the distribution of vegetation in floodplains.

In this chapter we explore how inundation patterns, valley setting, floodplain complexity and scaled processes influence the geomorphic template of Australian floodplains. We then draw these together to propose a framework for an interdisciplinary understanding of floodplain ecosystems based on the dynamics of change in the physical template.

## WHAT ARE FLOODPLAIN LANDSCAPES?

Floodplains are the relatively broad and smooth valley floors adjacent to river channels that are periodically inundated. Hydrologists and engineers have defined floodplains as the surface adjacent to a river channel that is inundated by a given return period of flow, e.g. 100 years (Ward and Trimble 2004); whereas, geomorphologists have emphasised the presence of alluvial deposits as defined by Nanson and Croke (1992) –'the largely horizontally-bedded alluvial landforms located adjacent to a river channel, separated from the channel by banks, and built of sediment transported by the present flow-regime'. In ecological terms, floodplains can be defined as areas adjacent to a river that are colonised by organisms adapted to wetting and drying. Floodplains provide habitat for a wide variety of plants and animals compared with other landforms; hence, they are also regarded as biological hot spots. In all these definitions, intermittent or periodic flooding is a common feature of floodplains. In broad terms, we define floodplains as 'areas of low-lying land composed of alluvial sediment that are subject to inundation by lateral overflow of water from rivers or lakes with which they are associated'.

## ELEMENTS OF THE FLOODPLAIN PHYSICAL TEMPLATE

### Inundation pattern

Floodplains are aquatic terrestrial transition zones that alternate between aquatic and terrestrial states (Junk *et al.* 1989). This view underpins the dominant conceptual model of floodplains as ecotones, where floods of different magnitude and duration pulse onto the floodplain, with the edge of the flood water forming a 'moving littoral' zone of higher productivity (Junk *et al.* 1989). This flood pulse concept assumes that floods expand and contract in a uniform and linear manner across floodplain surfaces with comparable changes in a range of environmental variables (Hughes 1997). For example, the characteristic lateral distribution of vegetation communities across floodplain surfaces has been shown in many studies to result from distinct hydro-geomorphic processes (Pinay *et al.* 1990; Petty and Douglas 2010). Different fluvial landforms occur at different elevations above the main channel and have different flow durations, wetting frequencies and soil conditions that change in a systematic manner from the main river channel to distal floodplain regions (Bridge 2003; Simons *et al.* 2004; Alvarez 2011). This physiographic approach to understanding hydro-geomorphic processes and vegetation distribution has dominated research and management perceptions of floodplain ecosystems for several decades. The flood pulse concept model is, however, limited to floodplains constrained within relatively confined valley settings, or in tropical and temperate climatic regions (Tockner and Stanford 2002).

There is emerging evidence that floodplains are organised as mosaics. Recent research in Australian unconstrained valley settings, where valley widths typically exceed 10 km, supports models of floodplains as internally heterogeneous landscapes (Thoms and Parsons 2011; Parsons and Thoms 2013). Stanford *et al.* (2005) termed this the 'shifting habitat mosaic', where flow, sediment movement, large wood deposition and riparian vegetation dynamics generate irregular and dynamic patch mosaics in floodplain landscapes. The composition and configuration of patches in a landscape influences the productivity and the flux of materials within and between landscape patches (Forman 1997). Applied to unconstrained floodplains, the spatial organisation of patches (like wetlands, scrolls, swales and abandoned channels) across a floodplain surface influences the flux of sediment (Thoms and Foster 2002) and associated nutrients (Southwell 2008) within and between different floodplain patches and the river channel. Under the shifting habitat mosaic paradigm it is not the flood pulse that dictates the types of plants, animals and processes found on the floodplain. Rather, floodplains are non-equilibrium systems in which patches are continuously redistributed in space and time by the flow regime, and organisms respond to this dynamic and heterogeneous mosaic (Latterell *et al.* 2006; Whited *et al.* 2007; Thorp *et al.* 2008).

Australian dryland floodplains in particular exhibit mosaic structure. Using a series of Landsat TM images, Murray *et al.* (2006) demonstrated strong anti-clockwise hysteresis patterns of inundation extent (Fig. 3.2) during a series of inundation events of the Narran floodplain – a large unconstrained dryland floodplain in Australia. Inundation resulted in a dynamic mosaic of wet patches rather than displaying a simple expansion and contraction of wetting across the floodplain surface. This complex response to inundation is mirrored in the spatial organisation of floodplain vegetation communities in the lower Balonne floodplain, of which the Narran floodplain ecosystem is part. Thoms and Parsons (2011) analysed a published 1:50 000 vegetation map of this 10 500 km$^2$ floodplain using a series of landscape

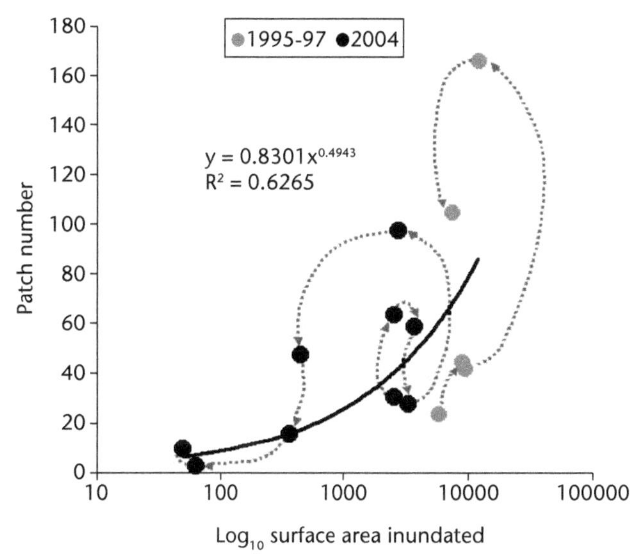


**Figure 3.2:** The relationship between floodplain inundation and discrete wet patches during two flood events in the Narran floodplain, NSW, Australia. After Murray *et al.* 2006.

metrics of the area, shape, interspersion and juxtaposition of eight types of vegetation community and land cover patches. Rather than forming a predictable linear gradient across the floodplain, vegetation communities of the lower Balonne floodplain landscape depict a heterogeneous mosaic (Fig. 3.3). Grassland forms the matrix element of the mosaic because it dominates the floodplain (40% of the total landscape area), occurs in large patches and is highly interspersed among the other vegetation community and land cover types (Fig. 3.3). *Eucalyptus coolabah* (coolibah) and *E. camaldulensis* (river red gum) trees occur along the major river courses and form the corridor element of the mosaic (Fig. 3.3). The other land cover types form the patch element of the mosaic, where each land cover type has a characteristic patch area, shape and interspersion or juxtaposition. The complex nature of biophysical patchiness in Australian floodplains suggests the need for an interdisciplinary approach that integrates ecological, hydrological and geomorphological elements contributing to the dynamics of floodplain landscapes.

## Valley setting

Regardless of the model used to describe floodplains, the valley setting (Hynes 1975) and physical dimensions of the floodplain (Gupta 2007) ultimately set the boundary conditions for floodplain character and the spatial organisation of patches within floodplain landscapes. In particular, the eco-geomorphological character of relatively constrained floodplain landscapes (i.e. those floodplains contained within valleys <2 km wide) differs from those that are unconfined (Thoms 2003). The geomorphic effectiveness of overbank flows, an important driver of floodplain processes, is influenced by the nature of the valley floor setting. Miller (1995) demonstrated that higher maximum

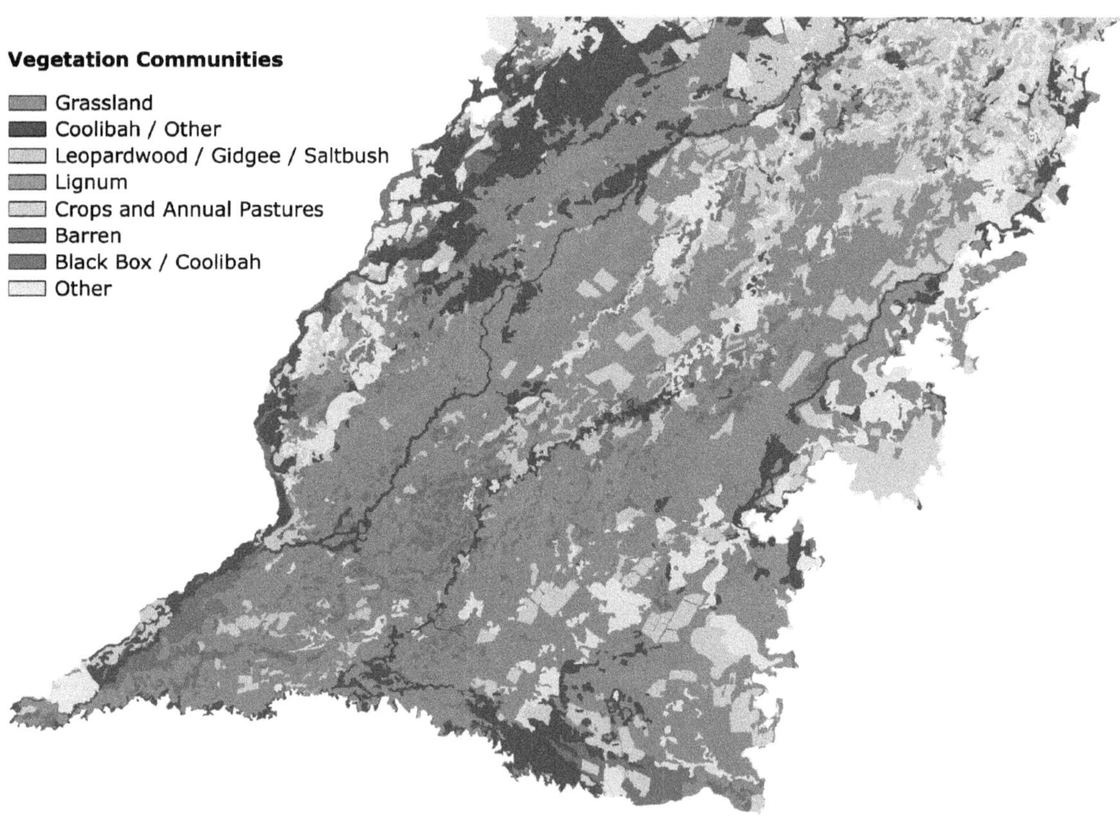


**Figure 3.3:** Vegetation and land cover patches in the lower Balonne floodplain, NSW, Australia. Modified from NFRPC (2004).

shear stresses are experienced on floodplain surfaces in narrower valleys than those that occur in association with floods of comparable size in broader valleys. This is primarily the result of differences in stage-velocity relationships between the two valley-floodplain settings (Thoms *et al.* 2000). These hydraulic differences have been used to explain variations in the distribution of sediment-associated nutrients (Thoms *et al.* 2000), the diversity and spatial organisation of morphological units (Thoms 2003) and the evolutionary sequence of these morphological units across floodplain surfaces in different valley floor settings (Tooth *et al.* 2010). Indeed, Mertes *et al.* (1995) used satellite imagery to describe the complexity of floodplain hydrology, geomorphology and riparian vegetation patterns across the lower Amazon floodplain using three satellite images, demonstrating considerable variation in the spatial heterogeneity of this floodplain between reaches, hypothesised to result from subtle differences in boundary conditions.

Evidence for differences in the physical character of floodplain templates between valley settings in Australia is increasing (Thoms *et al.* 2000; Cheetham *et al.* 2010; Thoms and Parsons 2011). Distinct floodplain reaches along Widden Brook in NSW, for example, were described by Cheetham *et al.* (2010). Despite some floodplain reaches being adjacent to one another, they were markedly different in terms of their stratigraphy, sedimentology and evolutionary processes. The Baramul floodplain sequences, located in upstream-constricted valleys, have four different surfaces composed of coarse to medium sands, with the lowermost surface being organically enriched. In contrast, Widden floodplain sequences, consisting of three floodplain surfaces, contain highly altered, vertically accreted, sandy sediments associated with those wider valley sections. In other relatively constricted valleys, the Kewarra floodplain sequences also had three surfaces but these can be differentiated in terms of age and the degree of sediment diagenesis (i.e. post-depositional changes in physical and chemical character). Differences in floodplain character and behaviour associated with variations in valley setting have also been reported by Warner (1992), Ferguson and Brierley (1999) and Thoms *et al.* (2000).

The complexity of the floodplain landscape in Widden Brook highlights the influence of larger-scale or top-down factors on contemporary floodplains. It was hypothesised that local-scale processes, preconditioned by the valley setting (in this case the relative width of the valley floor trough) dominate the formation of each floodplain reach, resulting in a complex physical floodplain landscape (Cheetham *et al.* 2010). Schumm (1973) discussed the relationship between valley floor gradient and valley floor stability, with failure of the valley floor occurring with steepening valley floor slope. This relationship has been further explored in terms of gully erosion and the propensity for landscape change in respect to slope stability thresholds (Patton and Schumm 1975; Schumm 1977). Channel sinuosity and gradient, along with sediment supply, are also discussed as geomorphic thresholds (Schumm 1977). These thresholds can be exceeded through natural progressive landscape evolution or forced by catastrophic events (Melville and Erskine 1986; Nanson and Erskine 1988) and human activities. Effectively, each terrace sequence can influence changes to the next terrace sequence in this way. For example, the exceedance of a local geomorphic threshold (such as a gradient) in one sequence can cause erosion of sediment from that sequence and deposition in the next (Erskine and Melville 2008). This process, in turn, contributes to the exceedance of another local geomorphic threshold (such as sediment supply or gradient) in the next sequence, leading to yet another floodplain abandonment event and so on. Localised, complex morphology and stratigraphy, like that of Widden Brook, is indicative of a complex morphological and stratigraphic response to the exceedance of one or more geomorphic thresholds (Erskine 2008). This complexity, when combined with the resultant discontinuous nature of remnant terrace pockets (Lewin and Macklin 2003), often prevents the linking of a specific geomorphic threshold to a particular landscape response. Nevertheless, the Widden

Catchment does provide an ideal landscape to discriminate regional from local controls on terrace formation.

## Floodplain complexity

Floodplains are often viewed as flat or nearly flat, featureless landforms adjacent to river channels (cf. Alvarez 2011). This view has prevailed for many years and results, in part, from limited topographical information of these landscapes. Digital elevation models representing the detailed surface character of various landscapes are now routinely created for a variety of landscape studies (Fonstad *et al.* 2013). Digital elevation models are generated from a variety of sources, including spot heights or elevation data from contours acquired from

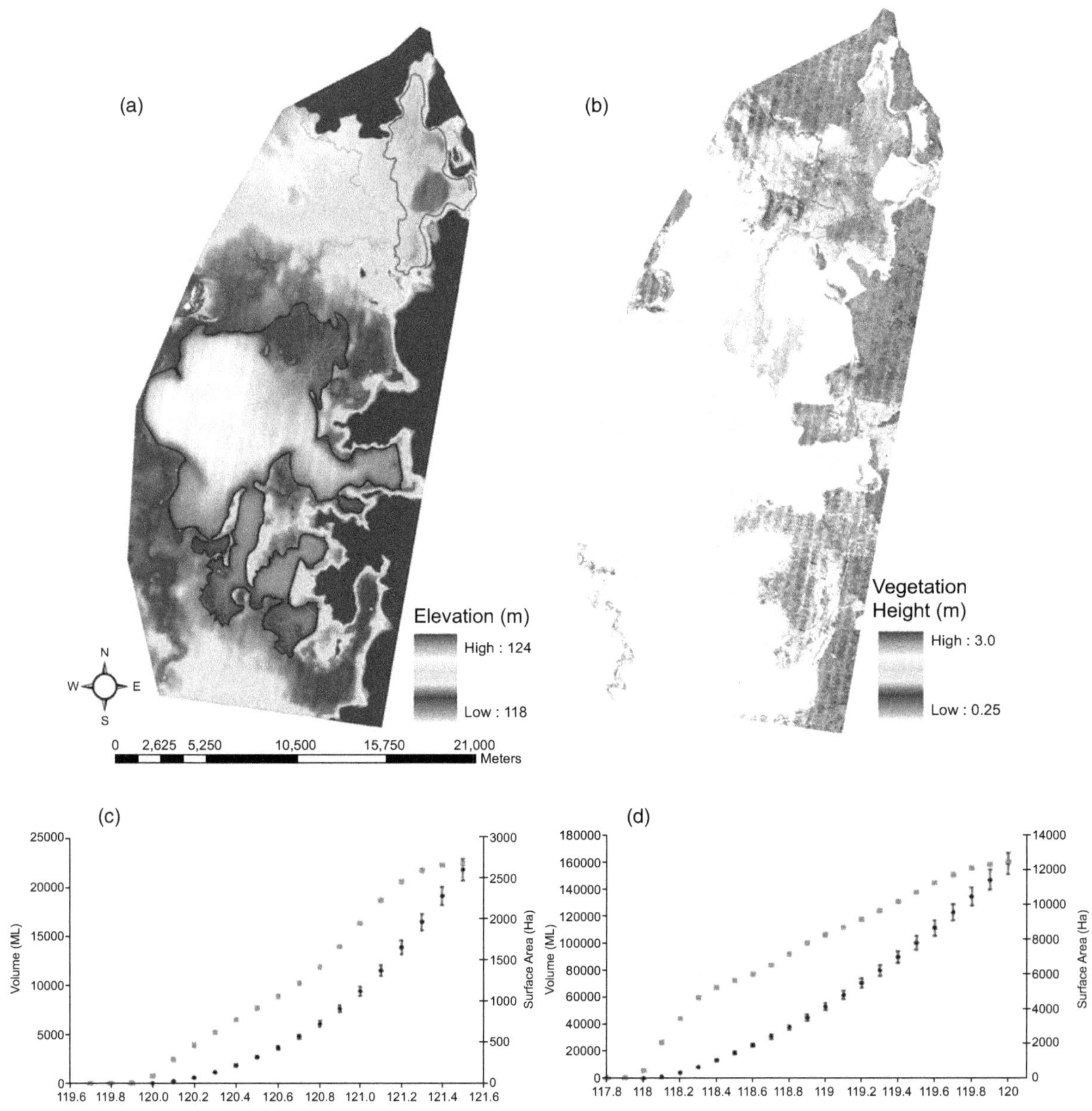


**Figure 3.4:** The surface morphology of the Narran floodplain ecosystem as derived via LiDAR: (a) the topography of the Narran system; (b) vegetation heights; (c) the isometric relationship for Clear Lake; and (d) the isometric relationship for Narran Lake.

topographic maps or aerial photographs (Webster *et al.* 2006).

A detailed digital elevation model of the Narran floodplain ecosystem in Australia highlights its complex topography (Fig. 3.4a, b). This digital elevation model was constructed from a LiDAR survey undertaken when the floodplain was dry and comprises a dataset of ~650 000 000 data points, with a spatial resolution of ~1 $m^2$, an absolute positional and vertical accuracy of 8–11 cm and an internal (point-to-point within flight lines) accuracy of 2–3 cm. The surface of the Narran floodplain ecosystem is characterised by several lakes, wetlands and a floodplain surface that is dissected by an extensive channel network. At the time of data collection, there were 804.5 km of floodplain channels with an average channel network density of 0.048 $km/km^2$. In addition, the two principal lakes within the Narran floodplain ecosystem, Narran Lake and the Northern Lakes, have their own unique isometric relationship between water depth and surface areas (Fig. 3.4c, d). This is because of several different sub-storages within the different lake systems being activated at different inundation levels. Thus, the topography of the Narran floodplain ecosystem does not resemble a flat or nearly flat, featureless landform, but instead displays a high degree of physical complexity (Scown *et al.* 2015).

Understanding complexity has become important to floodplain ecosystem science in several disciplines including ecology, geomorphology and hydrology. Characteristics of complex systems include: non-linearity, multiple feedbacks, emergent phenomena and heterogeneity (Fonstad *et al.* 2013). Theories of complexity imply that in floodplain landscapes, patterns and processes may not be directly related in ways that we might assume. For example, inundation may not result in a relatively simple expansion and contraction of floodwaters across floodplain surfaces, as proposed in the flood pulse concept (Junk *et al.* 1989). The enhanced topographic complexity of the Narran floodplain ecosystem, because of the presence of an extensive floodplain channel network, presumably contributes to its complex inundation patterns (Rayburg *et al.* 2009). The non-linear behaviour of the wetting and drying regime of this floodplain will have implications for the structure and function of floodplain vegetation at multiple scales in this system.

## Scaled processes

The hierarchical nature of fluvial processes is a fundamental tenet of geomorphology (Schumm 1977; Knighton 1984; Church 2002). Interrelated geomorphological factors sit within a hierarchy of influence, where certain factors set the conditions within which others can form (Schumm 1991; de Boer 1992). Geology and climate sit at the top of the hierarchy (Schumm and Lichty 1965) and act to control the physiography of a catchment, the types of vegetation and soils that are present in a catchment, and the uses to which humans put the land. In turn, catchment physiography, vegetation, soils and land use control channel-forming factors such as stream discharge, sediment input, sediment transport, sediment calibre, bank characteristics and valley slope. These channel-forming factors subsequently control the factors that occur at the bottom of the hierarchy, such as channel slope, water velocity, planform and cross-sectional channel dimensions, and the arrangement of bedforms in the channel (Knighton 1984). Thus, in accordance with principles of hierarchy theory, attributes operating at one level of a geomorphic hierarchy constrain the formation of attributes at successively lower levels. Associated with the geomorphic hierarchy is an inherent relationship between space and time (Schumm and Lichty 1965). For example, geomorphological processes that occur over long time scales ($10^6$–$10^3$ years), such as planation and denudation, manifest larger spatial-scale units such as drainage networks (Frissell *et al.* 1986). Conversely, geomorphological processes that occur over short time scales ($10^1$–$10^0$ years), such as small bank failures, flood scouring and deposition and minor bedform changes, generally manifest smaller spatial-scale units such as riffle-pool systems (Frissell *et al.* 1986).

The process-based interrelationship of factors within a geomorphological hierarchy forms the basis of many stream classification schemes.

Common among them is the premise that smaller-scale systems develop within the constraints set by larger-scale systems, and as a result of these hierarchical links, the spatial and temporal boundaries of each scale in the classification can be defined by a set of physical factors. There are many hierarchical stream classification schemes, each with different levels. Perhaps one of the most familiar of these is the scheme of Frissell *et al.* (1986) with five nested scales: stream systems, segment systems, reach systems, riffle-pool systems and microhabitat systems. Many other hierarchical stream classifications exist, and have been reviewed by Kondolf *et al.* (2003). The units within stream classification schemes have been used to guide stream management activities and restoration (e.g. Rosgen 1996). Importantly, stream classifications with a hierarchical process basis can help to identify the correct scale of observation for the geomorphic or ecological process of interest (Dollar *et al.* 2007; Parsons and Thoms 2007).

Most stream classification systems are focused on the channel environment. While stream classification schemes do take into account the larger-scale factors controlling the expression of physical features at a particular scale, the purpose of stream classification is generally to organise the instream channel environment. Thus, levels such as river networks, functional process zones, reaches, functional sets, functional units and mesohabitats dominate stream classifications (Table 3.1). The floodplain is largely excluded from classification schemes, although floodplains are sometimes considered as a process-domain or process-form unit (e.g. Montgomery 1999). The hydro-geomorphic processes shaping floodplain environments occur hierarchically and just as river systems can be organised hierarchically, so can floodplains. Floodplains have been classified on the ability of the main river channel to entrain and transport sediment and the erosional resistance of the floodplain sediment. For example, Nanson and Croke (1992) derived a genetic classification with three classes of floodplain being recognised – high-energy non-cohesive, medium-energy non-cohesive, and low-energy cohesive floodplains – and 13 derivative orders and suborders. Floodplain orders range from confined, coarse-grained, non-cohesive floodplains in high-energy environments to unconfined, fine-grained, cohesive floodplains in low-energy environments. Each floodplain class has distinct geomorphological features (such as scroll bars or extensive back swamps) that distinguish each other in terms of genesis and resulting morphology. However, many floodplain classifications do not take into account the hierarchical relationships between large and small-scale floodplain forming processes.

We propose that floodplains are hierarchically organised into nested scales of valley setting, functional process zones, functional sets, functional units and mesohabitats (Table 3.1). Valley setting sits at the top of the hierarchy and this is related to larger-scale landscape shaping processes such as glaciation, tectonics and weathering regime. Processes operating at commensurate spatial and temporal scales result in floodplains contained within constrained and unconstrained settings (Table 3.1). At the second level of the hierarchy is the functional process zone, which is related to processes of sediment storage and residence time. Floodplains forming at this level are associated with anastomoses and meandering of river channels. The third level of the hierarchy is the functional set level, which is controlled by sediment supply and flow regime. Floodplains at this level can be differentiated by whether they are formed by vertical or lateral accretion or through complex cut and fill processes. A large part of the Nanson and Croke (1992) floodplain classification scheme sits at this level of organisation. The fourth level of the hierarchy is the functional unit level of organisation, controlled by sediment deposition and erosion. Functional units commonly found on floodplains include wetlands, benches, floodplain channels, levees and scrolls. At the bottom of the hierarchy, mesohabitats are created by different sediment properties and interactions with vegetation and other organisms. Examples of mesohabitats include sediment patches of different nutrient and moisture properties and different vegetation types.

Classification systems such as proposed here for floodplains are useful for many purposes, especially in interdisciplinary settings. They provide a vehicle to improve communication between the disciplines (Kondolf *et al.* 2003) and when presented in a hierarchical framework, reduce the potential for scale mismatch between different disciplinary studies (Dollar *et al.* 2007).

## A FRAMEWORK FOR UNDERSTANDING THE ROLE OF THE PHYSICAL TEMPLATE IN FLOODPLAIN ECOSYSTEMS

The complexity, diversity and variability of floodplains challenge attempts to define them as functional ecosystems in a precise and widely applicable way (Nanson and Croke 1992). The form and behaviour of floodplains express many internal

**Table 3.1.** Typical levels in an instream river classification hierarchy (top) and proposed levels for a floodplain classification hierarchy (bottom)

River classification levels are compiled from Schumm (1977), Frissell *et al.* (1986), Petts and Amoros (1996) and Thorp *et al.* (2008).

| River hierarchy | | | | |
|---|---|---|---|---|
| **Level** | **Defining processes** | **Spatial scale (m²)** | **Temporal scale (yr)** | **Examples** |
| Stream network | Geologic processes (plate tectonics, uplift)<br>Glaciations<br>Planation and denudation | $10^6$ | $10^5$–$10^6$ | Watersheds, catchments |
| Functional process zone | Valley setting<br>Flow and sediment regimes | $10^4$–$10^6$ | $10^3$–$10^5$ | Braided, meandering |
| Functional set | Landform morphology<br>Lateral and vertical fluxes<br>Elevational change | $10^2$–$10^4$ | $10^1$–$10^3$ | Channel set (pools, riffles, backwaters)<br>Abandoned channel set<br>Aggradational set |
| Functional unit | Sedimentation<br>Flood frequency<br>Biomass production<br>Organic matter accumulation | $10^1$–$10^2$ | $10^{-1}$–$10^1$ | Gravel bars<br>Wetlands |
| Mesohabitat | Organic matter transport<br>Sediment scour and deposition | $10^{-1}$–$10^1$ | $10^{-2}$–$10^{-1}$ | Gravel patches<br>Sand bars<br>Scour holes<br>Submerged plant stands |
| **Floodplain hierarchy** | | | | |
| **Level** | **Defining processes** | **Spatial scale (m²)** | **Temporal scale (yr)** | **Examples** |
| Valley setting | Glaciation<br>Tectonics<br>Topography<br>Weathering regime | $10^6$ | $10^5$–$10^6$ | Constrained, unconstrained, high energy, low energy |
| Functional process zone | Sediment storage and residence time | $10^4$–$10^6$ | $10^3$–$10^5$ | Floodplains associated with anastomosing, braided, meandering and wandering river systems |
| Functional set | Sediment supply<br>Flow regime | $10^2$–$10^4$ | $10^1$–$10^3$ | Vertical, lateral accreted floodplain, cut and fill floodplain |
| Functional unit | Sediment deposition and erosion processes | $10^1$–$10^2$ | $10^{-1}$–$10^1$ | Billabongs, levees, scrolls, swales, benches, channels, wetlands |
| Mesohabitat | Sediment stability<br>Soil chemistry and moisture<br>Sediment texture<br>Vegetation<br>Microtopography | $10^{-1}$–$10^1$ | $10^{-2}$–$10^{-1}$ | Sediment patches, vegetation community |

and external geomorphological, hydrological and ecological influences, all of which interact closely. While the importance of these interactions is recognised (e.g. Phillips 1995), solutions to common floodplain problems tend to combine unconnected inputs from the different disciplines of geomorphology, hydrology and ecology rather than taking an interdisciplinary approach. Successful interdisciplinary science requires that separate disciplines gain a common understanding of the nature of the problem at hand, and identify the scales of relevant subsystem components, the underlying processes or phenomena, and the important variables involved, thus joining diverse disciplines and areas of understanding into a single conceptual-empirical structure. Frameworks are useful tools for achieving this as they aid different disciplines to work together in an integrated manner by ordering phenomena (Dollar *et al.* 2007).

The interdisciplinary study of floodplains lacks a conceptual framework on which to develop commonality and integration. There are floodplain models explaining, among other things, the influences of vertical and lateral river stability on floodplain construction processes (Leopold *et al.* 1964), and the importance of resource patches on floodplain mosaics (Pringle *et al.* 1988). However, these present the perspectives of single disciplines only and do not broadly serve the multi-dimensional decision-making environment of interdisciplinary floodplain science. Individually, they have value but do not provide a basis for ecologists, hydrologists and geomorphologists to integrate their thinking, concepts and data collection.

The generalised ecosystem framework of Pickett *et al.* (2003) identifies the key components of heterogeneity and their inter-linkages in space and over time. Applied to floodplains this framework suggests these ecosystems can be viewed as the interaction between an agent and the template or substrate (Fig. 3.5a) where agents create, maintain or transform the structural or functional features of a system and the template or substrate on which an agent acts is the spatial arrangement of physical or biological features of interest. For floodplains, the hydrological regime represents a key agent while different morphological units, such as scrolls, swales and floodplain channels, can be viewed as the template on which the physical agent of inundation acts, resulting in significant influences on floodplain ecosystem structure. The action of an agent on a template can result in the transition of a template from one state to another (state 2 in Fig. 3.5a). This transition to a new state can be affected by biotic and abiotic controllers, which influence the spatial extent and dispersion of materials as well as the intensity of the effect of an agent on the template. The combined action of an agent and various controllers modifies the existing structure of the template (e.g. if the character of state 2 differs from state 1 in Fig. 3.5a). The transition of a template from one state to another, because of the combined action of an agent and controller, can produce an ecological response, highlighting their functional relevance. If no ecological response occurs then they are functionally neutral and have no influence on the heterogeneity of the ecosystem. This conceptual framework has been used in various studies of landscape dynamics in both terrestrial and riverine ecosystems (e.g. Parsons *et al.* 2005).

The application of this framework to the physical template of Australian floodplains, and its functional significance to floodplain vegetation is outlined in Figure 3.5b. The template is represented by the floodplain vegetation landscape and the main responder is the spatial distribution of floodplain vegetation communities across floodplain surfaces. The flow regime and valley setting are the key agents that provide the structural and functional features of Australian floodplains and associated vegetation landscapes. The size, in particular the magnitude, frequency and duration of each inundation event, combined with the physical complexity of the floodplain surface controls not only the extent of wetting and drying but also the spatial pattern of the expansion and contraction of floodwaters. It is these interactions that are responsible for creating the altered floodplain template. Vegetation responds to this altered template in the shorter term by increasing biomass production and advancing life history stages such as flowering, seed production and recruitment. Vegetation responds in the longer term by forming mosaics of vegetation populations and communities in the floodplain landscape that can be characterised by measures of diversity, stand age, recruitment success, vegetation condition and spatial distribution. Thus, the interactions between the template, the controller and the altered template

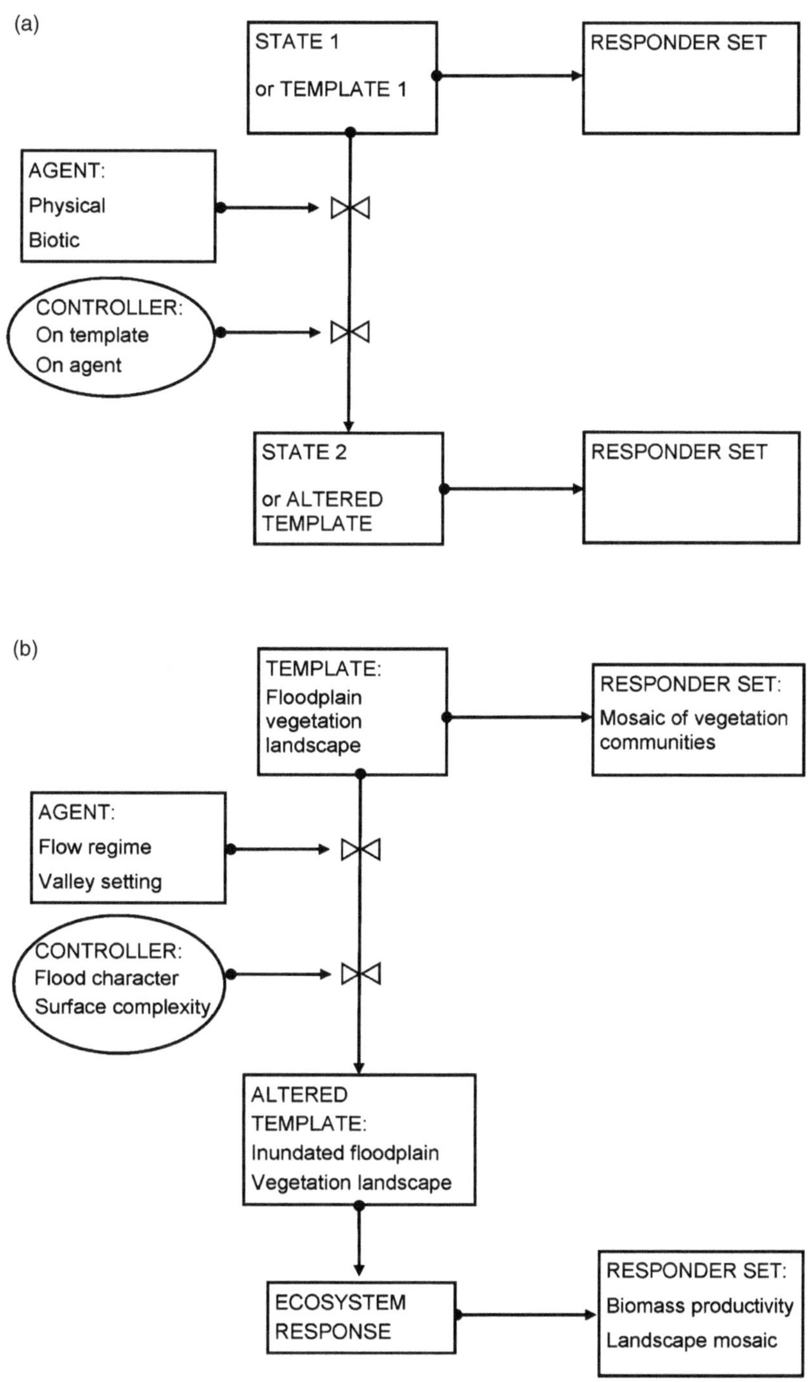


**Figure 3.5:** The ecosystems framework of Pickett *et al.* (2003): (a) a general framework illustrating a general representation of the generation of heterogeneity and biotic response to heterogeneity; and (b) applied to unconstrained floodplain ecosystems.

drive the dynamic nature of floodplain vegetation landscapes. The framework can be used in a hierarchical manner to allow integration between processes operating at different scales (e.g. Table 3.1). This structure accommodates feedback responses, allowing biotic consequences to contribute to the altered state. It also allows for consideration of downward constraints by higher levels and upward integration of processes from lower levels – an important factor in hierarchically organised systems like floodplain ecosystems.

## CONCLUSIONS

Large unconstrained floodplain ecosystems, set within valleys that can extend over 50 km in width, are a characteristic feature of Australia's riverine landscapes (Thoms and Sheldon 2000). The study of these ecosystems is not as well advanced as that of floodplains located in other climatic regions (cf. Fig. 3.1). We argue that our understanding of floodplains gained from research undertaken in temperate and tropical regions may not apply to Australian dryland floodplain ecosystems. This is especially so from an interdisciplinary perspective where the focus is on the interactions between the physical floodplain template and its associated vegetation landscape. Floodplains have long been characterised as aquatic terrestrial transition zones, with linear gradients in the expansion and contraction of floodwaters during overbank flows and the corresponding spatial organisation of vegetation communities from near channel to distal floodplain environments. Inundation across unconstrained floodplains produces a mosaic of wet patches, especially during the drying phase with intense fragmentation of floodwaters. The nonlinear pattern of wetting and drying in floodplains results, in part, from their complex surface topography. Processes that create this complexity are hierarchical in nature. Floodplains are not flat, featureless surfaces, but rather exhibit a high physical geodiversity, usually being dissected by a network of channels and containing numerous morphological features, including billabongs, ridges and swales. Each of these morphological features has a distinct substrate character in terms of their sediment texture, nutrient content and soil moisture properties.

The complexity of floodplain ecosystems challenges many traditional scientific methods. Their multi-causal, multiple-scale character often restricts the use of a conventional reductionist falsification scientific approach, except when applied at very small scales and within limited domains. The framework developed here for the interdisciplinary understanding of floodplain ecosystems provides a map that enables links between different disciplines to be made at appropriate scales. It recognises subsystems that contribute to the study of floodplains and the relevant levels of organisation within each, assigns appropriate scales and variables within the different subsystem organisational hierarchies, and describes interactions between the agents of change, templates and controllers.

## REFERENCES

Alvarez MC (2011) *Floodplains: Physical Geography, Ecology and Societal Interactions.* Nova Science Publications, New York, USA.

Anderson MG, Walling DE, Bates PD (1996) *Floodplain Processes.* Wiley, Chichester, UK.

Bridge JS (2003) *Rivers and Floodplains: Forms, Processes and Sedimentary Record.* Blackwell, Oxford, UK.

Cheetham M, Bush RT, Keene AF, Erskine WD, Fitzsimmons KE (2010) Longitudinal correlation of late Quaternary terrace sequences of Widden Brook, southeastern Australia. *Australian Journal of Earth Sciences* **57**, 97–109. doi:10.1080/08120090903416229

Church M (2002) Geomorphic thresholds in river landscapes. *Freshwater Biology* **47**, 541–557. doi:10.1046/j.1365-2427.2002.00919.x

de Boer DH (1992) Hierarchies and spatial scale in process geomorphology: a review. *Geomorphology* **4**, 303–318. doi:10.1016/0169-555X(92)90026-K

Dollar ESJ, James CS, Rogers KH, Thoms MC (2007) A framework for interdisciplinary understanding of rivers as ecosystems. *Geomorphology* **89**, 147–162. doi:10.1016/j.geomorph.2006.07.022

Erskine WD (2008) Channel incision and sand compartmentalization in an Australian sandstone basin subject to high flood variability. *International Association of Hydrological Sciences* **325**, 283–290.

Erskine WD, Melville MD (2008) Geomorphic and stratigraphic complexity: Holocene alluvial history of upper Wollombi Brook, Australia. *Geografiska Annaler* **90**, 19–35. doi:10.1111/j.1468-0459.2008.00331.x

Ferguson R, Brierley GJ (1999) Downstream changes in valley confinement as a control on floodplain morphology, Lower Tuross River, New South Wales, Australia: a constructivist approach to floodplain analysis. In *Varieties of Fluvial Form.* (Eds AJ Miller and A Gupta) pp. 377–407. Wiley, Chichester, UK.

Fonstad MA, Dietrich JT, Courville BC, Jensen JL, Carbonneau PE (2013) Topographic structure from motion: a new development in photogrammetric measurement. *Earth Surface Processes and Landforms* **38**, 421–430. doi:10.1002/esp.3366

Forman RTT (1997) *Land Mosaics: the Ecology of Landscapes and Regions.* Cambridge University Press, Cambridge, UK.

Frissell CA, Liss WJ, Warren CE, Hurley MD (1986) A hierarchical framework for stream habitat classification: viewing streams in a watershed context. *Environmental Management* **10**, 199–214. doi:10.1007/BF01867358

Gupta A (2007) *Large Rivers.* Wiley, Chichester, UK.

Gurnell AM, Petts GE (2002) Island-dominated landscapes of large floodplain rivers, a European perspective. *Freshwater Biology* **47**, 581–600. doi:10.1046/j.1365-2427.2002.00923.x

Hughes FMR (1997) Floodplain biogeomorphology. *Progress in Physical Geography* **21**, 501–529. doi:10.1177/030913339702100402

Hynes HBN (1975) The stream and its valley. *Verhandlungen der Internationalen Vereinigung für Theoretische und Angewandte Limnologie* **19**, 1–15.

Junk WJ, Bayley PB, Sparks RE (1989) The flood-pulse concept in river-floodplain ecosystems. *Canadian Special Publication of Fisheries and Aquatic Sciences* **106**, 110–127.

Klassen GJ, Van Urk A (1985) Resistance to flow of floodplains with grasses and hedges. In *Proceedings of the 21st International Association for Hydro-Environment Engineering and Research Congress*, pp. 469–473.

Knighton D (1984) *Fluvial Forms and Processes.* Edward Arnold, London.

Kondolf GM, Montgomery DR, Piegay H, Schmitt L (2003) Geomorphic classification of rivers and streams. In *Tools in Fluvial Geomorphology.* (Eds GM Kondolf and H Piegay) pp. 171–204. Wiley, Chichester, UK.

Latterell JJ, Bechtold JS, O'Keefe TC, Van Pelt R, Naiman RJ (2006) Dynamic patch mosaics and channel movement in an unconfined river valley of the Olympic Mountains. *Freshwater Biology* **51**, 523–544. doi:10.1111/j.1365-2427.2006.01513.x

Leigh C, Sheldon F, Kingsford RT, Arthington AH (2010) Sequential floods drive 'booms' and wetland persistence in dryland rivers: a synthesis. *Marine and Freshwater Research* **61**, 896–908. doi:10.1071/MF10106

Leopold LD, Wolman MG, Miller JP (1964) *Fluvial Processes in Geomorphology.* W.H. Freeman and Company, New York, NY.

Lewin J, Macklin MG (2003) Preservation potential for late Quaternary river alluvium. *Journal of Quaternary Science* **18**, 107–120. doi:10.1002/jqs.738

McCarthy TS, Tooth S, Kotze DC, Collins NB, Wandrag G, Pike T (2010) The role of geomorphology in evaluating remediation options for floodplain wetlands: the case of Ramsar-listed Seekoeivlei, eastern South Africa. *Wetlands Ecology and Management* **18**, 119–134. doi:10.1007/s11273-009-9153-7

Melville MD, Erskine WD (1986) Sediment remobilization and storage by discontinuous gullying in humid southeastern Australia. *International Association of Hydrological Sciences* **159**, 277–286.

Mertes LAK, Daniel DL, Melack JM, Nelson B, Martinelli LA, Forsberg BR (1995) Spatial patterns of hydrology, geomorphology, and vegetation on the floodplain of the Amazon River in Brazil from a remote sensing perspective. *Geomorphology* **13**, 215–232. doi:10.1016/0169-555X(95)00038-7

Millennium Ecosystem Assessment (MEA) (2003) *Ecosystems and Human Well-being: a Framework for Assessment*. Island Press, Washington, DC.

Miller AJ (1995) Valley morphology and boundary conditions influencing spatial patterns of flood flow. In *Natural and Anthropogenic Influences in Fluvial Geomorphology: the Wolman Volume*. (Eds JE Costa, AJ Miller, KW Potter and PR Wilcock) pp. 57–82. American Geophysical Union, Washington, DC.

Montgomery DR (1999) Process domains and the river continuum. *Journal of the American Water Resources Association* **35**, 397–410. doi:10.1111/j.1752-1688.1999.tb03598.x

Murray O, Thoms MC, Rayburg S (2006) The diversity of inundated areas in semi-arid floodplain ecosystems. *International Association of Hydrological Sciences* **306**, 277–286.

Naiman RJ, Decamps H, McClain ME (2005) *Riparia: Ecology, Conservation and Management of Streamside Communities*. Elsevier, Amsterdam.

Nanson GC, Croke JC (1992) A genetic classification of floodplains. *Geomorphology* **4**, 459–486. doi:10.1016/0169-555X(92)90039-Q

Nanson GC, Erskine WD (1988) Episodic changes of channel and floodplains on coastal rivers of NSW. In *Fluvial Geomorphology of Australia*. (Ed. RF Warner) pp. 201–221. Academic Press, Sydney.

Northern Floodplains Regional Planning Committee (NFRPC) (2004) *Vegetation Communities of the Northern Floodplains, Western New South Wales. Volume 2: Brewarrina Shire*. Northern Floodplains Planning Committee, Walgett, NSW.

Parsons M, Thoms MC (2007) Hierarchical patterns of physical-biological associations in river ecosystems. *Geomorphology* **89**, 127–146. doi:10.1016/j.geomorph.2006.07.016

Parsons M, Thoms MC (2013) Patterns of vegetation greenness during flood, rain and dry resource states in a large, unconstrained floodplain landscape. *Journal of Arid Environments* **88**, 24–38. doi:10.1016/j.jaridenv.2012.07.023

Parsons M, McLoughlin CA, Kotschy KA, Rogers KH, Rountree MW (2005) The effects of extreme floods on the biophysical heterogeneity of river landscapes. *Frontiers in Ecology and the Environment* **3**, 487–494. doi:10.1890/1540-9295(2005)003[0487:TEOEFO]2.0.CO;2

Patton PC, Schumm SA (1975) Gully erosion, northwestern Colorado: a threshold phenomenon. *Geology* **3**, 88–90. doi:10.1130/0091-7613(1975)3<88:GENCAT>2.0.CO;2

Petts GE, Amoros C (1996) The fluvial hydrosystem. In *Fluvial Hydrosystems*. (Eds GE Petts and C Amoros) pp. 1–12. Chapman & Hall, London.

Petty AM, Douglas MM (2010) Scale relationships and linkages between woody vegetation communities along a large tropical floodplain river, north Australia. *Journal of Tropical Ecology* **26**, 79–92. doi:10.1017/S0266467409990319

Phillips JD (1995) Biogeomorphology and landscape evolution: the problem of scale. *Geomorphology* **13**, 337–347. doi:10.1016/0169-555X(95)00023-X

Pickett STA, Cadenasso ML, Benning TL (2003) Biotic and abiotic variability as key determinants of Savanna heterogeneity at multiple spatiotemporal scales. In *The Kruger Experience: Ecology and Management of Savanna Heterogeneity*. (Eds J du Toit, H Biggs and KH Rogers) pp. 22–40. Island Press, Washington, DC.

Pinay G, Decamps H, Chauvet E, Fustec E (1990) Functions of ecotones in fluvial systems. In *The Ecology and Management of Aquatic-Terrestrial Ecotones*. (Eds RJ Naiman and H Decamps) pp. 141–169. Man and the Biosphere Series, Volume 4. UNESCO, Paris.

Pringle CM, Naiman RJ, Bretschko G, Karr JR, Oswood MW, Webster JR, Welcomme RL, Winterbourne MJ (1988) Patch dynamics in lotic systems: the stream as a mosaic. *Journal of the North American Benthological Society* **7**, 503–524. doi:10.2307/1467303

Puckridge JT, Sheldon F, Walker KF, Boulton AJ (1998) Flow variability and the ecology of large rivers. *Marine and Freshwater Research* **49**, 55–72. doi:10.1071/MF94161

Rayburg SC, Thoms MC, Neave N (2009) A comparison of digital elevation models generated from different data sources. *Geomorphology* **106**, 261–270. doi:10.1016/j.geomorph.2008.11.007

Reid MA, Ogden R, Thoms MC (2011) The influence of flood frequency, geomorphic setting and grazing on plant communities and plant biomass on a large dryland floodplain. *Journal of Arid Environments* **75**, 815–826. doi:10.1016/j.jaridenv.2011.03.014

Rosgen DL (1996) *Applied River Morphology*. Wildland Hydrology, Fort Collins, CO.

Sabo JL, Sponseller R, Dixon M, Gade K, Harms T, Heffernan J, Jani A, Katz G, Soykan C, Watts J, Welter J (2005) Riparian zones increase regional species richness by harboring different, not more, species. *Ecology* **86**, 56–62. doi:10.1890/04-0668

Schumm SA (1973) Geomorphic thresholds and complex response of drainage systems. In *Fluvial Geomorphology*. (Ed. M Morisawa) pp. 299–310. Publications in Geomorphology, State University of New York, Binghamton, NY.

Schumm SA (1977) *The Fluvial System*. John Wiley & Sons, New York, NY.

Schumm SA (1991) *To Interpret the Earth: Ten Ways to Be Wrong*. Cambridge University Press, Cambridge, UK.

Schumm SA, Lichty RW (1965) Time, space and causality in geomorphology. *American Journal of Science* **263**, 110–119. doi:10.2475/ajs.263.2.110

Scown M, Thoms MC, De Jager NR (2015) An index of floodplain surface complexity. *Hydrology and Earth System Science* **12**, 4507–4540.

Simons A, Bennett SJ, Neary VS (2004) Riparian vegetation and fluvial geomorphology: problems and opportunities. In *Riparian Vegetation and Fluvial Geomorphology*. (Eds SJ Bennett and A Simon) pp. 1–10. American Geophysical Union, Washington, DC.

Southwell M (2008) Floodplains as dynamic mosaics: sediment and nutrient patches in a large lowland river landscape. PhD thesis. University of Canberra, ACT.

Southwell MR, Thoms MC (2011) Patterns of nutrient concentrations across multiple floodplain surfaces in a large dryland river system. *Geographical Research* **49**, 431–443. doi:10.1111/j.1745-5871.2011.00699.x

Southwood TRE (1977) Habitat, the templet for ecological strategies? *Journal of Animal Ecology* **46**, 337–365. doi:10.2307/3817

Southwood TRE (1988) Tactics, strategies and templets. *Oikos* **52**, 3–18. doi:10.2307/3565974

Stanford JA, Lorang MS, Wissmar RC (2005) The shifting habitat mosaic of river ecosystems. *Verhandlungen des Internationalen Verein Limnologie* **29**, 123–136.

Thoms MC (1995) Sedimentation in the Barmah Forest. *International Association of Hydrological Sciences* **230**, 121–130.

Thoms MC (2003) Floodplain-river ecosystems: lateral connections and the implications of human interference. *Geomorphology* **56**, 335–349. doi:10.1016/S0169-555X(03)00160-0

Thoms MC, Foster JM (2002) Multivariate statistical techniques and the interpretation of floodplain sedimentation. *International Association of Hydrological Sciences* **276**, 453–464.

Thoms MC, Olley JM (2004) The stratigraphy, mode of deposition and age of inset flood plains on the Barwon-Darling River, Australia. In *Sediment Transfer Through the Fluvial System*. (Eds V Golosov, V Belyaev and DE Walling) pp. 316–325. International Association of Hydrological Sciences publication 288. IAHS Press, Wallingford, UK.

Thoms MC, Parsons M (2011) Patterns of vegetation community distribution in a large, semi-arid floodplain landscape. *River Systems* **19**, 271–282. doi:10.1127/1868-5749/2011/019-0050

Thoms MC, Sheldon F (2000) Australian lowland river systems: an introduction. *Regulated Rivers: Research and Management* **16**, 375–383. doi:10.1002/1099-1646(200009/10)16:5<375::AID-RRR591>3.0.CO;2-#

Thoms MC, Parker CR, Simons M (2000) Heavy metal dispersal in the Hawkesbury-Nepean River. In *River Management: the Australian Perspective*. (Eds B Finlayson and S Brizga) pp. 197–219. Wiley, Chichester, UK.

Thorp JH, Thoms MC, Delong MD (2008) *The Riverine Ecosystem Synthesis: Toward Conceptual Cohesiveness in River Science.* Academic Press, London.

Tockner K, Stanford JA (2002) Riverine flood plains: present state and future trends. *Environmental Conservation* **29**, 308–330. doi:10.1017/S037689290200022X

van Coller AL, Rogers KH, Heritage GL (2000) Riparian vegetation-environment relationships: complimentarity of gradients versus patch hierarchy approaches. *Journal of Vegetation Science* **11**, 337–350. doi:10.2307/3236626

Walker KF, Thoms MC (1993) A case history of the environmental effects of flow regulation on a semi-arid lowland river: the River Murray, South Australia. *Regulated Rivers: Research and Management* **8**, 103–119. doi:10.1002/rrr.3450080114

Ward AD, Trimble SW (2004) *Environmental Hydrology.* CRC Press, Boca Raton, FL.

Wardle G (2010) The story of our land revealed through plants. In *Desert Channels: the Impulse to Conserve.* (Eds L Robin, C Dickman and M Martin) pp. 90–105. CSIRO Publishing, Melbourne.

Warner RF (1992) Floodplain evolution in a New South Wales coastal valley, Australia: spatial process variations. *Geomorphology* **4**, 447–458. doi:10.1016/0169-555X(92)90038-P

Wasson RJ (1996) Annual and decadal variation of sediment yield in Australia, and some global comparisons. *International Association of Hydrological Sciences* **224**, 268–280.

Webster TL, Murphy JB, Gosse JC (2006) Mapping subtle structures with light detection and ranging (LIDAR): flow units and phreatomagmatic rootless cones in the North Mountain Basalt, Nova Scotia. *Canadian Journal of Earth Sciences* **43**, 157–176. doi:10.1139/e05-099

Whited DC, Lorang MS, Harner MJ, Hauer FR, Kimball JS, Stanford JA (2007) Climate, hydrologic disturbance, and succession: drivers of floodplain pattern. *Ecology* **88**, 940–953. doi:10.1890/05-1149

# 4

# A history of Australia's riverine habitats and vegetation

*Michael Reid, Sophie Bickford, Peter Gell and Christine Kenyon*

## INTRODUCTION

Our understanding of vegetation in Australia's riverine landscapes is greatly enhanced by an understanding of its history and dynamics over long timescales. Individuals and cohorts of structurally dominant riverine plant species, such as *Eucalyptus camaldulensis* (river red gum), *E. largiflorens* (black box) and *E. coolabah* (coolibah), can live for several hundred years and possess numerous adaptations that allow them to persist through long periods of suboptimal conditions. Similarly, it can take many years for plant taxa to disperse into newly created or transformed habitats within temporally dynamic riverine landscapes. Consequently, it is not possible to fully understand current species distributions and associations without knowledge of past states of riverine vegetation communities as well as their rates and trajectories of change. In contrast, many plants within riverine wetlands respond extremely rapidly to changing water regimes. Drying can desiccate water plants in a matter of days and dry lake beds can be colonised by native and exotic weedy taxa within a few months. Seeds buried in riverine sediments can also respond rapidly to rewetting, providing the period of drying is not prolonged, enabling wetland plant communities to recover. Shifts in the prevalence of rewetting, however, can lead to loss of seed banks over decadal scales and affect the capacity of wetland plant communities to respond to flooding. Changes in water quality can also dramatically affect aquatic ecosystems with increased nutrients, salts and turbidity all capable of inducing vegetation responses.

Plant fossil and palynological (i.e. pollen) records in Australia extend back for millions of years to, and beyond, the time when Australia was part of the super continent Gondwana. In this chapter, however, we will focus on the character and development of Australian riverine vegetation over the past 30 000 years. There are both practical and conceptual reasons for this focus. First, there are very few reliable and continuous fossil records of vegetation that extend beyond the past 30 millennia and fewer still that can be used to reliably inform on riverine vegetation. Second, conditions further back in time are of decreasing relevance to contemporary conditions. The focus on the past 30 000 years also enables us to examine the state of riverine vegetation over several important periods in Australia's history including significant periods of climate change and human intervention.

The chapter begins with an examination of the evidence regarding the state of Australia's riverine vegetation from the height of the most recent glacial period at around 22 000 years ago. We then discuss how riverine vegetation responded to the transition to the warmer conditions of the Holocene, and adjustments to channel form in response to altered gradients due to the rise in sea level. Next, we examine subsequent developments through the relative stability of the Holocene. Finally, this chapter examines how Australia's riverine vegetation has been affected by human activities, beginning with those of the first Australians and concluding with the dramatic changes that followed settlement by Europeans more than 200 years ago.

### Palaeo-reconstruction of past vegetation

There are several ways past vegetation communities can be reconstructed using currently available evidence. In general, however, fossil and subfossil records of plant remains that have been preserved in sedimentary records are analysed with respect to the timelines of sediment sequences, which can be established using a variety of techniques (e.g. radiocarbon dating). In this way, changes to vegetation over time can be revealed. The most widespread and complete of these records are provided by sequences of preserved pollen and spores.

A combination of features makes pollen and spores particularly valuable for palaeo-reconstruction of vegetation communities. First, pollen and spores are produced in large numbers by their parent plants and transported by wind over large distances, so even if sites of sediment accumulation are patchily distributed in the landscape the records preserved therein reflect, in an interpretable way, the extra-local and regional vegetation. Second, pollen and spores are highly refractory and are, thus, readily preserved in accumulated sediments. Finally, pollen and spores vary structurally along taxonomic lines and, although the level of taxonomic resolution that can be achieved via light microscopy varies across groups, Australian pollen and spores can largely be identified to ecologically meaningful levels.

Records of past vegetation can also be sourced from careful extraction of evidence from descriptions, accounts and anecdotes hidden in a wide range of historical documents. This can include the reporting of the occurrence of particular species by early botanists, and a subsequent lack of reports from more recent phases of intensive biological survey.

## RIVERINE VEGETATION OF THE LATE PLEISTOCENE AND LAST GLACIAL MAXIMUM

Pollen records extending back to the last glacial maximum (LGM) (i.e. around 20 000 years ago) are relatively rare because sites of sediment accumulation have limited lifespans. Sites may cease to accumulate sediments because drainage basins fill over time or changes in environmental conditions prevent material collecting in their basins. This is particularly important in riverine environments because rivers are so dynamic. The degree of environmental change experienced over the period leading up to, and following, the LGM was such that few riverine sites experienced continuous sediment accumulation over this period. Consequently, our understanding of riverine vegetation during this period is drawn from pollen records of sites that are not strictly speaking in riverine settings (that is, non-riverine sites of sediment accumulation that nonetheless preserve riverine pollen taxa from the landscape), along with information from the handful of riverine sites that do exist, as well as from other sources such as phylogeography and geomorphology. Based on these information sources, though, we can make some general inferences about riverine vegetation from the late Pleistocene and LGM.

Records from sites in the central highlands of Victoria (Fig. 4.1; McKenzie 1997; McKenzie 2002) and the southern Australian Alps (Kershaw *et al.* 2007) provide an overview of changes to riverine vegetation during and since the LGM. Not surprisingly, given the evidence of cooler and drier conditions during the LGM, these records suggest the treeline was substantially lower at this time. McKenzie (2002) suggests the small snow melt streams at these high altitudes lacked riparian trees and were instead fringed by semi-aquatic

**Figure 4.1:** Locations of sites and regions referred to in the text.

herbaceous taxa, such as Ranunculaceae, with alpine herbfields dominating the broader landscape. This interpretation is supported by the record from Caledonia Fen in the south-west portion of the Australian Alps (Kershaw *et al.* 2007).

At lower altitudes, several records suggest riparian zones provided important refugia for tree taxa. One of the longest records of continuous sediment accumulation in a riverine setting comes from Boulder Flat in East Gippsland (Fig. 4.1; Kenyon 1989). At ~140 m above sea level, this site is a riverine swamp formed on the margin of a small pocket floodplain of the Errinundra River which has been dated to over 26 000 years ago (Kenyon 1989). Although overall pollen concentrations are low in the basal section of the sediment retrieved from this site, the relative abundance of arboreal pollen in the record indicates that open forests, or at least woodlands, persisted in the area right through the LGM (Kenyon 1989). The persistence of substantial arboreal pollen representation in the Boulder Flat record through the LGM suggests that this riverine environment may have provided an important refuge for forest taxa during this period of harsh climatic conditions. However, other studies of late Pleistocene climates in East Gippsland suggest that the region escaped the harshest extremes of the LGM (Webb *et al.* 1992), perhaps partly as a result of the ameliorating effect of the warm east Australian current.

A record from Wyelangta Swamp, situated around 450 m above sea level in the Otway Ranges of Victoria (Fig. 4.1), tells a similar story for the LGM in south-east Australia. While the Otways generally appear to have been characterised by quite open vegetation with extensive alpine communities, including taxa now largely restricted to Tasmania, the sheltered riparian location at Wyelangta formed an apparent refuge for rainforest taxa during the late Pleistocene (McKenzie and Kershaw 2000). Importantly, radiocarbon dating provides evidence of a substantial hiatus in the sediment record corresponding to the LGM and the last glacial–interglacial transition (LGIT) (McKenzie and Kershaw 2000). Similarly, a record from Lake Baraba, situated ~100 km south-west of Sydney and 300 m above sea level (Fig. 4.1), also suggests that riparian environments provided important refugia for forest and woodland communities during the late Pleistocene (Black *et al.* 2006). At this site, *Casuarina* remained dominant from over 43 000 years ago right through to the early Holocene, when Myrtaceaous plants expanded, possibly due to the development of fringing *Melaleuca* communities. As with all data for this period, the record at Lake Baraba around the time of the LGM needs to be interpreted carefully as the radiocarbon date of around 23 060 years before present that provides the chronological control for the LGM is taken from the middle of a zone of poor pollen preservation. Chronological controls above and below this point are 7550 and >43 000 years before present, respectively (Black *et al.* 2006), so it is not possible to say definitively what the pollen assemblages of the LGM were.

Despite the uncertainties involved, there remains good evidence that riparian habitats at these lower altitudes (i.e. 150–500 m above sea level) were sheltered from extreme cold and experienced moisture subsidies from surface and subsurface drainage, thus providing important refugia for forest and woodland communities through the extreme climatic conditions of the LGM (Kenyon 1989; McKenzie and Kershaw 2000; Black *et al.* 2006; Kershaw *et al.* 2007). These habitats may also have remained moist as a result of the much-reduced evapotranspiration under cooler climates. For example, the relatively dry modern Brachina Creek in the Flinders Ranges (Fig. 4.1) was a perennial swamp through the LGM owing to substantially reduced evaporation loss and consequent elevated water tables (Williams *et al.* 2001).

Lowland rivers and their associated vegetation appear to have been more severely affected by the climate extremes of the late Pleistocene than those in montane or submontane environments. The Riverina Plain (Fig. 4.1) – which consists of the alluvial floodplains of the Murray, Murrumbidgee and Lachlan rivers and their tributaries in the southern Murray–Darling Basin (Page *et al.* 2009) – is distinguished by palaeochannels that display forms indicative of bedload or mixed load dominance with bankfull discharges substantially higher than those of today over several phases of fluvial activity since the last interglacial period (Bowler 1978; Page and Nanson 1996; Page *et al.* 2009). The higher bankfull discharges and coarse sediment loads, particularly during the coldest period leading up to and following the LGM, can be attributed to altered water balances and soil instability associated with reduced vegetation cover and periglacial processes at higher altitudes. Cooler conditions and a substantially lowered snow line would also have led to a larger winter snow pack and, hence, more seasonal discharge peaking in spring with the thaw (Page *et al.* 2009). Additionally, sea levels in the LGM were 130 m lower than present. Therefore, the gradient of all rivers was considerably greater during this period than in the Holocene. The South Australian reach of the Murray River, for example, is presently little more than 4 m above sea level. At the LGM, however, the river at Blanchetown (Fig. 4.1), site of lock 1, would have been over 130 m above the ocean baseline and flowing down a much steeper gradient even if the mouth was a further 200 km distant. It is also possible that channel instability may itself have contributed to the reduction in tree cover. In a recent synthesis, Page *et al.* (2009) speculated that the bankfull discharge to slope relationship associated with the late Pleistocene climate brought the rivers of the Riverina Plain close to the transition point between braided and meandering channel forms, resulting in inherently unstable channel forms.

There is little direct evidence of vegetation communities associated with Australia's lowland floodplains of the late Pleistocene and LGM. However, it has been suggested that riparian tree cover was substantially lower during the Late Pleistocene due to greater aridity (Bowler 1978), particularly at the LGM when fluvial activity appears to have been at its minimum (Page *et al.* 2009). Moreover, some researchers have suggested that reduced bank stability associated with lower riparian tree cover was a factor that contributed to the

**Table 4.1.** Summary of changes in Australian riverine environments since the late Pleistocene

| Period | Time span(years) | Climate conditions | Riverine zones | Environmental changes |
|---|---|---|---|---|
| Late Pleistocene to last glacial maximum | >50 000 to 20 000 | Increasingly cool and dry | Uplands | Lower tree lines, snowmelt streams fringed by semi-aquatic herbs and alpine herbfields |
| | | | Mid-elevation | Riparian zones likely refugia for trees and vegetation characteristic of moist environments, especially in areas close to the east coast, which experienced a moister climate relative to the surrounding landscape during the LGM |
| | | | Lowlands | Mixed and bedload rivers, broad and shallow with large meander wavelengths floodplains sparsely treed broad coastal plains |
| Pleistocene–Holocene transition | 20 000 to 10 000 | Period of rapid warming | Uplands and mid-elevation | Little information likely a dynamic period of adjustment to new hydrological regimes associated with climate amelioration and partial colonisation by woody vegetation to higher elevations |
| | | | Lowlands | Persistence of glacial river regimes in systems with high elevation catchments, but transition to suspended load sinuous channel rivers by 10 000 years ago in most river systems coastal lowland rivers subject to marine transgression changing river base-level likely to have contributed to adjustments in channel forms upstream |
| The Holocene | 10 000 to present | Relative climate stability, with warmest and wettest conditions occurring during the mid-Holocene | Uplands and mid-elevation | Development of upland, headwater swamps and bogs and 'chain of ponds' systems |
| | | | Lowlands | Establishment of 'modern' river channels relative stability in floodplains and floodplain vegetation development of estuaries and associated mangrove and saltmarsh communities, formation of coastal barrier systems |
| Human influence – Indigenous | >50 000 to present | Variable, including extremes of LGM and mid-Holocene | Uplands and mid-elevation | Burning of riparian vegetation |
| | | | Lowlands | Burning of floodplain vegetation to maintain open vegetation, encourage re-sprouting and aid harvesting of food plants harvesting of aquatic food plants construction of fish traps |
| Human influence – non-Indigenous | 220–present | Relative climate stability | Uplands | Clearing of land (incl. burning) resulting in loss of 'chain of ponds' systems and widespread gully erosion in eastern and southern Australia |
| | | | Mid-elevation | Clearing of land (incl. burning) sediment 'slugs' derived from upstream sheet and gully erosion sluicing for gold construction of dams exotic taxa (willows) |
| | | | Lowlands | Clearing of land (incl. burning), stock routes de-snagging elevated turbidity and associated loss of submerged macrophytes floodplain timber harvesting low-level weir construction irrigation |

palaeochannel forms observed (Bowler 1978; Hope 1994). The notion that floodplain tree cover was substantially reduced at the LGM is supported by phylogeographic evidence of river red gum distributions, which suggests that present day populations of this species in the Murray–Darling Basin are derived from northern populations that expanded southwards after the LGM as the climate became warmer and wetter; the original populations having been either eliminated or reduced to isolated fragments as the arid core of Australia expanded at the height of the last glacial period (Butcher *et al.* 2009).

In summary (Table 4.1), Australian temperate catchments during the LGM probably featured lower ground cover and heavily reduced tree cover as a result of cooler temperatures and/or greater aridity. These climatic conditions are likely to have created less stable catchment soils and therefore higher sediment supplies to streams and rivers. In addition, greater seasonality and higher runoff associated with reduced evapotranspiration may also have produced higher peak flows. Together with steeper channel gradients, these conditions are likely to have resulted in streams that carried higher volumes of sand and gravels and that were less stable than those of the more moderate climates of the Holocene.

## RIVERINE VEGETATION OF THE PLEISTOCENE–HOLOCENE TRANSITION

The period between 20 000 and 10 000 years ago was one of global climate warming. The climatic consequences of this global warming for the Australian region were complex, particularly in relation to effective precipitation and, by extension, runoff (Turney *et al.* 2006; Tibby 2012). It is likely that the complex patterns of climate change observed in the Australian region during the transition from glacial to interglacial global climate conditions reflect the effects of latitudinal movements of the subtropical high pressure systems and the southern hemisphere westerlies and associated low pressure systems, as well as variation in the strength of the Asian monsoon (Turney *et al.* 2006; Williams *et al.* 2009; Tibby 2012).

As for the LGM, there are just a handful of records from fluvial settings that provide a continuous record of changes in vegetation associated with the transition to Holocene climate (Tibby 2012). The small number of records can be attributed to instability in fluvial systems over this period. During the deglaciation, considerable channel readjustment is likely to have occurred in response to declines in the gradients of catchments associated with rises of more than 100 m in sea level. This would have been modulated by climate-driven changes in land cover runoff relations. Additionally, low altitude river lands would have become inundated and any records of change these environments may have contained are lost. Such inundation and channel readjustment combined would have eliminated some habitats, but generated others including transitional habitats along with those that persisted through to the Holocene. These changes may have occurred quite rapidly in conjunction with global sea level rise, which at the height of the deglaciation is estimated to have been around 50 cm per decade (Rahmstorf 2007). Many sites that have basal dates extending beyond the Holocene – such as Lake Baraba (Black *et al.* 2006), Tom Burns *Sphagnum* bog, Storm Creek, Snobs Creek (McKenzie 1997), Wyelangta (McKenzie and Kershaw 2000) and Delegate River (Fig. 4.1; Ladd 1979) – show evidence of discontinuities or hiatuses, perhaps associated with scouring of sediment, and have little chronological control around the critical period from 20 000 to 10 000 years ago, which incorporates the LGM and transition from glacial to interglacial climate conditions (Kenyon 1989; McKenzie 1997, 2002; McKenzie and Kershaw 2000; Black *et al.* 2006).

The collective picture that emerges from the records that cover this critical period is of instability in local riverine vegetation communities. While it is likely that this instability reflected changes in climate and their effects on catchment water balances and local fluvial processes (Ogden *et al.* 2001), it may also have resulted from intrinsic fluvial adjustments. The possible scouring of sediments, along with difficulties associated with disentangling the effects of climate, climate-driven fluvial adjustments and intrinsic fluvial adjustments, highlight how hard it can be to interpret records from fluvial settings (see Ladd 1979).

For the lowland rivers of the Murray–Darling Basin, the transition from late glacial high discharge, bedload rivers to modern narrow, sinuous channel forms carrying suspended load with reduced bankfull discharge appears to have occurred asynchronously across the basin (Ogden *et al.* 2001; Page *et al.* 2009). Ogden *et al.* (2001) found that the oldest modern channels occurred on the Goulburn River and Darling River by around 12 000 years ago, whereas the earliest evidence of modern channels on the Murray and Murrumbidgee rivers does not appear until after 10 000 years ago. Both Ogden *et al.* (2001) and Page *et al.* (2009) speculate that this difference may reflect differences in the relative elevations of the respective catchments. Accordingly, soils in the relatively low altitude catchments of the Goulburn and Darling stabilised earlier as the climate warmed and the upper catchments revegetated, while there was a delay in this process in the relatively higher altitude catchments of the Murray and Murrumbidgee. Similar scenarios were proposed to occur in the eastern Victorian highlands (Kenyon 1989).

Rapidly rising sea levels during the Pleistocene–Holocene transition also had dramatic effects on riverine vegetation. Most obviously, many late Pleistocene lowland riverine environments would have been consumed by the rising sea; others would have been transformed to estuarine environments. However, the rising sea levels are also likely to have indirectly impacted upstream fluvial processes (Nanson *et al.* 1995) and, consequently, riverine vegetation. By changing the base-level (i.e. the lowest point to which it can flow) of rivers discharging into the ocean, the rising sea levels would have altered river gradients to a varying degree. This effect would have been most dramatic in areas of the continent with narrow continental shelves, such as the south-east coast, where the 130 m+ increase in base level over as little as 20 km may have dramatically altered river long-profile slopes. Even for the Murray River, which during the LGM flowed across an additional 200 km or so of continental shelf, the 130 m increase in base level would have reduced the overall slope of the river and would have driven substantial changes in the transition to modern channel forms.

## RIVERINE VEGETATION OF THE HOLOCENE

Many records from upland areas commence in the early to mid-Holocene as a result of the development of bogs and swamp environments following climate amelioration and stabilisation along with fluvial adjustments to new water balances (McKenzie 1997, 2002; McKenzie and Kershaw 1997; Black *et al.* 2007; Nanson 2009). These changes to water balances provided conditions suited to more perennial surface waters, which enabled continuous sediment accumulation and were conducive to the preservation of fossil remains. Even through the relative stability of the Holocene, these records show dynamic riverine vegetation communities associated with hydroseral changes (i.e. wetland community succession associated with gradual depth changes due to sedimentation) (Nanson 2009). Nanson (2009) proposed a model of development for these upland floodplain systems whereby a gradual process of peat formation occurs within drainage lines to form 'chain of ponds' systems followed by rapid phases of incision caused by changes to the water balance in association with climate fluctuations or changes to catchment vegetation cover.

In lowland river settings, billabongs and other floodplain wetlands provide common sites of sediment accumulation and are used frequently in Holocene environmental reconstructions. These reconstructions generally focus on the aquatic ecosystems of the billabongs or wetlands themselves rather than riverine vegetation per se (Reid 2002, 2008; Gell *et al.* 2005a, 2005b, 2007; Kenyon 2005; Leahy *et al.* 2005; Reid *et al.* 2007; Fluin *et al.* 2010).

Nevertheless, several of these studies incorporate pollen and/or plant macrofossil records and many inform on changes in aquatic vegetation either directly, through pollen, and/or indirectly through changes in plant-associated taxa.

The dynamic nature of lowland floodplain systems, particularly in relation to the transition from late glacial to interglacial climate conditions, is such that most of these records are confined to the mid to late Holocene. The majority of sites formed through meander bend cut-off or avulsion, thus sediments record the initial transition from lotic to lentic conditions and, subsequently, the gradual transition to more terrestrial environments as these features filled with sediments. A common feature of these records is for aquatic pollen taxa to record a gradual transition from a low but persistent early abundance of Cyperaceae, presumably reflecting the presence of marginal emergent sedges, to later assemblages that also include submerged aquatic taxa, such as *Myriophyllum* and *Triglochin*, suggesting that more structurally and floristically diverse macrophyte communities developed over time (Reid 2002; Gell *et al.* 2005a; Kenyon 2005; Leahy *et al.* 2005; Reid *et al.* 2007). As discussed below, this successional process appears to have been interrupted at many sites by the disturbances created by European settlement.

In addition to providing records of local vegetation changes, floodplain wetlands also provide a record of broader floodplain vegetation communities. In contrast to the hydroseral succession suggested by aquatic pollen records, most pollen records suggest relative stability in floodplain vegetation over the past several thousand years and indicate that floodplain vegetation through much of the mid to late Holocene may have been dominated by the same broad taxonomic groups that dominate contemporary temperate floodplains: *Eucalyptus*, Poaceae, Chenopodiaceae and Asteraceae (Reid 2002; Kenyon 2005; Leahy *et al.* 2005; Reid *et al.* 2007), with taxa such as *Casuarina* (Bennett and Mooney 2003; Gell *et al.* 2005a; Black *et al.* 2006), *Callitris* (Reid *et al.* 2007) and *Duma* (Gell *et al.* 2005a) being locally important.

An exception to this general pattern is the Barmah–Millewa Forest (Fig. 4.1) where fossil pollen and geomorphic evidence suggest a highly dynamic environment through the Holocene. This dynamism reflects adjustments in channel form in response to tectonic activity in the late Pleistocene, subsequent climate change and channel development resulting in the effluent network and complex hydrology that supports the forest–wetland system of today (Kenyon 2005; Rutherfurd and Kenyon 2005; Stone 2006). Around 60 000 years ago, uplift of a section of Murray River floodplain north of modern-day Echuca known as the Cadell Tilt Block caused the ancestral course of the Murray to be diverted. The initial diversion took the river to the north of the Cadell Block along the course of Gulpa Creek and the Wakool River and later the Edward River. Lake Kanyapella formed, south of the Barmah region, in a fault-angle depression of the Cadell Fault ~30 000 years ago (Bowler and Harford 1966; Rutherfurd and Kenyon 2005). Lake Kanyapella was full during the Pleistocene, but dry by the mid-Holocene. The Lake Kanyapella lunette was subsequently breached and this event drained the Moira–Barmah palaeo-lake and a new channel formed to the south. The current southerly course of the Murray River in this region is the result of the most recent Holocene avulsion when the Murray River levee was breached near Picnic Point. As the new channel developed, it progressively built a levee and formed a digitate delta across the palaeo-lake floor, bisecting the flat, swampy Barmah and Moira lakes (Smith *et al.* 1989; Rutherfurd and Kenyon 2005). Pollen and sediment records from the Moira Marshes show a progression from lake to swamp to levee as the Barmah choke developed from around 4000 to 500 years ago (Kenyon 2005).

Coastal rivers appear to have undergone more dramatic changes during the course of the Holocene. Although the climate had reached its broad Holocene range by around 10 000 years ago, sea levels did not reach their approximate modern levels until around 7900–7700 years ago and continued to rise (Sloss *et al.* 2007). In the south-east, levels peaked at ~1.5 m higher than the present ~7400 years ago and remained around this level for several thousand years before declining to present levels ~2000 years ago (Sloss *et al.* 2007). These Holocene sea level fluctuations, along with interactions between fluvial and coastal processes, such as channel aggradation, barrier building and infilling of coastal valleys, meant that the levels of tidal influence and marine intrusion probably fluctuated greatly along many coastal rivers. These processes drove a continuous evolution in condition. At Lake Curlip on the lower Snowy River floodplain (Fig. 4.1), an open, active estuary featured conditions suited to *Ruppia*-dominated aquatic communities with fringing vegetation (MacGregor *et al.* 2005). As the developing barrier limited the tidal prism, the freshwater input from inland came to dominate and more diverse freshwater plant communities developed, dominated by *Potamogeton, Myriophyllum* and *Triglochin* (Ladd 1979). Ultimately, myrtaceous swamps prevailed, bringing tannic waters to the lake before European land use and transfer of water out of the catchment brought a return to more saline conditions.

## ABORIGINAL INFLUENCES ON RIVERINE VEGETATION

The history of riverine vegetation in Australia cannot be understood without taking into account that humans were using and shaping river environments for as long as Aboriginal people have occupied Australia, but especially over the past 5000 years (Head 1988; Balme 1995; Bandler 1995; Russell-Smith *et al.* 1997; Bowler *et al.* 2003). Connections between Aboriginal people and water are complex and the spiritual significance of water resources is reflected in numerous Dreamtime 'water' stories relating to the creation or return of water, the rivers, waterholes and wetlands. However lacking in detail our knowledge may be, there is sufficient evidence from our understanding of Aboriginal population levels and resource use to clearly indicate that Australian riverine landscapes were already actively managed environments when first encountered by European explorers and settlers.

The density of Aboriginal populations in riverine environments in pre-European Australia matched coastal regions (Lawrence 1971; Humphries 2007), and both were more densely occupied than other environments on the Australian mainland. In the Murray–Darling riverine area, population density of Aboriginal people appears to have been many times greater than in non-riverine inland areas (Meggitt 1966; Allen 1974; Coutts 1981; Kohen 1995). The Aboriginal population in the mid-Holocene along a 350-km stretch of the mid-Murray, for instance, was between 30 000 and 66 000 people, possibly representing the most densely populated area of Australia at this time (Pardoe 1998). Skeletal abnormalities indicate population pressure on natural resources in this region. Seasonal subsistence activities were determined by river flow and population densities along the rivers therefore fluctuated seasonally, with people moving onto adjacent plains or travelling down the river during dry periods and using rivers as routes for seasonal migrations (Bride 1898; Allen 1974; Flood 1989).

Water and fire regimes, key factors influencing vegetation composition and structure in riverine environments, were both manipulated by Aboriginal people to enhance the productivity of rivers, their tributaries, associated floodplains, swamplands and billabongs. Fire was used to keep floodplains and their surrounding vegetation open to aid hunting and passage. Consequently, floodplain forests were more open before European settlement than they are today due to regular burning. The explorer Charles Sturt described three inland Australian floodplain landscapes, one being 'plains of considerable extent, wholly destitute of timber' that 'invariably occur in the immediate neighbourhood of some river' (Gammage 2011). Swamp and wetland burning was judiciously but regularly undertaken on southern inland rivers to stimulate reed growth, harvest rhizomes, create open water for specific fish and plant species, and facilitate passage (Sturt 1838; Mitchell 1839; Curr 1863; Gott 1982). The following quote from

Edward Curr, an early squatter of the Murray River in the Moira region (Fig. 4.1), suggests that mosaic, rather than broadscale burning, of reed beds was undertaken – 'The reeds were by patches and strips of different hues and growth, in accordance with their ages and the periods at which they had last burnt' (Curr 1965). There is also strong evidence of consistent and thorough burning of floodplain and wetland vegetation by Aboriginal people on northern Australian rivers (Head 1994; Yibarbuk *et al.* 2001; Preece 2002). On the floodplain of the South Alligator River in western Arnhem Land (Fig. 4.1), for example, burning of floodplains was used to keep fuel loads low and slow-burning peat fires were used to remove thick stands of native grass *Hymenachne acutigluma* (hymenachne) and promote the growth of highly prized food species such as *Eleocharis dulcis* (water chestnut) (Russell-Smith *et al.* 1997).

Floodplains and adjacent woodlands were managed by Aboriginal people for a diverse range of plant foods. Along the mid reaches of the Murray River, for example, nutritious rhizomes, bulbs, corms and tubers of more than two hundred species of herbs, including lilies and orchids, were relied on in winter and spring when the new storage organs form (Curr 1965; Gott 1982, 2005; Zola and Gott 1992; Vanderwal 2001). Active cultivation of some species, such as *Microseris scapigera* (yam daisy) also occurred (Gott 1988; Zola and Gott 1992) and was likely to have influenced soil structures and the composition of understorey vegetation. Roots were gathered all year whereas seeds and fruits were seasonally available. Some plant parts were left in the ground and regular digging to obtain underground plant parts turned the litter and ash, fertilising the soil as well as thinning roots and loosening the soil, making it more absorbent (Buchan 1979; Gott and Conran 1991). Harvesting encouraged re-sprouting, enhanced growth or increased the extent of specific plants (Humphries 2007).

Local hydrological environments were also manipulated by Aboriginal people to enhance conditions for plants and animals, including fish. Engineering of river channels was undertaken in rivers throughout the country (Mathews 1903; Dargin 1976; Head 1989). In the eastern Riverina and south Gippsland, for example, waterways were blocked to extend swamps (Gilmore 1934). Engineering of wetlands was also carried out on northern tropical rivers, as observed by Gunn (1908, quoted in Gammage 2011) – 'Being so shallow and wide-spreading, the lagoons would dry up early in the Dry were it not that the black are able to refill them at will from the river, for here the Roper indulges in a third "duck under" so curious that with a few logs and sheets of bark the blacks can block the way of its waters and overflow them into the lagoons, thereby ensuring a plentiful larder to a host of wild fowl, and, incidentally to themselves'.

Manipulation of river hydrology and vegetation by Aboriginal people occurred against a template of larger-scale vegetation patterns determined by natural flow regimes and river morphologies that were markedly different from those of today. The historical record indicates that impacts early in European settlement wrought massive changes to the geomorphic character and behaviour of rivers, dramatically altering the composition and extent of riverine vegetation communities. Early surveyor and explorer descriptions led Gammage (2011) to describe a typical Australian river as one that was 'spreading and slow, easily forded, rich in swamps and reeds, with deep holes for fish, yabbies and platypus and lined with possum trees, marsupial grassland and bird and reptile shelter'. The Murrumbidgee and Molongolo rivers (Fig. 4.1) were 'chains of ponds' with frequent fords and dry crossings like so many of the other rivers surveyed by Hovell from 1832 to 1835 between Yass and Michalago. In 1824, Hamilton Hume and William Hovell traversed the Murray then named the Hume River, near Albury, and made the following description: 'we suddenly arrived at the bank of a very fine river – at least 200 feet wide, apparently deep, and the stream running at three knots' (Hovell in Webb 1984: 9). The banks were 'clothed with verdure to the water's edge; their general height various, but seldom either more or less than eight or nine feet, inclined or precipitous, as they happen, by the bendings of the stream, to be more or less exposed to the action of the current' (Hovell 1971: 38). Along the floodplain either side of the river was 'a perpetual succession of lagoons, extending generally in length from one to two miles, and about a quarter of a mile in breadth'. These lagoons were 'furnished with an inlet from the river, and an outlet into it' (Hovell 1971: 38). These areas were 'thickly wooded (the trees consisting principally of the blue gum, mostly of a large growth) ... overgrown with vines of various descriptions, and the fern, the peppermint, flax-plant, and currajong' (Bland 1831). Detailed geomorphic studies of a southern coastal river (Brierley and Fryirs 1998) support historical descriptions synthesised by Gammage (2011) showing that river courses through the middle to upper parts of the Bega catchment were discontinuous at the time of European settlement (Fig. 4.1), with extensive swamps along the middle and upper courses, and a continuous, low-capacity channel along the lowland plain (Brooks and Brierley 1997). Historical explorer descriptions indicate the rivers further north in New South Wales and Queensland were of similar character at the time of European settlement (Mitchell 1839).

## RIVERINE VEGETATION SINCE EUROPEAN SETTLEMENT

European arrival in Australia resulted in a dramatic reduction in numbers of Aboriginal people inhabiting riverine environments. Diseases introduced with European settlement ravaged Indigenous populations (Sturt 1838; Curr 1863; Beveridge 1883) and many people were displaced. Rivers were areas of high contest between their original inhabitants and the new settlers, with their water and surrounding grasslands and forests highly sought after by the first squatters. Aboriginal river management practices were swiftly abandoned and replaced with European exploitation.

### Squatters

The first phase of European settlement was the squatting phase, and by 1840 squatters and their stock occupied rivers in a belt from Port Phillip up through the Lachlan River and the coast to the Darling Downs (Fig. 4.1; Scott and Olley 2003). Not having tenure, squatters generally took up the open grassy plains for their stock and rather than using the axe, their main tool was fire, which was used almost annually to encourage grass growth for the purpose of running stock. European squatters often lit fires in different seasons from those set by Aboriginal people. The initial impacts of stock on riverine environments were likely great. Stocking rates on riverine holdings by early graziers were in many places very high compared with stocking rates of today. Inland rivers were especially subject to intensive grazing pressure in the summer months and during periods of drought when fodder and water were scarce and river bends were used as cattle holding yards (Newell 1970; Coulson 1979; Clark 2000), leading to degradation of floodplain vegetation and their associated wetlands (Robertson 1996, 1997). Early settlers' descriptions from this time focused on the economically valuable grasslands and indicate they were rich in herbs at the time of settlement (Hamilton 1845; Curr 1965; Rolls 1981), but that selective grazing, altered fire regimes and the introduction of weed species changed the species mix of grasses and depleted their herb component (Beadle 1948; Moore 1953; Hodgkinson 1856; Tremont and McIntyre 1994). An 1842 description from near the Murray River at Tongala, in northern Victoria, provides an illustration: 'the interstices [of kangaroo grass] were filled with luxuriant herbage; the yam generally, and in some cases the myrnong, or native carrot, prevailing' ... 'the inferior grasses and weeds – now [1863] unfortunately abundant – were almost, if not entirely absent' (Curr 1965). Not only were rivers the focus of large squatting runs, but they were used as freeways to move stock between the new colonies by 'overlanders'. At times the numbers of stock moved were great, with mobs of up to 30 000 head being moved at once (Roberts 1964; Pearson and Lennon 2010).

## Gold rush

Disturbance to catchments and rivers was greatly intensified from the 1850s due to the gold rush on the western slopes of New South Wales and in central and eastern Victoria. Mining practices rapidly degraded streams and surrounding catchments through sediment working and large-scale vegetation clearance. The impact of mining on sediment loads has been estimated for the Ovens River, where the entire 2100 ha of floodplain between Porepunkah and Myrtleford was worked to a depth of tens of metres (Evans 1975; Ladson 2000; Cottingham *et al.* 2001). Ladson (2000) estimated that, in 1913, 35 million $m^3$ of sediment was washed into the Ovens River. An absence of vegetation records from the time allows only inference on its impact on riverine vegetation (Newell 1970; Walker 1971). Large areas of box and ironbark (*Eucalyptus* spp.) forests were removed and several introduced weeds probably also appeared at this time (Newell 1970). Ethnohistoric evidence for intensive mining in the forests of East Gippsland reveal that regions afforded wilderness status today have in fact had large volumes of timber removed for mines and wood fires, only to recover to seemingly pristine condition (Gell *et al.* 1993).

Grazing and mining activities during this period appear to have resulted in wide-scale sheet and gully erosion in catchments (Convey 1980; Prosser *et al.* 2001; Olley and Scott 2002). Further downstream, the sediment transported from upstream disturbed catchments resulted in the deposition of 'sand slugs' in the middle reaches of streams, reducing the complexity and stability of stream beds and likely causing loss of habitat for aquatic macrophytes (Rutherfurd 2000; Prosser *et al.* 2001). Finer sediments were transported further downstream and contributed to order of magnitude increases in rates of sediment accumulation in lowland floodplain wetlands and billabongs (Peterson 1996; Gell *et al.* 2009). The effects of this sediment influx on aquatic plants are likely to have been dramatic. Regional surveys of billabongs through the middle reaches of the Murray show that submerged aquatic macrophytes are largely absent from these environments today (Ogden 2000). Palaeo-records from these environments suggest that this is a recent phenomenon caused by the activities of early European settlers, with some pollen records indicating declines in submerged aquatic plants around the time of European settlement (Gell *et al.* 2005a; Reid *et al.* 2007); indeed, records commonly show declines in plant-associated taxa such as benthic algae (Reid *et al.* 2007; Reid 2008, 2010), along with microcrustaceans (Ogden 2000), at the time of European settlement. Together, these data suggest a regional decline in submerged aquatic plants, probably due to attenuated light penetration caused by suspended sediments derived from eroded catchment soils (Ogden 2000; Reid *et al.* 2007). An exception is the Murray River flats at Swanport in South Australia (Fig. 4.1), where rapid sediment accumulation has shallowed the wetland allowing macrophytes to proliferate (Gell 2010).

## Closer settlement: catchment disturbance and river transformation

From the 1860s, the Closer Settlement Land Acts in New South Wales and Victoria brought closer settlement and the onset of more intensive farming practices as well as increased vegetation clearance. Rivers were again highly sought after by land selectors and their floodplains were favoured for cultivation. Water was diverted to irrigate dry floodplains and swamps were drained. Extensive reed beds that occurred in some places along inland rivers were often considered a hindrance to river trade and flow and so were removed, while channels were subject to straightening by artificial meander cut-off, de-snagging and dragline channel widening (Bushby 1980; Mulham 1994; Reinfelds *et al.* 1995). The removal of reed beds were major modifications that would have considerably changed flow patterns and flood regimes downstream and on surrounding floodplains, with implications for their vegetation community composition and function. Land holdings were much smaller than squatters' runs and overstocking was common. High value timber, including *Toona ciliate* (red cedar), was harvested along rivers on the eastern seaboard and ringbarking began in earnest from the 1870s and continued through to the 20th century. River red gum was heavily harvested in southern catchments resulting in alteration of forest structure from dominance by large, wide-spreading trees to mixed-aged stands of smaller pole trees (Mac Nally *et al.* 2011).

Catchments in high rainfall areas were cleared for dairies and cultivation. An 1865 article on mining in *The Argus* newspaper noted that several rivers had become drains even in the absence of mining and that these 'great changes' were the result of 'pastoral activities' (Powell 1989). In the Bega catchment by the 1860s the metamorphosis of many of its river environments had occurred with the clearance of riparian and floodplain vegetation and drainage of wetlands and back swamps triggering widening of the main river channel near the Bega township from around 40 m to 140 m (Brierley and Fryirs 1998). The highly variable flows of inland rivers were seen as an impediment to the development of agriculture and so settlers built weirs, dams, wells and water diversions to increase the reliability of water supply (Powell 1993; Kingsford 2000). Water was pumped from the Murray River system in substantial volumes from the 1890s, using wood-fired steam pumps for gravity-fed irrigation through an extensive system of privately constructed channels (McKinney 1889; Grant 1970). This water extraction impacted greatly on local river levels and altered floodplain hydrology, and the uncontrolled floodplain irrigation resulted in rising water tables. Groundwater levels in the region were recorded as being 8–10 m below the surface to 4 m in 1890 and 1 m in 1900, with salt problems evident in the Kerang region by the 1920s (Fig. 4.1; Powell 1993). Intensive stocking around a waypost near Salt Creek in the south-west of NSW induced increased billabong salinity from as early as 1880 (Fig. 4.1; Gell *et al.* 2007).

## River regulation

A phase of large-scale dam building in Australia began in the 1920s, with lock 1 at Blanchetown commissioned in 1922 and the construction of the Hume Dam on the Murray River. This phase of water resource development greatly escalated from the 1950s to the 1980s (Smith 1998; Bonyhady 2000). While water diversion had occurred from the 1860s in many rivers in southern Australia, with significant impacts on riverine vegetation, this relatively recent phase of gross modification of river flows and the resultant reduction in frequency, duration and timing of flooding is considered to have wrought the most change to the nature and function of all aspects of river ecology, including riverine vegetation (Walker 1985; Kingsford 2000). Furthermore, this has occurred in conjunction with rising groundwater tables and increased salinity and, more recently, the effects of climate change increasing drought and extreme rain events.

Rivers in the Murray–Darling Basin have been the most highly regulated and diverted, with 87% of the basin's divertible water extracted, primarily for irrigation. Rivers in the basins of north and south-eastern Australia, Tasmania, south-west Western Australia and South Australia have also been highly regulated through damming and diversion (Kingsford 2000). In the Murray–Darling Basin, dams with large holding capacities were constructed with the intention of radically altering the flow of rivers to service demand for irrigation (Walker 1985; Maheshwari *et al.* 1995). Water is stored in the winter months and released when demand is high in the summer months, resulting in a reduction in variability and long-term average flows as well as a seasonal redistribution of flows (Walker 1985). An aim of these schemes has been to minimise 'water wastage' by reducing water flow to floodplain wetlands, aquifers or the sea (Kingsford 2000). Through this activity, many floodplains and wetlands have been permanently, or near permanently, alienated from rivers and are reverting to terrestrial systems. In river red gum floodplain forests on the Murray River, for example, freshwater meadows have been depleted by

60%, shallow freshwater marshes by 61% and deep freshwater marshes by 55% (Ward 1991; VEAC 2008). In Barmah Forest, the loss of the *Pseudoraphis spinescens* (Moira grass) plains to encroaching river red gum, and the invasion of *Juncus ingens* (giant rush) has resulted from changes in flood regime since the commencement of regulated flows in the 1940s (Bren 1992). Die-back of floodplain vegetation as a whole on the Murray River has been widespread and was estimated to be almost 80% in 2011 (Mac Nally *et al.* 2011). Many remaining wetland and floodplain communities are in poor health and are undergoing changes in their composition, growth and regeneration in response to altered flooding regimes (e.g. Chesterfield 1986; Bren 1992; Jolly *et al.* 1993; Stokes *et al.* 2010; Mac Nally *et al.* 2011). Where water storages have been built, some wetlands have become permanently inundated, and environments near weirs permanently flow. In these situations, the expansion of plant species favouring stable water levels, such as introduced willows (*Salix* spp.) and *Typha*, has occurred in drowned riparian communities that developed under periodic dry–wet conditions (Gell 2012).

Hydrological changes to swamp vegetation have also been attributed to changing fire regimes and land use, particularly in relation to intensive forestry activities in East Gippsland (Gell *et al.* 1993). Post-European increases in burning, attributed to squatters, saw a decline in fire-sensitive *Tasmannia* (a sub-alpine shrub belonging to the family Winteraceae), which recovered under a regime of fire suppression imposed after the enquiry into the 1939 wildfires. Indeed, the most substantial changes in the recent pollen records from riverine wetlands in East Gippsland, even when compared with the 12 000-year record of Ladd (1979), were the prolific sporing of *Sphagnum* and flowering of *Myriophyllum pedunculatum* attributed to drying following post-harvesting forest regrowth on the Delegate River (Wilby and Gell 1994).

## CONCLUSIONS

Australia's riverine vegetation communities are dynamic, reflecting the variability of riverine landscapes generally. However, the current state of Australia's riverine vegetation is itself a legacy of longer-term natural and anthropogenic changes that have occurred over past millennia. During the last 30 000 years, riverine vegetation has responded to the complex and interactive effects on fluvial processes of the dramatic global climate transition from the LGM to the Holocene, substantial sea level changes, subsequent Holocene climate fluctuations, the influences of Aboriginal people and, perhaps most significantly, the profound land use and hydrological changes wrought by agri-industrial society in the past 200 years.

## REFERENCES

Allen H (1974) The Bagundji of the Darling Basin: cereal gatherers in an uncertain environment. *World Archaeology* **5**, 309–322. doi:10.1080/00438243.1974.9979576

Balme J (1995) 30 000 years of fishery in western New South Wales. *Archaeology in Oceania* **30**, 1–21. doi:10.1002/j.1834-4453.1995.tb00324.x

Bandler H (1995) Water resources exploitation in Australian prehistory environment. *The Environmentalist* **15**, 97–107. doi:10.1007/BF01901293

Beadle NCW (1948) *The Vegetation and Pastures of Western New South Wales with Special Reference to Soil Erosion*. NSW Government Printer, Sydney.

Bennett P, Mooney SD (2003) A palynological reconstruction of pre-European riparian vegetation at Wollombi, New South Wales and its application to stream bank management and revegetation. *Ecological Management & Restoration* **4**, S69–S78. doi:10.1046/j.1442-8903.4.s.9.x

Beveridge P (1883) Of the Aborigines inhabiting the great lacustrine and riverine depressions of the Lower Murray, Lower Murrumbidgee, Lower Lachlan and Lower Darling. *Journal and Proceedings of the Royal Society of New South Wales* **17**, 19–74.

Black MP, Mooney SD, Martin HA (2006) A >43 000-year vegetation and fire history from Lake Baraba, New South Wales, Australia. *Quaternary Science Reviews* **25**(21, 22), 3003–3016. doi:10.1016/j.quascirev.2006.04.006

Black MP, Mooney SD, Haberle SG (2007) The fire, human and climate nexus in the Sydney Basin, eastern Australia. *The Holocene* **17**(4), 469–480. doi:10.1177/0959683607077024

Bland W (1831) *Journey of Discovery to Port Phillip, New South Wales by Messrs WH Hovell and Hamilton Hume in 1824 and 1825.* A Hill, Sydney. [Facsimile edition, Libraries Board of South Australia, Adelaide, 1965.]

Bonyhady T (2000) *The Colonial Earth.* Melbourne University Press, Melbourne.

Bowler JM (1978) Quaternary climate and tectonics in the evolution of the riverine plain, southeastern Australia. *Landform Evolution in Australia.* (Eds JL Davies and MAJ Williams) pp. 70–112. Australian National Uuniversity Press, Canberra.

Bowler JM, Harford LB (1966) Quaternary tectonics and the evolution of the riverine plain near Echuca, Victoria. *Journal of the Geological Society of Australia* **13**(2), 339–354. doi:10.1080/00167616608728617

Bowler JM, Johnston H, Olley J, Prescott JR, Roberts RG, Shawcross W, Spooner NA (2003) New ages for human occupation and climatic change at Lake Mungo, Australia. *Nature* **421**, 837–840. doi:10.1038/nature01383

Bren LJ (1992) Tree invasion of an intermittent wetland in relation to changes in the flooding frequency of the River Murray, Australia. *Australian Journal of Ecology* **17**, 395–408. doi:10.1111/j.1442-9993.1992.tb00822.x

Bride TF (1898) *Letters from Victorian Pioneers.* 1983 edn. Currey O'Neil, Melbourne.

Brierley GJ, Fryirs K (1998) A fluvial sediment budget for upper Wolumla Creek, south coast, New South Wales, Australia. *The Australian Geographer* **29**, 107–124. doi:10.1080/00049189808703206

Brooks AP, Brierley GJ (1997) Geomorphic responses of lower Bega River to catchment disturbance, 1851–1926. *Geomorphology* **18**, 291–304. doi:10.1016/S0169-555X(96)00033-5

Buchan R (1979) Aborigines along the Murray. *Forest and Timber* **15**, 15–18.

Bushby JEP (1980) *Saltbush Country: History of the Deniliquin District.* Privately published with assistance of the Library of Australian History.

Butcher PA, McDonald MW, Bell JC (2009) Congruence between environmental parameters, morphology and genetic structure in Australia's most widely distributed eucalypt, *Eucalyptus camaldulensis. Tree Genetics & Genomes* **5**, 189–210.

Chesterfield EA (1986) Changes in the vegetation of the river red gum forest at Barmah, Victoria. *Australian Forestry* **49**, 4–15. doi:10.1080/00049158.1986.10674458

Clark ID (Ed.) (2000) *The Journals of George Augustus Robinson, Chief Protector, Port Phillip Aboriginal Protectorate, Volume 1: January 1839–30 September 1840.* Clarendon Heritage Matters, Ballarat, Victoria.

Convey T (1980) *The Days of Gold – Mining in the Tallangatta District.* Thomson Printing, Albury, NSW.

Cottingham P, Hannan G, Hillman T, Koehn J, Metzeling L, Roberts J, Rutherfurd I (2001) 'Report of the Ovens scientific panel on the environmental condition and flows of the Ovens River.' Technical report 9/2001. Cooperative Research Centre for Freshwater Ecology, University of Canberra, Canberra.

Coulson H (1979) *Echuca-Moama Murray River Neighbours.* McCabe Prints, Wangaratta, Victoria.

Coutts PJF (1981) *The Victorian Aboriginals 1800 to 1860.* Victorian Archaeological Survey, Ministry of Conservation, Victoria.

Curr EM (1863) *Recollections of Squatting in Victoria from 1841 to 1851.* Abridged with a forward and notes by Harley W. Forster. Melbourne University Press, Carlton, Victoria.

Curr EM (1965) *Recollections of Squatting in Victoria, Then Called the Port Phillip District (From 1841–1851).* Abridged with a foreword and notes by Harley W. Forster. Melbourne University Press, Carlton, Victoria.

Dargin P (1976) *Aboriginal Fisheries of the Darling-Barwon Rivers.* Development & Advisory Publications of NSW, Dubbo, NSW.

Evans E (1975) *Diary of a Welsh Swagman 1869–1894.* Abridged and annotated by William Evans. Sun Books, Melbourne, Victoria.

Flood JM (1989) *Archaeology of the Dreamtime: the Story of Prehistoric Australia and Her People*. Collins, Sydney.

Fluin J, Tibby J, Gell P (2010) The palaeolimnological record from Lake Cullulleraine, lower Murray River (south-east Australia): implications for understanding riverine histories. *Journal of Paleolimnology* **43**(2), 309–322. doi:10.1007/s10933-009-9333-8

Gammage B (2011) *The Biggest Estate on Earth: How Aborigines Made Australia*. Allen & Unwin, Sydney.

Gell P (2010) With the benefit of hindsight: the utility of palaeoecology in wetland condition assessment and identification of restoration targets. In *Ecology of Industrial Pollution*. (Eds L Batty and K Hallberg) pp. 162–188. Ecological Review Series, Cambridge University Press and the British Ecological Society, UK.

Gell PA (2012) Palaeoecology as a means of auditing wetland condition. In *Peopled Landscapes (Terra Australis 34): Archaeological and Biogeographic Approaches to Landscapes*. (Eds SG Haberle and B David) pp. 445–457. Australian National Uuniversity Press, Canberra.

Gell PA, Stuart I-M, Smith JD (1993) The response of vegetation to changing fire regimes and human activity in the Delegate River Catchment, East Gippsland, Victoria. *The Holocene* **3**(2), 150–160. doi:10.1177/095968369300300206

Gell PA, Bulpin S, Wallbrink P, Hancock G, Bickford S (2005a) Tareena Billabong – a palaeolimnological history of an ever-changing wetland, Chowilla floodplain, lower Murray–Darling Basin, Australia. *Marine and Freshwater Research* **56**(4), 441–456. doi:10.1071/MF04107

Gell P, Tibby J, Fluin J, Leahy P, Reid M, Adamson K, Bulpin S, MacGregor A, Wallbrink P, Hancock G, Walsh B (2005b) Accessing limnological change and variability using fossil diatom assemblages, south-east Australia. *River Research and Applications* **21**, 257–269. doi:10.1002/rra.845

Gell P, Tibby J, Little F, Baldwin D, Hancock G (2007) The impact of regulation and salinisation on floodplain lakes: the lower River Murray, Australia. *Hydrobiologia* **591**, 135–146. doi:10.1007/s10750-007-0806-3

Gell P, Fluin J, Tibby J, Hancock G, Harrison J, Zawadzki A, Haynes D, Khanum S, Little F, Walsh B (2009) Anthropogenic acceleration of sediment accretion in lowland floodplain wetlands, Murray–Darling Basin, Australia. *Geomorphology* **108**(1, 2), 122–126. doi:10.1016/j.geomorph.2007.12.020

Gilmore M (1934) *Old Days: Old Ways*. Angus & Robertson, Sydney.

Gott B (1982) Ecology of root use by the Aborigines of southern Australia. *Archaeology in Oceania* **17**, 59–67. doi:10.1002/j.1834-4453.1982.tb00039.x

Gott B (1988) Murnong – *Microseris scapigera*: a study of a staple food of Victorian Aborigines. *Australian Aboriginal Studies* **2**, 2–18.

Gott B (2005) Aboriginal fire management in south-eastern Australia: aims and frequency. *Journal of Biogeography* **32**, 1203–1208. doi:10.1111/j.1365-2699.2004.01233.x

Gott B, Conran J (1991) *Victorian Koorie Plants*. National Library Australia, Canberra.

Grant EJ (1970) *Walking With Time: the Story of the Wakool Country*. Abacada, Melbourne.

Hamilton G (1845) *The Journal of an Overlander, or, a Narrative of Journies in New South Wales and South Australia From 1836 to 1845 by George Hamilton*. National Library of Australia, MS 4299C, 1–73.

Head L (1988) Environment as artefact. *Australian Journal of Ecology* **13**, 21–49. doi:10.1111/j.1442-9993.1988.tb01415.x

Head L (1989) Using palaeoecology to date Aboriginal fish-traps at Lake Condah, Victoria. *Archaeology in Oceania* **24**, 110–115. doi:10.1002/j.1834-4453.1989.tb00220.x

Head L (1994) Landscapes socialised by fire: post-contact changes in Aboriginal fire use in northern Australia, and implications for pre-history. *Archaeology in Oceania* **29**, 172–181.

Hodgkinson C (1856) 'Report on the Murray River district, in reference to its geology, soil, prospective resources, proposed means of intercommunication.' Submitted to the Honourable Surveyor General in accordance with his instructions.

Hope GS (1994) Quaternary vegetation. In *History of the Australian Vegetation: Cretaceous to Recent*. (Ed. RS Hill) pp. 368–389. Cambridge University Press, Cambridge, UK.

Hovell W (1971) Journey of discovery to Port Phillip, New South Wales; by Messrs. W.H. Hovell and Hamilton Hume in 1824 and 1825. In *Australiana Facsimile Editions No. 12*. Libraries Board of South Australia, Adelaide. A. Hill, Sydney.

Humphries P (2007) Historical Indigenous use of aquatic resources in Australia's Murray–Darling Basin, and its implications for river management. *Ecological Management & Restoration* **8**, 106–113. doi:10.1111/j.1442-8903.2007.00347.x

Jolly ID, Walker GR, Thorburn PJ (1993) Salt accumulation in semi-arid floodplains soils with implications for forest health. *Journal of Hydrology* **150**, 589–614.

Kenyon CE (1989) A late Pleistocene and Holocene palaeoecological record from Boulder Flat, East Gippsland. Honours thesis. Monash University, Melbourne.

Kenyon CE (2005) Vegetation, fire and Aboriginal impact on the mid-Holocene Moira Marshes, New South Wales, Australia. *Proceedings of the Royal Society of Victoria* **117**(1), 41–59.

Kershaw AP, McKenzie GM, Porch N, Roberts RG, Brown J, Heijnis H, Orr ML, Jacobsen G, Newall PR (2007) A high-resolution record of vegetation and climate through the last glacial cycle from Caledonia Fen, southeastern highlands of Australia. *Journal of Quaternary Science* **22**(5), 481–500. doi:10.1002/jqs.1127

Kingsford R (2000) Ecological impacts of dams, water diversions and river management on floodplain wetlands in Australia. *Austral Ecology* **25**, 109–127. doi:10.1046/j.1442-9993.2000.01036.x

Kohen JL (1995) *Aboriginal Environmental Impacts*. UNSW Press, Sydney.

Ladd PG (1979) Past and present vegetation on the Delegate River in the highlands of eastern Victoria. II Vegetation and climatic history from 12,000 B.P. to present. *Australian Journal of Botany* **27**, 185–202. doi:10.1071/BT9790185

Ladson AR (2000) A multi-component indicator of stream condition for waterway managers: balancing scientific rigour with the need for utility. PhD thesis. Department of Civil and Environmental Engineering, University of Melbourne, Melbourne.

Lawrence R (1971) Habitat and economy: a historical perspective. In *Aboriginal Man and Environment in Australia*. (Eds DJ Mulvaney and J Golson) pp. 249–261. Australian National University Press, Canberra.

Leahy PJ, Tibby J, Kershaw AP, Heijnis H, Kershaw JS (2005) The impact of European settlement on Bolin Billabong, a Yarra River floodplain lake, Melbourne, Australia. *River Research and Applications* **21**(2, 3), 131–149. doi:10.1002/rra.837

Mac Nally R, Cunningham SC, Baker PJ, Horner GJ, Thomson JR (2011) Dynamics of Murray–Darling floodplain forests under multiple stressors: the past, present, and future of an Australian icon. *Water Resources Research* **47**(12), 1–11. doi:10.1029/2011WR010383.

MacGregor AJ, Gell PA, Wallbrink PJ, Hancock G (2005) Natural and post-European settlement variability in water quality of the lower Snowy River floodplain, eastern Victoria, Australia. *River Research and Applications* **21**, 201–213.

Maheshwari BL, Walker KF, Mcmahon TA (1995) Effects of regulation on the flow regime of the River Murray, Australia. *Regulated Rivers: Research and Management* **10**(1), 15–38. doi:10.1002/rrr.3450100103

Mathews RH (1903) The Aboriginal fisheries at Brewarrina. *Journal and proceedings of the Royal Society of NSW* **37**, 146–156.

McKenzie GM (1997) The late Quaternary vegetation history of the south-central highlands of Victoria, Australia. I. Sites above 900 m. *Australian Journal of Ecology* **22**, 19–36. doi:10.1111/j.1442-9993.1997.tb00638.x

McKenzie GM (2002) The late Quaternary vegetation history of the south-central highlands of Victoria, Australia. II. Sites below 900 m. *Austral Ecology* **27**(1), 32–54. doi:10.1046/j.1442-9993.2002.01155.x

McKenzie GM, Kershaw AP (1997) A vegetation history and quantitative estimate of Holocene climate from Chapple Vale, in the Otway region of Victoria, Australia. *Australian Journal of Botany* **45**(3), 565–581. doi:10.1071/BT96051

McKenzie GM, Kershaw AP (2000) The last glacial cycle from Wyelangta, the Otway region of Victoria, Australia. *Palaeogeography, Palaeoclimatology, Palaeoecology* **155**(1, 2), 177–193. doi:10.1016/S0031-0182(99)00100-5

McKinney HG (1889) Irrigation in its relation to the pastoral industry of New South Wales. *Proceedings of Royal Society of New South Wales* **23**, 75–96.

Meggitt MJ (1966) Gadjari among the Walbiri Aborigines of central Australia. *Oceania* **36**, 173–213. doi:10.1002/j.1834-4461.1966.tb00286.x

Mitchell TL (1839) *Three Expeditions Into the Interior of Eastern Australia.* Two volumes. Boone, London.

Moore CWE (1953) The vegetation of the south-eastern Riverina, New South Wales. I. The climax communities. *Australian Journal of Botany* **1**, 485–547. doi:10.1071/BT9530485

Mulham WE (1994) *The Best Crossing Place.* Reliance Printing and Stationary, Deniliquin, NSW.

Nanson RA (2009) The evolution of peat-swamp channels and organic floodplains, Barrington Tops, New South Wales, Australia. *Geographical Research* **47**(4), 434–448. doi:10.1111/j.1745-5871.2009.00596.x

Nanson GC, Barbetti M, Taylor G (1995) River stabilization due to changing climate and vegetation during the late Quaternary in Western Tasmania, Australia. *Geomorphology* **13**(1–4), 145–158. doi:10.1016/0169-555X(95)00040-C

Newell JW (1970) 'Soils and land use in the Ovens and Buffalo river valleys, Victoria.' Technical bulletin no. 21. Department of Agriculture, Melbourne.

Ogden RW (2000) Modern and historical variation in aquatic macrophyte cover of billabongs associated with catchment development. *Regulated Rivers: Research and Management* **16**, 497–512. doi:10.1002/1099-1646(200009/10)16:5<497::AID-RRR600>3.0.CO;2-Y

Ogden R, Spooner N, Reid M, Head J (2001) Sediment dates with implications for the age of the conversion from palaeochannel to modern fluvial activity on the Murray River and tributaries. *Quaternary International* **83–85**, 195–209. doi:10.1016/S1040-6182(01)00040-4

Olley JM, Scott A (2002) 'Sediment supply and transport in the Murrumbidgee and Namoi rivers since European settlement.' CSIRO Land & Water technical report 9/02. CSIRO Land & Water, Canberra.

Page KJ, Nanson GC (1996) Stratigraphic architecture resulting from late Quaternary evolution of the riverine plain, south-eastern Australia. *Sedimentology* **43**, 927–945. doi:10.1111/j.1365-3091.1996.tb01512.x

Page KJ, Kemp J, Nanson GC (2009) Late Quaternary evolution of riverine plain paleochannels, south-eastern Australia. *Australian Journal of Earth Sciences* **56**, S19–S33. doi:10.1080/08120090902870772

Pardoe C (1998) The cemetery as symbol: the distribution of pre-historical Aboriginal burial grounds in southeastern Australia. *Archaeology in Oceania* **23**, 1–16.

Pearson M, Lennon J (2010) *Pastoral Australia: Fortunes, Failures and Hard Yakka: a Historical Overview 1788–1967*. CSIRO Publishing, Melbourne.

Peterson L (1996) *Reading the Landscape: Documentation and Analysis of a Relict Feature of Land Degradation in the Bendigo District, Victoria.* Monash publications in geography no. 48. Department of Geography and Environmental Science, Monash University, Melbourne.

Powell JM (1989) *Watering the Garden State: Water, Land and Community in Victoria.* Allen & Unwin, Sydney.

Powell JM (1993) *The Emergence of Bioregionalism in the Murray–Darling Basin.* Murray–Darling Basin Commission, Canberra.

Preece N (2002) Aboriginal fires in monsoonal Australia from historical accounts. *Journal of Biogeography* **29**, 321–336. doi:10.1046/j.1365-2699.2002.00677.x

Prosser IP, Rutherfurd ID, Olley JM, Young WJ, Wallbrink PJ, Moran CJ (2001) Large-scale patterns of erosion and sediment transport in river networks, with examples from Australia. *Marine and Freshwater Research* **52**, 81–99. doi:10.1071/MF00033

Rahmstorf S (2007) A semi-empirical approach to projecting future sea-level rise. *Science* **315**, 368–370. doi:10.1126/science.1135456

Reid MA (2002) A diatom-based palaeoecological study of two billabongs on the Goulburn River floodplain, southeast Australia. In *Proceedings of the 15th International Diatom Symposium*. (Ed. J John) pp. 237–253. Gantner Verlag, Ruggell, Liechtenstein.

Reid MA (2008) Evidence for catastrophic shifts in the trophic structure of floodplain lakes associated with soil erosion. In *Sediment Dynamics in Changing Environments*. (Eds J Schmidt, T Cochrane, C Phillips, S Elliot, T Davies and L Basher) pp. 584–590. IAHS Press, Wallingford, UK.

Reid M (2010) *Establishing Benchmarks and Trajectories of Change in the Lower Murrumbidgee Floodplain*. Department of Environment, Climate Change and Water, Sydney, NSW.

Reid MA, Sayer CD, Kershaw AP, Heijnis H (2007) Palaeolimnological evidence for submerged plant loss in a floodplain lake associated with accelerated catchment soil erosion (Murray River, Australia). *Journal of Paleolimnology* **38**, 191–208. doi:10.1007/s10933-006-9067-9

Reinfelds I, Rutherfurd I, Bishop P (1995) History and effects of channelisation on the Latrobe River, Victoria. *Australian Geographical Studies* **33**, 60–76.

Roberts SH (1964) *The Squatting Age in Australia, 1835–1847*. 2nd edn. Melbourne University Press, Melbourne.

Robertson AI (1996) The effect of livestock on wetlands. In *Wetlands in a Dry Land: Understanding for Management*. (Ed. WD Williams) pp. 195–204. Environment Australia and The Land and Water Resources Research and Development Corporation, Canberra, Australia.

Robertson AI (1997) Land-water linkages in floodplain river systems: the influence of domestic stock. In *Frontiers in Ecology: Building the Links*. (Eds N Klomp and ID Lunt) pp. 207–218. Elsevier Science, Oxford, UK.

Rolls EC (1981) *A Million Wild Acres: 200 Years of Man and an Australian Forest*. Thomas Nelson, Melbourne.

Russell-Smith J, Lucas D, Gapindi M, Gapindi M, Kapirigi N, Namingum G, Lucas K, Guiliani P, Chaloupka G (1997) Aboriginal resource utilization and fire management practice in western Arnhem Land, monsoonal northern Australia: notes for prehistory, lessons for the future. *Human Ecology* **25**, 159–195. doi:10.1023/A:1021970021670

Rutherfurd I (2000) Some human impacts on Australian stream channel morphology. In *River Management: the Australasian Experience*. (Eds S Brizga and B Finlayson) pp. 11–49. John Wiley & Sons, Chichester, UK.

Rutherfurd ID, Kenyon CE (2005) Geomorphology of the Barmah–Millewa Forest. *Proceedings of the Royal Society of Victoria* **117**(1), 23–39.

Scott A, Olley J (2003) *Settlement, Erosion and Muddy Waters: Lessons from the Past*. Murray–Darling Basin Commission, Canberra.

Sloss CR, Murray-Wallace CV, Jones BG (2007) Holocene sea-level change on the southeast coast of Australia: a review. *The Holocene* **17**(7), 999–1014. doi:10.1177/0959683607082415

Smith DI (1998) *Water in Australia, Resources and Management*. Oxford University Press, Melbourne.

Smith ND, Cross TA, Dufficy JP, Clough SR (1989) Anatomy of an avulsion. *Sedimentology* **36**(1), 1–23. doi:10.1111/j.1365-3091.1989.tb00817.x

Stokes K, Ward K, Colloff M (2010) Alterations in flood frequency increase exotic and native species richness of understorey vegetation in a temperate floodplain eucalypt forest. *Plant Ecology* **211**, 219–233.

Stone T (2006) The late-Holocene origin of the modern Murray River course, southeastern Australia. *The Holocene* **16**(5), 771–778. doi:10.1191/0959683606hl971rr

Sturt C (1838) *The Unpublished Journal of Charles Sturt's 1838 Overlanding Expedition to Adelaide Following the Hume River*. Charles Sturt Papers MS 9025 (photocopy), held at the State Library of Victoria.

Tibby J (2012) The Younger Dryas: relevant in the Australian region? *Quaternary International* **253**, 47–54. doi:10.1016/j.quaint.2012.01.003

Tremont RM, McIntyre S (1994) Natural grassy vegetation and native forbs in temperate Australia: structure, dynamics and life-histories.

*Australian Journal of Botany* **42**, 641–658. doi:10.1071/BT9940641

Turney CSM, Kershaw AP, Lowe JJ, van der Kaars S, Johnston R, Rule S, Moss P, Radke L, Tibby J, McGlone MS, Wilmshurst JM, Vandergoes MJ, Fitzsimons SJ, Bryant C, James S, Branch NP, Cowley J, Kalin RM, Ogle N, Jacobsen G, Fifield LK (2006) Climatic variability in the southwest Pacific during the last termination (20–10 kyr BP). *Quaternary Science Reviews* **25**(9, 10), 886–903. doi:10.1016/j.quascirev.2005.09.007

Vanderwal R (Ed.) (2001) *John Bulmer's Recollections of Victorian Aboriginal Life 1855–1908.* Compiled by A. Campbell. Museum Victoria, Melbourne.

VEAC (2008) 'River red gum forests investigation: final report.' Victorian Environmental Assessment Council, Melbourne.

Walker MH (1971) *Come Wind, Come Weather: a Biography of Alfred Howitt.* Melbourne University Press, Melbourne.

Walker KF (1985) A review of the ecological effects of river regulation in Australia. *Hydrobiologia* **125**, 111–129.

Ward KA (1991) 'Investigation of the flood requirements of the Moira grass plains in Barmah Forest, Victoria.' Department of Conservation and Natural Resources, Shepparton, Victoria.

Webb M (1984) *Table Top: a History of the District 1824–1984.* Table Top History Committee, Albury, NSW.

Webb JA, Fabel D, Finlayson BL, Ellaway M, Shu L, Speirtz H-P (1992) Denudation chronology from cave and river terrace levels: the case of the Buchan karst, southeastern Australia. *Geological Magazine* **129**(3), 307–317. doi:10.1017/S0016756800019245

Wilby RL, Gell PA (1994) The impact of forest harvesting on water yield: modelling hydrological changes detected by pollen analysis. *Hydrological Sciences Journal/Journal des Sciences Hydrologiques* **39**(5), 471–486.

Williams M, Prescott JR, Chappell J, Adamson D, Cock B, Walker K, Gell P (2001) The enigma of a late Pleistocene wetland in the Flinders Ranges, South Australia. *Quaternary International* **83–85**, 129–144. doi:10.1016/S1040-6182(01)00035-0

Williams M, Cook E, van der Kaars S, Barrows T, Shulmeister J, Kershaw P (2009) Glacial and deglacial climatic patterns in Australia and surrounding regions from 35 000 to 10 000 years ago reconstructed from terrestrial and near-shore proxy data. *Quaternary Science Reviews* **28**(23, 24), 2398–2419. doi:10.1016/j.quascirev.2009.04.020

Yibarbuk D, Whitehead PJ, Russell-Smith J, Jackson D, Godjuwa C, Fisher A, Cooke P, Choquenot D, Bowman DMJS (2001) Fire ecology and Aboriginal land management in central Arnhem Land, Northern Australia: a tradition of ecosystem management. *Journal of Biogeography* **28**, 325–343. doi:10.1046/j.1365-2699.2001.00555.x

Zola N, Gott B (1992) *Koorie Plants Koorie People: Traditional Aboriginal Food, Fibre and Healing Plants of Victoria.* Koorie Heritage Trust, Melbourne.

# Section 2

Riverine plants

# 5

# Macroalgae, charophytes and bryophytes

*Michelle Casanova and Lucy Nairn*

## INTRODUCTION

*Sphagnum* bogs in alpine headwaters, moss gardens near waterfalls, leafy mats of liverworts, great skeins of filamentous green algae, beds and clumps of *Chara* and *Nitella* with bright orange reproductive organs, delicate, feathery red algae, the amazing emergence of life in once-desolate saltpans after a flood and the slime on rocks that makes keeping your footing in a stream so difficult – macroalgae, charophytes and bryophytes are found in all parts of Australia's riverine landscapes.

Bryophytes are essentially terrestrial; however, many species require the presence of free water only for transport of motile sperm cells and fertilisation of the gametangia borne on the haploid gametophyte. Algae (macroalgae and charophytes), in contrast, usually require free water, or at least a humid environment, for all stages of their life cycle. Bryophytes and algae are highly water dependent, and as such have the potential to be good indicators of water regime, water chemistry and the physical consequences of flow and water presence in riverine systems. Despite this reliance on water, these groups are also great survivors. Some groups of algae are characterised by desiccation-resistant spores or resting cells and some mosses and liverworts can be revived from the smallest shrivelled scrap of tissue through the addition of water.

In Australia's riverine ecosystems there are over 400 species of macroalgae. These species are confined to flowing or still water or, in the case of charophytes, as oospores when systems are dry (Day *et al.* 1995; Casanova 2007, 2009a). Twenty species of Australian bryophytes occur either wholly or partly under water, a further 24 species in boggy headwaters or marshes and an additional 40–50 species near streams in forests (Meagher and Fuhrer 2003). The study of these taxa in riverine systems has been hindered by their inconspicuous nature as most are small and noticeable only in large populations. Until quite recently there has been a lack of accessible taxonomic treatments of these taxa and a paucity of interest and expertise in their recognition and identification. Despite this, bryophytes and macroalgae play a significant role in many riparian systems and have great potential as indicators of riverine health, connectivity and resilience.

## DESCRIPTION

### Macroalgae

Algae are split between two kingdoms, Eubacteria and Eukaryota or Protista (Raven *et al.* 1992; van den Hoek *et al.* 1995), and the word 'algae' is therefore a term of convenience, referring to an artificial taxonomic group. The algae *sensu lato* are split into many different divisions, most of which are allocated to the kingdom Eukaryota/Protista (Huisman and Saunders 2007). In spite of evolutionary differences, the divisions share similar growth forms and morphologies (Stevenson 1996). Freshwater macroalgae (i.e. algae that can be seen with the naked eye) are most commonly found in the divisions Cyanophyta (the blue-greens), Rhodophyta (the reds), Heterokontophyta (the golden browns) and Chlorophyta (the greens).

Charophytes can also be considered to be macroalgae; however, some phylogenetic studies recognise their distinct characteristics and place them, along with the bryophytes and all 'higher' plants, in division Streptophyta (Karol *et al.* 2001; Huisman and Saunders 2007). Other taxonomists also consider the Zygnematales, Klebsormidiales and Coleochaetales to be part of this group rather than the Chlorophyta (Guiry 2012). For the purposes of this chapter, we discuss the charophytes separately in recognition of their very different role in Australia's riverine landscapes.

Macroalgae have a diversity of life histories. Cyanophyta reproduce by simple cell division, and some produce akinetes (thick-walled resting cells) that allow them to persist and disperse. Some form symbiotic relationships with liverworts and lichens in riparian habitats (Larkum 2007). Macroscopic colonies of *Rivularia*, *Gloeotrichia* and *Nostoc* species can be obvious attached to rocks or vegetation, or floating in still water (Baker and Fabbro 1999). *Stigonema*, *Anabaena*, *Coleodesmium*, *Loefgrenia*, *Lyngbya*, *Microcoleus*, *Nodularia* and *Microcystis* are likely to be present as part of a phytoplankton, epiphytic or epilithic community, but can also form macroscopic aggregations, blooms or paint-like scums in riparian systems, especially when flow rates are low (Entwisle *et al.* 1997).

Freshwater Rhodophyta have complicated life histories, with no motile or flagellated stages. Dispersal occurs via carpospores or fragmentation of thalli (Huisman and Kraft 2007). The absence of any known desiccation-resistant stages in their life cycles appears to restrict the genera *Batrachospermum*, *Sirodotia*, *Compsopogon* and *Psilosiphon* to permanently flowing water (Entwisle *et al.* 2007; Entwisle 2007). Rhodophytes living in estuaries also rely on the permanent presence of water, but can tolerate varying salinities (Riddin and Adams 2008).

In the Heterokontophyta, species in the genera *Tribonema*, *Botrydium* and *Vaucheria* produce macroscopic growths in streams, pools, seepages and damp soil in riparian areas and salt marshes. Some heterokonts produce endogenous cysts impregnated with silica, which allow them to recolonise habitats after disturbance (Entwisle 2007).

Chlorophytes are by far the most speciose of the riparian macroalgae (Day *et al.* 1995), and these include the familiar filamentous forms in the genera *Spirogyra*, *Zygnema* and *Mougeotia* in family Zygnemaceae (Casanova and Lewis 2007); *Oedogonium* and *Bulbochaete* in family Oedogoniaceae; *Cladophora* and *Rhizoclonium* in family Cladophoraceae; and *Chaetophora*, *Draparnaldia* and *Stigeoclonium* in family Chaetophoraceae, all of which can form macroscopic aggregations or 'skeins' of filaments in riparian systems (Entwisle *et al.* 2007). The life histories of Chlorophytes are varied (Haig 2010), and some can produce desiccation-resistant spores, oospores or hypnozygotes, allowing them to survive drought and colonise recently inundated streams and floodplains (Lewis and Entwisle 2007; Skinner and Entwisle 2007).

## Charophytes

Charophytes (streptophyte algae in family Characeae) are an important and distinctive group of algae of rivers, streams and wetlands. These plants are the closest living relatives to the ancestors of land plants (including the bryophytes; Karol *et al.* 2001) and they have many ultrastructural characteristics in common with them (Pickett-Heaps 1975; van den Hoek *et al.* 1995; McCourt *et al.* 2004). Charophytes have a haplobiontic life history in which the water-dependent vegetative plant represents the haploid generation (gametophyte), and the oospore alone represents the diploid generation (sporophyte) (Casanova 2007). Growing plants are

**Figure 5.1:** *Chara globularis* (charophyte) in shallow water at the edge of a riparian zone in southern Australia.

anchored in the sediment via colourless rhizoids and many can survive instream flows of several metres per second. The thallus has no waxy cuticle or epidermis and dehydrates quickly out of water, so plants are dependent on the presence of at least a film of free water. Reproduction relies on the movement of motile sperm cells, so free water is necessary for the production of oospores. Oospores can then survive months, or even decades, of complete desiccation. Thus, although several species can be found in fast-flowing streams, the greatest abundance and diversity of charophytes develops in slower-flowing systems and low-energy sites such as back-waters, waterholes and floodplain wetlands (Fig. 5.1).

## Bryophytes

Bryophytes are thought to have evolved from an early lineage of streptophyte green algae and are among the oldest vascular plants (Kenrick and Crane 1997). Originally considered to belong to a single division, bryophytes are now recognised as having several evolutionary lineages and are usually divided into three divisions: the Bryophyta (mosses), Marchantiophyta (liverworts) and Anthocerotophyta (hornworts). Some mosses and liverworts are found submerged in aquatic habitats; however, hornworts are found only rarely in such habitats (Scott 1994; J. Milne, pers. comm.).

**Figure 5.2:** A species of *Riccia* (liverwort) that grows in dry channels soon after drying down.

Bryophytes have a haplodiploid life history and the leafy haploid generation (gametophyte) is the most obvious growth stage. The diploid stage (sporophyte) is represented by a spore-producing capsule (with or without a stalk or seta). Bryophyte spores are present in floodplain and wetland soils, and some gametophytes, either leafy or thallose, can survive long periods of almost complete desiccation (Fig. 5.2). They are thus ideally suited to survival in Australia's hydrologically variable riparian systems.

# ECOLOGY

There are few ecological studies on riparian bryophytes or algae (macro or micro) in Australia. Entwisle's (1989, 1990) work revealed how algal communities changed over time in cold temperate streams and Mosisch looked at the effects of flow rates, nutrients and shading on algal biomass in a subtropical system (Mosisch and Bunn 1997; Mosisch *et al.* 1999, 2001; Mosisch 2001). Robson worked on biofilms in ephemeral streams (including some macroalgal taxa) (Robson 2000; Robson and Matthews 2004). Casanova (1993) studied the distribution and abundance of charophytes in permanent and temporary streams and wetlands on the New England Tablelands of New South Wales, and Nairn (2006) examined the community dynamics of macroalgae and bryophytes in several streams near Sydney. Aquatic bryophytes have fared even worse. Their habitats are occasionally noted in natural history studies (e.g. Ramsay and Cairns 2004), but we are only aware of two Australian ecological studies (Downes *et al.* 2003; Nairn 2006).

Spatial and temporal variability in macroalgal and bryophyte communities is considerable, and occurs at different scales. Consideration of historic and phylogenetic patterns as well as ecological factors is vital when attempting to explain present-day distributions of organisms (Crisci *et al.* 2006). At the broadest scale, macroalgal distribution reflects historic factors, while at small scales environmental and habitat variables are more likely to exert influence. Bryophytes and charophytes tend to be affected by environmental variables at larger spatial and longer temporal scales than non-charophyte macroalgae, which are very dynamic communities that vary between sites on the same river and over time frames as short as weeks (Biggs 1996; Nairn 2006).

In the headwaters, where temperatures are cooler, water is more reliably available and shading from riparian vegetation more complete, bryophytes are abundant. Even in open situations at the very top of catchments, bryophytes are characteristic

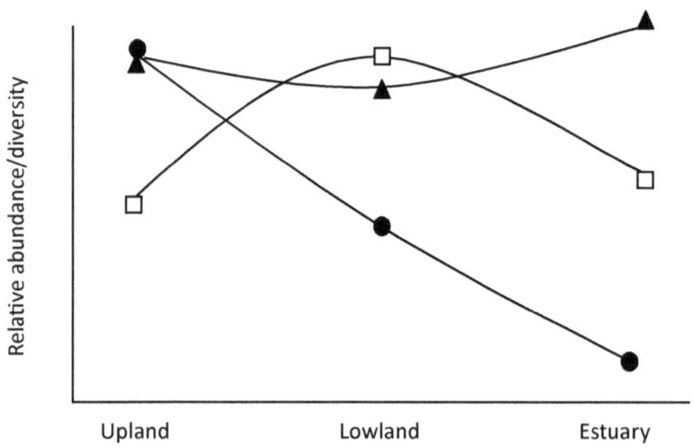


**Figure 5.3:** The hypothetical relative abundance and diversity of bryophytes (●), charophytes (□) and macroalgae (▲) in different regions of a riparian system.

components of bogs and springs (Whinam *et al.* 2003). Through the lowland middle reaches where streamflow is slower, temperatures are higher and nutrients are abundant, habitat for byrophytes is reduced and macroalgae and charophytes find a footing. There are few bryophytes in estuaries, but macroalgae are important in estuarine processes (Lirman *et al.* 2008). Charophytes are somewhat better studied in all parts of river systems (Casanova 1993, 2004; Casanova and Nicol 2009), and can be both speciose and abundant particularly where flow rates are low and water is clear (Casanova 2004). Figure 5.3 shows the hypothetical relative abundance and diversity of bryophytes, charophytes and macroalgae.

Upland areas are characterised by high precipitation, low nutrient inputs and low light in forested areas, but high light intensities above the treeline. Water availability can vary, depending on the type of headwaters, but many upland streams are somewhat permanent. Lowland areas are characterised by higher nutrient availability, substrate mobility and temperatures than upland areas, as well as more variability in water availability and permanence. The middle reaches of the longest Australian rivers flow through arid or semi-arid zones, which imposes particular characteristics and constraints on riparian systems. Equally important is the anthropogenic modification of flows and habitat in these systems. Finally, estuaries represent the end points in riparian systems, where the river enters the sea, and the sea invades the river. Estuaries are dynamic systems, subject to tidal fluxes, seasonal opening and closing and variable freshwater flows. There are some bryophytes associated with these systems, mostly in relation to the bordering salt marshes and sand dunes (Meagher and Fuhrer 2003); however, the charophytes and macroalgae (along with some 'seagrasses') tend to dominate. Instead of strictly freshwater species, marine chlorophytes and rhodophytes that can tolerate salinity variation, such as *Ulva*, *Enteromorpha* and *Gracillaria* can be important constituents of benthic estuarine communities (Riddin and Adams 2008).

Various models have been proposed to explain the distribution of macroalgal, charophyte and bryophyte communities. At large spatial scales, the determinants of assemblages of these groups are the same in riverine habitats as elsewhere: particularly climate (e.g. flood frequency and seasonality), catchment geology and land use (Biggs 1995; Suren 1996; Biggs *et al.* 1998). Historic events and changes in these determinants have also influenced each group, affecting groups differently depending on their 'evolutionary age'.

## Macroalgae

Algal communities are notoriously heterogeneous in time and space (Stevenson 1997). Availability of light, nutrient load, substratum texture, presence of grazers and flow variability all influence the patterns of macroalgae in streams (Biggs 1996). Some species exhibit seasonal variation while others do not (Entwisle 1990; Necchi *et al.* 1991; Necchi 1993). Coupled with the inherently unstable nature of stream environments, this can result in algal communities comprising seemingly haphazard assemblages (Blum 1959; Sheath and Burkholder 1985). In lotic systems, this heterogeneity has been attributed to broadscale factors such as climate, geology and land use, with smaller-scale factors such as riparian shading and flooding acting within these overarching effects (e.g. Stevenson 1997; Biggs *et al.* 1998).

Differences in abundance and species composition of algal communities have been reported between riffle, run and pool habitats in streams (Blum 1956; Biggs 1996). The processes driving these differences, however, are not well understood and there have been few published works on the subject. Water velocity appears to play a role, but its influence is confounded by interactions with nutrient availability. In enriched streams, the highest algal biomass is found in pools and runs with low flow velocity. In contrast, it is the higher velocity riffles that support the most periphyton in unenriched streams (Biggs 1996; Biggs and Stokseth 1996). Very high algal abundances or 'blooms' tend to be associated with eutrophic waters (e.g. Biggs and Price 1987), with the nutrient enrichment associated with agricultural and urban pollution often leading to a drop in diversity (e.g. Biggs 1995; Pan *et al.* 1999; Biggs and Smith 2002). Entwisle (1989) found an increase in species numbers with a moderate increase in nutrient enrichment in his study of the Yarra River Basin in Victoria, while Biggs and Smith (2002) found, in a study of streams in New Zealand, that streams with the lowest levels of nitrogen supported the highest diversity of algae. These authors suggest that, in nitrogen-limited streams, algae that can fix nitrogen are the first colonisers followed by nitrogen-limited taxa as microhabitat conditions change. In other studies, streams with moderate levels of nutrients often have the highest diversity (for a review see Borchardt 1996). Areas with frequent floods have also been found to have greater benthic algal diversity than less flood-prone streams (Clausen and Biggs 1997); however, the mechanisms by which this occurs are poorly understood.

The response of macroalgae to drought has been less well studied than flood effects (Peterson 1996), in spite of the potential importance of low flows in structuring stream communities (Resh *et al.* 1988; Peterson 2007). Some algae have been shown to be resistant to desiccation (Davis 1972; Morison and Sheath 1985) and dry algal biofilms can be important in the recolonisation of streams following rewetting (e.g. Robson 2000). Equally, refuges such as pools or patches of bryophytes can provide colonisation sources (Biggs 1996). Intermittent streams have received the most attention (e.g. Dodds *et al.* 1996; Robson 2000; Robson and Matthews 2004). Many streams in Australia, however, experience extreme low flows without ceasing to flow altogether (McMahon and Finlayson 2003). Along with desiccation, periods of low flow can influence algal communities through the accumulation of silt and leaf litter (Mosisch 2001) and the associated physical and chemical effects.

## Charophytes

Most studies on the distribution of charophytes use easily measured physical parameters such as pH (Imahori 1954; Spence 1967, 1982), conductivity

(Brock and Lane 1983; García and Chivas 2006), nutrient concentration (Forsberg 1965; John *et al.* 1982) and water clarity (Pip 1979; Vant *et al.* 1986) to explain the distribution of species among habitats (Corillion 1957; García 1999; García and Chivas 2006). The distribution of species groups (freshwater, mesohaline, saline and hypersaline species groups) is strongly influenced by salinity and, in general, the diversity of charophytes declines with increasing salinity (Brock and Lane 1983; García and Chivas 2006). Temperature and pH, however, can vary on short temporal scales in freshwater systems and, when charophytes are adequately sampled within a region, most species appear to have a broad tolerance to physio-chemical characteristics of water. The best predictors of charophyte occurrence in Australia appear to be time since inundation and the nutrient content of the water. Charophytes are rarely found in permanent waterbodies with high nutrient concentrations. Casanova and Kraft (2004) found 21 species of *Chara*, *Lamprothamnium* and *Nitella* in deep, permanent waterbodies, but those waters were characterised by high water clarity (regardless of nutrient concentration) or fluctuating salinity (coincident with high water clarity). High inorganic turbidity restricts light availability for charophyte growth (Schwarz *et al.* 1996), but several species of *Nitella* occur in highly turbid conditions in inland river wetlands and waterholes (Casanova and Porter 2013), and turbidity per se does not appear to hinder charophyte growth (Douglas-Hill 1997; Casanova *et al.* 2003).

Within habitats, the easily measured parameters of depth (Wood 1950) and water clarity (Middelboe and Markager 1997) have been used to explain the distribution of species. In the few studies where other variables have been assessed (Proctor 1990; Casanova 1993), shoot abundance of many species is positively correlated with time since inundation of the habitat as well as water temperature and nutrient concentrations (Casanova 1993). Studies of northern hemisphere charophytes indicate that they are outcompeted by angiosperms or phytoplankton under high nutrient conditions (Scheffer 1998; Blindow *et al.* 2002).

## Bryophytes

Little work has been done concerning physical variables affecting the distribution of aquatic bryophytes (Ceschin *et al.* 2012). Much of the work on bryophytes has instead focussed on effects of heavy metals and other organic pollutants (Whitton and Kelly 1995), or the role of bryophytes as habitat and food for grazers (e.g. Suren 1991, 1992; Suren and Winterbourn 1991). Those studies that have been conducted show that bryophytes tend to be persistent at the sites in which they become established (Suren 1996) and are less vulnerable to high flow events than macroalgae, provided the substratum they are attached to is stable (Sheath *et al.* 1986; Steinman and Boston 1993; Suren 1996). Stability of substratum is a major determinant of both the presence of bryophyte assemblages in a stream (Suren 1996) and the species composition (Muotka and Virtanen 1995). Working in New Zealand streams, Suren (1996) found that epilithic communities in streams were influenced by catchment geology, water quality, land use and the number of high-flow events.

Many bryophytes can survive under very low light conditions (e.g. Martin and Churchill 1982), although there are also cases where bryophytes survive high light intensities and aridity (Glime and Vitt 1984). All bryophytes are susceptible to desiccation, but especially aquatic ones (Glime 1971), which might explain why they can persist in shaded reaches of streams with reduced flows compared with similar reaches with higher light levels (Nairn 2006). Acidic waters are a common feature of alpine streams, where bryophytes are frequently abundant (e.g. Suren and Ormerod 1998). Surveys of bedrock streams in the southern highlands of New South Wales, however, found more bryophytes at sites with neutral pH than at acidic sites (Nairn 2006).

The occurrence of aquatic bryophytes has been found to concur with a 'hump-backed' model of diversity *sensu* Grime (1979), with diversity increasing with greater biomass to a point and then dropping off again (Muotka and Virtanen 1995). Virtanen *et al.* (2001) suggest that this relationship is only likely to occur in stable streams that contain boulders large enough to result in the zonation of bryophyte communities.

## HABITATS

In a cross-section of a waterway and its riparian zone, charophytes, macroalgae and some bryophytes are commonly found growing in the stream channel. Where stream flow rates are high and the bedrock is scoured, only macroalgae or bryophytes that can attach to rocks can persist (e.g. *Batrachospermum* spp., *Cladophora* spp., *Dendrocryphaea tasmanica*, *Fissidens* spp.). Boulders provide stable substrates for both macroalgae and bryophytes, while rolling cobbles gather only algae. Splash zones near waterfalls support bryophytes (e.g. *Bryum blandum*) and some macroalgae, particularly cyanobacteria (Meagher and Fuhrer 2003).

Where flow rates are lower, or channels are less scoured, charophytes and some bryophytes (*Fissidens*, *Riccardia*, *Anthoceros* and *Aneura* spp.) can establish in deposits of sediment or sand. During cease-to-flow events and in river backwaters, free-floating filamentous algae (*Spirogyra*, *Zygnema* spp.) and floating liverworts (*Riccia multifida*, *R. duplex*, *Ricciocarpus natans*) can become abundant. On in-channel benches and terraces, charophytes can grow when there is water flowing or free water present. Such habitats are often colonised by liverworts (*Riccia* spp.) as they dry out. Undercut banks can provide a moist, shaded habitat for mosses, liverworts (*Marchantia*, *Lunularia*) and hornworts (*Megaceros*), as can fallen timber and the soil surface within the standing vegetation on the floodplain.

Charophyte oospores can be present in the seed bank of the floodplain itself, and in water-retaining areas on the floodplain (e.g. billabongs and flood-runners), ready to germinate when the floodplain is inundated. In exceptional circumstances, charophytes can be found growing stranded or emergent on the banks of rivers (e.g. on the Daly River in the Northern Territory, Christmas Creek in central Queensland; M.T. Casanova, pers. obs; Fig. 5.4). This can occur where water seeping out of the bank maintains a film of moisture over the charophyte thallus. The same habitat can also be occupied by mosses and liverworts.

**Figure 5.4:** 'Emergent' charophytes on the bank of the Daly River in the Northern Territory. Note the *Eucalyptus* seedling that has germinated among the charophytes, indicating the relative dryness of the site and the accumulation of leaf litter, signifying the absence of flow.

At least half of the charophyte specimens in Australian herbaria have been collected from streams, rivers and creeks. The type specimens, from which each species is described, of ~37% of Australian species were collected in flowing waters. This is at least partly a consequence of the fact that the majority of collectors have been travellers (from Mueller in the 1840s to Casanova in the 2010s), and have collected at bridges and fords across Australia.

## BIOGEOGRAPHY

In spite of a commonly held view that most freshwater algae are cosmopolitan in distribution, a relatively high proportion of Australian algae have been found to be endemic to particular locations (Tyler 1996). In the case of red algae, up to 80% of species might be endemic (T. Entwisle, pers. comm.). The origins and affinities of these algae are beyond the scope of this chapter, but at least some taxa seem to have persisted on this continent for tens of thousands of years if not longer. A few genera in family Characeae produce lime-shells, or gyrogonites, around their oospores, which can be preserved as fossils in sediments. Fossils of *Lamprothamnium* species have been found in Australian Pleistocene and Holocene sediments (Burne *et al.* 1980; García and Chivas 2004), and fossils of *Lychnothamnus barbatus* have been collected from Pleistocene (40 000-year-old and 14 000-year-old) sediments in the Gulf of Carpentaria (García and Chivas 2006). *Lamprothamnium* species are presently abundant in brackish and saline habitats (e.g. the Avon River in Western Australia), and *Lychnothamnus* still occurs in freshwater riparian habitats in Queensland (McCourt *et al.* 1999; García and Chivas 2006). The genus *Lamprothamnium* is thought to have originated in the Australian–Gondwanan region (García and Casanova 2003), based on the persistence of dioecious members of that genus on this continent (Proctor 1980; Casanova 2013). Similarly, the monospecific red algal stream genus *Psilosiphon* is likely to have originated in Gondwana before the separation of New Zealand (Entwisle *et al.* 2000).

A small number of macroalgae and bryophytes can be considered truly cosmopolitan. *Mougeotia parvula, M. quadrangulata, Spirogyra communis, S. corrugata, Chara globularis* and *Nitella hyalina* are found on all continents (Lewis and Entwisle 2007), and *Chara zeylanica* has a pan-tropical distribution (Wood and Imahori 1965). Many more macroalgal species are widespread on two or more continents (Lewis and Entwisle 2007). Several taxa are shared with south-east Asia including India (e.g. *Mougeotia sinensis, Spirogyra brunnea, Nitella stuartii, Chara erythrogyna* and *Chara* section *Protochara*; Casanova and Karol 2014) and there is a group of south-eastern Australian endemic species with relationships to New Zealand and circumpolar island species (Entwisle *et al.* 2000; Casanova *et al.* 2007; Lewis and Entwisle 2007). Among the bryophytes there are a few truly cosmopolitan taxa, but most of the affinities among species are circumpolar or Gondwanan. The number of undisputed endemic taxa is relatively small, but this relates more to lack of research than anything else (E.A. Brown, pers. comm.; http://www.anbg.gov.au/bryophyte/bryogeography-aust-elsewhere.html).

Data on regional biogeography are deficient for most macroalgal species due to a paucity of collections. Charophytes can be segregated into summer-rainfall and winter-rainfall species, with some overlap of species where rainfall is less distinctly seasonal (Casanova 1993). Where data are sufficient (i.e. for a few species) there is evidence of restricted ranges, for example: *Batrachospermum diatyches* in Tasmania, *Thorea conturba* in New South Wales (Entwisle and Foard 2007), *Chara submollusca* in tropical Northern Territory and Queensland (Wood and Imahori 1965), *Nitella tasmanica* var. *afoliata* in Tasmania and Kangaroo Island (van Raam 1995; M. T. Casanova, pers. obs.), *Chara ecklonii* var. *albaniensis* in Western Australia (Wood 1972), and *Nitella partita* and *N. parooensis* in arid and semi-arid zone rivers (Casanova *et al.* 2003; Casanova and Porter 2013).

## DIVERSITY

A study of herbarium records of stream macroalgae from eastern Australia (Nairn 2006) found that the number of infra-generic macroalgal taxa recorded from eastern Australia was remarkably high when compared to that found by a similar study in North America. Sheath and Cole (1992) reported 259 infra-generic taxa from 1000 stream segments across the entire continent of North America (a latitude range of 10° to 70°) compared with the 302 recorded from eastern Australia in an area less than 15% of the size and with a latitude range of –10° to –48°. Similarly, the occurrence and diversity of charophytes in Australian streams exceeds that recorded for other regions (Wood 1972; van Raam 1995). Why the Australian macroalgal flora is so diverse is not immediately obvious. If diversity of habitat was the sole driver of taxonomic diversity, then the North American flora should be considerably richer than eastern Australia's. Sheath and Cole (1992) sampled habitats from tundra to tropical rainforest and desert streams, while the majority of collections made in eastern Australia are from two similar habitats: eucalypt-dominated forests in either temperate or subtropical climates.

Theories about the very high diversity of Australia's vascular flora might provide an explanation for the high diversity of macroalgae, charophytes and bryophytes. The most recent period of glaciation left North America and Europe with very low species numbers of vascular plants and animals. In contrast, Australia has a very rich flora (Mummery and Hardy 1994), with ~85% of the vascular plants being endemic (ESDWGBD 1991). If a large proportion of the algal taxa have also evolved here, one would expect the macroalgal flora to be similarly highly diverse. Australian soils are nutrient poor and, along with frequent fires and unpredictable rainfall, this is believed to have promoted speciation and prevented dominance by any one taxon (Wisheu *et al.* 2000). Australian streams also have low nutrient levels on a world scale (Harris 2001), and this low nutrient availability might be one factor contributing to the high diversity of aquatic flora in Australia. The extremely variable nature of hydrological regimes in Australia (McMahon and Finlayson 2003) has also almost certainly played a role. The highly variable nature of flows in Australian rivers (Lake 1995; McMahon and Finlayson 2003) might lead to communities being frequently 'reset', preventing one life history strategy from dominating.

## ECOSYSTEM SERVICES

Macroalgae, charophytes and bryophytes contribute to a wide range of critical ecosystem services, especially supporting and regulating services that underpin the provision of other ecosystem goods and services. In riverine habitats, the relative importance and total contribution to ecosystem services of macroalgae, charophytes and bryophytes typically vary along the run of a river or stream (Fig. 5.1; Auderset Joye *et al.* 2002; Ot'ahel'ová *et al.* 2007; Rothrock *et al.* 2008; Ceschin *et al.* 2012).

### Primary production

Stream bryophytes, macroalgae and charophytes are the main primary producers in upland streams (Minshall 1978; Naiman 1983). Algae are important primary producers in most streams (Minshall 1978), and can be the main energy source in many of them (Vannote 1980). Studies on bryophytes have shown that, where present, they can have rates of primary production that are similar to or higher than periphyton (e.g. Naiman 1983; Steinman and Boston 1993). There are difficulties, however, in estimating the primary production attributable to bryophytes, partly because their composition is unsuitable for assessment by many commonly used methods and partly because they host so many algal epiphytes (Ress *et al.* 2009).

In lowland systems, in-stream primary production can be inconspicuous. In rivers of eastern inland Australia the water is often turbid and charophytes and macroalgae can be difficult to see. Seed bank studies demonstrate the presence of charophytes in arid-zone rivers (Capon and Dowe 2007; M.T. Casanova, pers. obs.), but they are rarely collected from that habitat. Aquatic cyanobacteria and filamentous chlorophytes can be abundant at the water's edge in billabongs and waterholes, e.g. in an algal 'bath-tub ring'. Where such a zone occurs, it has been shown to be a major source of carbon in waterhole food webs (Bunn *et al.* 2003, 2006). These algae can form a significant proportion of the diets of fish in these systems, both when the rivers are flowing and when they consist of isolated pools (Sternberg *et al.* 2008). Charophytes and macroalgae are significant primary producers in the Daly River (NT) – *Spirogyra fluviatilis* establishes in the early dry season (April) after flow velocities recede, reaches peak biomass in July and is removed from the river with the first major wet-season flow (November) (Schult *et al.* 2007).

In estuaries, the main primary producers (excluding submerged angiosperm seagrasses) are charophytes and algae (Wehrenberg 2010).

## Bed and breakfast

Bryophyte beds in stream ecosystems support assemblages of invertebrates, often at much higher densities than on surrounding rocks (Stream Bryophyte Group 1999 and references therein). This includes rocks that are covered with periphyton. Working in Australian streams Downes *et al.* (2000) found that cobbles with a covering of the red alga *Audoiunella hermanii* had higher concentrations of macroinvertebrates than comparable smooth cobbles (but not as high as rough cobbles).

Bryophytes can provide habitat for epiphytic algae in upland streams, and the community that develops is related to the species of bryophyte and the shelter from, or exposure to, adverse environmental conditions in the site (Ress *et al.* 2010). The provision of habitat and food can contribute to the large numbers of invertebrates mentioned above. Macroalgae and charophytes provide food for macroinvertebrates in both upland and lowland streams (e.g. Feminella and Resh 1991; Bunn *et al.* 2003) and algae can directly form a significant proportion of the diets of fish in lowland systems, both when the rivers are flowing and when they consist of isolated pools (Sternberg *et al.* 2008). Bryophytes that colonise channels and floodplains as they dry down can provide food and shelter for soil invertebrates (Davidson *et al.* 1990).

## Nutrient processing

In-stream algae and bryophytes play a major role in nutrient spiralling (Mulholland 1996). Cyanobacteria can be responsible for the addition of nitrogen into riparian systems via their ability to fix atmospheric nitrogen, either as individual algal colonies, or as symbiotes with lichens or liverworts (e.g. Peterson and Grimm 1992). Up to 50% of riparian bryophytes can host epiphytic or endophytic cyanobacteria in temperate ecosystems, and many of those cyanobacteria are capable of fixing nitrogen (Adams and Duggan 2008). Bryophytes are also important in phosphorus uptake, in some cases more so than algae (Steinman and Boston 1993). It has been hypothesised that, in streams subject to spates, extensive bryophyte beds might provide a higher capacity for nutrient processing and retention than in those streams where filamentous algae dominate, due to their ability to persist under such conditions (Stream Bryophyte Group 1999).

In the Daly River, macroalgae and charophytes are likely to influence the availability of nutrients for phytoplankton growth. Nutrient limitation studies showed that additions of phosphorus alone increased cover of *Spirogyra fluviatilis* to 8–20% (compared with <5% with no additions) (Schult *et al.* 2007). The benthic macroalgae and charophytes in this system are effective scavengers of phosphorus where nitrogen is not limiting (Schult *et al.* 2007).

Charophytes in estuaries become vital nutrient sinks (Palma-Silva *et al.* 2002; Rodrigo *et al.* 2007), substrate for filamentous algal growth (Casanova 1994), food for herbivorous waterbirds (Noordhuis *et al.* 2002) and indicators of relative health and diversity (Appelgren and Mattila 2005). Charophyte beds can take up nitrogen at high rates (Rodrigo *et al.* 2007). Their morphology also promotes the settling of particulate material (Casanova *et al.* 2002) and their rhizomatous nature prevents re-suspension of sediments. Thus, healthy charophyte and macroalgal beds are indicators of good environmental condition in estuaries.

## Bank stabilisation

Plants that rely on an attachment to the soil (e.g. bryophytes and charophytes via rhizoids) can reduce sediment movement and erosion on floodplains and in channels. Bryophytes frequently colonise the banks and bed of river channels, drainage lines and depressions after draw-down or drying, and rosettes of *Riccia limbata*, *R. cavenosa*, *R. spongiformis* and tufts of *Ephemerum cristatum*, *Pottia drummondii* and *Bryum pachytheca* can occur on recently dried floodplain and channel soils (Eldridge and Tozer 1996). Bryophytes remain for a few weeks or months until they reproduce, dry out and die, leaving their spores in the soil (Casanova

2009b). While they grow they are macroscopic constituents of the soil crust community (along with microscopic algae and fungi) that stabilises newly exposed soil surfaces. The xanthophyte alga *Vaucheria* fulfils a similar role in some systems (Entwisle 1988). This soil crust community is quite different from the long-lived, delicate lichen-dominated soil crust community of arid-zone terrestrial soils (Eldridge and Tozer 1996).

When rivers are flowing, charophytes can also play a role in channel stabilisation. They can establish across the channel bed, accumulating sediment and epiphytic filamentous algae, reducing sediment movement and taking up nutrients (Bornette and Arens 2002).

## RESILIENCE

The term 'resilience' describes the capacity of an object or system to 'bounce-back' when it has been impacted or changed by a disturbance. Impacts in riparian systems include fire, drought and flooding, changes in nutrient availability, changes in physical substrate, and the introduction of novel competitors or consumers. Analysis of bryophyte, charophyte and macroalgal abundance, distribution and diversity can provide a measure of resilience when compared with other systems or undisturbed systems. Possession of a bank of spores or propagules in the soil can confer resilience, and many bryophytes, charophytes and macroalgae in Australian riparian zones possess such a bank (Casanova and Brock 1990; Brock 2011). A notable exception is the freshwater rhodophyta, members of which are restricted to permanent upland streams (Entwisle 2007). In contrast, rhodophytes like *Gracilariopsis* can persist under variable conditions in estuaries (Wehrenberg 2010) due to their capacity to withstand burying (up to 40 cm deep) and daily desiccation.

Riparian bryophyte assemblages exhibited greater resilience to the disturbance of logging than non-riparian bryophyte assemblages in a Swedish study (Dynesius *et al.* 2009), probably because there was less resource limitation and better habitat conditions in riparian zones. These characteristics, however, made riparian bryophyte communities more vulnerable to invasion. Australian macroalgal, charophyte and bryophyte communities and populations can generally withstand disturbance to some degree. Drought is survived via desiccation-resistant oospores (charophytes and macroalgae) and desiccation-resistant spores and thalli (bryophytes). High flows that restructure the habitat are survived via rapid re-establishment from fragments, gemmae and bulbils. *Chara australis* is a typical constituent of deep, cool, relatively permanent waters in eastern Australia (Tuross and Duea rivers on the south coast of NSW, Beardy Waters and Dumaresq Creek on the Northern Tablelands of NSW, Mt Emu Creek and Darlots Creek in western Victoria). Although the shoots can be removed by large flows, the rhizoids and bulbils can remain in the sediment and allow populations to re-establish after flood (Casanova 1994).

The abundance and representation of bryophytes and macroalgae have been used to interpret the effects of catchment disturbance in overseas studies (Ceschin *et al.* 2012) in a similar way to the use of invertebrates in Australian riparian studies, and bryophyte abundance and diversity have been informative in multivariate descriptions of riparian habitats at the landscape level (Manel *et al.* 2000). Standard methods for the detection and analysis of these non-motile elements of ecosystems (i.e. bryophytes, charophytes and macroalgae, and their propagules) are well established (e.g. Brock *et al.* 1994; Ot'ahel'ová *et al.* 2007; Rothrock *et al.* 2008; Ceschin *et al.* 2012), so the relative paucity of studies concerning these in Australian systems is seriously disappointing (Scott *et al.* 1997). The ongoing development of a more accessible taxonomy (e.g. Entwisle 1994; Scott 1994) for all these groups is likely to facilitate awareness of the role of these organisms in riparian processes.

## CONCLUSIONS

There can be little doubt that Australian riparian systems support a diversity of bryophytes, macroalgae and charophytes, and that these play a significant role in the ecology of these systems. Nevertheless, these elements have been little studied in Australia. Most of the information about their ecology comes as a digression from taxonomic surveys or from overseas studies. Given this group can be the basis of the food web, or provide significant habitat or resources to other elements of riparian communities, there is a real need to understand more than just the taxonomy and distribution of these groups. The role of bryophytes, charophytes and macroalgae in riparian processes is especially likely to be a rewarding subject for future study.

## REFERENCES


Adams DG, Duggan PS (2008) Cyanobacteria-bryophyte symbioses. *Journal of Experimental Botany* **59**, 1047–1058.

Appelgren K, Mattila J (2005) Variation in vegetation communities in shallow bays of the northern Baltic Sea. *Aquatic Botany* **83**, 1–13. doi:10.1016/j.aquabot.2005.05.001

Auderset Joye D, Castella E, Lachavanne J-B (2002) Occurrence of Characeae in Switzerland over the last two centuries (1800–2000). *Aquatic Botany* **72**, 369–385. doi:10.1016/S0304-3770(01)00211-X

Baker PD, Fabbro LD (1999) *A Guide to the Identification of Common Blue-green Algae (Cyanoprokaryotes) in Australian Freshwaters.* CRC for Freshwater Ecology, Thurgoona, NSW.

Biggs BJF (1995) The contribution of flood disturbance, catchment geology and land use to the habitat template of periphyton in stream ecosystems. *Freshwater Biology* **33**, 419–438. doi:10.1111/j.1365-2427.1995.tb00404.x

Biggs BJF (1996) Patterns in benthic algae of streams. In *Algal Ecology: Freshwater Benthic Ecosystems.* (Eds RJ Stevenson, ML Bothwell and RL Lowe) pp. 31–56. Academic Press, San Diego, CA.

Biggs BJF, Price GM (1987) A survey of filamentous algal proliferations in New Zealand rivers. *New Zealand Journal of Marine and Freshwater Research* **21**, 175–191. doi:10.1080/00288330.1987.9516214

Biggs BJF, Smith RA (2002) Taxonomic richness of stream benthic algae: effects of flood disturbance and nutrients. *Limnology and Oceanography* **47**, 1175–1186. doi:10.4319/lo.2002.47.4.1175

Biggs BJF, Stokseth S (1996) Hydraulic habitat suitability for periphyton in rivers. *Regulated Rivers: Research and Management* **12**, 251–261. doi:10.1002/(SICI)1099-1646(199603)12:2/3<251::AID-RRR393>3.0.CO;2-X

Biggs BJF, Stevenson RJ, Lowe RL (1998) A habitat matrix conceptual model for stream periphyton. *Archiv fuer Hydrobiologie* **143**, 21–56.

Blindow I, Hargeby A, Andersson G (2002) Seasonal changes of mechanisms maintaining clear water in a shallow lake with abundant *Chara* vegetation. *Aquatic Botany* **72**, 315–334. doi:10.1016/S0304-3770(01)00208-X

Blum JL (1956) The ecology of river algae. *Botanical Review* **22**, 291–341. doi:10.1007/BF02872474

Blum JL (1959) Algal populations in flowing waters. In *The Ecology of Algae – The Pymatuning Symposia in Ecology.* (Ed. CA Tyron) pp. 11–21. The University of Pittsburg, Pittsburg, PA.

Borchardt MA (1996) Nutrients. In *Algal Ecology: Freshwater Benthic Ecosystems.* (Eds RJ Stevenson, ML Bothwell and RL Lowe) pp. 184–227. Academic Press, San Diego, CA.

Bornette G, Arens M-F (2002) Charophyte communities in cut-off river channels: the role of connectivity. *Aquatic Botany* **73**, 149–162. doi:10.1016/S0304-3770(02)00017-7

Brock MA (2011) Persistence of seed banks in Australian temporary systems. *Freshwater Biology* **56**, 1312–1327. doi:10.1111/j.1365-2427.2010.02570.x

Brock MA, Lane JAK (1983) The aquatic macrophyte flora of saline wetlands in Western Australia in relation to salinity and permanence. *Hydrobiologia* **105**, 63–76. doi:10.1007/BF00025177

Brock MA, Theodore K, O'Donnel L (1994) Seed bank methods for Australian wetlands. *Australian Journal of Marine and Freshwater Research* **45**, 483–493.

Bunn SE, Davies PM, Winning M (2003) Sources of organic carbon supporting the food web of an arid zone floodplain river. *Freshwater Biology* **48**, 1–17.

Bunn SE, Thoms MC, Hamilton SK, Capon SJ (2006) Flow variability and dryland rivers: 'boom', 'bust' and the bits in between. *River Research and Applications* **22**, 179–186.

Burne RV, Bauld J, De Decker P (1980) Saline lake charophytes and their geological significance. *Journal of Sedimentary Petrology* **50**, 281–293.

Capon SJ, Dowe JL (2007) Diversity and dynamics of riparian vegetation. In *Principles for Riparian Lands Management*. (Eds S Lovett and P Price) pp. 13–32. Land and Water Australia, Canberra.

Casanova MT (1993) The ecology of charophytes in temporary and permanent wetlands: an Australian perspective. PhD thesis. University of New England, Armidale, NSW.

Casanova MT (1994) Vegetative and reproductive responses of charophytes to water-level fluctuations in permanent and temporary wetlands in Australia. *Australian Journal of Marine and Freshwater Research* **45**, 1409–1419.

Casanova MT (2004) 'A census of submerged plants of the Angas River and Tookayerta Creek catchments.' Report to the River Murray Catchment Water Management Board, Charophyte Services, Lake Bolac, Victoria.

Casanova MT (2007) Charophyceae. In *Algae of Australia: Introduction*. (Eds PM McCarthy and AM Orchard) pp. 368–373. Australian Biological Resources Study, Canberra and CSIRO Publishing, Melbourne.

Casanova MT (2009a) An overview of *Nitella* (Characeae, Charophyceae) in Australia. *Australian Systematic Botany* **22**, 193–218. doi:10.1071/SB08039

Casanova MT (2009b) 'Seed bank methodology and data for the Condamine River, Macquarie Marshes and Ovens River catchments.' Report to The Murray Darling Basin Commission. Charophyte Services, Lake Bolac, Victoria.

Casanova MT (2011) Using plant functional groups in investigate environmental water requirements. *Freshwater Biology* **56**, 2637–2652. doi:10.1111/j.1365-2427.2011.02680.x

Casanova MT (2013) *Lamprothamnium* in Australia (Characeae, Charophyceae). *Australian Systematic Botany* **26**, 268–290. doi:10.1071/SB13026

Casanova MT, Brock MA (1990) Charophyte germination and establishment from the seed bank of an Australian temporary lake. *Aquatic Botany* **36**, 247–254.

Casanova MT, Karol KG (2014) A revision of ecorticate species of *Chara* (subg. *Charopsis*, sect. *Protochara*, *comb. et stat. nov.*) (Characeae: Charophyceae). *Australian Systematic Botany* **27**, 23–27. doi:10.1071/SB13016

Casanova MT, Kraft G (2004) 'Charophytes in deep water lakes in Australia.' Final report to Australian Biological Resources Study, Canberra.

Casanova MT, Lewis SH (2007) Zygnemophyceae. In *Algae of Australia: Introduction*. (Eds PM McCarthy and AM Orchard) pp. 363–367. Australian Biological Resources Study, Canberra and CSIRO Publishing, Melbourne.

Casanova MT, Nicol JM (2009) *Chara canescens* in the southern hemisphere. *Charophytes* **2**, 55–60.

Casanova MT, Porter JL (2013) Two new *Nitella* species from arid-zone claypan wetlands in Australia (Characeae, Charophyceae). *Muelleria* **31**, 53–59.

Casanova MT, de Winton MD, Clayton JS (2002) Do charophytes clear turbid water? *Verhandlungen – Internationale Vereinigung für Theoretische und Angewandte Limnologie* **26**, 1440–1443.

Casanova MT, García A, Porter JL (2003) Charophyte rediscoveries in Australia: what and why? *Acta Micropalaeontologica Sinica* **20**, 129–138.

Casanova MT, de Winton MD, Karol KG, Clayton JS (2007) *Nitella hookeri* A. Braun (Characeae, Charophyceae) in New Zealand and Australia: implications for endemism, speciation and biogeography. *Charophytes* **1**, 2–18.

Ceschin S, Aleffi M, Bisceglie S, Savo V, Zuccarello V (2012) Aquatic bryophytes as ecological indicators of the water quality status in the Tiber River Basin (Italy). *Ecological Indicators* **14**, 74–81. doi:10.1016/j.ecolind.2011.08.020

Clausen B, Biggs BJF (1997) Relationships between benthic biota and hydrological indices in New Zealand streams. *Freshwater Biology* **38**, 327–342. doi:10.1046/j.1365-2427.1997.00230.x

Corillion R (1957) Les charophycées de France et d'Europe Occidentale. *Bulletin de la Société Scientifique de Bretagne* **32**, 1–471.

Crisci JV, Sala OE, Katinas L, Posadas P (2006) Bridging historical and ecological approaches in biogeography. *Australian Systematic Botany* **19**, 1–10. doi:10.1071/SB05006

Davidson AJ, Harborne JB, Longton RE (1990) The acceptability of mosses as food for generalist herbivores, slugs in the Arionidae. *Botanical Journal of the Linnean Society* **104**, 99–113.

Davis JS (1972) Survival records in the algae, and the survival role of certain algal pigments, fat and mucilaginous substances. *Biologist (Columbus, Ohio)* **54**, 52–93.

Day SA, Wickham RP, Entwisle TJ, Tyler PA (1995) *Bibliographic Checklist of Non-marine Algae in Australia*. Flora of Australia supplementary series 4. Australian Biological Resources Study, Canberra.

Dodds WK, Hutson RE, Eichem AC, Evans MA, Gudder DA, Fritz KM, Gray L (1996) The relationship of floods, drying, flow and light to primary production and producer biomass in a prairie stream. *Hydrobiologia* **333**, 151–159. doi:10.1007/BF00013429

Douglas-Hill A (1997) The role of turbidity in waterplant establishment. In *Australian Society for Limnology 36th Congress*. p. 49. Albury/Wodonga.

Downes BJ, Lake PS, Schreiber ESG, Glaister A (2000) Habitat structure, resources and diversity: the separate effects of surface roughness and macroalgae on stream invertebrates. *Oecologia* **123**, 569–581.

Downes BJ, Entwisle TJ, Reich P (2003) Effects of flow regulation on disturbance frequencies and in-channel bryophytes and macroalgae in some upland streams. *River Research and Applications* **19**, 27–42. doi:10.1002/rra.692

Dynesius M, Hylander C, Nilsson C (2009) High resilience of bryophyte assemblages in streamside compared to upland forests. *Ecology* **90**, 1042–1054. doi:10.1890/07-1822.1

Eldridge DJ, Tozer ME (1996) Distribution and floristics of bryophytes in soil crusts in semi-arid eastern Australia. *Australian Journal of Botany* **44**, 223–247.

Entwisle TJ (1988) A monograph of *Vaucheria* (Vaucheriaceae, Chrysophyta) in south-eastern mainland Australia. *Australian Systematic Botany* **1**, 1–77. doi:10.1071/SB9880001

Entwisle TJ (1989) Macroalgae in the Yarra River Basin: flora and distribution. *Proceedings of the Royal Society of Victoria* **101**, 1–76.

Entwisle TJ (1990) Macroalgae in the upper Yarra and Watts River catchments: distribution and phenology. *Australian Journal of Marine and Freshwater Research* **41**, 505–522. doi:10.1071/MF9900505

Entwisle TJ (1994) Macroalgae. In *Aquatic Cryptogams of Australia. Australian Society for Limnology Special Publication*. No. 10. (Ed. TJ Entwisle) pp. 29–105. Australian Society for Limnology, Abbotsford, Victoria.

Entwisle TJ (2007) Heterokontophyta: Xanthophyceae. In *Algae of Australia: Introduction*. (Eds PM McCarthy and AM Orchard) pp. 255–256. Australian Biological Resources Study, Canberra and CSIRO Publishing, Melbourne.

Entwisle TJ, Foard HJ (2007) Batrachospermales. In *Algae of Australia: Batrachosperales, Thoreales, Oedogoniales and Zygnemataceae*. (Eds TJ Entwisle, S Skinner, SH Lewis SH and HJ Foard) pp. 1–24. Australian Biological Resources Study, Canberra and CSIRO Publishing, Melbourne.

Entwisle TJ, Sonneman JA, Lewis SH (1997) *Freshwater Algae in Australia*. Sainty and Associates, Potts Point, Sydney.

Entwisle TJ, Vis ML, Foard HJ (2000) Biogeography of *Psilosiphon* (Batrachospermales, Rhodophyta) in Australia and New Zealand. *Cryptogamie. Algologie* **21**, 133–148. doi:10.1016/S0181-1568(00)00111-2

Entwisle TJ, Skinner S, Lewis SH, Foard HJ (Eds) (2007) *Algae of Australia: Batrachospermales, Thoreales, Oedogoniales and Zygnemataceae*. Australian Biological Resources Study, Canberra and CSIRO Publishing, Melbourne.

ESD Working Group on Biological Diversity (1991) 'The conservation of biological diversity as it relates to ecologically sustainable development.'

Report of working party to the Ecologically Sustainable Development Secretariat, Canberra.

Feminella JW, Resh VH (1991) Herbivorous caddisflies, macroalgae, and epilithic microalgae: dynamic interactions in a stream grazing system. *Oecologia* **87**, 247–256.

Forsberg C (1965) Environmental conditions for Swedish charophytes. *Symbolae Botanica Upsalienses* **18**, 5–62.

García A (1999) Charophyte flora of south-eastern South Australia and south-western Victoria, Australia: systematics, distribution and ecology. *Australian Journal of Botany* **47**, 407–426. doi:10.1071/BT97084

García A, Casanova MT (2003) *Lamprothamnium heraldii* sp. nov. (Charales, Charophyta) from Australia: the first dioecious representative of the genus. *Phycologia* **42**, 622–628. doi:10.2216/i0031-8884-42-6-622.1

García A, Chivas AR (2004) Quaternary and extant eryhaline Lamprothamnium Grovees (Chalales) from Australia: gyrogonite morphology and palaeolimnological significance. *Journal of Palaeolimnology* **31**, 321–341.

García A, Chivas AR (2006) Diversity and ecology of extant and Quaternary Australian charophytes (Charales). *Cryptogamie. Algologie* **27**, 323–340.

Glime JM (1971) Response of two species of *Fontinalis* to field isolation from stream water. *The Bryologist* **74**, 383–386. doi:10.2307/3241653

Glime JM, Vitt DH (1984) The physiological adaptations of aquatic musci. *Lindbergia* **10**, 41–52.

Grime JP (1979) *Plant Strategies and Vegetation Processes*. John Wiley & Sons, New York, NY.

Guiry MD (2012) How many species of algae are there? *Phycologia* **48**, 1057–1063.

Haig D (2010) What do we know about charophyte (streptophyta) life cycles? *Journal of Phycology* **46**, 860–867. doi:10.1111/j.1529-8817.2010.00874.x

Harris G (2001) A nutrient dynamics model for Australian waterways: land use, catchment biogeochemistry and water quality in Australian rivers, lakes and estuaries. In *Australian State of the Environment Second Technical Paper Series (Inland Waters)*. Department of Environment and Heritage, Canberra.

Huisman JM, Kraft GT (2007) Rhodophyta: Florideophyceae. In *Algae of Australia: Introduction*. (Eds PM McCarthy and AM Orchard) pp. 217–223. Australian Biological Resources Study, Canberra and CSIRO Publishing, Melbourne.

Huisman JM, Saunders GW (2007) Phylogeny and classification. In *Algae of Australia: Introduction*. (Eds PM McCarthy and AM Orchard) pp. 66–104. Australian Biological Resources Study, Canberra and CSIRO Publishing, Melbourne.

Imahori K (1954) *Ecology, Phytogeography and Taxonomy of the Japanese Charophyta*. Kanazawa University, Kanazawa, Japan.

John DM, Champ WST, Moore JA (1982) The changing status of Characeae in four marl lakes in the Irish midlands. *Journal of Life Science of the Royal Dublin Society* **4**, 47–71.

Karol KG, McCourt RM, Cimino MT, Delwiche CF (2001) The closest living relatives of land plants. *Science* **294**, 2351–2353.

Kenrick P, Crane PR (1997) *The Origin and Early Diversification of Land Plants: a Cladistic Study*. Smithsonian Institution Press, Washington, DC.

Lake PS (1995) Of floods and droughts: river and stream ecosystems of Australia. In *River and Stream Ecosystems*. (Eds CE Cushing, KW Cummins, GW Minshall) p. 817. Elsevier, Amsterdam.

Larkum AWD (2007) The major groups of algae: photosynthetic oxygenic prokaryotes. In *Algae of Australia: Introduction*. (Eds PM McCarthy and AM Orchard) pp. 198–206. Australian Biological Resources Study, Canberra and CSIRO Publishing, Melbourne.

Lewis SH, Entwisle TJ (2007) Zygnemaceae. In *Algae of Australia: Batrachosperales, Thoreales, Oedogoniales and Zygnemataceae*. (Eds TJ Entwisle, S Skinner, SH Lewis SH and HJ Foard) pp. 112–151. Australian Biological Resources Study, Canberra and CSIRO Publishing Melbourne.

Lirman D, Deangelo G, Serafy J, Hazra A, Smith-Hazra D, Herlan J, Luo J, Bellmund S, Wang J, Clausing R (2008) Seasonal changes in abundance and distribution of submerged aquatic vegetation in a highly managed coastal lagoon. *Hydrobiologia* **596**, 105–120. doi:10.1007/s10750-007-9061-x

Manel S, Buckton ST, Ormerod SJ (2000) Testing large-scale hypotheses using surveys: the effects of land use on the habitats, invertebrates and birds of Himalayan rivers. *Journal of Applied Ecology* **37**, 756–770. doi:10.1046/j.1365-2664.2000.00537.x

Martin CE, Churchill SP (1982) Chlorophyll concentrations and a/b ratios in mosses collected from exposed and shaded habitats in Kansas. *Journal of Bryology* **12**, 297–304. doi:10.1179/jbr.1982.12.2.297

McCourt RM, Karol KG, Casanova MT, Feist M (1999) Monophyly of genera and species of Characeae based on rbcL sequences, with special reference to Australian and European Lychnothamnus barbatus (Characeae: Charophyceae). *Australian Journal of Botany* **47**, 361–369.

McCourt RM, Delwiche CF, Karol KG (2004) Charophyte algae and land plant origins. *Trends in Ecology & Evolution* **19**, 661–666.

McMahon TA, Finlayson BL (2003) Droughts and anti-droughts: the low flow hydrology of Australian rivers. *Freshwater Biology* **48**, 1147–1160. doi:10.1046/j.1365-2427.2003.01098.x

Meagher D, Fuhrer B (2003) *A Field Guide to the Mosses and Allied Plants of Southern Australia*. Australian Biological Resources Study, Canberra.

Middelboe AL, Markager S (1997) Depth limits and minimum light requirements of freshwater macrophytes. *Freshwater Biology* **37**, 553–568. doi:10.1046/j.1365-2427.1997.00183.x

Minshall GW (1978) Autotrophy in stream ecosystems. *Bioscience* **28**, 767–771.

Morison MO, Sheath RG (1985) Responses to desiccation stress by *Klebsormidium rivulare* (Ulotrichales, Chlorophyta) from a Rhode Island stream. *Phycologia* **24**, 129–145. doi:10.2216/i0031-8884-24-2-129.1

Mosisch TD (2001) Effects of desiccation on stream epilithic algae. *New Zealand Journal of Marine and Freshwater Research* **35**, 173–179. doi:10.1080/00288330.2001.9516987

Mosisch TD, Bunn SE (1997) Temporal patterns of rainforest stream epilithic algae in relation to flow-related disturbance. *Aquatic Botany* **58**, 181–193. doi:10.1016/S0304-3770(97)00001-6

Mosisch TD, Bunn SE, Davies PM, Marshall CJ (1999) Effects of shade and nutrient manipulation on periphyton growth in a subtropical stream. *Aquatic Botany* **64**, 167–177. doi:10.1016/S0304-3770(99)00014-5

Mosisch TD, Bunn SE, Davies PM (2001) The relative importance of shading and nutrients on algal production in subtropical streams. *Freshwater Biology* **46**, 1269–1278. doi:10.1046/j.1365-2427.2001.00747.x

Mulholland PJ (1996) Role in nutrient cycling in streams. In *Algal Ecology: Freshwater Benthic Ecosystems*. (Eds RJ Stevenson, ML Bothwell and RL Lowe) pp. 609–639. Academic Press, San Diego, CA.

Mummery J, Hardy N (1994) *Australia's Biodiversity: an Overview of Selected Significant Components*. Department of the Environment, Sports and Territories, Canberra.

Muotka T, Virtanen R (1995) The stream as a habitat templet for bryophytes: species' distributions along gradients in disturbance and substratum heterogeneity. *Freshwater Biology* **33**, 141–160. doi:10.1111/j.1365-2427.1995.tb01156.x

Naiman RJ (1983) The annual pattern and spatial distribution of aquatic oxygen metabolism in boreal forest watersheds. *Ecological Monographs* **53**, 73–94. doi:10.2307/1942588

Nairn LC (2006) *Studies in the Ecology of Stream Bryophytes and Macroalgae from Eastern Australia: Scale, Pattern and Process*. University of Melbourne, Victoria.

Necchi O, Jr (1993) Distribution and seasonal dynamics of Rhodophyta in the Preto River Basin, southeastern Brazil. *Hydrobiologia* **250**, 81–90. doi:10.1007/BF00008229

Necchi O, Jr, Dip MR, Goes RM (1991) Macroalgae of a stream in southeastern Brazil: composition, seasonal variation and relation to physical and chemical variables. *Hydrobiologia* **213**, 241–250. doi:10.1007/BF00016425

Noordhuis R, van der Molen DT, van den Berg MS (2002) Response of herbivorous water-birds to the return of *Chara* in Lake Veluwemeer, the Netherlands. *Aquatic Botany* **72**, 349–367. doi:10.1016/S0304-3770(01)00210-8

Ot'ahel'ová H, Valachovič M, Hrivnák R (2007) The impact of environmental factors on the distribution patterns of aquatic plants along the Danube corridor (Slovakia). *Limnologica* **37**, 290–302. doi:10.1016/j.limno.2007.07.003

Palma-Silva C, Albertoni EF, Esteves FA (2002) Clear water associated with biomass and nutrient variation during the growth of a charophyte stand after a drawdown, in a tropical coastal lagoon. *Hydrobiologia* **482**, 79–87. doi:10.1023/A:1021295811688

Pan Y, Stevenson RJ, Hill BH, Kaufmann PR, Herlihy AT (1999) Spatial patterns and ecological determinants of benthic algal assemblages in mid-Atlantic streams, USA. *Journal of Phycology* **35**, 460–468. doi:10.1046/j.1529-8817.1999.3530460.x

Peterson CG (1996) Response of benthic algae to natural physical disturbance. In *Algal Ecology: Freshwater Benthic Ecosystems*. (Eds J Stevenson, ML Bothwell and RL Lowe) pp. 375–402. Academic Press, San Diego, CA.

Peterson CG (2007) Ecology of non-marine algae: streams. In *Algae of Australia: Introduction*. (Eds PM McCarthy and AM Orchard) pp. 434–458. Australian Biological Resources Study, Canberra and CSIRO Publishing, Melbourne.

Peterson CG, Grimm NB (1992) Temporal variation in enrichment effects during Periphyton succession in a nitrogen-limited desert stream ecosystem. *Journal of the North American Benthological Society* **11**, 20–36. doi:10.2307/1467879

Pickett-Heaps JD (1975) *Green Algae: Structure, Reproduction and Evolution in Selected Genera*. Sinauer, Sunderland, MA.

Pip E (1979) Survey of the ecology of submerged aquatic macrophytes in central Canada. *Aquatic Botany* **7**, 339–357.

Proctor VW (1980) Historical biogeography of *Chara* (Charophyta): an appraisal of the Braun-Wood classification plus a falsifiable alternative for future consideration. *Journal of Phycology* **16**, 218–233. doi:10.1111/j.1529-8817.1980.tb03023.x

Proctor VW (1990) Charophytes of the Llano Estacado (Texas and adjacent New Mexico) playas. *Journal of Biogeography* **17**, 75–84. doi:10.2307/2845189

Ramsay HP, Cairns A (2004) Habitat, distribution and phytogeographical affinities of mosses in the Wet Tropics bioregion, north-east Queensland, Australia. *Cunninghamia* **8**, 371–408.

Raven PH, Evert RF, Eichhorn SE (1992) *Biology of Plants*. 5th edn. Worth Publishers, New York.

Resh VH, Brown AV, Covich AP, Gurtz ME, Hiram WL, Minshall GW, Reice SR, Sheldon AL, Wallace JB, Wissmar RC (1988) The role of disturbance in stream ecology. *Journal of the North American Benthological Society* **7**, 433–455. doi:10.2307/1467300

Ress J, Lowe R, Waite M (2010) Bryophytic algal communities from Nu'uanu Pali, O'ahu (Hawai'i, USA). No 04009. In *The Final Program and Abstracts of the joint conference of the American Society of Plant Biologists and the Phycological Society of America*. July 18–22, Honolulu, HI. American Society of Plant Biologists, USA.

Riddin T, Adams JB (2008) Influence of mouth status and water level on the macrophytes in a small temporarily open/closed estuary. *Estuarine, Coastal and Shelf Science* **79**, 86–92. doi:10.1016/j.ecss.2008.03.010

Robson BJ (2000) Role of residual biofilm in the recolonization of rocky intermittent streams by benthic algae. *Marine and Freshwater Research* **51**, 725–732. doi:10.1071/MF00012

Robson BJ, Matthews TG (2004) Drought refuges affect algal recolonization in intermittent streams. *River Research and Applications* **20**, 753–763. doi:10.1002/rra.789

Rodrigo MA, Rojo C, Álvarez-Cobelaz M, Cirujano S (2007) *Chara hispida* beds as a sink of nitrogen: evidence from growth, nitrogen uptake and decomposition. *Aquatic Botany* **87**, 7–14. doi:10.1016/j.aquabot.2007.01.007

Rothrock PE, Simon TP, Stewart PM (2008) Development, calibration and validation of a littoral zone plant index of biotic integrity (PIBI) for lacustrine wetlands. *Ecological Indicators* **8**, 79–88. doi:10.1016/j.ecolind.2007.01.002

Scheffer M (1998) *Ecology of Shallow Lakes*. Chapman & Hall, London.

Schult J, Townsend S, Douglas M, Webster I, Skinner S, Casanova M (2007) *Recommendations for Nutrient Resource Condition Targets for the Daly River*. Charles Darwin University, Darwin, Australia.

Schwarz A, Hawes I, Howard-Williams C (1996) The role of photosynthesis/light relationships in determining lower depth limits of characeae in South Island, New Zealand lakes. *Freshwater Biology* **35**, 69–80. doi:10.1046/j.1365-2427.1996.00481.x

Scott GAM (1994) Liverworts and mosses. In *Aquatic Cryptogams of Australia*. (Ed. TJ Entwisle) Special Publication No. 10. pp. 106–141. Australian Society for Limnology, Abbotsford, Victoria.

Scott G, Entwisle TJ, May T, Stevens N (1997) *A Conservation Overview of Australian Non-marine Lichens, Bryophytes, Algae and Fungi*. Environment Australia, Canberra.

Sheath RG, Burkholder JM (1985) Characteristics of softwater streams in Rhode Island II. Composition and seasonal dynamics of macroalgal communities. *Hydrobiologia* **128**, 109–118. doi:10.1007/BF00008730

Sheath RG, Cole KM (1992) Biogeography of stream macroalgae in North America. *Journal of Phycology* **28**, 448–460. doi:10.1111/j.0022-3646.1992.00448.x

Sheath RG, Burkholder JM, Hambrook JA, Hogeland AM, Hoy E, Kane ME, Morison MO, Steinman AD, Van Alstyne KL (1986) Characteristics of softwater streams in Rhode Island. III. Distribution of macrophytic vegetation in a small drainage basin. *Hydrobiologia* **140**, 183–191. doi:10.1007/BF00007573

Skinnner S, Entwisle TJ (2007) Chlorophyta: Chlorophyceae. In *Algae of Australia: Introduction*. (Eds PM McCarthy and AM Orchard) pp. 334–337. Australian Biological Resources Study, Canberra and CSIRO Publishing, Melbourne.

Spence DHN (1967) Factors controlling the distribution of freshwater macrophytes with particular reference to the lochs of Scotland. *Journal of Ecology* **55**, 147–170.

Spence DHN (1982) The zonation of plants in freshwater lakes. *Advances in Ecological Research* **12**, 37–125. doi:10.1016/S0065-2504(08)60077-X

Steinman AD, Boston HL (1993) The ecological role of aquatic bryophytes in a woodland stream. *Journal of the North American Benthological Society* **12**, 17–26. doi:10.2307/1467681

Sternberg D, Balcombe SR, Marshall JC, Lobegeiger J (2008) Food resource variability in an Australian dryland river: evidence from the diet of two generalist native fish species. *Marine and Freshwater Research* **59**, 137–144.

Stevenson RJ (1996) An introduction to algal ecology in freshwater benthic habitats. In *Algal Ecology: Freshwater Benthic Ecosystems*. (Eds RJ Stevenson, ML Bothwell and RL Lowe) pp. 3–30. Academic Press, San Diego, CA.

Stevenson RJ (1997) Scale-dependent determinants and consequences of benthic algal heterogeneity. *Journal of the North American Benthological Society* **16**, 248–262. doi:10.2307/1468255

Suren AM (1991) Bryophytes as invertebrate habitat in two New Zealand alpine streams. *Freshwater Biology* **26**, 399–418. doi:10.1111/j.1365-2427.1991.tb01407.x

Stream Bryophyte Group (1999) Roles of bryophytes in stream ecosystems. *Journal of the North American Benthological Society* **18**, 151–184.

Suren AM (1992) Enhancement of invertebrate food resources by bryophytes in New Zealand alpine headwater streams. *New Zealand Journal of Marine and Freshwater Research* **26**, 229–239. doi:10.1080/00288330.1992.9516518

Suren AM (1996) Bryophyte distribution patterns in relation to macro-, meso- and micro-scale variables in South Island, New Zealand streams. *New Zealand Journal of Marine and Freshwater Research* **30**, 501–523. doi:10.1080/00288330.1996.9516738

Suren AM, Ormerod SJ (1998) Aquatic bryophytes in Himalayan streams: testing a distribution model in a highly heterogeneous environment.

*Freshwater Biology* **40**, 697–716. doi:10.1046/j.1365-2427.1998.00366.x

Suren AM, Winterbourn MJ (1991) Consumption of aquatic bryophytes by alpine stream invertebrates in New Zealand. *New Zealand Journal of Marine and Freshwater Research* **25**, 331–343. doi:10.1080/00288330.1991.9516487

Tyler PA (1996) Endemism in freshwater algae with special reference to the Australian region. *Hydrobiologia* **336**, 127–135. doi:10.1007/BF00010826

Vannote RL (1980) The river continuum concept. *Canadian Journal of Fisheries and Aquatic Sciences* **37**, 130–137. doi:10.1139/f80-017

van den Hoek C, Mann DG, Jahns HM (1995) *Algae: an Introduction to Phycology*. Cambridge University Press, Cambridge, UK.

van Raam JC (1995) *The Characeae of Tasmania*. Cramer, Berlin.

Vant WN, Davies-Colley RJ, Clayton JS, Coffey BT (1986) Macrophyte depth limits in North Island lakes of differing clarity. *Hydrobiologia* **137**, 55–60. doi:10.1007/BF00004172

Virtanen R, Muotka T, Saksa M (2001) Species richness-standing crop relationship in stream bryophyte communities: patterns across multiple scales. *Journal of Ecology* **89**, 14–20. doi:10.1046/j.1365-2745.2001.00513.x

Wehrenberg ML (2010) Alternative life history of *Gracilariopsis* sp. (Rhodophyta) within the dynamic substrate of a central Californian estuary. No. P04004. In *The Final Program and Abstracts of the Joint Conference of the American Society of Plant Biologists and the Phycological Society of America*. July 18–22 Honolulu, HI. American Society of Plant Biologists, USA.

Whinam J, Hope GS, Clarkson BR, Buxton RP, Alspach PA, Adam P (2003) *Sphagnum* peatlands of Australasia: their distribution, utilisation and management. *Wetlands Ecology and Management* **11**, 37–49. doi:10.1023/A:1022005504855

Whitton BA, Kelly MG (1995) Use of algae and other plants for monitoring rivers. *Australian Journal of Ecology* **20**, 45–56. doi:10.1111/j.1442-9993.1995.tb00521.x

Wisheu IC, Rosenzweig ML, Olsvig-Whittaker L, Shmida A (2000) What makes nutrient-poor Mediterranean heathlands so rich in plant diversity? *Evolutionary Ecology Research* **2**, 935–955.

Wood RD (1950) Stability and zonation of the Characeae. *Ecology* **31**, 642–647. doi:10.2307/1931582

Wood RD (1972) *Characeae of Australia*. Cramer, Lehre, Germany.

Wood RD, Imahori K (1965) *A Revision of the Characeae*. Cramer, Weinheim, Germany.

# 6

# Aquatic vascular macrophytes of Australia's riverine systems

*Stephen Mackay and Cassandra James*

## INTRODUCTION

Aquatic plants are an integral component of Australian streams and rivers. Macrophytes provide habitat, spawning sites and refuges to many native aquatic vertebrate and invertebrate species, including many species of conservation significance (Kemp 1984; Humphries 1995; Armstrong and Booth 2005). Their contribution to aquatic food webs through direct herbivory or as detritus is generally thought to be relatively minor (Bunn *et al.* 1997; Douglas *et al.* 2005), although this is subject to some debate (Watson and Barmuta 2011). However, the complex epiphytic assemblages of invertebrates, bacteria, protozoans, microalgae and fungi found on macrophyte surfaces are an important food source for a wide variety of aquatic organisms (Biggs and Close 1989). Aquatic macrophytes shape their physical and chemical environment through interactions with the water column and its suspended materials, as well as by affecting their substrates (Madsen *et al.* 2001). Human uses of rivers are influenced significantly by the presence of macrophytes. Indigenous Australians use macrophytes as food and material for making mats, baskets, fishing nets and string (Mitchell 1986; Jackson *et al.* 2012). Macrophytes are exploited by the aquarium and nursery industries. However, excessive macrophyte growth can impede downstream delivery of water for human users, clog water intake pipes and increase flood risks by reducing channel volume. The recreational and aesthetic values of waterways may be reduced by extensive plant growth, which restricts recreational fishing, entangles boat propellers and is hazardous to swimmers (Mitchell 1974, 1978). Aquatic macrophytes can thus be considered 'ecosystem engineers' in riverine landscapes capable of influencing both the natural ecosystem and many of the river functions for which humans value rivers.

Providing a clear definition of an aquatic plant, or macrophyte, is surprisingly difficult. Part of the problem lies with the term 'macrophyte', which has no precise taxonomic definition and loosely means 'large leaf' (Sculthorpe 1967). Macrophytes can be found in permanent waterbodies, temporary aquatic habitats that undergo periodic inundation and drying, and terrestrial habitats. Hence, aquatic macrophytes can be species that are permanently or infrequently submerged. Indeed, most macrophyte species are considered to be semi-aquatic or amphibious (Sand-Jensen and Frost-Christensen 1999), which may reflect the relatively recent terrestrial origins of aquatic plants (Sculthorpe 1967; Cook 1999). Relatively few taxa can actually be considered truly aquatic in the sense that surface water is required for successful life cycle completion.

Aquatic macrophytes are generally defined as plants that can be identified with the naked eye and grow in water, in soil covered with water, or on damp ground (Weaver and Clements 1938, cited in

Sculthorpe 1967; Jacobs and Wilson 1996). Broadly speaking, the term macrophyte is often taken to include macroalgae (particularly the charophytes *Chara* and *Nitella*), bryophytes (mosses and liverworts), pteridophytes (ferns) and non-woody angiosperms. There are no aquatic gymnosperms (Sculthorpe 1967).

In this chapter we provide a broad overview of aquatic macrophyte diversity and ecology in Australian riverine landscapes. We use the term macrophyte to describe aquatic vascular flora (i.e. plants with conductive tissues) that grow in or on water, or on waterlogged soils. Macroalgae and bryophytes are covered in Chapter 5. We review current knowledge on some major Australian macrophyte groups of riverine landscapes, discuss some of their common adaptations to life in water and explore principal environmental drivers of spatial and temporal patterns in macrophyte community structure. Where possible we provide examples from Australian studies, although much of the work on aquatic macrophyte ecology is focussed on Europe and Asia. There is a relative wealth of knowledge relating to wetland flora of Australia's inland and coastal systems (e.g. Brock 1994), discussed in Chapters 9, 10 and 13. There is still much to learn, however, about the ecology of Australia's aquatic flora in riverine contexts, which remain poorly understood, particularly in neotropical regions (Padial *et al.* 2008).

## DIVERSITY

### Taxonomic diversity

Aquatic (vascular) macrophytes are believed to be derived from terrestrial ancestors, with the transition from terrestrial to aquatic habitats probably occurring numerously and at different times (Cook 1999). It is estimated that there are 2500 macrophyte species worldwide, which is less than 1% of the total global vascular flora (Chambers *et al.* 2008). Aquatic flora is dominated by monocotyledons, in contrast to terrestrial flora, which is dominated by dicotyledons (Hutchinson 1975). Sculthorpe (1967) lists ten families of aquatic dicotyledons (51 genera) and 18 families of aquatic monocotyledons (61 genera) while Cook (1999) lists 407 genera from 87 plant families as being aquatic. Estimates of the number of macrophyte species found in Australia vary. Chambers *et al.* (2008) estimated 439 macrophyte species for the Australasian region compared with 553 species listed by Jacobs and Wilson (1996) for the Australian continent alone. This latter estimate is certainly conservative as Jacobs and Wilson's study was not intended to provide a definitive aquatic plant inventory. Clearly, any aquatic macrophyte inventory is dependent on the ecological criteria used to define a macrophyte, convention (Jacobs 1999) and the current state of understanding regarding systematics and species phylogeny.

On a global scale, the Australasian aquatic vascular flora is relatively depauperate, ranking sixth of eight biogeographic regions in terms of total species (Chambers *et al.* 2008). For example, the cosmopolitan genus *Potamogeton* (pondweeds) is represented in Australia by 11 species whereas 80–90 *Potamogeton* species occur globally (Cook 1996; Papassotiriou *et al.* 2011). However, the Australasian flora is relatively speciose in Aponogetonaceae, Menyanthaceae (e.g. *Nymphoides* spp.) and Haloragaceae (e.g. *Myriophyllum* spp.) (Chambers *et al.* 2008). Endemism in the Australian macrophyte flora is also relatively low (Cowie *et al.* 2000), in contrast to the Australian flora as a whole (Marshall 1988), although low endemism is a general feature of aquatic plants globally (Santamaría 2002). Chambers *et al.* (2008) suggest that endemism in the aquatic flora of Australasia is intermediate compared with other major biogeographic regions, with the highest levels of endemism found in the afrotropics and neotropics. Notably, however, Australia is a centre for endemism for some largely aquatic genera including *Myriophyllum* (>37 endemic species; Moody and Les 2010) and *Aponogeton* (11 endemic species; Hellquist and Jacobs 2011).

In addition to the native aquatic flora, a large number of macrophytes have been introduced to Australia. These include species introduced deliberately as ornamental plants (e.g. *Eichhornia crassipes*, water hyacinth) or as ponded pasture species (e.g. *Urochloa mutica*, para grass), or introduced accidently through attachment on other imported goods, or as a result of misidentification. The inventory of macrophytes introduced to Australia is long and subject to the same vagaries as estimates of native species regarding the definition of macrophyte. Nevertheless, this list includes many of those species considered the world's worst weeds, with many introduced species now considered fully naturalised. Among the 32 listed Australian weeds of national significance, there are six macrophytes – *Hymenachne amplexicaulis*, *Salvinia molesta*, *Cabomba caroliniana*, *Alternanthera philoxeroides*, *Sagittaria platyphylla* and *Eichhornia crassipes* (http://www.weeds.org.au/WoNS/). The latter two were added in April 2012 (Australian Weeds Committee 2012).

## Morphological diversity

Morphological classification of macrophytes into growth or life forms has a long history (Sculthorpe 1967). Many classification schemes distinguish between free-floating macrophytes and those that are rooted or attached to the substrate. Genera such as *Lemna*, *Azolla* and the introduced species *Eichhornia crassipes* are free-floating macrophytes. Submerged free-floating macrophytes are rarer but include *Ceratophyllum demersum* (hornwort), *Aldrovanda vesiculosa* (waterwheel plant) and some aquatic species of the genus *Utricularia* (bladderworts). 'Attached' macrophytes are often subdivided into groups based on the location of their photosynthetic biomass (i.e. stems and leaves) in relation to the water surface: submerged, emergent or floating. Submerged species include many members of the family Hydrocharitaceae (e.g. *Egeria*, *Elodea* and *Hydrilla*) that have long flexuous stems and can produce dense canopies close to the water surface, although reproductive organs may be submerged, floating or borne above the water surface. These caulescent growth forms (i.e. possessing a stem) are often distinguished from the rosette growth forms of some species of *Isoetes* and *Vallisneria*, which have short stems and leaves attached to the base of the plant. Emergent species have their photosynthetic biomass protruding above the water surface, or grow on waterlogged soils, and include a large number of sedges (*Baumea*, *Cyperus*, *Schoenoplectus*), rushes (*Juncus*), knotweeds (*Persicaria*) and grasses. Plants with floating leaves that remain rooted in the substrate include waterlilies (*Nymphaea*), marshworts (*Nymphoides*) and water ribbons (*Triglochin*).

The high levels of polymorphism and phenotypic plasticity exhibited by many macrophyte species confounds attempts to develop a clear-cut classification system based on morphological attributes alone (Sculthorpe 1967). For example, the widely distributed *Myriophyllum verrucosum* (red water milfoil) exhibits different growth forms when submerged than when growing in terrestrial habitats on damp soils. The submerged form has green pinnate (i.e. comb-like) leaves and flexuous stems up to 1.5 m long. Plants stranded on damp mud have short stems (up to 10 cm long) with red-purple leaves that may have entire or weakly pinnate margins (Orchard 1985). Heterophylly (i.e. having more than one type of leaf form on the same plant) is also a common feature of several macrophyte species. The native pondweed *Potamogeton tricarinatus*, for example, produces both thin, submerged leaves and floating, ovate or elliptical leaves with a waxy surface (Aston 1977). Submerged and floating leaves may vary in their photosynthetic attributes (Frost-Christensen and Sand-Jensen 1995). Macrophytes that produce floating leaves are often able to survive on damp ground in a terrestrial form (Kadono 1984).

Growth plasticity may also be manifested in more subtle changes in the allocation of biomass between and within different plant components (i.e. roots, stems, leaves and reproductive organs) in response to environmental variation. In low-nutrient environments and environments subject to frequent flooding (Barrat-Segretain 2001), for example, a greater proportion of plant biomass may be allocated to roots or root structure may be modified (Xie *et al.* 2007). Similarly, a greater proportion of plant biomass may be allocated to leaves in low-light environments (Blanch *et al.* 1998).

High morphological plasticity is a typical feature of many aquatic macrophytes, with species exhibiting variation in growth forms in response to variation in environmental conditions that allow them to occur over a wide range of conditions (Lacoul and Freedman 2006).

## ADAPTATIONS TO LIFE IN WATER

Life in water presents vastly different constraints to those encountered by plants on land. Water is one thousand times denser than air and resistance to the diffusion of gases (e.g. carbon dioxide and oxygen) is high. Plants may be subject to stresses associated with low oxygen (i.e. anoxia) and associated chemical toxicities in waterlogged soils. Macrophytes of lotic waters are also subject to quite different mechanical and physical stresses than the gravitational effects experienced by land plants. Hence, macrophytes have developed some unique ways of dealing with life in water, many of which are themes repeated throughout the aquatic flora (Sculthorpe 1967; Cook 1999).

### Growth traits

Macrophytes exhibit a range of morphological or structural and physiological adaptations to life in water. The buoyancy provided by water allows the development of a plant architecture relatively unencumbered by the structural support requirements that influence terrestrial plants. Hence, plants of completely submerged habits often have significantly less complex structural compounds, such as lignin and hemicellulose, than terrestrial plants. Nearly all macrophytes have a system of lacunae or gas chambers which both facilitate internal gas movement and aid buoyancy. Air spaces (or aerenchyma) in roots and stems facilitate the exchange of gases (in particular oxygen) to submerged plant parts. The development of adventitious roots (i.e. roots arising from parts of the plant other than the root base) in many macrophytes also assists the acquisition and internal transport of oxygen. Connective xylem tissues that normally transport water from roots may be severely reduced in macrophytes (Pedersen and Sand-Jensen 1993). Submerged macrophytes may lack features to control water loss, such as a thick waxy cuticle and stomata. Features such as hydrophobic hairs on leaf surfaces (e.g. *Salvinia molesta*) and waxy surfaces (e.g. *Potamogeton sulcatus*) found on the leaves of some floating and emergent plants prevent leaves from becoming waterlogged.

Sessile aquatic organisms anchored in flowing water must tolerate the prevailing conditions. For macrophytes this requires a compromise between reducing the drag forces exerted on the plant by flowing water and the need to intercept light for photosynthesis and growth. To avoid being uprooted, lotic macrophytes show several morphological or structural adaptations that minimise drag forces on the plant. A low, prostrate growth form, as seen, for example, in

**Figure 6.1:** *Cladopus queenslandicus* (Podostemaceae) from Behana Creek, Mulgrave River Catchment, north Queensland.

*Cladopus queenslandicus*, an endemic species that clings to rocks and can be found in fast-flowing headwater streams of northern Queensland (Fig. 6.1), is subject to relatively low drag compared with other growth forms (e.g. Suren *et al.* 2000). Species with long strap-like leaves are also subject to less drag than those with broader leaves and more branching growth forms. Species may reconfigure their morphology in response to increasing water velocities, with leaves compressing against the stem and stems and/or leaves bending over to form a more streamlined shape, lowering the drag and minimising physical damage to the plant (Sand-Jensen 2003).

### Reproductive traits

Clonal growth and vegetative reproduction are extremely common among aquatic flora (Philbrick and Les 1996) and even small stem fragments may be capable of establishing new plants (Barrat-Segretain *et al.* 1998). Vegetative spread may occur through plant stem fragments, rhizomes, stolons, tubers and specialised vegetative organs (e.g. turions) in some species. Vegetative reproduction is an extremely efficient method of plant dispersal and colonisation at small and large spatial scales. For example, the spread of *Elodea canadensis* across Europe is thought to have been primarily through vegetative reproduction (Sculthorpe 1967). Vegetative dispersal probably accounts for the spread of many introduced species in Australia. Macrophytes have also devised some curious ways of reproducing and dispersing sexually in water (for an overview see Sculthorpe 1967). Flowers, fruits or seeds, for example, may possess floatation aids that enable them to rise to the water surface and disperse, or aerial flowering and fruiting in species that are otherwise completely submerged. Flowers of *Vallisneria nana* develop underwater but pollination occurs on the surface. Male flowers are free-floating and released from the base of the plant, but female flowers are borne on peduncles. The peduncle elongates to allow the female flower to reach the water surface. Upon fertilisation, the peduncle of the female flower coils to pull the flower below the water surface, where the fruit develops (Aston 1977).

## DISTRIBUTIONS AND DYNAMICS

### Biogeography

A cosmopolitan distribution is a notable feature of aquatic macrophytes (Santamaría 2002), with many species having intercontinental distributions. Within Australia, classification of aquatic floras at the genus and species levels shows that the tropics form a distinct group from temperate and subtropical regions (Jacobs and Wilson 1996), and similar patterns have been noted elsewhere (Cook 1985; Crow 1993). Families exhibiting strong tropical affinities include Podostemaceae, Aponogetonaceae and Nymphaeaceae (Crow 1993). Those with generally temperate distributions include Potamogetonaceae, Haloragaceae and Ranunculaceae (Aston 1977; Crow 1993). This disjunction can also be seen within individual genera, for example, within the genus *Vallisneria*: *V. nana* (ribbonweed) has a generally tropical–subtropical distribution, but is replaced by *V. australis* (= *V. americana)* in temperate regions (Jacobs and Frank 1997).

Climatic factors probably govern broadscale patterns in aquatic plant species distributions and are thought to be largely responsible for the differences observed between the tropical floras and those elsewhere (Jacobs and Wilson 1996). The role of climate in determining species distributions is generally thought to have a more limited role in broad climatic zones (Santamaría 2002). However, in Australia the strong climatic gradients observed between the coast and inland areas may still be important in governing macrophyte distributions at even relatively local scales. In the temperate climate of north-eastern New South Wales, for example, Growns (2009) found macrophyte assemblage structure to be most strongly related to longitude, which is likely to be a proxy for several climatic parameters including annual rainfall, and maximum and minimum temperatures. Chessman and Royal (2010) found, however, that the influence of climatic drivers of aquatic plant distributions in eastern Australia were hard to disentangle from non-climatic variables and, hence, at these smaller spatial scales the effects of climate may be difficult to isolate from other local environmental influences.

Distance barriers may play a significant role in the development of aquatic flora, particularly at the generic level, with geographically closer areas having more similar floras (Jacobs 1999). However, long-distance dispersal by waterbirds is thought to be an important factor in the distribution of many aquatic plants and may account for the cosmopolitan distribution of many species (Les *et al.* 2003). Viable diaspores from aquatic genera such as *Lemna*, *Typha*, *Myriophyllum* and *Nitella* have been found in the faecal samples of waterbirds in the Macquarie Marshes, a complex of temporary to semi-permanent wetlands in New South Wales (Green *et al.* 2008). Nomadic waterfowl may be instrumental in the long-distance dispersal of aquatic plants, although at broader spatial scales (i.e. 100–1000 km) plants may be dispersal-limited (Raulings *et al.* 2011). Divergence times between phylogenetically related taxa suggest that continental drift is unlikely to account for the cosmopolitan distribution of many aquatic macrophytes as divergence between taxa is too recent to be accounted for by this mechanism (Les *et al.* 2003).

## Local-scale factors controlling distributions in lotic systems

At a local scale, such as the scale of rivers and reaches within rivers, the distribution and biomass of macrophytes in Australian lotic systems tends to be highly spatially and temporally variable (Hughes 1990; Rea *et al.* 2002; Mackay *et al.* 2003, 2010; Chessman *et al.* 2008). This variability is due to environmental variability, recovery time from disturbance, the patchy nature of disturbance and the likelihood that plant propagules can successfully establish at a given point in the river network. The environmental factors that influence the distribution and growth of macrophytes in lotic environments have been extensively investigated in temperate systems and include flow (particularly flow extremes), water velocity, water depth, substrate composition, substrate nutrient concentrations, water quality (including nutrient concentration, pH, turbidity), riparian shading, competition and herbivory (Biggs 1996; Carr *et al.* 1997; Franklin *et al.* 2008). Aquatic vegetation itself, however, has a significant effect on river channel geomorphology, river flows and nutrient dynamics (reviewed in Clarke 2002). Such complex synergistic relationships present significant challenges to understanding relationships between environmental factors and macrophytes. By classifying these environmental variables as disturbances (i.e. factors associated with biomass loss) or resources (i.e. factors associated with biomass gain) a much simplified understanding of the relationships between environmental factors and macrophytes can be obtained (Biggs 1996; Riis and Biggs 2001). The key factors characterising disturbance regimes are temporal variability in river hydrology that influence hydraulic regimes (i.e. changes in water velocity, depth and substrate mobility). The key factors characterising resources are light, nitrogen, phosphorus and carbon (Riis and Biggs 2001).

## Hydrological drivers

The hydrological regime is considered a primary orchestrator and determinant of lotic species distributions, community composition and structure. Hydrology plays a fundamental role in macrophyte distribution, composition and structure, and the limited information available on macrophytes of Australian lotic systems provides further evidence to support this (Hughes 1990; Rea *et al.* 2002; Mackay 2007; Duivenvoorden 2008; Chessman and Royal 2010). Australian streams and rivers vary considerably in their hydrologic attributes, from the highly variable and intermittent inland streams that remain dry for months and even years to the more perennial coastal streams of the tropics and temperate zones (Kennard *et al.* 2010). Macrophytes are found across all these systems, but their distribution, composition and structure vary.

In Australian lotic ecosystems, flow extremes (i.e. floods and droughts) and hydraulic forces (i.e. the forces exerted by moving water on the stream bed and plants) are major causes of plant death and reductions in plant biomass (Biggs 1996). Low- and no-flow conditions causing periodic drying out of river channels are typical of many inland streams and rivers of Australia's arid and semi-arid regions. Macrophytes are found in these systems that periodically dry out, although their abundances are generally relatively low in such systems. With very low flow, or the cessation of flow, substantial alterations to water quality occur with concomitant changes in nutrient, oxygen and dissolved carbon concentrations, temperatures and conductivity – factors that all impact on aquatic plant physiology and growth. During drying, plants in littoral reaches may become exposed and, if they remain dry, suffer substantial mortality (Duivenvoorden 2008). A persistent drought-tolerant seed bank may enable some species to survive through periods of desiccation and re-establish on wetting. Persistent seed banks in the river channel, however, may lack submerged species altogether (James *et al.* 2007). Other species, for example, *Aponogeton queenslandicus*, have dormant tubers or, in the case of the aquatic fern *Marsilea drummondii* (nardoo), sporocarps that can tolerate drought. For species without a propagule bank or those with a short-lived transient propagule bank, ready dispersal from permanent waterbodies is usually a prerequisite for long-term persistence in ephemeral waterbodies.

Floods are important to macrophytes as, arguably, the principal source of disturbance (i.e. cause of biomass loss) in riverine landscapes (Fig. 6.2). Macrophyte communities may require many months to recover from floods that cause complete uprooting of below-ground structures, as colonisation must be achieved by immigration of seeds or propagules (Henry *et al.* 1996; Riis *et al.* 2004; Duivenvoorden 2008; Sousa *et al.* 2010). A secondary effect of flooding on macrophytes is biomass reduction or mortality due to reductions in photosynthesis caused by turbid floodwaters (Mitchell and Rogers 1985; Shannon *et al.* 2001; Duivenvoorden 2008).

**Figure 6.2:** Flooding in Obi Obi Creek (Mary River, south-east Queensland) resulted in the deposition of *Egeria densa* (centre) in riparian trees.

The constraints of flooding on colonisation and establishment may limit macrophytes to streams where the time between floods capable of substrate mobilisation is long enough to allow macrophytes to establish. Riis and Biggs (2003) found that macrophytes were absent from New Zealand streams with more than 13 floods per year. Attempts to find similar relationships in south-east Queensland streams have yielded mixed results, perhaps because of the variation in hydraulic forces in channels with differing morphologies (Mackay 2007).

The hydraulic environment is the mechanism by which flow influences macrophytes during both floods and base flow conditions, and describes the forces exerted by flowing water on plants and the stream bed. Biggs (1996) described increases in water velocity and shear stress as the cause of biomass loss during the rising limb of floods. Sufficiently high water velocities can cause plucking and tearing (i.e. fragmentation) of plant biomass, and sufficiently high shear stresses can mobilise sediments which can cause abrasion of plant material or uprooting. In flume studies, Riis and Biggs (2003) found that the biomass lost by eight macrophyte species with varying morphologies when exposed to water velocities of up to 1.5 m/s was less than 1% of the original biomass of each species. This suggests that in minor floods uprooting or sediment abrasion would be more likely to cause biomass loss than plucking and tearing. Mackay (2007) concluded that a combination of sediment abrasion or uprooting and fragmentation accounted for changes in macrophyte biomass after minor flooding at 11 sites in south-east Queensland. The relative importance of plucking or tearing, uprooting and sediment abrasion may depend on the species present, their biomass, acclimation to previous flow events and local-scale sediment characteristics (Biggs 1996).

Floods have significant impacts on macrophyte assemblages, but are also an important mechanism for the dispersal of propagules into new habitats (Gurnell *et al.* 2008), including off-stream (lentic) habitats that may be periodically disconnected from the main river channel. As many submersed macrophytes use vegetative reproduction as the primary means of dispersal, flows that uproot plants or create stem fragments aid dispersal and colonisation of new habitats. Vegetative dispersal is effective in long-range (continental) dispersal, but dispersal of fragments into new habitats is dependent on connectivity between the source and the sink, and retention of water for a sufficient period for fragments to sink and establish.

Flow regimes are clearly important to the spatial and temporal variability of aquatic macrophyte communities. In practice, it is difficult to identify consistent relationships between flow regime attributes and assemblage characteristics (Arthington *et al.* 2012). In addition to climate and catchment characteristics, flow regime attributes at a given point in the stream network are influenced by channel geomorphology such as channel width and bed slope (Gordon *et al.* 2005). Researchers are placing greater importance on the hydraulic environment as a means of describing the habitat of instream aquatic biota, including aquatic plants (Biggs 1996; Riis and Biggs 2003; Sand-Jensen 2008).

The longitudinal distribution of macrophytes along rivers is generally related to variations in hydraulic variables such as water velocity and substrate composition. Macrophyte biomass is often shown to peak at low (not zero) to medium water velocities (less than 0.3 m/s), with aquatic vascular macrophytes typically limited to water velocities of less than 1.0 m/s (Chambers *et al.* 1991; French and Chambers 1996). Bryophytes generally dominate at high water velocities, particularly on coarse substrates (Mackay *et al.* 2010). Although water velocity may impose growth constraints on macrophytes, low water velocities can reduce the diffuse boundary layers surrounding submerged plant surfaces and increasing the mass transfer of essential nutrients and carbon to the plant (Biggs 1996).

### Effects of flow alteration

Flow alterations through river impoundment and water abstraction have altered patterns of hydrologic and hydraulic disturbances in many, if not most, Australian streams and rivers (see Chapter 14). Changes in river flow leading to increased frequency and duration of low-flow periods have probably facilitated the establishment and spread of some aquatic macrophytes, for example, *Egeria densa* in the Hawkesbury-Nepean River, NSW (Cummins *et al.* 1997). Similarly, constant year-round base flow in the highly regulated lower Murray River led to the colonisation and growth of aquatic plants in the main river channel (Walker *et al.* 1994) and off-channel habitats (e.g. Grundell *et al.* 2012). Some regional studies, however, have failed to find any impact of water abstraction on macrophyte assemblages. Chessman *et al.* (2008) attributed the lack of impact from water abstraction on aquatic macrophytes in their study in north-eastern New South Wales to the naturally high spatial variability of macrophyte communities in streams in the study region and the natural resilience of those communities to drought. The extremely high spatial and temporal hydrologic variability of Australian lotic environments may limit our ability to detect the responses (if any) of macrophytes to altered flow regimes.

## RESOURCES

### Light

Light is one of the principal factors regulating plant growth in streams and rivers (Carr *et al.* 1997).

**Figure 6.3:** Shade provided by intact riparian zones may limit, but not necessarily exclude, growth of alien semi-aquatic species in Wet Tropics streams. (a) Riparian zone of the Little Mulgrave River. (b) Riparian zone of Woopen Creek, a tributary of the Russell River.

In addition to controlling rates of photosynthesis (Kirk 1994), light also has an important influence on biomass allocation, physiology and plant morphology, the production and germination of vegetative and sexual propagules, and rates of macronutrient uptake and assimilation (Bolduan *et al.* 1994; Cedergreen and Madsen 2003; Jian *et al.* 2003). The main influences on the light environment in rivers and streams are shading by riparian canopy cover and turbidity. In smaller streams and rivers, shading by overhanging riparian vegetation affects both the quantity and quality of light reaching macrophytes. These affects are also dependent on forest geometry, weather, season and time of day. In forested headwater streams, narrow stream widths combined with high levels of riparian canopy cover effectively limit macrophyte growth, and biomass in heavily shaded headwater streams is typically low or absent (Fig. 6.3). Riparian shading is often seen as a management tool for controlling excessive aquatic plant growth. Attempts to control unwanted macrophyte growth through riparian shading, however, may be confounded by the tolerance of low light levels by some species, especially submerged species. Riparian shading may reduce the biomass of nuisance species but may not totally eliminate these species.

The effects of riparian shading are reduced as stream width increases and water quality and transparency progressively become more important, particularly for submerged macrophytes. Light attenuation in water can be rapid, even in very clear waters, due to absorption, refraction and scattering by suspended particles (Kirk 1994). In Australian desert rivers, however, photic depths are very shallow (<0.4 m; Fellows *et al.* 2009) due to fine clays in suspension. Both the low light environment and the ephemeral, temporary nature of river flows are thought to preclude substantial macrophyte bed development in these systems (e.g. Bunn *et al.* 2003), although beds often develop in associated temporary floodplain wetlands and littoral habitats (Ogden 2000; Reid et al. 2007; Reid 2008).

## Nutrients

For the majority of macrophytes, macronutrients (i.e. nitrogen, phosphorus, carbon) are derived from either the substrate and/or directly from the water column. The degree to which plants use these two sources has been a matter of debate for some time (Denny 1972), and it is likely to depend both on the relative availability of nutrients in the water column and substrate and the species itself (Best *et al.* 1996).

The extent to which nitrogen and phosphorus limit aquatic plant growth is a contentious issue. In flowing waters, aquatic plants have a continuous supply of nitrogen and phosphorus from upstream sources. Indeed, some authors believe that background nutrient levels in many streams are sufficient for aquatic plant growth (Madsen and Adams 1989). Furthermore, some aquatic plants are able to store phosphorus in tissues in excess of their requirements ('luxury consumption'; Madsen and Cedergreen 2002). While laboratory studies have shown that nutrient additions can increase growth and influence competitive outcomes between species, nutrient concentrations are not always correlated or associated with aquatic plant growth (Canfield and Hoyer 1988; Feijoó *et al.* 1996; Carr *et al.* 2003; Hoyer *et al.* 2004). Macrophytes themselves can also play a key role in nutrient cycling. Decomposition of macrophytes may represent an important pathway for nutrients to enter aquatic food webs (Shilla *et al.* 2006).

The supply of carbon can be a major limitation on photosynthesis and growth of aquatic macrophytes due to the slow diffusion of carbon dioxide in water. This is particularly so for slow-moving and still waters, where transport of dissolved inorganic carbon occurs mainly via diffusion in contrast to the advective transport of dissolved inorganic carbon in flowing waters. Aquatic macrophytes have several methods for acquiring carbon in water. The lack of a thick waxy cuticle enables nutrients and gases to be absorbed directly from the water while a high surface area to biomass ratio, typical of plants with highly dissected or thin leaves (e.g. *Ceratophyllum demersum*), increases the area for interception of resources. Some aquatic macrophyte species can utilise sediment $CO_2$ to varying extents (Winkel and Borum 2009). Many submerged macrophytes exhibit physiological mechanisms that enhance uptake of carbon dioxide, such as crassulacean acid metabolism or a $C_4$-like fixation (Keeley 1998; Maberly and Madsen 2002). *Hydrilla verticillata*, a species found throughout the eastern seaboard of Australia, can shift to a $C_4$-like carbon fixation pathway (Maberly and Madsen 2002). $C_4$ fixation is advantageous under conditions of high oxygen concentration, high pH and low inorganic carbon concentration (Maberly and Madsen 2002).

Many submerged macrophytes can utilise bicarbonate ($HCO_3^-$) as an alternative carbon source to carbon dioxide. In freshwater, the relative proportions of different aqueous carbon fractions varies strongly with ambient pH. In waters with low pH, carbon will be available primarily in the form of free $CO_2$ and carbonic acid ($H_2CO_3$). As pH increases the relative proportion of bicarbonate increases and, at a pH of 6.0–9.2, bicarbonate is the major source of carbon in solution (Riis and Biggs 2001). Submerged macrophytes found in Australian freshwaters such as *Ceratophyllum demersum* and *Potamogeton* species are thought to be capable of utilising bicarbonate. Bicarbonate utilisation, however, probably incurs energy costs to the plant (Jones 2005), in contrast to the uptake of free $CO_2$ that occurs by diffusion and consequently has no direct energy costs (Raven and Lucas 1985).

## CONCLUSIONS

Aquatic macrophytes are an essential, but not always highly visible, component of riverine ecosystems. Their prominence in aquatic ecosystems is in direct contrast to our knowledge of their ecology and taxonomy in Australian streams and rivers. They contribute in numerous ways to the overall structural complexity and heterogeneity of riverine ecosystems. Their physical presence provides structural complexity in the water column and slows water velocity, causing fine sediments and organic material to settle and reduce turbidity. Substantial alterations to water chemistry in and surrounding macrophyte beds occur as a result of plant photosynthesis and respiration, with changes in dissolved gases, pH and nutrients. Macrophytes provide habitat, spawning sites and refuges for many native aquatic vertebrate and invertebrate species, including several species of conservation significance. Their contribution to aquatic food webs through direct herbivory or as detritus is generally thought to be relatively minor, although this is subject to some debate. However, the complex epiphytic assemblages of invertebrates, bacteria, protozoans, microalgae and fungi found on macrophyte surfaces are an important food source for a wide variety of aquatic organisms (Biggs and Close 1989). Collectively these mechanisms contribute to maintaining aquatic biodiversity and support key ecological functions of riverine ecosystems.

## REFERENCES

Armstrong G, Booth DT (2005) Dietary ecology of the Australian freshwater turtle (*Elseya* sp.: Chelonia: Chelidae) in the Burnett River, Queensland. *Wildlife Research* **32**, 349–353.

Arthington AH, Mackay SJ, James CS, Rolls RJ, Sternberg D, Barnes A, Capon SJ (2012) 'Ecological limits of hydrologic alteration: a test of the ELOHA framework in south-east Queensland.' Waterlines report, National Water Commission, Canberra.

Aston HI (1977) *Aquatic Plants of Australia*. Melbourne University Press, Melbourne.

Australian Weeds Committee (2012) *Weeds of National Significance 2012*. Department of Agriculture, Fisheries and Forestry, Canberra.

Barrat-Segretain M-H (2001) Biomass allocation in three macrophyte species in relation to the disturbance level of their habitat. *Freshwater Biology* **46**, 935–945. doi:10.1046/j.1365-2427.2001.00728.x

Barrat-Segretain M-H, Bornette G, Hering-Vilas-Boas A (1998) Comparative abilities of vegetative regeneration among aquatic plants growing in disturbed habitats. *Aquatic Botany* **60**, 201–211. doi:10.1016/S0304-3770(97)00091-0

Best EPH, Woltman H, Jacobs FHH (1996) Sediment-related growth limitation of *Elodea nuttallii* as indicated by a fertilization experiment. *Freshwater Biology* **36**, 33–44. doi:10.1046/j.1365-2427.1996.d01-474.x

Biggs BJF (1996) Hydraulic habitat of plants in streams. *Regulated Rivers: Research and Management* **12**, 131–144. doi:10.1002/(SICI)1099-1646(199603)12:2/3<131::AID-RRR385>3.0.CO;2-X

Biggs BJF, Close ME (1989) Periphytic biomass dynamics in gravel bed rivers: the relative effects of flows and nutrients. *Freshwater Biology* **22**, 209–231. doi:10.1111/j.1365-2427.1989.tb01096.x

Blanch SJ, Ganf GG, Walker KF (1998) Growth and recruitment in *Vallisneria americana* as related to average irradiance in the water column. *Aquatic Botany* **61**, 181–205. doi:10.1016/S0304-3770(98)00065-5

Bolduan BR, van Eeckhout GC, Quade HW, Gannon JE (1994) *Potamogeton crispus* – the other invader. *Lake and Reservoir Management* **10**, 113–125. doi:10.1080/07438149409354182

Brock MA (1994) Aquatic vegetation of inland wetlands. In *Australian Vegetation*. 2nd edn. (Ed. RH Groves) pp. 437–466. Cambridge University Press, Cambridge, UK.

Bunn SE, Davies PM, Kellaway DM (1997) Contributions of sugar cane and invasive pasture grass to the aquatic food web of a tropical lowland stream. *Marine and Freshwater Research* **48**, 173–179. doi:10.1071/MF96055

Bunn SE, Davies PM, Winning M (2003) Sources of organic carbon supporting the food web of an arid zone floodplain river. *Freshwater Biology* **48**, 619–635. doi:10.1046/j.1365-2427.2003.01031.x

Canfield DE, Jr, Hoyer MV (1988) Influence of nutrient enrichment and light availability on the abundance of aquatic macrophytes in Florida streams. *Canadian Journal of Fisheries and Aquatic Sciences* **45**, 1467–1472. doi:10.1139/f88-171

Carr GM, Duthie HC, Taylor WD (1997) Models of aquatic plant productivity: a review of the factors that influence growth. *Aquatic Botany* **59**, 195–215. doi:10.1016/S0304-3770(97)00071-5

Carr GM, Bod SAE, Duthie HC, Taylor WD (2003) Macrophyte biomass and water quality in Ontario rivers. *Journal of the North American Benthological Society* **22**, 182–193. doi:10.2307/1467991

Cedergreen N, Madsen TV (2003) Nitrate reductase activity in roots and shoots of aquatic macrophytes. *Aquatic Botany* **76**, 203–212. doi:10.1016/S0304-3770(03)00050-0

Chambers PA, Prepas EE, Hamilton HR, Bothwell ML (1991) Current velocity and its effect on aquatic macrophytes in flowing waters. *Ecological Applications* **1**, 249–257. doi:10.2307/1941754

Chambers PA, Lacoul P, Murphy KJ, Thomaz SM (2008) Global diversity of aquatic macrophytes in freshwater. *Hydrobiologia* **595**, 9–26. doi:10.1007/s10750-007-9154-6

Chessman BC, Royal MJ (2010) Complex environmental gradients predict distributions of river-dependent plants in eastern Australia. *Aquatic Sciences* **72**, 431–441. doi:10.1007/s00027-010-0146-7

Chessman BC, Royal MJ, Muschal M (2008) Does water abstraction from unregulated streams affect aquatic macrophyte assemblages? An evaluation based on comparisons with reference sites. *Ecohydrology* **1**, 67–75. doi:10.1002/eco.8

Clarke SJ (2002) Vegetation growth in rivers: influences upon sediment and nutrient dynamics. *Progress in Physical Geography* **26**, 159–172. doi:10.1191/0309133302pp324ra

Cook CDK (1985) Range extensions of aquatic vascular plant species. *Journal of Aquatic Plant Management* **23**, 1–6.

Cook CDK (1996) *Aquatic Plant Book*. SPB Academic Publishing, Amsterdam.

Cook CDK (1999) The number and kinds of embryo-bearing plants which have become aquatic: a survey. *Perspectives in Plant Ecology, Evolution and Systematics* **2**(1), 79–102. doi:10.1078/1433-8319-00066

Cowie ID, Short PS, Osterkamp Madsen M (2000) *Floodplain Flora: a Flora of the Coastal Floodplains of the Northern Territory, Australia*. Flora of Australia Supplementary Series Number 10. Australian Biological Resources Study, Canberra.

Crow GE (1993) Species diversity in aquatic angiosperms: latitudinal patterns. *Aquatic Botany* **44**, 229–258. doi:10.1016/0304-3770(93)90072-5

Cummins S, Roberts D, Church A, Ling J (1997) Invertebrate associations in macrophyte assemblages within the Hawkesbury-Nepean River, NSW. Science and technology in the management of the Hawkesbury-Nepean Catchment. *Geographical Society NSW* **14**, 36–40.

Denny P (1972) Sites of nutrient absorption in aquatic macrophytes. *Journal of Ecology* **60**, 819–829. doi:10.2307/2258568

Douglas MM, Bunn SE, Davies PM (2005) River and wetland food webs in Australia's wet-dry tropics: general principles and implications for management. *Marine and Freshwater Research* **56**, 329–342. doi:10.1071/MF04084

Duivenvoorden LJ (2008) Effects of water level fluctuations on *Vallisneria nana* in the Burnett River in southeast Queensland, Australia. *River Research and Applications* **24**, 1362–1376. doi:10.1002/rra.1157

Feijoó CS, Momo FR, Bonetto CA, Tur NM (1996) Factors influencing biomass and nutrient content of the submersed macrophyte *Egeria densa* Planch. in a pampasic stream. *Hydrobiologia* **341**, 21–26. doi:10.1007/BF00012299

Fellows CS, Bunn SE, Sheldon F, Beard NJ (2009) Benthic metabolism in two turbid dryland rivers. *Freshwater Biology* **54**, 236–253. doi:10.1111/j.1365-2427.2008.02104.x

Franklin P, Dunbar M, Whitehead P (2008) Flow controls on aquatic macrophytes: a review. *The Science of the Total Environment* **400**, 369–378. doi:10.1016/j.scitotenv.2008.06.018

French TD, Chambers PA (1996) Habitat partitioning in riverine macrophyte communities. *Freshwater Biology* **36**, 509–520. doi:10.1046/j.1365-2427.1996.00105.x

Frost-Christensen H, Sand-Jensen K (1995) Comparative kinetics of photosynthesis in floating and submerged *Potamogeton* leaves. *Aquatic Botany* **51**, 121–134. doi:10.1016/0304-3770(95)00455-9

Gordon ND, McMahon TA, Finlayson BL, Gippel GJ, Nathan RJ (2005) *Stream Hydrology: an Introduction for Ecologists*. 2nd edn. John Wiley & Sons, Chichester, UK.

Green AJ, Jenkins KM, Bell D, Morris PJ, Kingsford RT (2008) The potential role of waterbirds in dispersing invertebrates and plants in arid Australia. *Freshwater Biology* **53**, 380–392.

Growns I (2009) Differences in bioregional classifications among four aquatic biotic groups: implications for conservation reserve design and monitoring programs. *Journal of Environmental Management* **90**, 2652–2658. doi:10.1016/j.jenvman.2009.02.002

Grundell R, Gell P, Mills K, Zawadzki A (2012) Interaction between a river and its wetland: evidence from the Murray River for spatial variability in diatom and radioisotope records. *Journal of Paleolimnology* **47**, 205–219. doi:10.1007/s10933-011-9572-3

Gurnell A, Thompson K, Goodson J, Moggridge H (2008) Propagule deposition along river margins: linking hydrology and ecology. *Journal of Ecology* **96**, 553–565. doi:10.1111/j.1365-2745.2008.01358.x

Hellquist CB, Jacobs SWL (2011) Aponogetonaceae. In *Flora of Australia. Vol. 39. Alismatales to Arales.* (Ed. A Wilson) pp. 44–52. CSIRO Publishing and the Australian Biological Resources Study, Canberra.

Henry CP, Amoros C, Bornette G (1996) Species traits and recolonization processes after flood disturbances in riverine macrophytes. *Vegetatio* **122**, 13–27. doi:10.1007/BF00052812

Hoyer MV, Frazer TK, Notestein SK, Canfield DE (2004) Vegetative characteristics of three low-lying Florida coastal rivers in relation to flow, light, salinity and nutrients. *Hydrobiologia* **528**, 31–43. doi:10.1007/s10750-004-1658-8

Hughes JMR (1990) Lotic vegetation dynamics following disturbance along the Swan and Apsley rivers, Tasmania, Australia. *Journal of Biogeography* **17**, 291–306. doi:10.2307/2845125

Humphries P (1995) Life history, food and habitat of southern pygmy perch, *Nannoperca australis*, in the Macquarie River, Tasmania. *Marine and Freshwater Research* **46**, 1159–1169. doi:10.1071/MF9951159

Hutchinson GE (1975) *A Treatise on Limnology. Vol. 3. Limnological Botany.* Wiley, New York, NY.

Jackson S, Finn M, Featherston P (2012) Aquatic resource use by Indigenous Australians in two tropical river catchments: the Fitzroy River and Daly River. *Human Ecology* **40**, 893–908. doi:10.1007/s10745-012-9518-z

Jacobs SWL (1999) Terrestrial wetlands and water-plants. In *Flora of Australia. Vol. 1.* 2nd edn. (Eds AE Orchard and HS Thompson) pp. 403–435. CSIRO Publishing and the Australian Biological Resources Study, Canberra.

Jacobs SWL, Frank KA (1997) Notes on *Vallisneria* (Hydrocharitaceae) in Australia, with descriptions of two new species. *Telopea* **7**, 111–118. doi:10.7751/telopea19971001

Jacobs SWL, Wilson KL (1996) Biogeographical analysis of the freshwater plants of Australasia. *Australian Systematic Botany* **9**, 169–183. doi:10.1071/SB9960169

James C, Capon S, White M, Rayburg S, Thoms M (2007) Spatial variability of the soil seed bank in a heterogeneous ephemeral wetland system in semi-arid Australia. *Plant Ecology* **190**, 205–217. doi:10.1007/s11258-006-9201-9

Jian Y, Li B, Wang J, Chen J (2003) Control of turion germination in *Potamogeton crispus*. *Aquatic Botany* **75**, 59–69. doi:10.1016/S0304-3770(02)00165-1

Jones JI (2005) The metabolic cost of bicarbonate use in the submerged plant *Elodea nuttallii*. *Aquatic Botany* **83**, 71–81. doi:10.1016/j.aquabot.2005.05.002

Kadono Y (1984) Comparative ecology of Japanese *Potamogeton*: an extensive survey with special reference to growth form and life cycle. *Japanese Journal of Ecology* **34**, 161–172.

Keeley JE (1998) C4 photosynthetic modifications in the evolutionary transition from land to water in aquatic grasses. *Oecologia* **116**, 85–97. doi:10.1007/s004420050566

Kemp A (1984) Spawning of the Australian lungfish, *Neoceratodus forsteri* (Krefft), in the Brisbane River and in Enoggera Reservoir, Queensland. *Memoirs of the Queensland Museum* **21**, 391–399.

Kennard MJ, Pusey BJ, Olden JD, Mackay SJ, Stein JL, Marsh N (2010) Classification of natural flow regimes in Australia to support environmental flow management. *Freshwater Biology* **55**, 171–193. doi:10.1111/j.1365-2427.2009.02307.x

Kirk JTO (1994) *Light and Photosynthesis in Aquatic Environments.* Cambridge University Press, Cambridge, UK.

Lacoul P, Freedman B (2006) Environmental influences on aquatic plants in freshwater ecosystems. *Environmental Reviews* **14**, 89–136. doi:10.1139/a06-001

Les DH, Crawford DJ, Kimball RT, Moody ML, Landolt E (2003) Biogeography of discontinuously distributed hydrophytes: a molecular appraisal of intercontinental disjunctions. *International Journal of Plant Sciences* **164**, 917–932. doi:10.1086/378650

Maberly SC, Madsen TV (2002) Freshwater angiosperm carbon concentrating mechanisms: processes and patterns. *Functional Plant Biology* **29**, 393–405. doi:10.1071/PP01187

Mackay SJ (2007) Drivers of macrophyte assemblage structure in southeast Queensland streams. PhD thesis. Griffith University, Australia.

Mackay SJ, Arthington AH, Kennard MJ, Pusey BJ (2003) Spatial variation in the distribution and abundance of submersed macrophytes in an Australian subtropical river. *Aquatic Botany* **77**, 169–186. doi:10.1016/S0304-3770(03)00103-7

Mackay SJ, James CS, Arthington AH (2010) Macrophytes as indicators of stream condition in the wet tropics region, northern Queensland, Australia. *Ecological Indicators* **10**, 330–340. doi:10.1016/j.ecolind.2009.06.017

Madsen JD, Adams MS (1989) The distribution of submerged aquatic macrophyte biomass in a eutrophic stream, Badfish Creek: the effect of environment. *Hydrobiologia* **171**, 111–119. doi:10.1007/BF00008171

Madsen TV, Cedergreen N (2002) Sources of nutrients to rooted submerged macrophytes growing in a nutrient-rich stream. *Freshwater Biology* **47**, 283–291. doi:10.1046/j.1365-2427.2002.00802.x

Madsen JD, Chambers PA, James WF, Koch EW, Westlake DF (2001) The interaction between water movement, sediment dynamics and submersed macrophytes. *Hydrobiologia* **444**, 71–84. doi:10.1023/A:1017520800568

Marshall LG (1988) Extinction. In *Analytical Biogeography: an Integrated Approach to the Study of Animal and Plant Distribution*. (Eds AA Myers and PS Giller) pp. 219–254. Chapman & Hall, London.

Mitchell DS (1974) The effects of excessive aquatic plant populations. In *Aquatic Vegetation and its Use and Control*. (Ed. DS Mitchell) pp. 50–56. UNESCO, Paris.

Mitchell DS (1978) *Aquatic Weeds in Australian Inland Waters, Department of Environment, Housing and Community Development*. Australian Government Publishing Service, Canberra.

Mitchell DS (1986) Aquatic macrophytes and man. In *Limnology in Australia*. (Eds P De Dekker and WD Williams) pp. 588–598. CSIRO Publishing, Melbourne and Dr W Junk Publishers, Dordrecht, Netherlands.

Mitchell DS, Rogers KH (1985) Seasonality/aseasonality of aquatic macrophytes in southern hemisphere inland waters. *Hydrobiologia* **125**, 137–150. doi:10.1007/BF00045931

Moody ML, Les DH (2010) Systematics of the aquatic angiosperm genus *Myriophyllum* (Haloragaceae). *Systematic Botany* **35**, 121–139. doi:10.1600/036364410790862470

Ogden RW (2000) Modern and historical variation in aquatic macrophyte cover of billabongs associated with catchment development. *Regulated Rivers: Research and Management* **16**, 497–512. doi:10.1002/1099-1646(200009/10)16:5<497::AID-RRR600>3.0.CO;2-Y

Orchard AE (1985) *Myriophyllum* (Haloragaceae) in Australasia. II. The Australian species. *Brunonia* **8**, 173–291. doi:10.1071/BRU9850173

Padial AA, Bini LM, Thomaz SM (2008) The study of aquatic macrophytes in neotropics: a scientometrical view of the main trends and gaps. *Brazilian Journal of Biology* **68**, 1051–1059. doi:10.1590/S1519-69842008000500012

Papassotiriou SE, Jacobs SWL, Hellquist CB (2011) Potamogetonaceae. In *Flora of Australia. Vol. 39. Alismatales to Arales*. pp. 84–95. Australian Biological Resources Study, Canberra and CSIRO Publishing, Melbourne.

Pedersen O, Sand-Jensen K (1993) Water transport in submerged macrophytes. *Aquatic Botany* **44**, 385–406. doi:10.1016/0304-3770(93)90079-C

Philbrick CT, Les DH (1996) Evolution of aquatic angiosperm reproductive systems. *Bioscience* **46**, 813–826. doi:10.2307/1312967

Raulings E, Morris K, Thompson R, Mac Nally R (2011) Do birds of a feather disperse plants together? *Freshwater Biology* **56**, 1390–1402. doi:10.1111/j.1365-2427.2011.02576.x

Raven JA, Lucas WJ (1985) Energy costs of carbon acquisition. In *Inorganic Carbon Uptake by Aquatic Photosynthetic Organisms*. (Eds WJ Lucas and JA

Berry) pp. 305–324. The American Society of Plant Physiologists, Rockville, MD.

Rea N, Dostine PL, Cook S, Webster I, Williams D (2002) 'Environmental water requirements of *Vallisneria nana* in the Daly River, Northern Territory.' Report no. 35/2002. Department of Infrastructure, Planning and Environment, Darwin.

Reid MA, Sayer CD, Kershaw AP, Heijnis H (2007) Palaeolimnological evidence for submerged plant loss in a floodplain lake associated with accelerated catchment soil erosion (Murray River, Australia). *Journal of Paleolimnology* **38**, 191–208.

Reid MA (2008) Evidence for catastrophic shifts in the trophic structure of floodplain lakes associated with soil erosion. In *Sediment Dynamics in Changing Environments. Vol. 325.* (Eds J Schmidt, T Cochrane, C Phillips, S Elliot, T Davies, L Basher) pp. 584–509. IAHS Press, Wallingford, UK.

Riis T, Biggs BJF (2001) Distribution of macrophytes in New Zealand streams and lakes in relation to disturbance frequency and resource supply – a synthesis and conceptual model. *New Zealand Journal of Marine and Freshwater Research* **35**, 255–267. doi:10.1080/00288330.2001.9516996

Riis T, Biggs BJF (2003) Hydrologic and hydraulic control of macrophyte establishment and performance in streams. *Limnology and Oceanography* **48**, 1488–1497. doi:10.4319/lo.2003.48.4.1488

Riis T, Biggs BJF, Flanagan M (2004) Colonisation and temporal dynamics of macrophytes in artificial stream channels with contrasting flow regimes. *Archiv fuer Hydrobiologie* **159**, 77–95. doi:10.1127/0003-9136/2004/0159-0077

Sand-Jensen K (2003) Drag and reconfiguration of freshwater macrophytes. *Freshwater Biology* **48**, 271–283. doi:10.1046/j.1365-2427.2003.00998.x

Sand-Jensen K (2008) Drag forces on common plant species in temperate streams: consequences of morphology, velocity and biomass. *Hydrobiologia* **610**, 307–319. doi:10.1007/s10750-008-9446-5

Sand-Jensen K, Frost-Christensen H (1999) Plant growth and photosynthesis in the transition zone between land and stream. *Aquatic Botany* **63**, 23–35. doi:10.1016/S0304-3770(98)00107-7

Santamaría L (2002) Why are most aquatic plants widely distributed? Dispersal, clonal growth and small-scale heterogeneity in a stressful environment. *Acta Oecologica* **23**, 137–154. doi:10.1016/S1146-609X(02)01146-3

Sculthorpe CD (1967) *The Biology of Aquatic Vascular Plants.* Edward Arnold, London.

Shannon JP, Blinn DW, McKinney T, Benenati EP, Wilson KP, O'Brien C (2001) Aquatic food base response to the 1996 test flood below Glen Canyon Dam, Colorado River, Arizona. *Ecological Applications* **11**, 672–685. doi:10.1890/1051-0761(2001)011[0672:AFBRTT]2.0.CO;2

Shilla DA, Asaeda T, Kian S, Lalith R, Manatunge J (2006) Phosphorus concentration in sediment, water and tissues of three submerged macrophytes of Myall Lake, Australia. *Wetlands Ecology and Management* **14**, 549–558. doi:10.1007/s11273-006-9007-5

Sousa WTZ, Thomaz SM, Murphy KJ (2010) Response of native *Egeria najas* Planch. and invasive *Hydrilla verticillata* (L.f.) Royle to altered hydroecological regime in a subtropical river. *Aquatic Botany* **92**, 40–48. doi:10.1016/j.aquabot.2009.10.002

Suren AM, Smart GM, Smith RA, Brown SLR (2000) Drag coefficients of stream bryophytes: experimental determinations and ecological significance. *Freshwater Biology* **45**, 309–317. doi:10.1111/j.1365-2427.2000.00621.x

Walker KF, Boulton AJ, Thoms MC, Sheldon F (1994) Effects of water-level changes induced by weirs on the distribution of littoral plants along the River Murray, South Australia. *Australian Journal of Marine and Freshwater Research* **45**, 1421–1438. doi:10.1071/MF9941421

Watson A, Barmuta LA (2011) Feeding-preference trials confirm unexpected stable isotope analysis results: freshwater macroinvertebrates do consume macrophytes. *Marine and Freshwater Research* **62**, 1248–1257. doi:10.1071/MF10298

Weaver JE, Clements FE (1938) *Plant Ecology*. 2nd edn. McGraw-Hill, New York, NY.

Winkel A, Borum J (2009) Use of sediment $CO_2$ by submersed rooted plants. *Annals of Botany* **103**, 1015–1023. doi:10.1093/aob/mcp036

Xie Y, Luo W, Ren B, Li F (2007) Morphological and physiological responses to sediment type and light availability in roots of the submerged plant *Myriophyllum spicatum*. *Annals of Botany* **100**, 1517–1523. doi:10.1093/aob/mcm236

# 7

# Riparian herbs

*Samantha Capon*

## INTRODUCTION

Australia's riverine landscapes encompass an enormous diversity of riparian habitats, broadly defined here as lands that 'adjoin, directly influence or are influenced by a body of water' (Price and Tubman 2007). Predominantly, these are the parts of the landscape that are periodically inundated by overbank floods from the waterbodies with which they are associated and therefore also exist for at least some of the time as relatively dry, semi-terrestrial habitats. While trees and shrubs are present in many, though not all, of Australia's riparian habitats, virtually all riparian areas support herbaceous plants at one time or another. Referred to in this chapter as 'riparian herbs', these relatively small, mostly non-woody plants represent a diverse array of groundcover and understorey species that typically contribute to the bulk of species richness in Australia's riparian habitats. While riparian herb assemblages can include submerged as well as free and attached floating aquatic plants which occur mainly in the presence of surface water, these are discussed in Chapters 5 and 6. Instead, this chapter focuses on the amphibious and terrestrial herbaceous plants that are present in riparian habitats during damp or dry phases in the absence of standing or flowing water (although these plants sometimes persist during wet phases as well). Emergent plants, typically considered to be a major group of aquatic plants (e.g. Sainty and Jacobs 2003), are also included given their prevalence in riparian habitats.

In this chapter we provide a brief overview of riparian habitats in Australia and the major environmental factors influencing the character and condition of their herbaceous vegetation. The diversity of Australian riparian herbs is then discussed with particular reference to key plant groups, their taxonomic diversity and their distribution, both regionally and among riparian habitats. A synthesis of plant traits follows with an emphasis on those growth and life history characteristics that enable riparian herbs to persist in habitats that are characterised by variable flooding and drying regimes. Major spatial and temporal patterns in the composition and structure of riparian herbaceous vegetation in relation to wetting and drying patterns are discussed. The final section highlights the provision of ecosystem services by riparian herbs and their economic and cultural significance to Australian people.

## RIPARIAN HABITATS: FROM A HERB'S PERSPECTIVE

There is considerable variation among Australia's riparian environments with respect to their physical, chemical and biological characteristics as well as their levels of anthropogenic modification and influence, both regionally and within catchments. Many of these regional and landscape factors play an important role in determining the suitability of riparian habitats for the establishment and growth of herbs. As in other plant habitats, climate, soil and land use are particularly significant drivers of plant distributions and vegetation dynamics (Sainty and Jacobs 2003). The distinguishing feature of riparian habitats, however, is their periodic connection with, and inundation by, surface waters from the waterbodies they adjoin. Additionally, riparian habitats tend to be cooler with higher relative humidity than adjacent uplands (Brosofske *et al.* 1997; Danehy and Kirpes 2000).

Abiotic and biotic components and processes of riparian ecosystems are driven primarily by the variable wetting and drying regimes they experience as a result of periodic overbank flooding (Naiman *et al.* 2010). Flooding represents a 'pulse' disturbance in riparian habitats (Lake 2000), during which water and nutrient availability are increased, light and temperature conditions are altered, and soil anoxia and potentially toxicity occur (Blom and Voesenek 1996). Flooding also provides a vector for the dispersal of plant propagules via hydrochory (Nilsson *et al.* 1991) and can structurally support plant bodies during periods of submergence. Swiftly flowing waters, however, may cause mechanical damage to plants or displace plants through scouring (Menges 1986; Naiman and Décamps 1997; Young *et al.* 2001). Additionally, sediments deposited by floodwaters may bury small plants, limiting emergence, growth and reproduction (Sluis and Tandarich 2004; Lowe *et al.* 2010). Fluvial processes associated with flooding can create bare areas of substrate for colonisation by riparian herbs (Stromberg 2001), but can also compact soils, making them more difficult for roots to penetrate (Blom and Voesenek 1996).

As a result of periodic inundation, riparian habitats often have greater soil moisture and nutrient content than adjacent upland areas (Megonigal *et al.* 1997). Additionally, riparian habitats tend to exhibit relatively high levels of environmental heterogeneity at local scales due to spatial and temporal variation in patterns of wetting and drying and their interaction with the local topography. Flooding can further contribute to habitat heterogeneity by influencing other disturbances, such as fire through its effects on the accumulation and distribution of fuels (see Chapter 16) (Stromberg 2001).

## Habitat types

Despite the considerable variation in Australia's riparian habitats at regional, landscape and local scales, a few broad habitat types can be distinguished (*sensu* Brock *et al.* 2006): channel beds, channel banks, floodplains and wetlands. These vary in their distribution and abundance both regionally and within catchments in relation to topographic position. In dryland regions, for instance, these features are often difficult to distinguish and channel banks can be virtually imperceptible where channels are associated with very low gradient rivers. In more mesic regions, however, all of these habitat types are likely to be present in lowland areas while channel and channel bank habitats are usually the only riparian habitats in upstream reaches of Australian river catchments (Capon and Dowe 2007). Riparian vegetation itself is an important influence on habitat structure through its impact on, for example, river channel form and stability through sometimes quite complex feedback mechanisms. In many parts of Australia grazing, fire regimes and salinity also play an important role in shaping herb assemblages in riparian habitats (see Chapters 15, 16 and 17).

### Channel bed

The channel bed is typically defined as the area between the banks of a river ordinarily covered by water. Many of Australia's inland rivers, however, are intermittent or ephemeral and much of their active channel beds can be exposed for varying durations, during which time they provide habitats for colonisation by riparian herbs. Channel habitats can also include topographic features such as islands and depositional bars, which provide comparatively stable surfaces for herb establishment (Hupp and Osterkamp 1985). During dry phases, riparian herb assemblages may be influenced by the stability and character of bed sediments, as well as canopy cover (Mackay *et al.* 2003; Fritz *et al.* 2004). The duration of exposure will also determine the scope for vegetation development in dry channel habitats. Some herbs (e.g. emergent sedges and grasses) may persist in channel habitats when surface water flows occur. Hydrologic (e.g. flood frequency) and hydraulic characteristics, especially water velocity, will then be important determinants of vegetation composition and structure (Mackay *et al.* 2003).

## Channel banks

Channel bank habitats are those that occur directly adjacent to active channel beds, including levee banks, and are therefore subject to varying degrees of fluvial action in relation to their elevation and distance from the channel bed (Fig. 7.1). Gradients of moisture availability and fluvial disturbance provide important controls on the distribution, abundance and diversity of plants in these habitats (Merritt *et al.* 2010). Sediment characteristics of channel banks can influence flood effects on processes of erosion and deposition as well as affecting soil chemistry (Horowitz 1991; Bechtold and Naiman 2006) and soil moisture capacity (Hillel 1980). Moisture levels within the permeable layer of the river bed and bank are strongly dependent on the hydraulic conductivity of the soils, with highly permeable subsurface soils unable to maintain a water table (Rassam *et al.* 2006). Levee banks, for instance, often comprise coarser material and may drain faster than low-lying water ponding areas (Naiman and Décamps 1997). In lowland river reaches, flow rates can be reduced and therefore more alluvium with higher nutrient content tends to be deposited in these channel bank habitats (Sainty and Jacobs 2003). Where banks are steep and channel flows swift, exposure to wave action and currents may exert further influences on riparian herbs in these habitats (Roberts and Ludwig 1991). Light availability may also be important in structuring bank vegetation where light levels reaching the ground storey are low such as in tropical forests.

**Figure 7.1:** Riparian herb community growing on the channel bank of a waterhole in the Lake Eyre Basin. (Photo: Cassandra James.)

## Floodplains

Floodplain habitats encompass those parts of riverine landscapes beyond active channel beds and their banks that are also inundated regularly or occasionally by overbank flooding, but to a lesser degree (Hupp and Osterkamp 1985). Junk and Welcomme (1990) define floodplains as 'areas of low lying land that are subject to inundation by lateral overflow water from rivers with which they are associated'. In Australia, floodplains often occupy vast areas, particularly in arid and semi-arid inland catchments where low topographic gradients promote widespread shallow flooding in response to high flows. In the Cooper Creek Catchment, for instance, floodplains extend to widths of up to 60 km, the periphery of which may flood less than once a decade (Capon 2003, 2005). In contrast, the expansive floodplains of tropical northern Australia flood annually with relative predictability. Differences in flood frequency, timing, duration and rates of rise and drawdown are major determinants of

**Figure 7.2:** Herbs growing in the drying lake bed of Clear Lake, Narran Lakes following recession of floodwaters in 2004.

floodplain vegetation composition and structure at both regional and landscape scales (Capon 2003).

### Wetlands

Wetlands are areas inundated with slow-moving or still water on a temporary or permanent basis. Riverine landscapes in Australia contain a range of wetland habitats that may be associated with channel beds, channel banks or floodplains, including billabongs, marshes, swamps and lakes, both freshwater and saline (Fig. 7.2). Connected with river channels to varying degrees by overbank flooding, riverine wetlands may receive further water from local rainfall and runoff as well as upwelling of groundwater. In addition to hydrologic characteristics, water quality (e.g. salinity, turbidity) can also be an important determinant of riparian herb assemblages in wetland habitats (Porter 2000). Within wetlands themselves, a range of distinctive plant habitats akin to channels and channel banks (or bed and fringing habitats) may also be discernible.

## PLANT DIVERSITY

Riparian herb assemblages in Australian riverine landscapes include a diverse array of plant life forms and taxonomic groups ranging from obligate aquatic to terrestrial species. While the term 'herb' is used in a variety of ways, here it refers to relatively small, non-woody, vascular plants. As there are no herbaceous gymnosperms (Judd *et al.* 1999), riparian herbs can be angiosperms, ferns or fern allies. For the most part, this group has terrestrial ancestry with varying degrees of modification of terrestrial structures facilitating re-adaptation to amphibious or aquatic conditions (Brock 1994). Australia's angiosperms, especially semi-aquatic herbs, tend to be cosmopolitan, with many northern species originating from the tropics, Asia or Malaysia (Brock 1994). Endemism is more common in southern parts of the continent, with many endemic species present among cosmopolitan genera (Beadle 1981; Brock 1994). A large proportion of herb species present in Australia's riparian habitats are also widely distributed in terrestrial habitats and are often only restricted to riparian habitats at the drier extremes of their ranges (Hancock *et al.* 1996). The major herbaceous life forms present in Australian riparian habitats are grasses, sedges and rushes, forbs, and ferns and fern allies.

### Grasses

Grasses (i.e. members of the Poaceae family) occur in a wide range of riparian habitats throughout Australia and often dominate the fringes of channels and wetlands as well as the understorey of woody vegetation in channel bank and floodplain habitats. Grasslands comprising one to several dominant species can also occupy vast areas of floodplain and wetland habitat (e.g. lake beds) during dry phases where they may provide the major structural vegetation in the riverine landscape (Sainty and Jacobs 2003). Both perennial and annual grasses are present in Australian riparian habitats, though dominant species tend to be perennial in most places (Roberts and Marston 2011).

Many grass species present in riparian habitats also occur in terrestrial habitats, reflecting the capacity of grasses to adapt to a range of conditions. In particular, tussock grasses (i.e. grasses that grow in clumps or tufts) tend to exhibit few adaptations to flooding but can often tolerate brief periods of waterlogged conditions. A common tussock grass present in riparian habitats of south-eastern Australia is *Paspalidium jubiflorum* (Warrego grass; Roberts and Marston 2011), while *Eragrostis* spp. are widespread in drier parts of inland Australia. Tussock grasslands dominated by *Astrebla* spp. (Mitchell grass) are also extensive in inland and northern Australia (Groves and Whalley 2002) as are *Dicanthium* spp. (bluegrass) on heavier clays in tropical regions.

Some grasses can be considered riparian specialists, capable of persisting under fluctuating surface water levels. *Phragmites australis* (common reed), for example, is a common emergent grass present throughout much of coastal and inland Australia, with the exception of the north-west, that occurs in a wide range of riverine habitats including channels, wetland fringes, floodplains and as an understorey to riparian woodlands (Brock 1994; Sainty and Jacobs 2003; Roberts and Marston 2011). Other aquatic grass species favour more frequently inundated or even submerged habitats, for example, *Amphibromus fluitans* (river swamp wallaby grass) in south-eastern Australia, *Oryza australiensis* (native rice) in northern Australia and *Echinochloa turneriana* (channel millet) in the channel country of south-west Queensland. The semi-aquatic grass *Pseudoraphis spinescens* (spiny mud grass or Moira grass), distributed in all mainland states of Australia and a common species of wetlands and floodplains in the southern Murray–Darling Basin as well as the Northern Territory, persists under wet and dry conditions with distinctive life history phases in each (i.e. floating and turf, respectively) (Roberts and Marston 2011). Other common and widespread mat-forming grasses present in shallow or occasionally flooded riparian habitats include *Cynodon dactylon* (couch), *Paspalum distichum* (water couch) and *Sporobolus mitchellii* (rat's-tail couch), all of which occur in all mainland states of Australia (Sainty and Jacobs 2003; Roberts and Marston 2011).

## Sedges and rushes

Sedges (i.e. members of the Cyperaceae family) and rushes (i.e. members of the Juncacae and Restionaceae families) comprise a conspicuous component of riparian herb assemblages in a diverse range of riverine habitats (Fig. 7.3). Along with sedge-like (e.g. *Lomandra* spp.) and rush-like (e.g. *Typha* spp.) plants, sedges and rushes are common in fringing habitats of channels and wetlands throughout Australia and can form monocultures in many temporary and ephemeral wetlands. Sedgelands and rushlands can also dominate low-lying areas of channel banks and floodplains where water ponds (Roberts and Marston 2011). *Baumea* spp. and *Eleocharis* spp. (spike-rushes) are particularly common emergent sedges of shallow waters and wetland edges of Australia (Brock 1994).

**Figure 7.3:** Sedges (mostly *Eleocharis* spp.) growing in a wetland in Yanga National Park, Lowbidgee, after an environmental water allocation in 2008.

Sedges and rushes tend to have erect, emergent habits often forming tussocks, for example, *Carex tereticaulis* (terete club-sedge), *Cyperus gymnocaulos* (spiny sedge) and *Juncus ingens* (giant rush), or dense clonal stands such as *Typha* spp. (cumbungi), *Schoenoplectus validus* (river club-rush) and *Eleocharis sphacelata* (tall spike-rush) (Roberts and Marston 2011). Some species (e.g. *Isolepis fluitans*; floating club-rush) may also exhibit elongated or creeping forms in submerged conditions (Sainty and Jacobs 2003). Other common emergents of Australian channel and channel bank habitats include *Bolboschoenus fluviatilis* (marsh club-rush) and *Schoenoplectus mucronatus* (Sainty and Jacobs 2003).

### Forbs

Forbs are herbaceous plants that are not graminoids (i.e. not grasses, sedges or rushes) and encompass a very diverse array of plant habits, life histories and taxonomic groups. In many Australian riparian habitats, forbs can account for a high proportion of plant species richness (Reid and Capon 2011), although they may only represent a small portion of the biomass, especially where grasses, sedges and rushes are present (Capon 2003). Diverse forb assemblages are often particularly prevalent during moist conditions following the drawdown of floodwaters, and low-growing forbs also tend to be common in wetland fringing habitats (Brock 1994; Reid and Capon 2011). Many forb species are widely distributed, occurring in arid, semi-arid, subtropical and more temperate regions in a range of riverine and terrestrial habitats. Common families represented in riparian forb assemblages in Australia include Amaranthaceae, Apiaceae, Asteraceae, Brassicaceae, Calittrichaceae, Campanulaceae, Caryophyllaceae, Chenopodiaceae, Convolvulaceae, Crassulaceae, Epacridaceae, Euphorbiaceae, Fabaceae, Lythraceae, Oxalidadaceae, Plantaginaceae and Polygonacae.

### Ferns and fern allies

Ferns and fern allies (i.e. members of the Psilotaceae, Lycopodiaceae, Selaginallaceae, Isoetaceae and Equistaceae families) are relatively uncommon components of riverine vegetation in much of Australia. A few species, however, are widely distributed throughout Australia and occur frequently in riverine habitats. In particular, *Marsilea* spp. (nardoo) are amphibious rhizomatous ferns that vary in habit depending on the presence of surface water. Displaying floating stems and leaves during flooding, nardoo forms mats with some erect stems under damp conditions and often dominates low-lying depressions in inland wetlands and floodplains following floodwater recession (Sainty and Jacobs 2003). *Azolla pinnata* and *A. filiculoides* are also common free-floating ferns of waterbodies in northern, eastern, south-eastern and south-western Australia, as is the noxious introduced *Salvinia molesta* (Sainty and Jacobs 2003).

## PLANT TRAITS

Riparian herbs are a varied group and exhibit a wide range of traits that enable them to persist in the face of the fluctuating environmental conditions imposed by variable riparian wetting and drying regimes. These traits can be broadly divided between growth traits, which allow plants to survive through flooding or drying, and life history traits, which enable plants to avoid these disturbances either temporally or spatially. Riparian herbs can also possess traits that facilitate their survival under a range of other stressors and disturbances to which they may be subject (e.g. high salinity levels, grazing) (see Chapters 15 and 17). While this chapter focuses on traits in relation to water regimes, traits that favour persistence under hydrologic variability (e.g. long-lived soil seed banks) may also confer an ability to cope with other stressors (e.g. fire).

### Growth traits

To persist through periods of flooding, riparian herbs must be able to cope with the reduced levels of oxygen that occur as a result of soil saturation. Flood-intolerant plants may survive such conditions for short periods (i.e. days), but will eventually wilt and die due to a lack of oxygen and an inability to sustain energy production (Cronk and Fennessy 2001). Some flood-tolerant plants, however, may switch to alternative metabolic pathways during flooding to enable respiration to continue under anoxic conditions (Hook 1984). Many riparian herbs (and aquatic plants in general) also exhibit structural traits that enhance their uptake and internal transport of oxygen.

One of the most common means by which riparian herbs facilitate gas exchange and increase the aeration of roots and rhizomes during flooding involves the formation of aerenchyma in stems, leaves and roots (Blom and Voesenek 1996). Aerenchyma is porous tissue comprising large lacunae (i.e. gas-filled spaces) that enhance gas flow (Cronk and Fennessy 2001). *Phragmites australis*, for example, is well adapted to tolerating fluctuating water levels at higher elevations due to its hollow stem anatomy. *Typha domingensis* (cumbungi) also forms aerenchyma, but has the additional capacity to alter its anatomy in response to water level so that resistance to gas flow is lessened when plants grow at lower elevations, making it better suited across a range of depths but with static water levels (White and Ganf 2002). Aerenchyma formation in response to flooding can also induce hypertrophy, in which the stem base swells as a result of cell rupture and separation, further contributing to the porosity of stems and enhancing gas movement (Cronk and Fennessy 2001). While internal gas movement in most plants occurs via passive diffusion, some riparian herbs exhibit other mechanisms that may favour growth during submergence in deep water or that enable deeper root extension in waterlogged soils (Sorrell *et al.* 1997). Convective gas flows, driven by temperature and humidity gradients between the atmosphere and internal gas in leaf and shoot tissues, enable many wetland plants to grow where diffusion is unable to supply sufficient oxygen to roots. This is especially prevalent among emergent plants (e.g. *Eleocharis sphacelata*) (Sorrell and Tanner 2000). In *P. australis*, for instance, internal gas flow to roots and rhizomes occurs through pressurised ventilation generated in leaf sheaths and culms (Brix *et al.* 1992).

The development of aerenchyma can enable riparian herbs to float during flooding, thereby increasing their access to light, carbon dioxide and water (Cronk and Fennessy 2001). In some riparian herbs, particularly sedges and rushes, flooding also induces rapid elongation of stems and petioles, which helps plants reach the water surface (Roberts and Marston 2011). Australian riparian herbs in which flood-induced stem elongation has been observed include species of *Ranunculus*, *Rumex*, *Typha* and *Scirpus* (Cronk and Fennessy 2001). In general, emergent riparian herbs tend to allocate greater biomass to above-ground vegetative structures in response to flooding and increasing water depths (Smith and Brock 2007).

Many plants develop adventitious roots (i.e. roots arising from stems or leaves) in response to flooding (Blom *et al.* 1990), a response that may be facilitated by hypertrophy (Arteca 1997). Adventitious roots enhance oxygen supply to flooded plants and also contribute to water and nutrient uptake (Cronk and Fennessy 2001). Australian riparian herbs that develop adventitious roots include species of *Typha*, *Rumex* (docs and sorrels) and *Rorippa* (watercress). More flood-tolerant riparian herbs also tend to exhibit a concentration of roots in the shallower, better-oxygenated part of the soil profile. During flooding, root growth in *P. australis*, for example, occurs mainly near the soil surface (Weisner and Strand 1996).

Other growth traits exhibited by riparian herbs that confer a capacity to persist under variable wetting and drying regimes include pliable stems and leaves. Flexible stems, for example, contribute to the ability of the fern *Marsilea* spp. to cope with changes in water level, while the supple stems and leaves of *P. australis* also enable growth in fringing habitats exposed to swift flows and wave action (Weisner 1991; Coops and Van der Velde 1996). During periods of drought, some riparian herbs (e.g. *P. australis*) may shed leaves to reduce water demands (Pagter *et al.* 2005). Some more aquatic riparian herbs also exhibit heterophylly, in which leaf form differs under submerged and emergent conditions. Typically, heterophyllous amphibious plants exhibit smaller, rounder leaves during dry phases and longer, more dissected leaves during flooding that increase the ratio of their surface area to volume thereby enhancing the interception of light and uptake of nutrients and carbon dioxide (Wetzel 1983). Heterophyllous Australian riparian herbs include species of *Sagittaria*, *Callitriche*, *Ludwigia* and *Ranunculus*.

## Life history traits

Riparian herbs display a variety of life history traits that allow plant populations to persist in hydrologically variable habitats. Many of these life history traits provide a means by which riparian herbs can avoid the stresses associated with flooding or drought and re-establish when favourable conditions return. Such traits are particularly important for species that lack growth traits that facilitate tolerance of hydrologic disturbance. Some life history traits enable riparian herb species to capitalise on the subsidies provided by flooding in riparian habitats. For instance, many riparian herbs have buoyant seeds or fruits that allow dispersal by stream flow and floodwaters in a process known as hydrochory (Nilsson *et al.* 1991). In some other species (e.g. *Eleocharis acuta*, common spike-rush) the germinability of seeds is increased as a result of passing through the guts of waterbirds (Bell 2000; Green *et al.* 2008).

The dominant means of reproduction among wetland plants globally is vegetative regeneration (Cronk and Fennessy 2001). Notwithstanding genetic mutations, reproduction by 'cloning' essentially results in new individuals in which the genetic identity of parent plants is conserved, providing a means of relative trait stability that may be advantageous where habitat conditions are reasonably constant (e.g. semi-permanent wetlands). Vegetative regeneration can occur from a range of plant tissues including modified shoots, roots and leaf buds or from plant fragments (e.g. stem fragments and leaves) that separate from adults, disperse and regenerate at other locations. The formation of seeds from maternal cells without fertilisation (apomixis) also occurs in at least two Australian salt marsh species (e.g. *Halosarcia indica* and *H. pergranulata*), but is typically rare among wetland plants (Grace 1993; Cronk and Fennessy 2001). *Glossostigma cleistanthum* (mud mat), a native amphibious herb of central and eastern Australia, can also propagate via cleistogamy (i.e. automatic self-pollination) in which tiny, non-opening and self-fertilising flowers are produced underwater. This species also produces open or chasmogamous flowers (Les *et al.* 2006).

Common mechanisms of vegetative regeneration among riparian herbs include stem layering, in which arching stems, on contact with soil, develop adventitious roots that can then produce new individuals. *Alternanthera philoxeroides* (alligator weed), one of Australia's most significant national weeds, spreads via stem layering. Other stem modifications that provide a means of asexual reproduction include stolons (i.e. specialised shoots that grow horizontally above ground) and rhizomes (i.e. underground horizontally growing stems). New individuals can arise at the nodes of both stolons and rhizomes. Rhizomes are particularly advantageous to riparian herbs as they can persist through disturbances such as drought and fire in relatively protected conditions, providing reservoirs for starch reserves and meristematic tissue from which plants can later regenerate in response to flooding or rainfall. Among riparian herbs, rhizomes are prevalent in perennial grasses and other emergents, especially sedges and rushes. The sedge *Eleocharis acuta*, for example, displays robust rhizome growth even as a seedling (Bell 2000) and rhizomes can survive for between five and ten years in moist soils (Roberts and Marston 2011). In Europe and Japan, rhizomes of *P. australis* may persist for up to six years (Clevering and van der Toorn 2000; Karunaratne *et al.* 2004; Asaeda *et al.* 2006).

Modified shoot bases can also provide riparian herbs with a means of persistence and subsequent regeneration. *Crinum* spp. (riparian lilies), for example, survive as dormant bulbs that are modified stem bases from which plants grow. In *E. acuta*, shoot bases from which new plants can regenerate are often protected during dry periods by dead stems that are retained on the surface (Bell 2000). Riparian herbs may also have a range of root modifications (e.g. tubers) that, while used primarily for carbohydrate storage, provide an additional means of dispersal and asexual regeneration (Cronk and Fennessy 2001). Regeneration from dispersed plant fragments, including shoots and roots, is common among many widespread wetland plants (e.g. *Myriophyllum* spp. and *Elodea* spp.). Brittle older stems of *Pseudoraphis spinescens*, for example, easily break away from parent plants and can then root at nodes to produce clones (Ward 1992; Cowie *et al.* 2000). Both *Juncus* subgenus *genuine* and *Phalaris aquatica* have been observed to establish from rhizome fragments in soils of riparian zones in south-eastern Australia (Williams *et al.* 2008).

Sexual reproduction pathways can also facilitate the persistence of riparian herbs in hydrologically variable habitats and are often particularly important where flooding and drying are very unpredictable (e.g. arid areas) (Capon and Brock 2006).

The production of large, long-lived dormant soil seed banks, for instance, allows some species to remain at a location in a dormant state for prolonged periods during which conditions are unsuitable for plant establishment and growth. Most riparian herbs exhibit multiple reproductive pathways (i.e. vegetative and sexual), the prevalence of which varies spatially and temporally in relation to environmental conditions. *Pseudoraphis spinescens*, for example, shifts to vegetative growth via rooting of stems as floodwaters recede while plants flower and produce seeds when inundated (Ward 1992). At larger scales, *P. australis* usually spreads vegetatively but can also regenerate following major disturbances (e.g. extreme drought) from small persistent soil seed banks.

Flowering in some riparian herbs (e.g. *Eragrostis australasica*, cane grass) is triggered by rainfall or flooding or can be seasonal to coincide with predictable periods of rainfall or flow (Brock *et al.* 2006; Roberts and Marston 2011). Water level can also influence the timing of life history events. Flowering in *Eleocharis acuta*, for example, occurs in plants grown in damp soil before those in shallow water (Bell 2000). Seed production among riparian herbs ranges from the production of high numbers of small seeds (e.g. *Juncus* spp.) to the production of a few large seeds (e.g. *Glyceria australis*) (Smith and Brock 2007). Dispersal of riparian herb seeds (and vegetative propagules) occurs via a range of vectors including wind, water and animals (e.g. ants and waterbirds).

Many riparian herb species produce large, long-lived soil seed banks, some of which are likely to persist for a decade or more (Capon and Brock 2006; Capon 2007). Germination of riparian herbs from soil seed banks can depend on a range of cues relating to soil moisture, temperature and light availability, but these are often associated with flooding (Brock *et al.* 2006). Germination rates of *Pseudoraphis spinescens* seeds, for example, are around ten times greater in saturated soils than under shallow inundation (Finlayson *et al.* 1990). In general, most riparian herbs do not germinate during flooding, but favour the waterlogged or damp conditions that follow the recession of floodwaters (Capon and Brock 2006; Roberts and Marston 2011). Typically, only a proportion of seeds residing in riparian soil seed banks germinates at any one time, ensuring a source for later regeneration should successful recruitment not occur (Capon 2007). Establishing individuals of riparian herbs, both sexual and asexual, also tend to have rapid life cycles, responding quickly to favourable conditions and allowing the production of new seeds or vegetative propagules to occur before further disturbance events. Regeneration of *E. acuta*, for example, has been observed from rhizomes after seven days in response to wetting along the Murray River (Siebentritt 2003). Rhizome production in *E. acuta* and *Typha* spp. can also commence when seedlings are only around seven weeks old (Nicol and Ganf 2000). Reduced plant size among seedlings may also be a response to rapid drying conditions (Baker 1982).

## Functional groups

Riparian herbs have been classified into a variety of functional groups in relation to their traits and responses to flooding and drying. Naiman and Décamps (1997), for example, divide riparian plants into four groups depending on their life history strategy with respect to flood disturbance. Under this scheme, 'invaders' include those species that produce large numbers of wind- and water-dispersed seeds that colonise riparian habitats during favourable conditions. 'Endurers' are those plants that can re-sprout following damage or burial wrought by flooding (or grazing). 'Resisters' include riparian plants with growth traits that enable tolerance of flooding for considerable durations (i.e. many aquatic and amphibious plants), while 'avoiders' account for the remaining species that may be present at times but do not have specific adaptations to flood disturbances. A widely adopted functional plant classification for water plants in Australia was developed by Brock and Casanova (1997) and recently amended by Casanova (2011). Under this scheme, plants are classified based on their life history traits and their

**Table 7.1.** Water plant functional groups (FG) (adapted from Brock and Casanova 1997; Casanova 2011)

| FG code | FG name | Definition | Australian examples |
|---|---|---|---|
| Tdr | Terrestrial dry | Essentially terrestrial but can invade or persist in riparian areas and wetlands | *Bidens pilosa* (cobbler's pegs), *Cynodon dactylon* (couch), *Oxalis* spp. |
| Tda | Terrestrial damp | Germinate and establish on saturated or damp ground but not flood-tolerant | *Gratiola peruviana* (Australian brooklime) |
| ATl | Amphibious fluctuation tolerator – low growing | Germinate on saturated or flooded soil, can grow underwater but need to be exposed to flower and set seed | *Hydrocotyle verticillata* (shield pennywort), *Rorippa nasturtium-aquaticum* (watercress) |
| ATe | Amphibious fluctuation tolerator – emergent | Emergents that survive in saturated soil or shallow water but need most photosynthetic parts to be exposed | *Eleocharis acuta* (common spike-rush), *Ranunculus amphitrichus* (small river buttercup) |
| ATw | Amphibious fluctuation tolerator – woody | Woody perennial plants that germinate in shallow water or in drying soils but require near-continuous water in root zone | *Eucalyptus camaldulensis* (river red gum) |
| ARp | Amphibious fluctuation responder – plastic | As per ATlFG but exhibit a morphological response to changes in water level (e.g. shoot elongation or change in leaf type) | *Myriophyllum* spp. (water milfoil), *Potamogeton tricarinatus* (pondweed) |
| ARf | Amphibious fluctuation responder – floating | Float on water surface or grow underwater with floating leaves, some can compete life cycles on mud | *Lemna disperma* (common duck-weed); *Marsilea drummondii* (nardoo) |
| Se | Perennial – emergent | Woody and monocotyledonous emergent plants that require permanent water in root zone, favour stable water levels | *Triglochin procerum* (water ribbons); *Typha* spp. (cumbungi) |
| Sk | Submerged – k-selected | True aquatic species that require flooding for at least six months for germination or sexual reproduction | *Potamogeton crispus* (curly pondweed); *Vallisneria spiralis* (ribbonweed) |
| Sr | Submerged – r-selected | Fast-growing and rapidly reproducing species that colonise recently flooded habitats and require drying to stimulate high germination rates | *Lepilaena australis* (Austral water mat); *Zannichellia pallustris* (horned pondweed) |

response to the presence of surface water (Table 7.1).

## DISTRIBUTIONS AND DYNAMICS

At regional scales, the composition of riparian herb assemblages primarily reflects the greater species pool that typically includes plants adapted to the specific climatic, edaphic and other environmental conditions of that area. Riparian vegetation, however, tends to comprise different suites of herb species from the plant assemblages present in upland areas, though not necessarily more species. In Western Australia, for example, riparian vegetation in the floristically diverse heathlands and plateaus can be comparatively depauperate in terms of species richness (Hancock *et al.* 1996). The dissimilarity of riparian and upland vegetation depends largely on the degree of differentiation of riparian habitats from upland areas, mainly in relation to hydrology. In wetter landscapes, for instance, such as the northern highlands of Queensland's wet tropics, riparian vegetation in the upper and middle reaches of a catchment may be quite similar to the surrounding vegetation. In contrast, riparian vegetation in arid areas can be very distinct from upland vegetation and is often considerably more diverse both structurally and in terms of species numbers (e.g. Capon 2005). The diversity of riparian herb assemblages also tends to be higher in the absence of other stressors and disturbances (Paijmans *et al.* 1985). As the environment becomes more extreme, riparian herb assemblages skew towards species with traits that allow them to survive in such conditions. In desert floodplains, for example, herbaceous vegetation is typically dominated by ephemeral species with long-lived soil seed banks that are able to persist in the face of hydrologic unpredictability (Brock *et al.* 2006).

At a catchment scale, the composition and diversity of riparian herb assemblages usually vary in relation to topographic position as this influences flow. Lowland reaches, for instance, typically support more species (Sainty and Jacobs 2003). Similarly, spatial patterns in riparian vegetation at local scales predominantly reflect gradients of flood disturbance and therefore tend to relate to elevation or distance from channels or waterbodies. Herbaceous vegetation in many riparian habitats is characterised by distinctive zonation from wetter to drier areas as a result of hydrologic influences on processes of plant establishment, growth and mortality. Abiotic factors and levels of flood tolerance among plants tend to be the primary determinants of vegetation composition and structure at lower elevations where flooding occurs more frequently and intensely. Riparian habitats subject to prolonged and frequent inundation often support fewer species and lower overall biomass, particularly among amphibious and terrestrial species (Casanova and Brock 2000). Frequently flooded habitats may also support more annual, short-lived species that are able to respond rapidly to favourable conditions and complete their life cycles before further flooding (Capon 2003, 2005). Biotic factors (e.g. competition) tend to be more important at drier ends of flooding gradients where herb assemblages usually comprise a greater proportion of competitive and terrestrial species.

Riparian herb assemblages also tend to be very heterogeneous at a local scale due to the interactions of topography and patterns of wetting and drying and their influence on plant establishment and growth. Riparian soil seed banks are typically characterised by high levels of local heterogeneity and this can contribute to fine-scale differences in the composition and structure of herb assemblages (Webb *et al.* 2006; James *et al.* 2007). Other factors that affect the establishment of herb communities in riparian habitats, such as canopy cover, litter cover and grazing, are also influenced by surface water hydrology and tend to be spatially patchy as a result (Stromberg *et al.* 2007). Canopy cover, for instance, can influence the composition and abundance of seeds reaching sites and entering soil seed banks as well as germination via effects on light and temperature (Williams *et al.* 2008).

The composition and structure of herb assemblages in most riparian habitats in Australia are highly dynamic in time as well as space, exhibiting significant shifts in biomass and diversity mainly in response to patterns of wetting and drying. During dry periods, terrestrial species generally dominate riparian herbaceous vegetation, but since these tend to have limited flood tolerance, they are often quickly extirpated on flooding. Submerged and floating aquatic plants establish during periods of inundation and are often the only herbs able to germinate under the anoxic conditions associated with flooding. Some of these species persist as floodwaters recede, although obligate aquatic species do not tolerate drying. Germination and establishment of most riparian herb species occur during the moist conditions following floodwater recession. Such conditions also tend to promote high levels of productivity in some species (e.g. *Eleocharis acuta*) and can trigger reproduction in others (e.g. *Glossostigma* spp.) (Blanch and Brock 1994; Brock *et al.* 2006; Robertson and James 2007). The abundance and species diversity of riparian herb assemblages are therefore usually greatest under these conditions. The rate of drawdown of floodwaters can influence the composition of developing herb communities during this phase as rapid drying can favour the establishment of terrestrial species (Nicol *et al.* 2003). Flow timing can also be important as it will influence the types and rates of responses in processes such as hydrochory, germination and growth (Greet *et al.* 2011). The duration of dry periods intervening floods can also be significant as this will affect the availability and condition of propagules and adult plants present to respond to wetting. In general, however, high species richness and biomass of riparian herb assemblages are promoted by frequent short floods (Casanova and Brock 2000).

## ECOSYSTEM SERVICES

Australia's riparian herbs provide many important goods and services for people and have a wide range of economic as well as non-market values. Riparian herbs have long been used by Indigenous Australians for food and medicinal purposes and have also provided materials for making items such as woven baskets (Gott 2008). Rhizomes and leaf bases of *Typha* spp. (cumbungi), for instance, provided a staple food source while fibres from these plants were used to make string (Gott 1999). Seeds of many grasses common to riparian habitats in central Australia (e.g. *Panicum decompositum*, native millet; *Dactyloctenium radulans*, button grass; and *Eragrostis* spp.) also traditionally provided a staple food source (Latz 1999).

The provision by riparian herbs of fodder and grazing pastures is a particularly important ecosystem service of many Australian riverine landscapes with high economic value. The native floodplain pastures of the channel country in the Lake Eyre Basin, for example, are considered to be among the best cattle-fattening pastures in the world (Capon 2003).

Riparian herbs have significant supporting and regulating services in riverine landscapes. Groundcover and understorey vegetation can be especially significant for minimising bank erosion and regulating processes associated with soil and water quality (e.g. sediment transport) (Zierholz *et al.* 2001; Micheli and Kirchner 2002) and in-stream nutrient cycling (e.g. Esslemont *et al.* 2007). Herbaceous vegetation in riparian areas also provides important habitat and food resources for terrestrial and aquatic biota (e.g. grasses overhanging channels) as well as material input into aquatic ecosystems (e.g. leaves). A wide range of cultural services are also supported by riparian herbs, which contribute significantly to the aesthetic and wilderness values of many riverine landscapes as well as their value for recreational, tourism and scientific uses.

## CONCLUSIONS

Riparian herbs represent a significant component of the vegetation in Australia's riverine landscapes. Typically accounting for a large proportion of the floristic diversity present in riparian habitats, amphibious and terrestrial herbs may also be the only plants present where conditions limit the establishment and growth of woody vegetation or aquatic species. Exhibiting a wide range of growth and life history traits, the reproduction and growth of Australia's riparian herbs tend to be shaped predominantly by flooding and drying. Consequently, surface water hydrology has an overriding influence on riparian herb assemblages throughout the continent.

## REFERENCES


Arteca RN (1997) Flooding. In *Plant Ecophysiology.* (Ed. MNV Prasad) pp. 151–171. John Wiley & Sons, New York, NY.

Asaeda T, Manatunge J, Roberts J, Hai DN (2006) Seasonal dynamics of resource translocation between the aboveground organs and age-specific rhizome segments of *Phragmites australis. Environmental and Experimental Botany* **57**(1, 2), 9–18. doi:10.1016/j.envexpbot.2005.03.006

Baker WR (1982) Evolution and biogeography of arid Australian Scrophulariaceae. In *Evolution of the Flora and Fauna of Arid Australia.* (Eds WR Baker and PJM Greenslade) pp. 286–350. Peacock Publications, Adelaide.

Beadle NCW (1981) *The Vegetation of Australia.* Gustav Fischer Verlag, Stuttgart, Germany.

Bechtold JS, Naiman RJ (2006) Soil texture and nitrogen mineralization potential across a riparian toposequence in a semi-arid savanna. *Soil Biology & Biochemistry* **38**, 1325–1333. doi:10.1016/j.soilbio.2005.09.028

Bell D (2000) The ecology of co-existing *Eleocharis* species. PhD thesis. University of New England, NSW.

Blanch SJ, Brock MA (1994) Effects of grazing and depth on two wetland plant species. *Marine and Freshwater Research* **45**(8), 1387–1394. doi:10.1071/MF9941387

Blom CWPM, Voesenek LACJ (1996) Flooding: the survival strategies of plants. *Trends in Ecology & Evolution* **11**(7), 290–295. doi:10.1016/0169-5347(96)10034-3

Blom CWPM, Bögemann GM, Laan P, Van der Sman AJM, Van de Steeg HM, Voesenek LACJ (1990) Adaptations to flooding in plants from river areas. *Aquatic Botany* **38**(1), 29–47. doi:10.1016/0304-3770(90)90097-5

Brix H, Sorrel BK, Orr PT (1992) Internal pressurization and convective gas flow in some emergent. *Limnology and Oceanography* **37**(7), 1420–1433. doi:10.4319/lo.1992.37.7.1420

Brock MA (1994) Aquatic vegetation of inland wetlands. In *Australian Vegetation*. 2nd edn. (Ed. RH Groves) pp. 437–466. Cambridge University Press, Cambridge, UK.

Brock MA, Casanova MT (1997) Plant life at the edge of wetlands: ecological responses to wetting and drying patterns. In Frontiers in Ecology: Building the Links (Eds N Klomp and I Lunt) pp. 181–192. Elsevier Science, Oxford, UK.

Brock M, Capon S, Porter J (2006) Disturbance of plant communities dependent on desert rivers. In *Ecology of Desert Rivers*. (Ed. RT Kingsford) pp. 100–132. Cambridge University Press, Cambridge, UK.

Brosofske KD, Chen J, Naiman RJ, Franklin JF (1997) Harvesting effects on microclimatic gradients from small streams to uplands in western Washington. *Ecological Applications* **7**(4), 1188–1200. doi:10.1890/1051-0761(1997)007[1188:HEOMGF]2.0.CO;2

Capon SJ (2003) Plant community responses to wetting and drying in a large arid floodplain. *River Research and Applications* **19**(5, 6), 509–520. doi:10.1002/rra.730

Capon SJ (2005) Flood variability and spatial variation in plant community composition and structure on a large arid floodplain. *Journal of Arid Environments* **60**(2), 283–302. doi:10.1016/j.jaridenv.2004.04.004

Capon SJ (2007) Effects of flooding on seedling emergence from the soil seed bank of a large desert floodplain. *Wetlands* **27**(4), 904–914. doi:10.1672/0277-5212(2007)27[904:EOFOSE]2.0.CO;2

Capon SJ, Brock MA (2006) Flooding, soil seed bank dynamics and vegetation resilience of a hydrologically variable desert floodplain. *Freshwater Biology* **51**(2), 206–223. doi:10.1111/j.1365-2427.2005.01484.x

Capon SJ, Dowe JL (2007) Diversity and dynamics of riparian vegetation. In *Principles for Riparian Lands Management*. Land and Water Australia, Canberra.

Casanova MT (2011) Using water plant functional groups to investigate environmental water requirements. *Freshwater Biology* **56**(12), 2637–2652. doi:10.1111/j.1365-2427.2011.02680.x

Casanova MT, Brock MA (2000) How do depth, duration and frequency of flooding influence the establishment of wetland plant communities? *Plant Ecology* **147**(2), 237–250. doi:10.1023/A:1009875226637

Clevering OA, van der Toorn J (2000) Observations on the colonization of a young polder area in the Netherlands with special reference to the clonal expansion of *Phragmites australis*. *Folia Geobotanica* **35**(4), 375–387. doi:10.1007/BF02803550

Coops H, Van der Velde G (1996) Effects of waves on helophyte stands: mechanical characteristics of stems of *Phragmites australis* and *Scirpus lacustris*. *Aquatic Botany* **53**(3, 4), 175–185. doi:10.1016/0304-3770(96)01026-1

Cowie ID, Short PS, Osterkamp Madsen M (2000) Floodplain flora: a flora of the coastal floodplains of the Northern Territory, Australia. *Flora of Australia Supplementary Series, Vol. 10.*

Cronk JK, Fennessy MS (2001) *Wetland Plants: Biology and Ecology*. CRC Press, Boca Raton, FL.

Danehy RJ, Kirpes BJ (2000) Relative humidity gradients across riparian areas in eastern Oregon and Washington forests. *Northwest Science* 74(3), 224–233.

Esslemont G, Maher W, Ford P, Lawrence I (2007) Riparian plant material inputs to the Murray River, Australia. *Journal of Environmental Quality* **36**(4), 963–974. doi:10.2134/jeq2006.0318

Finlayson CM, Cowie ID, Bailey BJ (1990) Characteristics of a seasonally flooded freshwater system in monsoonal Australia. In *Wetland Ecology and Management: Case Studies.* (Eds DF Whingham, RE Good, J Kvet) pp. 141–162. Springer, Dordrecht, Netherlands.

Fritz KM, Evans MA, Feminella JW (2004) Factors affecting biomass allocation in the riverine macrophytes *Justicia americana. Aquatic Botany* **78**, 279–288. doi:10.1016/j.aquabot.2003.11.003

Gott B (1999) Cumbungi, *Typh*a species: a staple Aboriginal food in southern Australia. Australian Aboriginal Studies 1999(1), 33.

Gott B (2008) Indigenous use of plants in southeastern Australia. *Telopea* **12**(2), 215–226.

Grace JB (1993) The adaptive significance of clonal reproduction in angiosperms: an aquatic perspective. *Aquatic Botany* **44**, 159–180. doi:10.1016/0304-3770(93)90070-D

Green AJ, Jenkins KM, Bell D, Morris PJ, Kingsford RT (2008) The potential role of waterbirds in dispersing invertebrates and plants in arid Australia. *Freshwater Biology* **53**, 380–392.

Greet J, Webb JA, Cousens RD (2011) The importance of seasonal flow timing for riparian vegetation dynamics: a systematic review using causal criteria analysis. *Freshwater Biology* **56**(7), 1231–1247. doi:10.1111/j.1365-2427.2011.02564.x

Groves R, Whalley RD (2002) Grass and grassland ecology in Australia. *Flora of Australia.* Vol. 43. *Poaceae.* (Eds K Mallett and AE Orchard) pp. 157–182. Australian Biological Resources Study and CSIRO, Australia.

Hancock CN, Ladd PG, Froend RH (1996) Biodiversity and management of riparian vegetation in Western Australia. *Forest Ecology and Management* **85**(1–3), 239–250. doi:10.1016/S0378-1127(96)03761-9

Hillel D (1980) *Fundamentals of Soil Physics.* Academic Press, New York, NY.

Hook DD (1984) Adaptation to flooding with fresh water. In *Flooding and Plant Growth.* (Ed. TT Kozlowski) pp. 265–294. Academic Press, Orlando, FL.

Horowitz AJ (1991) *A Primer on Sediment-Trace Element Chemistry.* CRC Press, Lewis, MI.

Hupp CR, Osterkamp WR (1985) Bottomland vegetation distribution along Passage Creek, Virginia, in relation to fluvial landforms. *Ecology* **66**, 670–681. doi:10.2307/1940528

James CS, Capon SJ, White MG, Rayburg SC, Thoms MC (2007) Spatial variability of the soil seed bank in a heterogeneous ephemeral wetland system in semi-arid Australia. *Plant Ecology* **190**(2), 205–217. doi:10.1007/s11258-006-9201-9

Judd WS, Campbell CS, Kellogg EA, Stevens PF, Donoghue MJ (1999) Plant systematics: a phylogenetic approach. *Ecologia Mediterranea* **25**(2), 215–216

Junk WJ, Welcomme RL (1990) Floodplains. In *Wetlands and Shallow Continental Water Bodies. Vol. 1.* (Ed. BC Patten) pp. 491–524. SPB Academic Publishing, The Hague, Netherlands.

Karunaratne S, Aseaeda T, Yutani K (2004) Shoot regrowth and age-specific rhizome storage dynamics of *Phragmites australis* subjected to summer harvesting. *Ecological Engineering* **22**(2), 99–111. doi:10.1016/j.ecoleng.2004.02.006

Lake PS (2000) Disturbance, patchiness, and diversity in streams. *Journal of the North American Benthological Society* **19**(4), 573–592. doi:10.2307/1468118

Latz P (1999) *Pocket Bushtucker: a Field Guide to the Plants of Central Australia and their Traditional Uses.* Jukurrpa Books, Alice Springs, NT.

Les DH, Capers RS, Tippery NP (2006) Introduction of *Glossostigma* (Phrymaceae) to North America: a taxonomic and ecological overview. *American Journal of Botany* **93**(6), 927–939. doi:10.3732/ajb.93.6.927

Lowe BJ, Watts RJ, Roberts J, Robertson A (2010) The effect of experimental inundation and sediment deposition on the survival and growth of two herbaceous riverbank plant species. *Plant Ecology* **209**(1), 57–69. doi:10.1007/s11258-010-9721-1

Mackay SJ, Arthington AH, Kennard MJ, Pusey BJ (2003) Spatial variation in the distribution and abundance of submersed macrophytes in an Australian subtropical river. *Aquatic Botany* **77**(3), 169–186. doi:10.1016/S0304-3770(03)00103-7

Megonigal JP, Conner WH, Kroeger S, Sharitz RR (1997) Aboveground production in southeastern floodplain forests: a test of the subsidy-stress hypothesis. *Ecology* **78**(2), 370–384.

Menges ES (1986) Environmental correlates of herb species composition in five southern Wisconsin floodplain forests. *American Midland Naturalist* **115**, 106–117. doi:10.2307/2425840

Merritt DM, Scott ML, Poff NL, Auble GT, Lytle DA (2010) Theory, methods and tools for determining environmental flows for riparian vegetation: riparian vegetation-flow response guilds. *Freshwater Biology* **55**, 206–225. doi:10.1111/j.1365-2427.2009.02206.x

Micheli ER, Kirchner JW (2002) Effects of wet meadow riparian vegetation on streambank erosion. 2. Measurements of vegetated bank strength and consequences for failure mechanics. *Earth Surface Processes and Landforms* **27**(7), 687–697.

Naiman RJ, Décamps H (1997) The ecology of interfaces: riparian zones. *Annual Review of Ecology and Systematics* **28**, 621–658. doi:10.1146/annurev.ecolsys.28.1.621

Naiman RJ, Decamps H, McClain ME (2010) *Riparia: Ecology, Conservation, and Management of Streamside Communities*. Elsevier Academic Press, Burlington, MA.

Nicol JM, Ganf GG (2000) Water regimes, seedling recruitment and establishment in three wetland plant species. *Marine and Freshwater Research* **51**(4), 305–309. doi:10.1071/MF99147

Nicol JM, Ganf GG, Pelton GA (2003) Seed banks of a southern Australian wetland: the influence of water regime on the final floristic composition. *Plant Ecology* **168**(2), 191–205. doi:10.1023/A:1024430919811

Nilsson C, Gardfjell M, Grelsson G (1991) Importance of hydrochory in structuring plant communities along rivers. *Canadian Journal of Botany* **69**(12), 2631–2633. doi:10.1139/b91-328

Pagter M, Bragato C, Brix H (2005) Tolerance and physiological responses of *Phragmites australis* to water deficit. *Aquatic Botany* **81**(4), 285–299. doi:10.1016/j.aquabot.2005.01.002

Paijmans K, Galloway RW, Faith DP, Fleming PM, Haantjens HA, Heyligers PC, Kalma JD, Loffler E (1985) *Aspects of Australian Wetlands.* CSIRO Publishing, Melbourne.

Porter J (2000) Effects of salinity, turbidity and water regime on arid zone wetland seed banks. *Internationale Vereinigung für Theoretische und Angewandte Limnologie* **28**, 1468–1471.

Price P, Tubman W (2007) Structure and characteristics of riparian lands. In *Principles for Riparian Lands Management.* (Eds S Lovett and P Price) pp. 1–12. Land and Water Australia, Canberra.

Rassam DW, Fellows CS, De Hayr R, Hunter H, Bloesch P (2006) The hydrology of riparian buffer zones; two case studies in an ephemeral and a perennial stream. *Journal of Hydrology* **325**, 308–324. doi:10.1016/j.jhydrol.2005.10.023

Reid M, Capon S (2011) Role of the soil seed bank in vegetation responses to environmental flows on a drought-affected floodplain. *River Systems* **19**(3), 249–259. doi:10.1127/1868-5749/2011/019-0022

Roberts J, Ludwig JA (1991) Riparian vegetation along current-exposure gradients in floodplain wetlands of the River Murray. *Australian Journal of Ecology* **79**, 117–127. doi:10.2307/2260787

Roberts J, Marston F (2011) *Water Regime for Wetland and Floodplain Plants: a Source Book for the Murray–Darling Basin*. National Water Commission, Canberra.

Robertson HA, James KR (2007) Plant establishment from the seed bank of a degraded floodplain wetland: a comparison of two alternative management scenarios. *Plant Ecology* **188**(2), 145–164. doi:10.1007/s11258-006-9153-0

Sainty GR, Jacobs SW (2003) *Waterplants in Australia.* 4th edn. Sainty and Associates, Potts Point, NSW.

Siebentritt M (2003) The influence of water regime upon vegetation in lower River Murray wetlands: implications for restoration. PhD thesis.

Department of Environmental Biology, University of Adelaide, Adelaide.

Sluis W, Tandarich J (2004) Siltation and hydrologic regime determine species composition in herbaceous floodplain communities. *Plant Ecology* **173**(1), 115–124. doi:10.1023/B:VEGE.0000026335.44232.1c

Smith RGB, Brock MA (2007) The ups and downs of life on the edge: the influence of water level fluctuations on biomass allocation in two contrasting aquatic plants. *Plant Ecology* **188**(1), 103–116. doi:10.1007/s11258-006-9151-2

Sorrell BK, Tanner CC (2000) Convective gas flow and internal aeration in *Eleocharis sphacelata* in relation to water depth. *Journal of Ecology* **88**(5), 778–789. doi:10.1046/j.1365-2745.2000.00493.x

Sorrell BK, Brix H, Orr PT (1997) *Eleocharis sphacelata*: internal gas transport pathways and modelling of aeration by pressurized flow and diffusion. *New Phytologist* **136**(3), 433–442. doi:10.1046/j.1469-8137.1997.00769.x

Stromberg JC (2001) Restoration of riparian vegetation in the south-western United States: importance of flow regimes and fluvial dynamism. *Journal of Arid Environments* **49**(1), 17–34. doi:10.1006/jare.2001.0833

Stromberg JC, Beauchamp VB, Dixon MD, Lite SJ, Paradzick C (2007) Importance of low-flow and high-flow characteristics to restoration of riparian vegetation along rivers in arid south-western United States. *Freshwater Biology* **52**(4), 651–679. doi:10.1111/j.1365-2427.2006.01713.x

Ward KA (1992) *Investigation of the Flood Requirements of the Moira Grass Plains in Barmah Forest, Victoria*. Department of Conservation & Environment, Benalla, Victoria.

Webb M, Reid M, Capon S, Thoms M, Rayburg S, James C (2006) Are floodplain-wetland plant communities determined by seed bank composition or inundation periods? *IAHS Publication* **306**, 241–248.

Weisner SEB (1991) Within-lake patterns in depth penetration of emergent vegetation. *Freshwater Biology* **26**(1), 133–142. doi:10.1111/j.1365-2427.1991.tb00515.x

Weisner SE, Strand JA (1996) Rhizome architecture in Phragmites australis in relation to water depth: implications for within-plant oxygen transport distances. *Folia Geobotanica* **31**(1), 91–97.

Wetzel RG (1983) *Limnology*. Saunders College Publishing, Philadelphia, PA.

White SD, Ganf GG (2002) A comparison of the morphology, gas space anatomy and potential for internal aeration in *Phragmites australis* under variable and static water regimes. *Aquatic Botany* **73**(2), 115–127. doi:10.1016/S0304-3770(02)00010-4

Williams L, Reich P, Capon SJ, Raulings E (2008) Soil seed banks of degraded riparian zones in southeastern Australia and their potential contribution to the restoration of understorey vegetation. *River Research and Applications* **24**(7), 1002–1017. doi:10.1002/rra.1123

Young WJ, Schiller CB, Harris JH, Roberts J, Hillman TJ (2001) River flow, processes, habitats an driver life. In *Rivers as Ecological Systems: the Murray–Darling Basin* (Ed. WJ Young) pp. 45–99. Murray–Darling Basin Commission, Canberra.

Zierholz C, Prosser IP, Fogarty PJ, Rustomji P (2001) In-stream wetlands and their significance for channel filling and the catchment sediment budget, Jugiong Creek, New South Wales. *Geomorphology* **38**(3), 221–235.

# 8

# Riverine trees and shrubs

*Samantha Capon, Cassandra James and Amy George*

## INTRODUCTION

From coastal estuaries to the vast floodplains of the arid interior, woody plants can be found in a wide range of riverine habitats throughout Australia. Although often absent from areas that experience frequent and prolonged flooding, as well as many saline wetlands of inland Australia (Specht 1990), trees and shrubs form a considerable diversity of riverine vegetation communities across the continent including closed and open forests, woodlands and shrublands, in addition to many areas dominated by herbaceous vegetation that also support scattered trees and shrubs. Despite this structural diversity, woody riparian vegetation in Australia is often locally dominated by one to a few species, many of which are also widespread across the continent (e.g. *Eucalyptus camaldulensis*, river red gum) (Capon and Dowe 2007). Among these are a handful of rheophytes (e.g. *Casuarina cunninghamiana*, river she-oak) adapted to growing in swiftly flowing waters as well as trees and shrubs characteristic of marginal floodplain habitats that may be inundated less frequently than once a decade (e.g. *E. largiflorens*, black box), but nevertheless rely on inundation for their survival. Wherever woody plants occur in riverine landscapes, however, and in whatever form, they play an integral role in shaping the environments of which they are a part – physically, ecologically and culturally.

In this chapter, we provide a broad survey of the floristics and biogeography of woody riparian plants across Australia. We review the common traits of these species that facilitate their growth and reproduction in habitats that are characterised by periodic flooding and drying. Adaptations exhibited by riparian trees and shrubs to other common disturbances and stresses (e.g. fire, grazing and salt) are considered in other chapters in this volume (see Chapters 15, 16 and 17). We also discuss the dynamics of woody vegetation in Australia's riverine landscapes with a focus on describing major temporal and spatial patterns at both population and community levels and the chief drivers of these. Finally, we present a brief appraisal of the major ecosystem functions, goods and services provided by riparian trees and shrubs in Australia.

## FLORISTICS AND BIOGEOGRAPHY

Australia's woody riparian flora can be considered quite depauperate in comparison with that of climatically similar regions of other continents (Bowman and Woinarski 1994; Bowman 2000). Riparian tree species richness in south-eastern Brazil, for example, is substantially higher, at over 220 species, than in tropical northern Australia, which is characterised by relatively simple associations of *Melaleuca* and *Eucalyptus* (Metzger *et al.* 1997). In Europe, woody species usually comprise between 20 and 40% of plant diversity in riparian areas, and even sites with comparatively stressful conditions (e.g. cold or frequently flooded) fall in the lower bounds of this range (Schnitzler *et al.* 2007). In contrast, woody riparian vegetation across much of Australia tends to be limited to one or a few dominant species (Cole 1986; Fielding and Alexander 1996), which typically account for a low proportion of local plant species richness. Furthermore, woody plants that do grow in Australia's riparian habitats often occur in terrestrial habitats as well. Tree and shrub species contributing to upland vegetation in wet or mesic areas, for instance, often occupy riparian habitats in drier areas (Hancock *et al.* 1996). Even the ubiquitous riparian species *Eucalyptus camaldulensis* can be found on hill slopes some distance from drainage channels in wetter regions (Specht 1990).

Australia's riparian plant diversity, at the level of family and genus, reflects broad floristic and biogeographic patterns across the continent. Species belonging to the diverse Myrtaceae family, which includes *Eucalyptus*, *Melaleuca* and *Callistemon* species, account for much woody riparian plant diversity throughout the continent, including some specialist riparian species (Capon and Dowe 2007). Eucalypts, in particular, dominate a large proportion of Australia's wooded riparian zones. Other families frequently represented by riparian trees and shrubs include Fabaceae (e.g. *Acacia* spp.), Proteaceae (e.g. *Banksia* spp., *Grevillea* spp.) and Chenopodiaceae (e.g. *Chenopodium* spp., *Sclerolaena* spp., *Atriplex* spp.). In northern Australia, the palm family Arecaceae (e.g. *Archontophoenix* and *Livistona*) is also well represented in riparian habitats (see Chapter 12; Capon and Dowe 2007). Trees and shrubs in Australia's estuarine and other coastal wetlands encompass a diverse range of mangrove and salt marsh families, which are discussed in Chapter 9 of this volume.

## Trees

Trees are tall, perennial plants with an elongated, usually single, stem that are typically woody, although palms are usually included despite lacking true wood. In Australia, the tree flora, including that of riverine landscapes and floodplains, also includes many multi-stemmed forms known locally as mallees, particularly with reference to *Eucalyptus* species that sprout multiple stems from below-ground lignotubers (i.e. a woody growth comprising buds and energy reserves just below or above the ground) (Parsons 1994). Specht (1990) identifies 21 major species that comprise Australian forested wetlands (nine eucalypts, ten melaleucas and two casuarinas), each with a distinctive distribution and often being replaced by functionally similar species in observationally comparable habitats beyond their range (Specht 1990).

Like the rest of Australia's woody vegetation, riparian trees throughout the continent are dominated by myrtaceous species, especially *Eucalyptus* (Barlow 1994). *Eucalyptus camaldulensis* is the most widely distributed large eucalypt of inland Australia, where it occurs along almost all major watercourses (Fig. 8.1). *Eucalyptus largiflorens* (black box) and *E. coolabah* (coolibah) are also common woodland trees of semi-arid and arid channel bank and floodplain habitats, the former occurring in the south-east and the latter in similar habitats of the north, with some overlap between the two species in central Australia (Fig. 8.1). *Eucalyptus ovata* (swamp gum) is widely distributed throughout northern and eastern Tasmania and in frequently inundated coastal areas on the mainland from South Australia, through Victoria to New South Wales (Fig. 8.1; Clucas and Ladiges 1979; Specht 1990). Other common coastal riparian eucalypts include *E. robusta* (swamp mahogany), *E. tereticornis* (forest red gum) and *E. elata* (river peppermint) along the east coast of mainland Australia, and *E. rudis* (Western Australian flooded gum) in the west (Fig. 8.1; Specht 1990).

Melaleucas are common representatives of the Myrtaceae family in riparian habitats of coastal regions and across northern Australia, and include some of the few Australian trees that are true rheophytes or obligate riparian species. *Melaleuca leucadendra* and *M. argentea* frequently fringe channels throughout northern Australia (Fig. 8.1; Specht 1990; Pettit *et al.* 2001) while northern monsoonal tropical floodplains often support paperbark

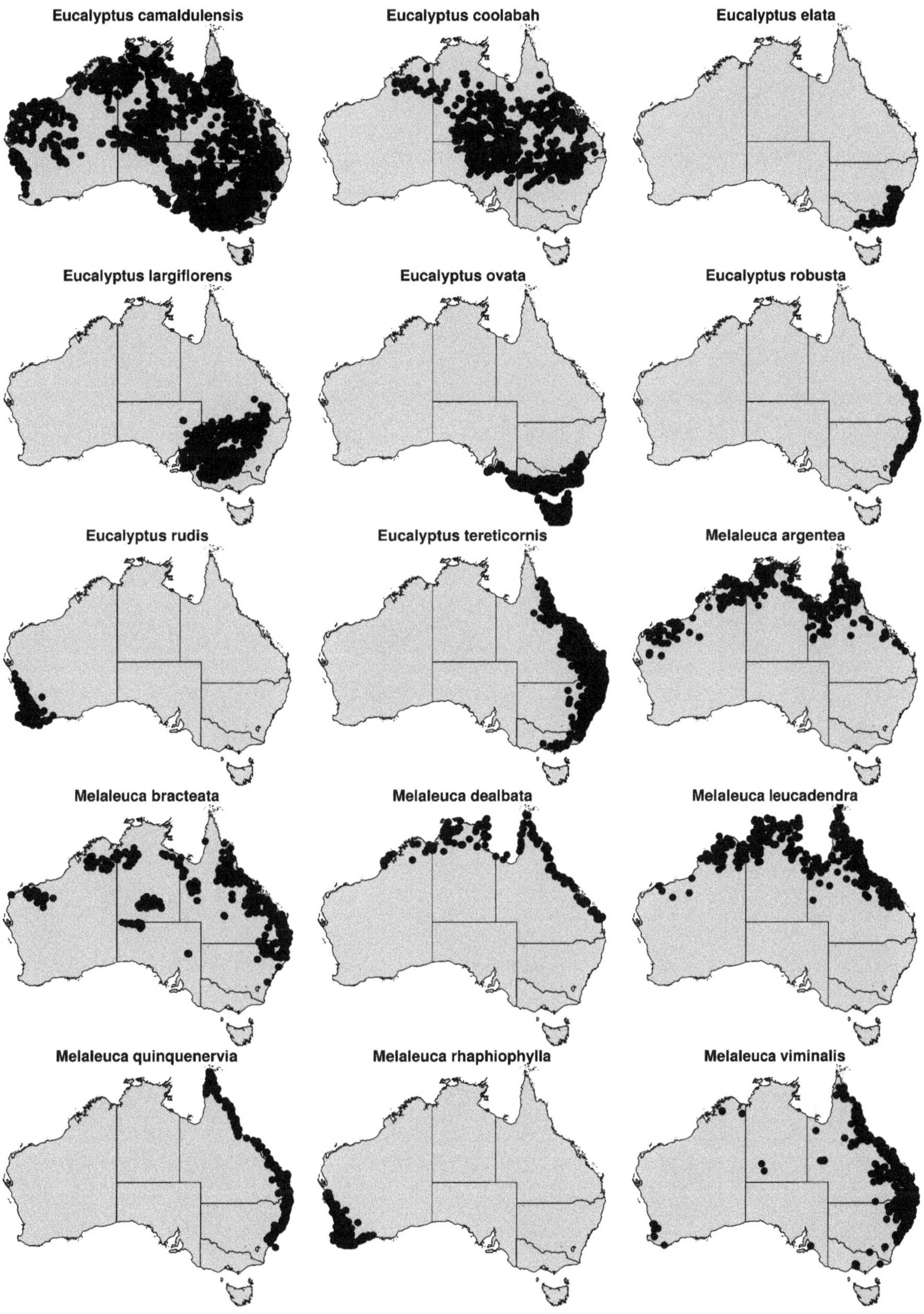


**Figure 8.1:** Distribution of common riverine *Eucalyptus* and *Melaleuca* species across Australia based on records from the Atlas of Living Australia (www.ala.org.au).

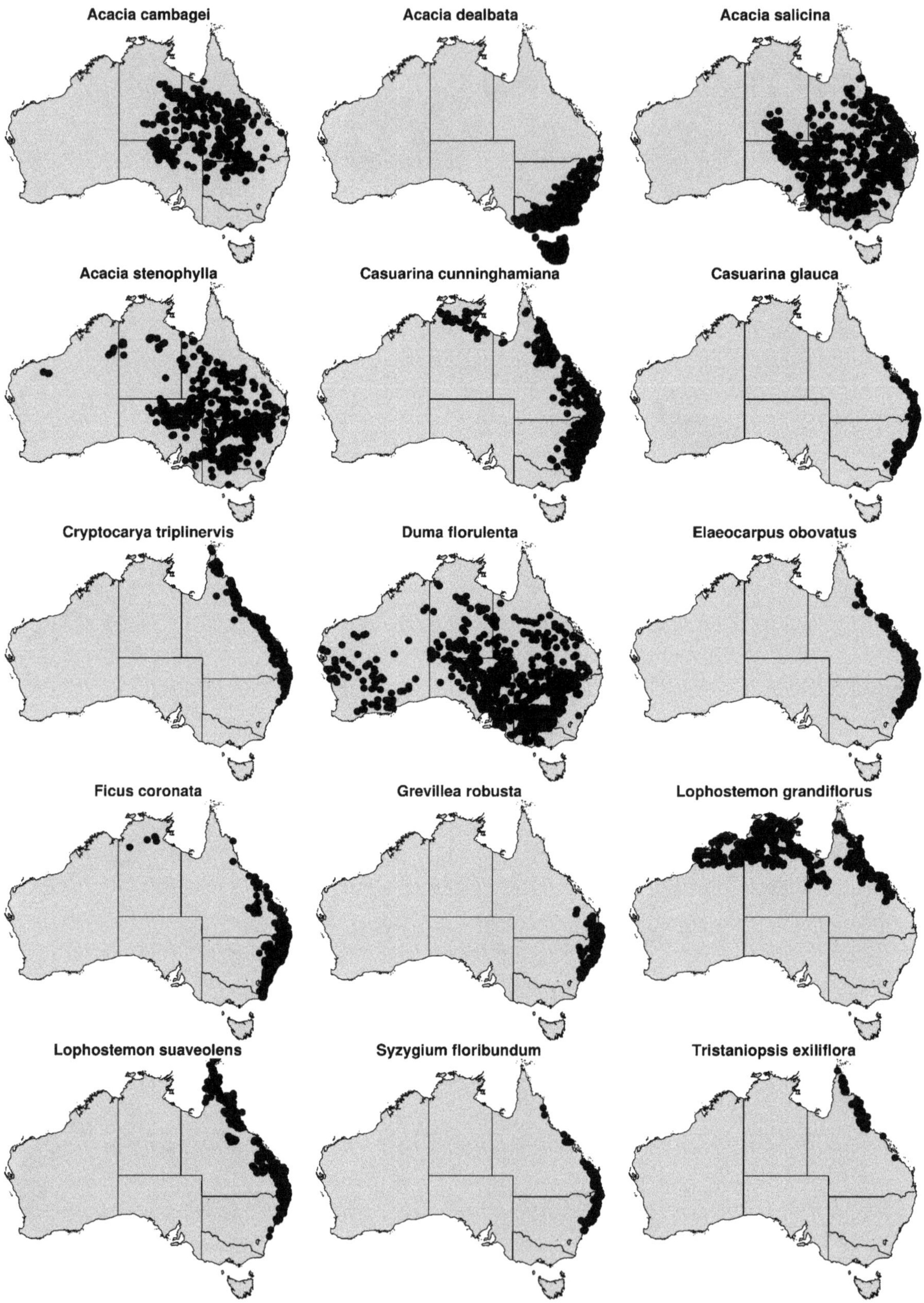


**Figure 8.2:** Distribution of some common woody riverine species based on records from the Atlas of Living Australia (www.ala.org.au).

swamp forests of *M. viridiflora* and *M. cajuputi* (Finlayson 2005). *Melaleuca bracteata* and *M. dealbata* are common in riverine habitats along Australia's central to northern east coast, as well as in isolated patches in northern and central Australia (Fig. 8.1). *Melaleuca quinquenervia* is a particularly dominant tree of coastal riverine habitats in New South Wales and Queensland, as are *M. rhaphiophylla* and *M. preissiana* in south-west Western Australia (Fig. 8.1; Pettit *et al.* 2011). *Melaleuca viminalis* (previously *Callistemon viminalis*) is very prevalent in channel and riparian habitats of south-east Queensland (Fig. 8.1).

Casuarinas are a significant component of riparian canopies of Australia's east coast as well as in the upper to mid reaches of many inland catchments (e.g. in the eastern Murray–Darling Basin). The moderately sized *C. cunninghamiana* is particularly prevalent along coastal rivers from southern New South Wales up to northern Queensland and into the Northern Territory where it forms belts along watercourses and can also extend onto floodplains. *Casuarina glauca* (swamp she-oak) is a smaller tree that forms pure stands in low-lying wetland habitats along the east coast, often in association with mangroves (Specht 1990).

A high diversity of rainforest species (e.g. *Lophostemon grandiflorus*, northern swamp mahogany; *L. suaveolens*, swamp box; and *Elaeocarpus obovatus*, hard quandong) also contribute to riparian canopies in sub-tropical eastern and tropical northern Australia, many of which have distributions that extend well beyond riverine habitats. In the wet tropics of Queensland, *Tristaniopsis exiliflora* (kanuka box; Figs 8.2, 8.3) is a defining species of rainforest riparian zones. In higher rainfall areas, swamp rainforests are often dominated by palms (e.g. *Archontophoenix* and *Livistona*). Gallery

**Figure 8.3:** Photos of some common woody riverine species: (a) *Tristaniopsis exiliflora* (kanuka box) lining a riverbank in the wet tropics of Queensland, (b) *Eucalyptus coolabah* (coolibah) on a floodplain (Lake Eyre Basin), (c) *Melaleuca leucadendra* (weeping paperbark) lining a waterhole in the dry tropics of northern Queensland, (d) *Duma florulenta* (tangled lignum) shrubs in an inundated ephemeral wetland in northern NSW.

**Figure 8.4:** Photos of some common woody riverine vegetation communities: (a) floodplain woodland in northern NSW comprising *Eucalyptus coolabah, Acacia stenophylla* and *Eremophila bignoniiflora*; (b) *Duma florulenta* (tangled lignum) shrubs growing along a dry river channel in northern NSW with woodland composed of *Acacia stenophylla* and *Eucalyptus coolabah*.

rainforests (i.e. riparian forests in sparsely wooded landscapes) can be very diverse and typically include a range of fig species (e.g. strangler and sandpaper figs). High numbers of woody species are also found in riparian habitats of south-east Queensland with common species including *Ficus coronata* (sandpaper fig) and *Syzygium floribundum* (weeping lilly pilly) and, in drier inland sites, *Grevillea robusta* (silky oak; Arthington *et al.* 2012) (Fig. 8.2). Members of the laurel family (e.g. *Cryptocarya triplinervis*; Fig. 8.2) are widely distributed along creeks and rivers of Australia's east coast.

Acacias tend to replace eucalypts as the dominant tree taxa in inland Australia, especially where mean annual rainfall is below 600 mm (Gillison 1994). Although eucalypts remain prevalent in riparian habitats of these drier landscapes, acacias usually make a significant contribution to woodland canopies and may be the only or dominant tree in some places, for example, along minor channels or swampy woodlands. Common species include *Acacia stenophylla* (river cooba), *A. salicina* (cooba), *A. cambagei* (gidgee) and *A. dealbata* (silver wattle) (Figs 8.2, 8.4).

## Shrubs

Shrubs are multi-stemmed woody plants of intermediate height. In riverine habitats, shrubs may form a mid-storey in woodlands, dominate the woody vegetation in swampy shrublands or appear as scattered plants in herbaceous vegetation communities. Shrubs often form dense bands fringing watercourses and wetlands, particularly in steeper, upland areas (Hancock *et al.* 1996; Roberts and Marston 2011). Channel features (e.g. bars and islands) also frequently support dense thickets of shrubs, which in drier regions often further colonise dry river beds (Tooth *et al.* 2008). Myrtaceous shrubs occur widely in riparian habitats throughout eastern and northern Australia (e.g. *Leptospermum* spp., tea trees; and *Melaleuca* spp., bottlebrushes). *Pomaderris apetala* (dogwood) is a common riparian shrub that only occurs in Tasmania (Daley and Kirkpatrick 2004). *Duma florulenta* (tangled lignum) is a particularly widespread and abundant shrub of riverine habitats of inland Australia (Figs 8.2, 8.3d, 8.4b). During drier periods, arid and semi-arid riparian, floodplain and wetland habitats also typically support a moderate diversity of chenopod shrubs and sub-shrubs (dwarf shrubs), including *Atriplex* spp., *Sclerolaena* spp. and *Maireana* spp., as well as *Acacia* shrubs (see Chapter 13).

## WOODY PLANT TRAITS

Woody plants growing in riverine habitats can be riparian specialists, with specific adaptations that have evolved in response to selection pressures particular to these environments (e.g. periodic flooding), or they may be generalists that possess traits that enable them to establish and survive, at least some of the time, in riparian habitats. The majority of Australia's woody riparian plants are likely to fall into the latter category as so many species have distributions that extend well beyond riverine habitats. Furthermore, Australia has few obligate rheophytes, or plants specialised to grow in swiftly flowing water (van Steenis 1981), but many that can be considered facultative rheophytes (Capon and Dowe 2007).

In general, trees and shrubs inhabiting riverine habitats must possess traits that enable them to cope with the environmental variability associated with periodic inundation, as well as drying, which occurs seasonally in some regions (e.g. northern Australia), but with high levels of unpredictability in others (e.g. central Australia). Indeed, many traits of trees and shrubs as a broad group are thought to be advantageous in variable environments, such as long life spans and generation times (Schaffer 1974). The broad distribution of many common Australian riparian trees and shrubs across multiple climatic zones (Figs 8.1, 8.2), also suggests considerable flexibility and adaptive capacity among these species. Nevertheless, to reproduce and persist in riparian habitats, trees and shrubs require traits that enable the successful completion of a suite of life history stages, as well as supporting growth, under a range of different conditions.

## Reproductive traits

Riparian and wetland trees and shrubs do not typically form persistent soil seed banks (e.g. Chong and Walker 2005; Capon 2007) and therefore depend on dispersal, germination and establishment traits that enable reproduction and persistence in habitats characterised by variable hydrological stresses.

### Sexual reproduction

Very little is known regarding the reproductive biology or ecology of many of Australia's riparian trees and shrubs (e.g. Roberts and Marston 2011). Sexual reproduction in trees of dryland riparian habitats, however, tends to be highly episodic and associated with disturbances such as floods (Brock *et al.* 2006). Here, we identify a range of features of breeding systems that may infer an advantage under conditions of hydrological variability including: (i) having generalist pollinators; (ii) an ability to 'self'; (iii) prolific flowering and seed setting; (iv) variable and protracted flowering; (v) precocious flowering and seed production; (vi) flood-induced flowering and seed set; and (vii) phenological coupling.

(i) *Generalist pollinators* – a tight coupling between pollen supply and pollinators may be a disadvantage in variable environments as conditions suitable for pollinators may differ from those suitable for pollen production. Adopting pollinators with generalist strategies may provide more flexibility because there is less dependency on the availability of a specific pollinator. Such a lack of specificity in pollination is typical of many eucalypts and acacias (Pryor 1976; Gibson *et al.* 2011).

(ii) *Ability to 'self'* – 'selfing' (i.e. fusing of male and female gametes from the same individual, as opposed to cross-fertilisation) provides reproductive assurance when pollinators and/or pollen sources are scarce (Lloyd 1979, 1992). Although selfing may have negative effects on the survival and growth of progeny (termed inbreeding depression), these negative effects may be outweighed by the benefits of being able to reproduce in hydrologically variable environments even in the absence of intraspecific pollen availability. The ability to self may also facilitate opportunistic reproduction during favourable conditions. Mixed mating strategies, with species showing both cross- and self-fertilisation, are typical among many of the tree and shrub species discussed in this chapter, for example, trees of the genus *Eucalyptus* such as *Eucalyptus camaldulensis* (McDonald *et al.* 1996). Dioecious species such as *Duma florulenta* and *Atriplex nummularia*, however, do not generally have the capacity to self, although in the case of *A. nummularia*, monoecious plants do occur (Sampson and Byrne 2012).

(iii) *Prolific flowering and seed setting* – prolific flowering is observed in many eucalypts, acacias and melaleucas but can be highly variable between years. *Eucalyptus camaldulensis*, for example, often only flowers en masse every couple of years. Such 'mast' years also occur in *Casuarina cunninghamiana* (Woolfrey and Ladd 2001). Advantages to masting may include increased efficiency of wind pollination and predator seed satiation. High flower production does not necessarily mean high seed production, however, as large numbers of flowers or fruits can be aborted before fertilisation or maturation, respectively. Tybirk (1997) suggests that typically less than 1% of *Acacia* flowers result in fruit. Numbers of viable seeds per capsule can also vary considerably in *Eucalyptus* species depending on factors such as position in the canopy, possibly in relation to the degree of outcrossing, which tends to be greater at the top of the canopy (Patterson *et al*. 2001).

(iv) *Flowering variability and protracted flowering* – variable timing and duration of flowering is apparent among many riparian trees and shrubs. Flowering of *E. largiflorens* in the lower Murray, for example, has been observed in all months apart from February and March (Jensen *et al*. 2008). A broad temporal spread of reproductive effort may indicate a response to water availability (Johns *et al*. 2009). Flowering in *Acacia stenophylla*, for instance, is typically observed between March and August, but can also occur at other times of the year (T. Doody, pers. comm., reported in Johns *et al*. 2009) suggesting an opportunistic response to water availability.

(v) *Precocious flowering and seed production* – melaleucas are reportedly able to reproduce within one year of germinating and can produce large numbers of flowers within only three years (Meskimen 1962 in Serbesoff-King 2003). Precocious flowering is also reported for 6-month-old seedlings of *E. camaldulensis* (Khan 1965), but whether this occurrence can produce any viable seed is unknown.

(vi) *Flood-induced flowering and seed set* – in hydrologically variable environments, key life history events may be timed to coincide with the availability of water (Brock *et al*. 2006). For phenological coupling to be effective in promoting sexual regeneration, however, plants must either be able to produce seed quickly enough to capitalise on higher moisture availability after flooding or high local rainfall (i.e. flood-triggered reproduction) or have a ready supply of viable seeds stored aerially or in the soil that can respond to favourable conditions (i.e. cued seed release). Examples of flood-triggered reproduction include *D. florulenta*, which flowers rapidly in response to wetting (Capon *et al*. 2009), although recruitment may not follow winter floods due to low temperatures. *Duma florulenta* also exhibits a rapid seed maturation time, with seeds produced between 14 and 30 days after anthesis, enabling this species to capitalise on high moisture conditions after flooding (Chong and Walker 2005).

(vii) *Phenological coupling (serotiny, timing)* – because of the relatively long time usually taken from anthesis to seed maturation, phenological coupling for many inland perennial riverine species is probably anticipatory (i.e. restricted to the cued release of seeds) rather than the reactive, flood-triggered seed production discussed above. Seasonality in flowering and seed set is observed widely in eucalypts, for example, with considerable geographic variation that may reflect differences in seasonal wetting regimes (e.g. Pettit and Froend 2001). The ability to retain at least a proportion of seeds produced in the canopy also allows some riverine species to couple seed release with conditions most suitable for seed germination and establishment. In fluctuating environments, a serotinous habit may be advantageous through facilitating early establishment in the growing season and protecting seeds from granivores and other influences that may result in early seed mortality on the ground. Bet-hedging via staggered seed dispersal through time may also be beneficial.

Serotinous floodplain species can either release their seeds at the time of maturity, retain mature seeds in their canopy, or do both. *Eucalyptus camaldulensis*, for example, may retain a portion of the seeds in the canopy. Dexter (1970) estimated that ~25% of seeds of *Eucalyptus camaldulensis* are retained. This differs from studies of *E. camaldulensis* in the Ord River where Pettit and Froend (2001) found that all seed was released in April and no serotiny was observed. In contrast, *E. rudis*, an ecologically and genetically similar species to *E. camaldulensis* found in temperate south-western Western Australia, retained seeds in the canopy for at least 12 months (Pettit and Froend 2001).

Seeds retained in the canopy can be released in response to favourable conditions – a form of bet-hedging with a trade-off between short-term gain and long-term risk reduction (Venable 2007). Christensen and Kimber (1975) noted an increase in the proportion of species exhibiting serotiny as reliability of rainfall decreased in south-western Australia. Eucalypt seed, however, is rarely held in the canopy for more than three or four years (House 1997). Serotiny is exhibited by *E. coolabah* in some locations (contrary to Rogers and Ralph 2011; S. Capon, pers. obs.). Roberts (1993) also notes old brown fruit on a large proportion (~50%) of trees surveyed, although whether these contained viable seed is unknown.

Factors triggering seed release from serotinous species of Australian riverine habitats are poorly understood, particularly in dryland regions (Johns *et al.* 2009; Hamilton-Brown *et al.* 2009). For eucalypts in general, capsules are released when the sap stream is cut off, but by this time most seed is thought to have already been released (Cremer 1965). Desiccation of the capsule alters the ovary wall, causing the loculi within the capsule to widen and valves to open, after which the seeds drop to the ground via gravity (Cremer 1965). Hence, the triggers that instigate seed release are likely to relate to hydrology and climate. In Dowd Morass, a Ramsar-listed, brackish-water wetland in south-eastern Australia, environmental triggers of seed release of *Melaleuca ericifolia*, a coastal swamp paperbark, include water level, salinity, air temperature and vapour pressure deficit, with a high pulse of seed release from this species observed during periods of water drawdown (Hamilton-Brown *et al.* 2009). Notably, however, drawdown during 2005 failed to elicit a similar seed release response to that observed after a large seed release event in the previous year, possibly because little mature seed remained in the canopy after the release of the previous year (Hamilton-Brown *et al.* 2009).

Although the mechanisms of seed release are unknown in most inland Australian floodplain species, observations of the key serotinous species suggest considerable geographic variation in seed release. Seed fall of *E. camaldulensis*, for instance, tends to be highest in spring when floods are often receding in the southern Murray–Darling Basin (Dexter 1978; George 2004), although peaks have been observed in winter in the Mount Lofty Ranges in South Australia (Pudney 1998). In a comparison between the riparian vegetation of the tropical Ord River and the temperate Blackwood River of Western Australia, Pettit and Froend (2001) found that *E. camaldulensis* and *E. rudis* both timed seed release to coincide with periods of the year when river water levels typically recede, i.e. at the end of the tropical wet season in April in the Ord River compared with the end of the temperate winter in the Blackwood River.

### Vegetative reproduction

Vegetative reproduction is common among many wetland plants and confers many advantages over sexual reproduction, including the provision of support from parent plants in early stages of establishment (Cronk and Fennessy 2001). Since asexual reproduction precludes the need for gamete exchange and seed development, such reproductive strategies can facilitate a rapid reproductive response to favourable conditions. Furthermore, when populations are very small with limited individuals available, reliance on sexual reproduction can be risky. The production of vegetative reproductive organs and propagules (e.g. root stock, tubers, rhizomes), however, can be expensive, requiring large storage reserves which may be incapable of surviving extended periods of stressful conditions such as drought. Vegetative reproduction is therefore most likely to occur and be beneficial in more hydrologically stable or predictable riverine habitats.

Among common Australian riverine trees and shrubs, a range of vegetative reproductive mechanisms are apparent. Layering, i.e. the development of new individuals from parent plant stems, occurs in *C. cunninghamiana* from parents damaged by floods (Woolfrey and Ladd 2001) and *D. florulenta* (Chong and Walker 2005). *Acacia stenophylla* also demonstrates suckering which enables moderate to fast growth of recruits, even at extremely stressful sites if roots are damaged. This species can also coppice (i.e. regrow from the stump or roots following damage) when young or under favourable conditions (Boxshall and Jenkyn 2001).

Intra-specific reliance on sexual and asexual reproduction for persistence can vary in relation to temporal and spatial environmental variability. Stored seed reserves, for instance, may enable regeneration of a population after particularly severe flood disturbances that damage a high proportion of the canopy. Similarly, local regeneration in *D. florulenta* appears to rely principally on vegetative mechanisms, but this species also produces vast quantities of relatively short-lived seeds which may enable dispersal and establishment in habitats further afield (Chong and Walker 2005; Capon *et al.* 2009).

## Propagule and dispersal traits

A range of traits among both sexual and asexual plant propagules may favour persistence of woody plant species in hydrologically variable riverine habitats. These include: (i) propagule longevity, via either seed dormancy or traits that infer longevity among asexual propagules; (ii) variable dormancy breaking and bet-hedging; and (iii) the ability to disperse via hydrochory and subsequently establish.

(i) *Propagule longevity* – physical seed dormancy, facilitated by water-impermeable seed coats, is common in shrubs of arid and semi-arid regions (Baskin and Baskin 2001). In members of the Mimosaceae family, including *Acacia stenophylla*, physical dormancy is enabled via hard seed coats, which typically require physical abrasion or high temperature treatments (e.g. submersion in boiling water) to break dormancy and initiate germination (Gunn 2001). Seed dormancy in acacias often develops in the absence of stimulation to germinate and while seeds are still young and soft (Tybirk 1997). Consequently, fresh or immature seeds may germinate without such dormancy-breaking treatments (Gunn 2001). High levels of recruitment among *A. stenophylla* along floodlines also suggests possible natural dormancy breakage in response to flooding and drying. In other riverine acacias (e.g. the exotic *A. nilotica*) prolonged seed submergence (up to 18 weeks) has been observed to break dormancy and promote germination (Warrag and Eltigani 2005). Physiological dormancy of *Atriplex* species seeds may be facilitated by bracteoles that enclose seeds and retard germination under conditions of accumulating salt (Beadle 1952), although other studies (e.g. Stevens *et al.* 2006) have found only a minor effect of bracteole removal on germination. Vegetative propagules of woody riverine plants may also exhibit considerable longevity. Rhizome banks and root stock of *D. florulenta*, for example, are likely to be able to survive for ten years or more in the absence of flooding (Roberts and Marston 2011).

(ii) *Variable dormancy breaking and bet-hedging* – variation in the timing of dormancy breakage and the initiation of plant establishment from both sexual and asexual propagules represents a means of bet-hedging and spreading the risk of reproductively unsuccessful germination events. For example, the germinablity of seeds of *Acacia ligulata*, a species that occurs in a variety of habitats including floodplains, differs between those ingested by birds and those dispersed by ants, with the former germinating readily while the latter tend to remain dormant until sufficiently scarified or exposed to heat (Letnic *et al.* 2000). The dimorphic seed of some *Atriplex* species, within which seed longevity varies between seed types, may also represent a bet-hedging strategy. In *Atriplex vesicaria*, for example, most seeds germinate readily, but a small percentage remains dormant (Hunt 2001).

(iii) *Ability to capitalise on hydrochory* – a wide range of morphological and/or physiological traits of plant propagules allows them to capitalise on hydrochory (i.e. dispersal via water) including the production of small and light seeds as well as particular seed shapes. Double-coated seeds of *E. camaldulensis*, for example, facilitate floatation for up to ten days (Pettit and Froend 2001). Propagule appendages such as plumes or wings can also enable seed dispersal across the water surface. Wings of *Casuarina cunninghamiana* seeds may be functionally similar to plumes of *Salix* and *Populus* seeds (Woolfrey and Ladd 2001). Similarly, the bracteoles surrounding *Atriplex* seeds probably assist in the dispersal of this genus by both water and wind.

Many woody plant propagules that disperse by hydrochory are also able to germinate while floating, including *E. camaldulensis*, *E. largiflorens* and *D. florulenta* (Chong and Walker 2005; Jensen *et al.* 2008). Seeds and very young seedlings of these species therefore appear to be capable of surviving at least short periods of submergence, although longer periods of submergence probably result in mortality. In general, waterlogging of greater than five months or so prohibits seedling phases in most trees and shrubs (Boomsma 1950).

## Seedling and recruitment traits

Seedlings and recruits of woody plants in riverine habitats typically suffer high levels of mortality and, consequently, recruitment represents a major population bottleneck for most species (Espelta *et al.* 1995; Jongejans *et al.* 2006; Lehmann *et al.* 2009), especially in unpredictable dryland regions (Streng *et al.* 1989; Hughes 1990; Cooper *et al.* 1999; Horton and Clark 2001). A range of traits that maximise survival in hydrologically variable habitats are exhibited by seedlings of riverine trees and shrubs at both population and individual levels, including: (i) recruitment by numbers; (ii) opportunistic germination and seedling growth; (iii) drought tolerance and avoidance; (iv) flood tolerance; and (v) an ability to re-sprout after disturbances.

(i) *Recruitment by numbers* – riparian species such as *E. camaldulensis* produce very large numbers of seed that contain few reserves but possess high initial viability that declines rapidly once on the ground. This enables the germination of large numbers of seedlings under favourable conditions thus promoting the survival of at least some recruits. This strategy of recruitment by numbers is similar to that described for other common riparian perennial species such as cottonwoods in North America (Mahoney and Rood 1998).

(ii) *Opportunistic germination and seedling growth* – seedlings of many riverine trees and shrubs (e.g. *Eucalyptus microtheca* and *D. florulenta*) exhibit extremely rapid growth which enables recruits to capitalise on favourable conditions and maximise the opportunity to gain sufficient stature and reserves to survive subsequent periods of stress such as drought (Chunyang and Tuomela 1997; Capon *et al.* 2009). Similarly, rapid germination in response to suitable cues provides young recruits with even longer durations under productive conditions in which to become established. *Duma florulenta* seeds, for example, responded to wetting within three days when exposed to fluctuating temperatures (Chong and Walker 2005).

(iii) *Drought tolerance and avoidance* – the development of morphological and physiological traits that enable recruits to tolerate or avoid drought are likely to be beneficial in most riverine habitats, but particularly those in highly variable or dryland environments. Lignotuber development, deep tap roots, stomatal aperture modulation, biomass adjustments (e.g. increasing root to shoot ratios), flexible water use, leaf surface wax, leaf shedding and leaf area alterations (e.g. via leaf shedding) all represent traits that facilitate drought tolerance or avoidance. Traits associated with drought tolerance typically include those that allow plant tissues to avoid dehydration through minimising water loss, maximising water uptake and/or by tolerating low tissue water potential (Chaves *et al.* 2003). Seedlings of many Australian woody riverine species instead demonstrate drought avoidance mechanisms such as extensive tap root system development and extension to track declining soil water levels. Increasing rooting depth in *E. camaldulensis* seedlings to avoid water deficiency, for example, has been observed in numerous situations (e.g. Pereira and Kozlowski 1976; White *et al.* 2000; Argus *et al.* 2015). Greenhouse studies on *E. coolabah* seedlings also suggest that allocation to roots in this species can increase in response to drought (Susiluoto and Berninger 2007). Similarly, rapid root extension and low shoot to root ratios have been observed in seedlings of several riparian species of North America and Europe (e.g. Brown and Archer 1990; Padilla and Pugnaire 2007; Woods *et al.* 2014).

High levels of above ground biomass production, contributing to carbohydrate reserves, may also be important for maintaining physiological function and enabling physiological adjustments to water stress (e.g. stomatal aperture) that allow seedlings to continue growth under drought conditions (Farrell *et al.* 1996). Minimising water loss while maintaining high tissue water potential through stomatal closure is often one of the earliest plant responses to drought, and is widely considered to be one of the main ways in which *E. camaldulensis* seedlings control transpirational water loss (Pereira and Kozlowski 1976; White *et al.* 2000). Other morphological adaptations that can reduce water loss include a reduction in specific leaf area or leaf abscission, also observed in *E. camaldulensis* seedlings (Gibson *et al.* 1994). Additionally, *E. camaldulensis* displays structural changes in leaves as seedling grow such that, with age, leaves become stronger, narrower and more vertical (i.e. sclerophyllous; Gibson and Bachelard 1994), and wax becomes thicker on abaxial, rather than adaxial leaf surfaces (Pereira and Kozlowski 1976).

Tolerance of aridity by seedlings can also be facilitated at cellular level via osmotic adjustment (e.g. by increasing cell solutes) and/or cell wall elasticity, both of which contribute to maintaining cell turgor and metabolic function under dry conditions. Modest osmotic adjustment is observed in *E. microtheca* (Li 1998) and *E. camaldulensis* (Lemcoff *et al.* 1994; Merchant *et al.* 2006, 2007). Cell wall elasticity (measured as the bulk modulus elasticity, ε) was also found to increase significantly in *E. camaldulensis* (Lemcoff *et al.* 2002) and *E. microtheca* (also probably *E. coolabah*) seedlings in response to drought (Li 1998). Other studies, however, have found these species to be relatively inelastic in response to drought (e.g. Tuomela 1997; White *et al.* 2000). This contrast in results could be a function of plant age or size, as cell wall elasticity is known to vary with ontogeny and leaf age and the studies of White *et al.* (2000) were performed on larger, and presumably older, plants.

(iv) *Flood tolerance* – stomatal closure can be one of the earliest responses of juvenile plants to flooding (Pereira and Kozlowski 1976) and is typically the first response among plants with amphistomatous leaves (i.e. with stomata on both leaf sides). Maintaining stomatal conductance during flooding could be an adaptation to allow species to capitalise on available water in hydrologically variable environments (Argus *et al.* 2015). Variation in the response of *E. camaldulensis* has been observed in stomatal conductance of seedlings after flooding. Van der Moezel *et al.* (1989) reported flooding induced rapid stomatal closure on upper, but not lower, leaf surfaces followed by reopening of stomata after a short period for *E. camaldulensis*. In contrast, Argus *et al.* (2015) found that stomatal conductance of seedlings was maintained for more than two weeks after flooding in the subspecies *E. camaldulensis* subsp. *refulgens* from the Pilbara region of north-west Australia. Basal stem thickening is also commonly observed among *E. camaldulensis* seedlings (e.g. Argus *et al.* 2015) in response to flooding and may represent a means of eliminating potentially toxic ethylene gas resulting from rapid, flood-induced production of this hormone as well as promoting the movement of oxygen to roots (Blake and Reid 1981).

Oxygen uptake by seedlings during flooding may be enhanced by the growth of adventitious roots, for example, in *E. camaldulensis* (Sena Gomes and Kozlowski 1980; Van der Moezel *et al.* 1988; Argus *et al.* 2014) and *D. florulenta* (Capon *et al.* 2009). Development of aerenchymatous (i.e. porous) tissue also facilitates greater internal gas exchange during flooding. This trait has been observed in seedlings of *E. camaldulensis* (Van der Moezel *et al.* 1988; Argus *et al.* 2014) but not *E. largiflorens* (Heinrich 1990; McEvoy 1992, reported in Jolly and Walker 1996). Instead, *E. largiflorens* seedlings may tolerate flooding via a capacity to absorb oxygen at very low concentrations and low rates (Heinrich 1990; Jolly and Walker 1996; Akeroyd *et al.* 1998). Root aerenchyma development in *Casuarina* species is also thought to confer high waterlogging tolerance on this species (Van der Moezel 1988, 1989).

(v) *Ability to re-sprout after disturbances* – seedlings and juvenile plants are likely to be more susceptible to a range of disturbances than adult plants due to their smaller size. *Eucalyptus camaldulensis* and *Melaleuca leucadendra*, for example, both exhibit a high capacity for shoot re-sprouting after disturbance from a very young age (i.e. five days post-emergence of cotyledons) (Chong *et al.* 2007). Many *Eucalyptus* seedlings also develop a lignotuber, which enables older recruits to recover after drought or grazing (McEvoy 1992).

## Adult traits

Mature trees and shrubs in Australia's riverine habitats typically exhibit traits that facilitate some degree of tolerance to both flooding and drying. Mechanisms of stress tolerance and responses to disturbance vary widely between species, however. Riparian shrubs, for instance, tend to be long-lived stress tolerators, but generally do not persist in the face of frequent disturbance (Pettit *et al.* 2001; Capon 2003). Many common woody plants of riverine habitats (e.g. *E. camaldulensis* and *D. florulenta*) also display opportunistic and flexible patterns of growth that allow plants to capitalise on favourable conditions and build sufficient reserves to survive subsequent periods of stress or disturbance (Argent *et al.* 2004). Consequently, few species demonstrate regular annual growth patterns.

Because of their greater stature and strength, adult trees and shrubs in riverine habitats are exposed to less stress and disturbance in relation to flooding and drying than their juvenile counterparts. Mature woody stems, for example, are less vulnerable to mechanical damage by flood waters than stems of seedlings, and deeper more established root systems can better facilitate water uptake during dry periods. Older plants are also more likely to have established mycorrhizal associations (i.e. symbiotic fungi) that enhance the uptake of water and nutrients and thus facilitate better growth under stressful conditions. Nevertheless, both prolonged inundation and drought can lead to declining health and mortality of mature trees and shrubs (Froend and van der Moezel 1994; Thornton and Briggs 1994; Siebentritt *et al.* 2004).

Drought tolerance and avoidance among mature riverine trees and shrubs can be facilitated by a range of traits including low transpiration rates, low canopy area, and a pendulous habit such as that demonstrated by *E. largiflorens* (Jolly and Walker 1996). Many riparian and floodplain trees (e.g. *E. coolabah* and *E. largiflorens*) develop deep root systems and are able to utilise groundwater, even if this is somewhat saline (Jolly and Walker 1996; Payne *et al.* 2006). *Eucalyptus camaldulensis* in floodplain habitats, for example, can access permanent water tables more than 10 m below ground (Mensforth *et al.* 1994). Similarly, *Casuarina cunninghamiana* is likely to be able to access water up to 6–8 m deep while melaleucas may be limited to water in the upper few metres (O'Grady *et al.* 2006; Hubble *et al.* 2010). Some riverine trees can be opportunistic and flexible water users, capable of switching between groundwater and surface water depending on conditions and availability (Brock *et al.* 2006). The capacity to recover from drought, as well as other disturbances (e.g. fire), via epicormic growth is also exhibited by riverine eucalypts as well as in eucalypt species more generally (Bachelard 1969).

Flood tolerance among mature woody plants is facilitated by similar traits to those exhibited by seedlings, for example, development of aerenchyma and adventitious roots. Trees and shrubs growing in lotic habitats also tend to display traits associated with stem flexibility. Melaleucas, for example, often exhibit a multi-stemmed reclining or prostrate habit, with weeping foliage and thick, spongy bark that reduce shear stress and deflect currents, promoting the accumulation of sand and gravel in formations parallel to the direction of flow, which serves to anchor roots and lower trunks through partial burial (Fielding and Alexander 2001). Similarly, the thin, flexible foliage of *Casuarina cunninghamiana* presents little resistance to flows (Woolfrey and Ladd 2001).

## WOODY VEGETATION DYNAMICS

Hydrology has an overriding influence on the outcome of each life history stage of woody plants in riverine habitats. Consequently, the presence, abundance, population structure and condition of trees and shrubs in these environments reflect the interaction of hydrological patterns and the species traits that determine plant responses to these. In many Australian riverine habitats, variable hydrological regimes therefore result in highly dynamic populations and vegetation communities.

Flooding can trigger a range of responses among woody plants including reproduction, dispersal, germination, establishment and growth. Depending on the flood tolerance of species, stresses associated with flooding (e.g. soil anoxia) may also result in reduced health of trees and shrubs and eventual mortality. Among seedlings, flood-induced mortality may serve to thin dense thickets of regeneration and enable seedling release (George *et al.* 2005; Horner *et al.* 2010). Drying also causes visible changes in woody riparian vegetation, for example, the dramatic declines in floodplain tree health observed throughout the Murray–Darling Basin during the recent 'millennium drought' (Capon 2014). More subtle effects of drought, such as reductions in reproductive effort or recruitment, however, may take longer to manifest in the visible structure and composition of the vegetation.

The effects of wetting and drying on riverine trees and shrubs are strongly dependent on hydrological features of flow and flood pulses as well as the characteristics of the intervening dry periods. At a particular location, the timing, duration, depth and frequency of wetting can all play a role in determining plant and vegetation responses (Brock *et al.* 2006). The seasonal timing of floods in relation to cued seed release, for example, can influence the potential for germination to occur (Pettit and Froend 2001) while recession of flood waters during hot summer months may expose establishing seedlings to rapid drying, reducing their likelihood of survival (Dexter 1970). Inundation depth is particularly important with regards to seedling stages, with deeper floods more likely to drown young plants. Flood-induced mortality of less tolerant species (e.g. many riparian shrubs) may also increase with longer durations of flooding (Capon 2003).

Over longer time scales, flood history also shapes woody vegetation dynamics within particular riverine habitats. Patterns of wetting and drying over periods up to a decade can shape recruitment including reproductive effort, germination and establishment outcomes (George *et al.* 2005). Riparian eucalypts, for example, may germinate in response to flooding in one year but then require several years of shallow flooding and favourable rainfall to successfully establish (Roberts and Marston 2011). Effects may also be cumulative, for example, declining tree condition due to drought may result in lower reproductive effort and subsequent reductions in recruitment (George 2004). Because of highly variable hydrological regimes, woody plant recruitment in many Australian riverine landscapes is often episodic with large cohorts of seedlings establishing in response to wetter periods, but subsequently thinning during dry periods (Woolfrey and Ladd 2001; Roberts and Marston 2011). In more predictable climates, such as the Mediterranean climate of south-west Western Australia, riparian tree recruitment may be more continuous (Pettit and Froend 2001).

Geomorphological processes associated with flooding and drying (e.g. scouring, burial and sediment redistribution) also have a significant influence on woody vegetation dynamics in riverine landscapes, especially in lotic habitats (Hupp and Osterkamp 1996). Some riparian shrubs may be favoured by frequent fluvial disturbance, for instance, via gap creation (Wintle and Kirkpatrick 2007). Particular sediment characteristics may also promote the establishment and growth of some woody species. Seedling establishment in *Casuarina cunninghamiana*, for example, is limited by sandy substrates but favoured on cobble banks (Woolfrey and Ladd 2001).

Because of spatial differences in patterns of flooding and drying, the distribution and character of woody vegetation across riverine landscapes largely reflects flooding and associated fluvial processes. These factors also shape the distinction between upland and riparian vegetation. Riparian vegetation is typically differentiated from upland vegetation with respect to both composition and structure and, while not always more diverse, riparian vegetation typically contains different mixtures of species than that of neighbouring land systems (Sabo *et al.* 2005). Riparian vegetation also tends to display more structural complexity than neighbouring upland vegetation, for example, more mid-storey layers may be present (e.g. Hancock *et al.* 1996). In landscapes where soil moisture regimes are not strongly differentiated, such as in wetter parts of south-western Western Australia and the wet tropics of Queensland, there is generally less distinction between riparian and upland vegetation. In contrast, woody riparian vegetation in arid regions may be the only perennial vegetation present in the landscape (see Chapter 13).

Within riverine landscapes, the stability of sediments is an important influence on the distribution of trees and shrubs, with very frequently disturbed surfaces typically devoid of woody vegetation. On floodplains, long-lived trees tend to occur on the most stable surfaces, with shorter-lived trees and shrubs growing on intermediately disturbed surfaces (Friedel *et al.* 1993). Despite often requiring inundation to sustain different life stages, most Australian freshwater forest species are unable to withstand long periods of inundation, as prolonged periods of waterlogging usually inhibit seedling establishment of tall trees and shrubs (Specht 1990). The distribution of woody species across flood frequency or elevation gradients therefore reflects the responses of both juvenile and mature plants to wetting and drying. In much of the Murray–Darling Basin, for example, channel banks are lined with *E. camaldulensis* which then grades into either *E. largiflorens* or *E. coolabah* on the floodplain as the flood frequency declines (see Chapter 10). In Australia's tropical monsoonal floodplains, major tree species exhibit clumped distributions across an elevation gradient with the greatest species richness occurring at the dry extreme (Bowman and McDonough 1991).

Australia's riverine forests tend to occur in frequently flooded but well drained riverine habitats. Woody vegetation structure tends to become more open in places that experience regular waterlogging, with woodlands typically occupying infrequently flooded riverine habitats (Specht 1990). In some riverine habitats, shrubs may comprise the dominant life form in terms of both species diversity and cover (e.g. Pettit *et al.* 2001). Shrublands can occupy both swampy, water-ponding habitats, as is the case with much *D. florulenta* shrubland, or rarely flooded, marginal floodplain areas. The latter often support mixed chenopod or samphire communities (Capon 2005). Riverine shrublands range from dense thickets to sparse open shrublands and the density of shrubs tends to vary in relation to flooding, often declining with decreasing flood frequency (Pettit *et al.* 2001; Capon *et al.* 2009).

Flood history also shapes the population structure, as well as the composition and density of woody vegetation in riverine habitats. Where variable hydrology results in episodic recruitment, populations often exhibit distinct peaks in mature tree size classes in relation to flood history (George *et al.* 2005). Patterns of tree establishment can also vary spatially, for example, along elevation gradients (e.g. Pettit and Froend 2001) or the watermark of receding floodwaters (Dexter 1970), or reflecting geomorphic features. In central Australia, for example, Roberts (1993) observed concentrations of young *E. coolabah* on top of steep channel banks (as well as sand mounds).

Within species, individual trees can also exhibit significantly different morphologies in relation to vegetation density and therefore, indirectly, hydrology. Trees in riverine forests, for example, tend to have straight boles and high branches while those in woodlands display an open spreading shape (Brooker and Kleinig 1999). The morphology of riverine shrubs (e.g. *D. florulenta*) can also vary with respect to plant density and flooding (Capon *et al.* 2009).

## ECOSYSTEM SERVICES

Woody vegetation is crucial to the ecological functioning of Australia's riverine landscapes and the catchments within which they are situated, providing a wide range of highly valued goods and services to people. Trees and shrubs represent critical structural habitat for both terrestrial and aquatic fauna as well as a source of food (e.g. leaves, fruit, seed) for many species. Woody vegetation can also influence habitat quality indirectly by altering microclimates (e.g. shading) or supplying wood and litter. Often containing the only continuous patches of woody vegetation in many landscapes, riparian zones offer important corridors for wildlife movement. Scattered trees in modified landscapes, however, are also disproportionately significant in terms of their influence on ecological structure and function (Fischer *et al.* 2010). Such habitat values of woody riparian vegetation are likely to be increasingly important as climate change escalates (Capon *et al.* 2013).

Riverine trees and shrubs can be involved in many critical physical and chemical ecological processes both locally and, due to their topographic position, at a catchment or landscape scale (e.g. water filtration, nutrient cycling). Even scattered trees, for example, can result in local increases in soil nutrient levels (Wilson 2002; Manning *et al.* 2006). Woody plants are particularly influential on the physical form of riverine landscapes and can play a key role in shaping channels and in-channel features (Tooth and Nanson 2000). Tree roots, for example, can minimise channel bank erosion due to mass failure (Hubble *et al.* 2010).

Woody vegetation in riverine landscapes has many use and non-use values for people, supplying a wide range of provisional and cultural ecosystem goods and services (Millennium Ecosystem Assessment 2005). Indigenous Australians have a long history of harvesting wood from riparian species for firewood and the production of artefacts such as boomerangs, canoes, spears, jewellery and ornaments. The bark, sap, fruit, seeds and leaves of some species also provide a food source or medicinal ingredients, and branches and bark (e.g. of river red gum) can be used in the construction of shelters or as floor mats. The roots of some plants (e.g. *E. coolabah*) may even provide a critical water source in drier regions (Latz 1999).

Timber harvesting continues to be an important use of many floodplain forests in the Murray–Darling Basin (Di Stefano 2002) and numerous Australian riparian species (e.g. river red gum) have become commercially significant on a global scale (Boland *et al.* 2006). Many cultural, aesthetic and recreational values of Australia's riverine landscapes also rely on the presence and health of trees and shrubs. The 'dig tree' (*E. coolabah*) on the banks of the Cooper Creek in south-west Queensland, for example, represents an iconic historical memorial of the ill-fated Burke and Wills expedition to traverse the continent (Murgatroyd 2002).

## CONCLUSIONS

Australia's woody riverine vegetation is an essential element of many of the continent's most iconic landscapes. From the majestic river red gum forests of the Murray River floodplains to wet tropic gallery rainforests and the scattered coolibahs of the arid interior, trees and shrubs in Australian riverine landscapes comprise a diversity of life forms and vegetation communities that are characterised by high levels of temporal and spatial heterogeneity. Variable patterns of wetting and drying are a major determinant of the species pools present in these systems and the range of plant traits that enable species to survive often harsh conditions and capitalise on times of plenty. The character and condition of forests, woodlands and shrublands across riverine landscapes map complex histories of flooding while simultaneously interacting with the landscape to shape its physical form and determine future hydrological regimes. Often distinctive at a catchment level, woody riverine vegetation supports a wide diversity of ecosystem goods and services and is critical to ecological functioning at multiple scales. Despite the importance and ubiquity of many riparian tree and shrub species, however, there are substantial knowledge gaps surrounding the biology and ecology of Australia's woody riverine vegetation. The conservation and management of these significant plants into the future depends on research that addresses these gaps as well as greater recognition of the significance of riverine forests, woodlands and shrublands to both ecological health and human well-being.

## ACKNOWLEDGEMENTS

The authors acknowledge the support of the Cotton Catchment Communities Cooperative Research Centre and the Cotton Research and Development Corporation, who both funded research projects that enabled significant portions of this review to be conducted.

## REFERENCES

Akeroyd MD, Tyerman SD, Walker GR, Jolly ID (1998) Impact of flooding on the water use of semi-arid riparian eucalypts. *Journal of Hydrology* **206**, 104–117. doi:10.1016/S0022-1694(98)00092-4

Argent RM, McMahon TA, Bowler JM, Finlayson BL (2004) The dendroecological potential of *Eucalyptus camaldulensis* Dehnhardt (river red gum) from the Barmah Forest, Victoria, Australia. *Australian Geographical Studies* **42**, 89–102. doi:10.1111/j.1467-8470.2004.00245.x

Argus RE, Colmer TD, Grierson PF (2015) Early physiological flood tolerance is followed by slow post-flooding root recovery in the dryland riparian tree *Eucalyptus camaldulensis* subsp. *refulgens*. *Plant, Cell & Environment* **38**, 1189–1199. doi:10.1111/pce.12473

Arthington AH, Mackay SJ, James CS, Rolls RJ, Sternberg D, Barnes A, Capon SJ (2012) 'Ecologic al-limits-of-hydrologic-alteration: a test of the ELOHA framework in south-east Queensland.' Waterlines report series 75. National Water Commission, Canberra.

Bachelard EP (1969) Studies on the formation of epicormic shoots on eucalypt stem segments. *Australian Journal of Biological Sciences* **22**, 1291–1296.

Barlow BA (1994) Phytogeography of the Australian region. In *Australian Vegetation.* 2nd edn (Groves edn). Cambridge University Press, Melbourne.

Baskin CC, Baskin JM (2001) *Seeds: Ecology, Biogeography, and Evolution of Dormancy and Germination*. Elsevier Academic Press, San Diego, CA.

Beadle NCW (1952) Studies in halophytes. I. The germination of the seed and establishment of the seedling of five species of *Atriplex* in Australia. *Ecology* **33**, 49–62. doi:10.2307/1931251

Blake TJ, Reid DM (1981) Ethylene, water relations and tolerance to waterlogging of three *Eucalyptus* species. *Australian Journal of Plant Physiology* **8**, 497–505. doi:10.1071/PP9810497

Boland DJ, Brooker MIH, Chippendale GM, Hall N, Hyland BPM, Johnston RD, Kleinig DA, McDonald MW, Turner JD (Eds) (2006) *Forest Trees of Australia*. CSIRO Publishing, Melbourne.

Boomsma CD (1950) The red gum (*E. camaldulensis* (Dehn)) association of Australia. *Australian Forestry* **14**, 99–110. doi:10.1080/00049158.1950.10675787

Bowman DMJS (2000) *Australia Rainforests: Islands of Green in a Land of Fire*. Cambridge University Press, Cambridge.

Bowman DMJS, McDonough L (1991) Tree species distribution across a seasonally flooded elevation gradient in the Australian monsoon tropics. *Journal of Biogeography* **18**, 203–212. doi:10.2307/2845293

Bowman DMJS, Woinarski JCZ (1994) Biogeography of Australian monsoon rainforest mammals: implications for the conservation of rainforest mammals. *Pacific Conservation Biology* **1**, 98–106.

Boxshall DJ, Jenkyn T (2001) *Eumong – River Cooba,* Acacia stenophylla. *Farm Forestry Species Profile for North Central Victoria*. Department of Primary Industries, Victoria.

Brock M, Capon S, Porter J (2006) Disturbance of plant communities dependent on desert rivers. In *Ecology of Desert Rivers*. (Ed. RT Kingsford) pp. 100–132. Cambridge University Press, Cambridge, UK.

Brooker MIH, Kleinig DA (1999) *Field Guide to Eucalypts: South-Eastern Australia*. Bloomings Books, Hawthorn, Victoria.

Brown JR, Archer S (1990) Water relations of a perennial grass and seedlings vs adult woody plants in a subtropical savanna, Texas. *Oikos* **57**, 366–374. doi:10.2307/3565966

Capon SJ (2003) Plant community responses to wetting and drying in a large arid floodplain. *River Research and Applications* **19**, 509–520. doi:10.1002/rra.730

Capon SJ (2005) Flood variability and spatial variation in plant community composition and structure on a large arid floodplain. *Journal of Arid Environments* **60**, 283–302. doi:10.1016/j.jaridenv.2004.04.004

Capon SJ (2007) Effects of flooding on seedling emergence from the soil seed bank of a large desert floodplain. *Wetlands* **27**, 904–914. doi:10.1672/0277-5212(2007)27[904:EOFOSE]2.0.CO;2

Capon S (2014) Adapting water resources management to drought and water scarcity in Australia's Murray–Darling Basin. In *Adaptation to Climate Change Through Water Resources Management: Capacity, Equity and Sustainability*. (Eds D Stucker and E Lopez-Gunn) pp. 110–133. Routledge, New York, NY.

Capon SJ, Dowe JL (2007) Diversity and dynamics of riparian vegetation. In *Principles for Riparian Lands Management*. (Eds S Lovett and P Price) pp. 13–32. Land and Water, Canberra.

Capon SJ, James CS, Williams L, Quinn GP (2009) Responses to flooding and drying in seedlings of a common Australian desert floodplain shrub: *Muehlenbeckia florulenta* Meisn. (tangled lignum). *Environmental and Experimental Botany* **66**, 178–185. doi:10.1016/j.envexpbot.2009.02.012

Capon SJ, Chambers LE, Mac Nally R, Naiman RJ, Davies P, Marshall N, Pittock J, Reid M, Capon T, Douglas M, Catford J, Baldwin DS, Stewardson M, Roberts J, Parsons M, Williams SE (2013) Riparian ecosystems in the 21st century: hotspots for climate change adaptation? *Ecosystems* **16**, 359–381. doi:10.1007/s10021-013-9656-1

Chaves MM, Maroco JP, Pereira JS (2003) Understanding plant responses to drought – from genes to the whole plant. *Functional Plant Biology* **30**, 239–264. doi:10.1071/FP02076

Chong C, Walker KF (2005) Does lignum rely on a soil seed bank? Germination and reproductive phenology of *Muehlenbeckia florulenta* (Polygonaceae). *Australian Journal of Botany* **53**, 407–415. doi:10.1071/BT04130

Chong C, Edwards W, Waycott M (2007) Differences in resprouting ability are not related to seed size or seedling growth in four riparian woody species. *Journal of Ecology* **95**, 840–850. doi:10.1111/j.1365-2745.2007.01241.x

Christensen PES, Kimber PC (1975) Effect of prescribed burning on the flora and fauna of south-west Australian forests. *Proceedings of the Ecological Society of Australia* **9**, 85–106.

Clucas RD, Ladiges PY (1979) Variations in populations of *Eucalyptus ovata* Labill., and the effects of waterlogging on seedling growth. *Australian Journal of Botany* **27**, 301–315. doi:10.1071/BT9790301

Cole MM (1986) *The Savannas: Biogeography and Geobotany.* Academic Press, London.

Cooper DJ, Merritt DM, Andersen DC, Chimner RA (1999) Factors controlling the establishment of Fremont cottonwood seedlings on the upper Green River, USA. *Regulated Rivers: Research and Management* **15**, 419–440. doi:10.1002/(SICI)1099-1646(199909/10)15:5<419::AID-RRR555>3.0.CO;2-Y

Cremer KW (1965) How eucalypt forests release their seed. *Australian Journal of Botany* **13**, 11–16. doi:10.1071/BT9650011

Cronk JK, Fennessy MS (2001) *Wetland Plants: Biology and Ecology*. CRC press, Boca Raton, FL.

Dexter BD (1970) *Regeneration of River Red Gum* Eucalyptus camaldulensis *Dehnh*. University of Melbourne, Melbourne.

Dexter BD (1978) Silviculture of the river red gum forests of the central Murray floodplain. *Proceedings of the Royal Society of Victoria* **90**, 175–194.

Di Stefano J (2002) River red gum (*Eucalyptus camaldulensis*): a review of ecosystem processes, seedling regeneration and silvicultural practice. *Australian Forestry* **65**, 14–22. doi:10.1080/00049158.2002.10674848

Espelta JM, Riba M, Javier R (1995) Patterns of seedling recruitment in west-Mediterranean *Quercus ilex* forest influenced by canopy development. *Journal of Vegetation Science* **6**, 465–472. doi:10.2307/3236344

Farrell RCC, Bell DT, Akilan K, Marshall JK (1996) Morphological and physiological comparisons of clonal lines of *Eucalyptus camaldulensis*: responses to drought and waterlogging. *Australian Journal of Plant Physiology* **23**, 497–507. doi:10.1071/PP9960497

Fielding CR, Alexander J (1996) Sedimentology of the upper Burdekin River of north Queensland, Australia – an example of a tropical, variable discharge river. *Terra Nova* **8**, 447–457. doi:10.1111/j.1365-3121.1996.tb00770.x

Fielding CR, Alexander J (2001) Fossil trees in ancient fluvial channel deposits: evidence of seasonal and longer-term climatic variability. *Palaeogeography, Palaeoclimatology, Palaeoecology* **170**, 59–80. doi:10.1016/S0031-0182(01)00227-9

Finlayson CM (2005) Plant ecology of Australia's tropical floodplain wetlands: a review. *Annals of Botany* **96**, 541–555. doi:10.1093/aob/mci209

Fischer J, Stott J, Law BS (2010) The disproportionate value of scattered trees. *Biological Conservation* **143**, 1564–1567. doi:10.1016/j.biocon.2010.03.030

Friedel MH, Pickup G, Nelson DJ (1993) The interpretation of vegetation change in a spatially and temporally diverse arid Australian landscape. *Journal of Arid Environments* **24**, 241–260. doi:10.1006/jare.1993.1022

Froend RH, van der Moezel PG (1994) The impact of prolonged flooding on the vegetation of Coomalbidgup Swamp. *Journal of the Royal Society of Western Australia* **77**, 15–22.

George AK (2004) Eucalypt regeneration on the lower River Murray floodplain, South Australia. PhD thesis. The University of Adelaide, SA.

George AK, Walker KF, Lewis MM (2005) Population status of eucalypt trees on the River Murray floodplain, South Australia. *River Research and Applications* **21**, 271–282. doi:10.1002/rra.846

Gibson A, Bachelard EP (1994) Relationships between site characteristics and survival strategies of *Eucalyptus camaldulensis* seedlings [northern Australia]. International Workshop, Zhangzhou. Australian Centre for International Agricultural Research, Canberra, ACT (Australia), Fujian Province (China).

Gibson A, Bachelard EP, Hubick KT (1994) Growth strategies of *Eucalyptus camaldulensis* Dehnh at three sites in northern Australia. *Australian Journal of Plant Physiology* **21**, 653–662. doi:10.1071/PP9940653

Gibson MR, Richardson DM, Marchante E, Marchante H, Rodger JG, Stone GN, Byrne M,

Fuentes-Ramírez A, George N, Harris C, Johnson SD, Le Roux JJ, Miller JT, Murphy DJ, Pauw A, Prescott MN, Wandrag EM, Wilson JRU (2011) Reproductive biology of Australian *Acacia* species: important mediator of invasiveness? *Diversity & Distributions* **17**, 911–933. doi:10.1111/j.1472-4642.2011.00808.x

Gillison AN (1994) Woodlands. In *Australian Vegetation.* (Ed. RH Groves) pp. 227–256. Cambridge University Press, Cambridge, UK.

Gunn B (2001) *Australian Tree Seed Centre: Operations Manual.* CSIRO Forestry and Forestry Products, Canberra.

Hamilton-Brown S, Boon PI, Robinson R, Morris K, Raulings EJ (2009) Aerial seed storage in *Melaleuca ericifolia* Sm. (swamp paperbark): environmental triggers for seed release. *Hydrobiologia* **620**, 121–133. doi:10.1007/s10750-008-9623-6

Hancock CN, Ladd PG, Froend RH (1996) Biodiversity and management of riparian vegetation in Western Australia. *Forest Ecology and Management* **85**, 239–250. doi:10.1016/S0378-1127(96)03761-9

Heinrich P (1990) 'The eco-physiology of riparian river red gum (*Eucalyptus camaldulensis*).' Final report, Australian Water Resources Advisory Council, The University of Melbourne, Melbourne.

Horner GJ, Baker PJ, Nally RM, Cunningham SC, Thomson JR, Hamilton F (2010) Forest structure, habitat and carbon benefits from thinning floodplain forests: managing early stand density makes a difference. *Forest Ecology and Management* **259**, 286–293. doi:10.1016/j.foreco.2009.10.015

Horton JL, Clark JL (2001) Water table decline alters growth and survival of *Salix gooddingii* and *Tamarix chinensis* seedlings. *Forest Ecology and Management* **140**, 239–247. doi:10.1016/S0378-1127(00)00314-5

House SM (1997) Reproductive biology of eucalypts. In *Eucalypt Ecology: Individuals to Ecosystems.* (Eds J Williams and J Woinarski) pp. 30–55. Cambridge University Press, Cambridge, UK.

Hubble TCT, Docker BB, Rutherfurd ID (2010) The role of riparian trees in maintaining riverbank stability: a review of Australian experience and practice. *Ecological Engineering* **36**, 292–304. doi:10.1016/j.ecoleng.2009.04.006

Hughes FMR (1990) The influence of flooding regimes on forest distribution and composition in the Tana River floodplain, Kenya. *Journal of Applied Ecology* **27**, 475–491. doi:10.2307/2404295

Hunt LP (2001) Low seed availability may limit recruitment in grazed *Atriplex vesicaria* and contribute to its local extinction. *Plant Ecology* **157**, 53–67. doi:10.1023/A:1014515122231

Hupp CR, Osterkamp WR (1996) Riparian vegetation and fluvial geomorphic processes. *Geomorphology* **14**, 277–295. doi:10.1016/0169-555X(95)00042-4

Jensen AE, Walker KF, Paton DC (2008) The role of seedbanks in restoration of floodplain woodlands. *River Research and Applications* **24**, 632–649. doi:10.1002/rra.1161

Johns C, Reid CJ, Roberts J, Sims N, Doody T, Overton I, McGinness HM, Rogers K, Campbell C, Gawne B (2009) 'Native trees of the River Murray floodplain: literature review and experimental designs to examine effects of flow enhancement and floodwater retention.' Final report for the MDBA, Murray–Darling Freshwater Research Centre, Wodonga.

Jolly ID, Walker GR (1996) Is the field water use of *Eucalyptus largiflorens* F. Muell. affected by short-term flooding? *Australian Journal of Ecology* **21**, 173–183. doi:10.1111/j.1442-9993.1996.tb00598.x

Jongejans E, Soons MB, de Kroon H (2006) Bottlenecks and spatiotemporal variation in the sexual reproduction pathway of perennial meadow plants. *Basic and Applied Ecology* **7**, 71–81. doi:10.1016/j.baae.2005.04.004

Khan MAW (1965) Flowering seedlings of *Eucalyptus camaldulensis* and *E. rudis*. *Indian Forester* **9**, 203.

Latz P (1999) *Pocket Bushtucker; a Field Guide to the Plants of Central Australia and Their Traditional Uses.* Jukurrpa Books, Alice Springs, NT.

Lehmann CER, Prior LD, Bowman DMJS (2009) Fire controls population structure in four dominant tree species in a tropical savanna. *Oecologia* **161**, 505–515. doi:10.1007/s00442-009-1395-9

Lemcoff JH, Guarnaschelli AB, Garau AM, Bascialli ME, Ghersa CM (1994) Osmotic adjustment and its use as a selection criterion in *Eucalytpus* seedlings. *Canadian Journal of Forest Research* **24**, 2404–2408. doi:10.1139/x94-310

Lemcoff JH, Guarnaschelli AB, Garau AM, Prystupa P (2002) Elastic and osmotic adjustments in rooted cuttings of several clones of *Eucalyptus camaldulensis* Dehnh. from southeastern Australia after a drought. *Flora* **197**, 134–142. doi:10.1078/0367-2530-00023

Letnic M, Dickman CR, McNaught G (2000) Bet-hedging and germination in the Australian arid zone shrub *Acacia ligulata. Austral Ecology* **25**, 368–374. doi:10.1046/j.1442-9993.2000.01047.x

Li CY (1998) Some aspects of leaf water relations in four provenances of *Eucalyptus microtheca* seedlings. *Forest Ecology and Management* **111**, 303–308. doi:10.1016/S0378-1127(98)00336-3

Chunyang L, Tuomela K (1997) Differentiation of seed germination and early seedling growth in ten provenances of *Eucalyptus microtheca. Journal of Forest Research* **8**, 13–19. doi:10.1007/BF02864932

Lloyd DG (1979) Some reproductive factors affecting the selection of self-fertilization in plants. *American Naturalist* **113**, 67–69. doi:10.1086/283365

Lloyd DG (1992) Self- and cross-fertilization in plants. II. The selection of self-fertilization. *International Journal of Plant Sciences* **153**, 370–380. doi:10.1086/297041

Mahoney JM, Rood SB (1998) Streamflow requirements for cottonwood seedling recruitment – an integrative model. *Wetlands* **18**, 634–645. doi:10.1007/BF03161678

Manning AD, Fischer J, Lindenmayer DB (2006) Scattered trees are keystone structures – implications for conservation. *Biological Conservation* **132**, 311–321. doi:10.1016/j.biocon.2006.04.023

McDonald MW, Bell JC, Butcher PA (1996) Effect of seed collection strategies on capturing genetic diversity in *Eucalyptus camaldulensis*. In *Innovations in Tropical Tree Seed Technology, Proceedings of the IUFRO Symposium of Project Group P.2.04.00 'Seed Problems'*. 7–10 September 1995, Arusha, Tanzania. (Ed. K Olesen) pp. 166–174. International Union of Forestry Research Organisations, Wien, Austria.

McEvoy PK (1992) Ecophysiology of three *Eucalyptus* species on the River Murray floodplain. Master's thesis. University of Melbourne, Melbourne.

Mensforth LJ, Thorburn PJ, Tyerman SD, Walker GR (1994) Sources of water used by riparian *Eucalyptus camaldulensis* overlying highly saline groundwater. *Oecologia* **100**, 21–28. doi:10.1007/BF00317126

Merchant A, Tausz M, Arndt SK, Adams MA (2006) Cyclitols and carbohydrates in leaves and roots of 13 *Eucalyptus* species suggest contrasting physiological responses to water deficit. *Plant, Cell & Environment* **29**, 2017–2029. doi:10.1111/j.1365-3040.2006.01577.x

Merchant A, Callister A, Arndt S, Tausz M, Adams M (2007) Contrasting physiological responses of six *Eucalyptus* species to water deficit. *Annals of Botany* **100**, 1507–1515. doi:10.1093/aob/mcm234

Meskimen GF (1962) *A Silvical Study of the Melaleuca Tree in South Florida*. University of Florida, Gainesville, FL.

Metzger JP, Bernacci LC, Goldenberg R (1997) Pattern of tree species diversity in riparian forest fragments of different widths (SE Brazil). *Plant Ecology* **133**(2), 135–152. doi:10.1023/A:1009791831294

Millennium Ecosystem Assessment (2005) *Millennium Ecosystem Assessment Findings*. Millennium Ecosystem Assessment. Island Press, Washington, DC .

Murgatroyd S (2002) *The Dig Tree*. Bloomsbury, London.

O'Grady AP, Eamus D, Cook PG, Lamontagne S (2006) Comparative water use by the riparian trees *Melaleuca argentea* and *Corymbia bella* in the wet-dry tropics of northern Australia. *Tree Physiology* **26**, 219–228. doi:10.1093/treephys/26.2.219

Padilla FM, Pugnaire FI (2007) Rooting depth and soil moisture control Mediterranean woody seedling survival during drought. *Functional Ecology* **21**, 489–495. doi:10.1111/j.1365-2435.2007.01267.x

Parsons RF (1994) *Eucalyptus* scrubs and shrublands. In *Australian Vegetation*. 2nd edn. (Ed. RH

Groves) pp. 291–319. Cambridge University Press, Cambridge, UK.

Patterson B, Vaillancourt RE, Potts BM (2001) Eucalypt seed collectors: beware of sampling seedlots from low canopy! *Australian Forestry* **64**, 139–142. doi:10.1080/00049158.2001.10676178

Payne EGI, Costelloe JF, Woodrow IE, Irvine EC, Western AW, Herczeg AL (2006) Riparian tree water use by *Eucalyptus coolabah* in the Lake Eyre Basin. In *30th Hydrology and Water Resources Symposium: Past, Present and Future.* pp. 692–697. Engineers Australia, Canberra.

Pereira JS, Kozlowski TT (1976) Leaf anatomy and water relations of *Eucalyptus camaldulensis* and *E. globulus* seedlings. *Canadian Journal of Botany* **54**, 2868–2880.

Pettit NE, Froend RH (2001) Availability of seed for recruitment of riparian vegetation: a comparison of a tropical and a temperate river ecosystem in Australia. *Australian Journal of Botany* **49**, 515–528. doi:10.1071/BT00059

Pettit NE, Froend RH, Davies PM (2001) Identifying the natural flow regime and the relationship with riparian vegetation for two contrasting western Australian rivers. *Regulated Rivers: Research and Management* **17**, 201–215. doi:10.1002/rrr.624

Pettit N, Townsend S, Dixon I, Wilson D (2011) Plant communities of aquatic and riverine habitats. In *Aquatic Biodiversity in Northern Australia: Patterns, Threats and Future.* (Ed. B Pusey) pp. 37–50. Charles Darwin University Press, Darwin.

Pryor LD (1976) *Biology of Eucalypts.* Edward Arnold, London.

Pudney S (1998) Seed dispersal in *Eucalyptus camaldulensis*: implications for natural regeneration. Honours thesis. The University of Adelaide, SA.

Roberts J (1993) Regeneration and growth of coolibah, *Eucalyptus coolabah* subsp. *arida*, a riparian tree, in the Cooper Creek region of South Australia. *Australian Journal of Ecology* **18**, 345–350. doi:10.1111/j.1442-9993.1993.tb00461.x

Roberts J, Marston F (2011) *Water Regime for Wetland and Floodplain Plants: a Source Book for the Murray–Darling Basin.* National Water Commission, Canberra.

Rogers K, Ralph TJ (Eds) (2011) Floodplain wetland biota in the Murray–Darling Basin: water and habitat requirements. CSIRO Publishing, Melbourne.

Sabo JL, Sponseller R, Dixon M, Gade K, Harms T, Heffernan J, Jani A, Katz G, Soykan C, Watts J, Welter J (2005) Riparian zones increase regional species richness by harboring different, not more, species. *Ecology* **86**, 56–62. doi:10.1890/04-0668

Sampson JF, Byrne M (2012) Genetic diversity and multiple origins of polyploidy *Atriplex nummularia* Lindl. (Chenopodiaceae). *Biological Journal of the Linnaean Society* **105**, 218–230. doi:10.1111/j.1095-8312.2011.01787.x

Schnitzler A, Hale BW, Alsum EM (2007) Examining native and exotic species diversity in European riparian forests. *Biological Conservation* **138**, 146–156. doi:10.1016/j.biocon.2007.04.010

Schaffer WM (1974) Selection for optimal life histories: the effects of age structure. *Ecology* **55**, 291–303.

Sena Gomes AR, Kozlowski TT (1980) Effects of flooding of *Eucalyptus camaldulensis* and *Eucalyptus globulus* seedlings. *Oecologia* **46**, 139–142. doi:10.1007/BF00540117

Serbesoff-King K (2003) *Melaleuca* in Florida: a literature review on the taxonomy, distribution, biology, ecology, economic importance and control measures. *Journal of Aquatic Plant Management* **41**, 98–112.

Siebentritt MA, Ganf GG, Walker KF (2004) Effects of an enhanced flood on riparian plants of the River Murray, South Australia. *River Research and Applications* **20**, 765–774. doi:10.1002/rra.794

Specht RL (1990) Forested wetlands in Australia. In *Ecosystems of the World.* (Eds AE Lugo, M Brinson, S Brown) pp. 387–406. Elsevier, Amsterdam.

Stevens JC, Barrett-Lennard EG, Dixon KW (2006) Enhancing the germination of three fodder shrubs (*Atriplex amnicola, A. nummularia, A. undulata*; Chenopodiaceae): implications for the optimisation of field establishment. *Australian Journal of Agricultural Research* **57**, 1279–1289. doi:10.1071/AR06031

Streng DR, Glitzenstein JS, Harcombe PA (1989) Woody seedling dynamics in an east Texas floodplain forest. *Ecological Monographs* **59**, 177–204. doi:10.2307/2937285

Susiluoto S, Berninger F (2007) Interactions between morphological and physiological drought responses in *Eucalyptus microtheca*. *Silva Fennica* **41**, 221–233. doi:10.14214/sf.292

Thornton SA, Briggs SV (1994) A survey of hydrological changes to wetlands of the Murrumbidgee River. *Wetlands* **13**, 1–13.

Tooth S, Nanson GC (2000) The role of vegetation in the formation of anabranching channels in an ephemeral river, northern plains, arid central Australia. *Hydrological Processes* **14**, 3099–3117. doi:10.1002/1099-1085(200011/12)14:16/17<3099::AID-HYP136>3.0.CO;2-4

Tooth S, Jansen JD, Nanson GC, Coulthard TJ, Pietsch T (2008) Riparian vegetation and the late Holocene development of an anabranching river: Magela Creek, northern Australia. *Geological Society of America Bulletin* **120**(7, 8), 1021–1035. doi:10.1130/B26165.1

Tuomela K (1997) Leaf water relations in six provenances of *Eucalyptus microtheca*: a greenhouse experiment. *Forest Ecology and Management* **92**, 1–10. doi:10.1016/S0378-1127(96)03961-8

Tybirk K (1997) 'Reproductive biology and evolution of the genus *Acacia*.' Bulletin no. 20 of the International Group for the Study of Mimosoideae – world wide wattle.

Van der Moezel PG, Watson LE, Pearce-Pinto GVN, Bell DT (1988) The response of six *Eucalyptus* species and *Casuarina obesa* to the combined effect of salinity and waterlogging. *Australian Journal of Plant Physiology* **15**, 465–474. doi:10.1071/PP9880465

Van der Moezel PG, Walton CS, Pearce-Pinto GVN, Bell DT (1989) Screening for salinity and waterlogging tolerance in five *Casuarina* species. *Landscape and Urban Planning* **17**, 331–337. doi:10.1016/0169-2046(89)90087-X

van Steenis CGGJ (1981) *Rheophytes of the World*. Sijthoff & Noordhoff, Alphen aan den Rijn, Netherlands.

Venable DL (2007) Bet hedging in a guild of desert annuals. *Ecology* **88**, 1086–1090. doi:10.1890/06-1495

Warrag EI, Eltigani MA (2005) Breaking seed coat dormancy of *Acacia nilotica* seeds under simulated natural habitat conditions in Sudan. *Tropical Ecology* **46**, 127–132.

White DA, Turner NC, Galbraith JH (2000) Leaf water relations and stomatal behavior of four allopatric *Eucalyptus* species planted in Mediterranean southwest Australia. *Tree Physiology* **20**, 1157–1165. doi:10.1093/treephys/20.17.1157

Wilson B (2002) Influence of scattered paddock trees on surface soil properties: a study of the Northern Tablelands of NSW. *Ecological Management & Restoration* **3**, 211–219. doi:10.1046/j.1442-8903.2002.00115.x

Wintle BC, Kirkpatrick JB (2007) The response of riparian vegetation to flood-maintained habitat heterogeneity. *Austral Ecology* **32**, 592–599. doi:10.1111/j.1442-9993.2007.01753.x

Woods SR, Archer SR, Schwinning S (2014) Seedling responses to water pulses in shrubs with contrasting histories of grassland encroachment. *PLoS ONE* **9**(1), e87278. doi:10.1371/journal.pone.0087278

Woolfrey AR, Ladd PG (2001) Habitat preference and reproductive traits of a major Australian riparian tree species (*Casuarina cunninghamiana*). *Australian Journal of Botany* **49**, 705–715. doi:10.1071/BT01009

# Section 3

Riverine vegetation

9

# Vegetation of coastal floodplains and wetlands in south-eastern Australia

*Paul Boon, David Keith and Elisa Raulings*

## INTRODUCTION

The coastline of south-eastern Australia between northern New South Wales and western Victoria supports a wide range of wetlands with diverse vegetation communities associated with estuaries, their inland river channels and floodplains. Beds of submerged angiosperms (i.e. seagrasses) occur in the subtidal and intertidal zones of many estuaries in this region (Coles *et al.* 2003). At slightly higher elevations, but still subject to daily inundation by tides, mangrove shrublands or forests are often present around estuaries, mostly on muddy substrata but occasionally on rocky substrata derived from basalt or granite (Duke 2006). Slightly higher, and subject to less frequent tidal inundation, are coastal salt marshes, areas of land that experience recurrent but infrequent low-energy tidal inundation and are vegetated by low-growing vascular plants, especially succulent chenopods and salt-tolerant grasses. In addition to these two well-known classes of intertidal wetland there are a variety of other types of coastal, but largely non-tidal, wetlands. They tend to occur in slightly more elevated areas, hence the lack of regular tidal inundation. They are, however, almost always exposed to various amounts of freshwater, either as surface runoff or as groundwater subsidies. These supratidal wetlands are often dominated by rushes, sedges or reeds, such as *Juncus kraussii, Gahnia trifida, Baumea juncea* and *Phragmites australis*, or by grasses such as *Paspalum distichum, Poa labillardierei, Themeda triandra, Austrodanthonia* spp., and *Distichlis distichophylla*. Woodlands of *Melaleuca* spp. (paperbarks), *Leptospermum lanigerum* (woolly tea tree) and *Casuarina glauca* (swamp she-oak) dominate wetlands that experience brackish-water influences on the lower parts of coastal floodplains, with the importance of each type varying geographically across south-eastern Australia. The highest elevations of coastal floodplains are influenced only by freshwater flows, and in these areas forested wetlands dominated by tall eucalypt species inter-grade with the terrestrial vegetation of the hinterland.

While there is now widespread recognition that wetlands provide valuable ecosystem services (Woodward and Yui 2001; Bateman *et al.* 2011), coastal wetlands in particular provide critical foraging and nursery habitat for many commercially and recreationally important species of fish, crustaceans and molluscs, and are critical to Australia's fishing industries (Creighton *et al.* 2015). Coastal wetlands are also important feeding, breeding and roosting areas for waterbirds, including many migratory species listed under treaties such as the *Japan–Australia Migratory Bird Agreement* (JAMBA), the *China–Australia Migratory Bird Agreement* (CAMBA) and the *Convention on the Conservation of Migratory Species of Wild Animals* (Bonn Convention). These treaties require that the habitats of listed taxa be protected, conserved, secured and restored. Even so, the degree to which adequate protection is afforded to coastal wetlands varies greatly across states and wetland types in south-eastern Australia (Boon *et al.* 2015). About one-third of Australia's Ramsar-listed wetlands are in the coastal south-east of the country, and in total they cover an area of over 420 000 ha (Table 9.1).

Despite the importance of coastal wetlands for the provision of ecosystem services, other than mangroves relatively few types have been subject to detailed study in Australia. Mangroves are well covered in the literature (e.g. Macnae 1966; Lear and Turner 1977; Saenger *et al.* 1977; Clough 1982; Hutchings and Recher 1982; Hutchings and Saenger 1987; Robertson and Alongi 1992; Claridge and Burnett 1993; Harty 1997; Duke 2006; Morrisey *et al.* 2010), whereas most other types of coastal wetland have been neglected by coastal-zone managers, ecologists and research funders (Boon 2012; Boon *et al.* 2015). In the case of coastal salt marshes, for example, until the recent publication of *Australian Saltmarsh Ecology* (Saintilan 2009a) and some chapters in *Coastal Wetlands: an Integrated Ecosystem Approach* (Perillo *et al.* 2009) and *Human Impacts on Saltmarshes: a Global Perspective* (Silliman *et al.* 2009), the most recent book with substantive sections on Australian coastal salt marsh is 25 years old: Adam's (1990) *Saltmarsh Ecology*. In fact, the presence of entries for mangroves (Bridgewater 1999) and seagrasses (Walker 1999) but not for salt marsh or other estuarine or intertidal wetlands in the introductory volume of *Flora of Australia* (Orchard 1999) says a great deal about the overall neglect of coastal salt marsh and other non-mangrove coastal wetlands in Australia. The paradox is that a number of early botanical studies in Australia were undertaken on estuarine wetlands, especially on mangroves and coastal salt marsh. Hamilton (1919) and Collins (1921), for example, studied salt marsh and mangrove vegetation in the Sydney region; Pidgeon (1940) reported on spatial zonation (and inferred

**Table 9.1.** Ramsar-listed wetlands in coastal south-eastern Australia
Source: Environment Australia (2001: 22, 23).

| Ramsar number and name | Area (ha) |
|---|---|
| 3. Moulting Lagoon (Tasmania) | 4507 |
| 4. Logan Lagoon (Tasmania) | 2257 |
| 5. Lavinia (Tasmania) | 7034 |
| 6. Pitt Water-Orielton Lagoon (Tasmania) | 3334 |
| 7. Apsley Marshes (Tasmania) | 880 |
| 8. East Coast Cape Barren Island Lagoons (Tasmania) | 4473 |
| 9. Flood Plain Lower Ringarooma River (Tasmania) | 3519 |
| 10. Jocks Lagoon (Tasmania) | 19 |
| 12. Little Waterhouse Lake (Tasmania) | 56 |
| 13. Corner Inlet (Victoria) | 67 186 |
| 18. Port Phillip Bay (Western Shoreline) and Bellarine Peninsula (Victoria) | 22 897 |
| 19. Western Port (Victoria) | 59 297 |
| 21. Gippsland Lakes (Victoria) | 60 015 |
| 23. Towra Point Nature Reserve (New South Wales) | 387 |
| 24. Hunter Estuary Wetlands (New South Wales) | 2971 |
| 25. Coorong and Lakes Alexandrina and Albert (South Australia) | 140 500 |
| 52. Myall Lakes (New South Wales) | 44 612 |
| 57. Edithvale-Seaford Wetlands (Victoria) | 261 |
| Total | 424 205 |

successional patterns) in mangroves and salt marshes along the central coast of New South Wales; Patton (1942) reported on salt marshes in Victoria; and Curtis and Somerville (1948) described coastal salt marshes in Tasmania.

This chapter draws together what is known about the ecology of vegetation in coastal floodplains and associated estuarine and adjacent intertidal areas in south-eastern Australia, with a focus on New South Wales, Victoria and Tasmania. For the purposes of the chapter, 'estuarine' is defined as a coastal area where ocean water is measurably diluted by freshwater (Tagliapietra *et al.* 2009). Thus, 'intertidal' covers a narrower set of conditions than 'estuarine', which in turn is narrower in scope than 'coastal'. The chapter focuses first on the hydrological factors that most strongly influence coastal wetland and floodplain vegetation: tidal inundation by seawater, episodic inputs of freshwater by surface flows, and freshwater inputs of groundwater. The major types of coastal wetland and floodplain vegetation are then considered, followed by a discussion of the principle threatening factors. Where available, current inventory information is included, as well as information on losses (or, in some cases, gains) following European colonisation.

## A PRIMER ON THE HYDROLOGY OF COASTAL FLOODPLAINS AND WETLANDS

Coastal floodplains are hydrologically unusual in that they receive water from at least two sources: the ocean (as seawater) and their parent river (as freshwater). Groundwater inputs also may be important in some cases, although the degree to which groundwater subsidises fringing vegetation in estuaries or along the coast more generally is very poorly known. Given this interplay between oceanic and freshwater influences, the relationship between the relative level of the sea and the relative level of the land is central to the formation and ecological functioning of coastal wetlands (Rogers *et al.* 2006, 2012; Woodroffe and Davies 2009; Day *et al.* 2011; Comeaux *et al.* 2012).

Tidal regimes not only delimit the upper boundary of coastal salt marshes and the commencement of truly terrestrial flora in the hinterland, but set the upper and lower limit of mangroves and the upper limit of seagrasses. In concert with patterning in freshwater inputs, tidal regimes also control the intricate positioning of vegetation in coastal wetlands according to the tolerance of various species to inundation and salinity (halophytic: salt-tolerant; glycophytic: salt-intolerant). Long-term changes in tidal amplitude can also have a profound role in structuring coastal plant communities (Bleakney 1972; Swinbanks 1982). Adam (1990), for example, noted that historical fluctuations in the dominance of Poaceae and Cyperaceae in a Long Island (USA) salt marsh were related to the 18.6-year tidal cycle, with sedges dominating over grasses when the tidal range was smallest.

Variability in the tidal regime experienced by coastal wetlands is strongly associated with a wide range of site-specific geomorphological and meteorological factors. Funnel-shaped estuaries, for example, usually give rise to an amplified tidal range, whereas the tidal range in estuaries with very constricted openings to the sea is often greatly attenuated. Prolonged onshore winds can raise tidal heights, and offshore winds can depress tidal heights; storm surges can greatly modify the amplitude of tidal inundation over short periods. In estuaries that experience large inputs of freshwater, river discharge can strongly influence the extent of saline intrusions and tidal extent. Within individual wetlands, especially if they are large, the periodicity and amplitude of inundation varies also according to elevation and to microtopographical relief (Raulings *et al.* 2010; Boon 2011).

Ranwell (1972) differentiated between 'intertidal' and 'high' marshes in the United Kingdom: the former were posited to experience frequent tidal inundation and have relatively constant soil moisture and salinity regimes, whereas the latter experienced unpredictable inundation and had highly variable edaphic conditions. A simple differentiation of coastal wetland vegetation along a putative elevational gradient from the ocean, however, does not fully reflect the microtopographic diversity at a given site. Salt marshes and mangroves, for example, are often criss-crossed with tidal channels which modify the penetration of saline intrusions (Perillo 2009). Recent modelling of the southern Queensland coast has shown how these subtle variations in topography not only affect the current distribution of wetland vegetation types, but will influence the responses of different species to rising sea levels under anthropogenic climate change (Traill *et al.* 2011).

Freshwater inputs also play a critical role in structuring the vegetation, fauna and ecological interactions of coastal wetlands (Boorman 2009). Coastal wetlands receive inputs of freshwater directly via rainfall, indirectly as runoff from the adjacent hinterland, and as groundwater flows. In the case of coastal floodplains, inundation can occur also when flood-swollen rivers overtop their banks. The relative ratio of inundation with tidally forced saline water and inundation with freshwater (of diverse sources) is thought to be a factor controlling many aspects of the ecology of coastal wetlands (Clarke and Hannon 1967, 1969, 1970; Adam 1990, 2002; Boorman 2009). In the upper parts of salt marshes, for example, swards of salt-tolerant grasses (e.g. *Distichlis* and *Sporobolus* spp.) often occur where freshwater inputs are large enough to significantly dilute soil salinities (Specht and Specht 1999). As salinities further decrease with reduced tidal influence, less salt-tolerant plant taxa such as *Juncus kraussii* often dominate. Under fully freshwater conditions, a range of mostly salt-sensitive emergent macrophytes, including *Bolboschoenus* and *Carex* spp., become the visually dominant taxa. Woody taxa are also affected by the interplay of marine and fresh waters. The germination and establishment of the swamp paperbark (*Melaleuca ericifolia*), for example, progressively declines as salinity increases (Robinson *et al.* 2006, 2008; Salter *et al.* 2007), and even temperate mangroves, long thought to be the model marine woody plant, germinate best in water less salty than straight seawater (Morrisey *et al.* 2010).

Many non-perennial estuarine plants also require periods of reduced salinity for their seeds to germinate, and seasonal relief of high salinities can prompt annual taxa to divert resources to reproductive structures and set seed (Ungar 1962, 1978). Ponds of stagnant rainwater, for example, are critical for the occurrence of many submerged plant taxa in coastal salt marshes of western and northern Europe (Beeftink 1977). Similarly, many intertidal wetlands along the south-east Australian coast possess large, shallow depressions that rapidly fill with rainwater after storms and are soon colonised by submerged angiosperm taxa such as *Ruppia* and *Lepilaena* species, which rapidly form extensive shallow aquatic meadows. Even small depressions constitute an important component of the environmental heterogeneity of coastal wetlands and are often valuable foraging sites for shorebirds, as well as breeding sites for insects such as mosquitoes and midges.

The influence of groundwater exchange on coastal vegetation is less well understood than that of seawater or surface inputs of freshwater, and is likely to vary among different types of coastal wetlands (Crowe and Shikaze 2004). Groundwater intrusions can affect (mostly to decrease) soil salinities and lower the position of the freshwater–brackish water interface in coastal salt marsh and other estuarine wetlands (Werner and Lockington 2006; Carter *et al.* 2008). Groundwater can also affect nutrient dynamics, especially through the supply of critical plant nutrients to otherwise nutrient-limited coastal ecosystems (Krest *et al.* 2000). There is increasing evidence that groundwater exports from coastal wetlands can have impacts on receiving waters as well; the specific case of acidic and metal-laden waters seeping from hydrologically modified coastal wetlands that overlay acid sulfate soils is discussed later. An example of linkages among groundwater, estuaries and coastal wetlands is provided by Ahern *et al.* (2006), who correlated the occurrence of cyanobacterial blooms with the efflux of groundwater from paperbark-dominated wetlands in coastal southern Queensland. They concluded that discharges of acidic groundwater increased the availability of nutrients to phytoplankton, thus initiating and prolonging algal blooms in the adjacent estuary.

# VEGETATION COMMUNITIES OF COASTAL FLOODPLAINS AND WETLANDS IN SOUTH-EASTERN AUSTRALIA

## Mangroves

### Description

Woodroffe and Davies (2009: 66) defined mangroves as 'trees, shrubs, or palms, exceeding 0.5 m in height that occur in the upper intertidal zone'. Duke (2006: 12) adopted a similar definition but was more prescriptive about tidal position, indicating that mangroves normally grew 'above mean sea level in the intertidal zone of marine coastal environments and estuarine margins'. Neither definition is entirely suitable for south-eastern Australia, because of the presence of stunted ('bonsai') mangrove plants less than 0.5 m tall at the southernmost limit of mangroves in Corner Inlet-Nooramunga, Victoria. Figure 9.1 shows how tall woody salt marsh taxa (in this case *Tecticornia arbuscula*) can be nearly as tall as the seaward-fringing mangroves (*Avicennia marina* subsp. *australasica*) in southern Victorian coastal wetlands.

In general, *Avicennia marina* subsp. *australasica* occurs along the south-eastern Australian coast as a dense, monospecific shrubland or low closed forest, with individuals growing as shrubs or small trees from 0.3 to 4 m tall (Harty 1997; Duke 2006). Plant densities, however, can vary from individual specimens growing sparsely on the shoreline, to dense, continuous belts of plants many hundreds of metres wide. Truly mixed communities are confined to far north-eastern New South Wales and Queensland, as a result of the increase in floristic diversity of mangrove species with decreasing latitude (Specht and Specht 1999; Duke 2006).

In most cases in south-eastern Australia, there is little or no angiosperm understorey in the mangrove community, except in the better-drained zones or in landward areas where mangroves become more scattered and where *Sarcocornia quinqueflora* and *Triglochin striata* may occur as a sparse ground layer. Nevertheless, much of the sediment in temperate mangroves may be covered by brown

**Figure 9.1:** Mangrove and salt marsh vegetation near Hastings, Western Port, Victoria. The mangrove *Avicennia marina* subsp. *australasica* occurs in the background as the dark-green band of shrubs and small trees; the foreground is dominated by the herb *Sarcocornia quinqueflora* and the shrub *Tecticornia* (formerly *Sclerostegia*) *arbuscula*, the latter often of a similar height to the surrounding mangroves. (Photo: Paul Boon.)

algae, including *Hormosira banksii*. Algal mats of *Bostrychia* and *Catanella* species are also often associated with pneumatophores and the lower stems of mangroves in Victoria (Morgan 1986).

### Distribution

Australia has the fourth highest species diversity of mangrove taxa (22 genera from 19 families; 41 species) of any country, after the Philippines, Indonesia and Papua New Guinea (Duke 2006). A strong north–south gradient in floristic diversity (Specht and Specht 1999; Duke 2006) is evident, with 39 mangrove species found in Queensland and five mangrove species north of Coffs Harbour in New South Wales (~30°S): *Bruguiera gymnorhiza*, *Rhizophora stylosa*, *Excoecaria agallocha*, *Aegiceras corniculata* and *Avicennia marina* subsp. *australasica*. Only *Avicennia marina* subsp. *australasica* extends further south than Merimbula in New South Wales (~37°S), and it is the only mangrove species that occurs in Victoria. Indeed, the highest latitude expression of mangroves is in southern Victoria, at Millers Landing (38°45′S). Mangroves are not present in Tasmania.

In addition to this latitudinal gradient, mangrove species are often distributed along elevational and/or tidal gradients. In those places where *Avicennia marina* subsp. *australasica* and *Aegiceras corniculata* co-occur in the central coast of New South Wales, for example, the latter is usually found towards the landward edge and further upstream, whereas the former tend towards the seaward edge or further downstream (Chapman and Underwood 1995). The difference in distribution along a coastal floodplain between the two species is also hinted at by their common names: *Aegiceras corniculata* is known colloquially as the river mangrove, whereas *Avicennia marina* subsp. *australasica* is commonly called the grey mangrove or the white mangrove.

Mangroves are present along much of the New South Wales coast, occurring in 69 of the 133 estuaries and small coastal creeks in the state (Duke 2006). Large stands are found in the Port Stephens system, and in the estuaries of the Hunter River and the Hawkesbury River, invariably dominated, at least in the lower reaches, by *Avicennia marina* subsp. *australasica*. These three estuarine systems account for 16%, 15% and 11%, respectively, of the total area of mangroves in New South Wales (Harty 1997).

Mangroves in Victoria have a discontinuous distribution, from the Barwon River in the west to the distal sections of Cunninghame Arm (Gippsland Lakes) in the east. They are particularly well developed around Western Port (including on French Island) and in the Corner Inlet-Nooramunga area of south Gippsland (Boon *et al.* 2015). As noted earlier, the most southerly occurrence of mangroves globally is in Victoria, at Millers Landing in Corner Inlet-Nooramunga.

Although mangroves in south-eastern Australia are usually found on muddy substrata, in a few locations they can be found growing on a rocky shoreline. The rocky substrata can be quite varied, in Victoria, for example, on basalt at The Jawbone Marine Sanctuary, on granite at Millers Landing, and on Silurian sedimentary rocks south of Hastings in Western Port (N. Rosengren, pers. comm.). Presumably because of the large expanses of muddy substrata available for colonisation, mangroves are common along the lower reaches of coastal rivers, from the Brisbane River in southern Queensland to the Barwon River near Geelong in Victoria. In contrast to these riverine–estuarine environments, mangroves occur also along the shoreline of coastal embayments that receive little or no direct freshwater input (e.g. in Corner Inlet-Nooramunga and in Western Port, in south Gippsland).

### Current area

There are few reliable estimates of the extent of mangroves in south-eastern Australia, and the older inventories are not in good agreement (e.g. see Galloway 1981, 1982; Bucher and Saenger 1991; OzCoasts 2009). The area of mangroves in Victoria has been recently quantified, on the basis of 1:5000 aerial photographs and intensive ground-truthing, at 51.77 km$^2$ (Boon *et al.* 2015). The most recent and reliable mangrove mapping in New South Wales was based on interpretation of 1:25 000 aerial photographs and extensive field reconnaissance by Williams *et al.* (2006), who updated earlier work by West *et al.* (1985). They estimated the total area of mangroves in New South Wales to be ~124 km$^2$.

## Coastal salt marsh

### Description

Coastal salt marsh presents a formidable problem in definition and spatial delimitation (Adam 1990; Winning 1991; Boon *et al.* 2015). The definition adopted by Boon *et al.* (2015) for a statewide assessment in Victoria was 'land that experiences recurrent low-energy inundation by seawater and not usually by substantial freshwater flows, which is vegetated by low-growing vascular plants (generally <1.5 m height), such as succulent chenopods and salt-tolerant monocots but excluding vegetation dominated by the mangrove *Avicennia marina*'.

A distinction between salt marshes and mangroves based on plant height and structure works well along the wetter subtropical and warm temperate coast of northern New South Wales and Queensland, where large mangrove trees form a dense shrubland or forest and salt marshes are mostly poorly developed. The differentiation, however, is unclear along many parts of the Victorian coast, where *Avicennia marina* subsp. *australasica* often forms a stunted shrubland with a lower stature than the *Tecticornia arbuscula* shrubs of the adjacent salt marshes and into which mangroves often recruit (Fig. 9.1). In such cases the distinction between mangroves and salt marshes is often based on floristics and convention (Adam 1990). Mangroves can be found also as sparsely growing individuals within what would appear to be a large expanse of salt marsh. In these cases, it is difficult to tell unequivocally where the mangrove zone ends and the salt marsh zone starts. In some mapping studies an arbitrary, but necessary, differentiation between the two wetlands types has been made on the basis of distance between individual mangrove plants (Wilton and Saintilan 2000).

Coastal salt marsh across the world is dominated by plants belonging to relatively few families, presumably due to the evolutionary difficulties of overcoming the physiologically harsh conditions. The vast majority of plant species belong to the families Chenopodiaceae, Cyperaceae, Juncaceae and Poaceae, with some representatives also from the Amaranthaceae, Apiaceae, Frankeniaceae and Plumbaginaceae (Adam 1990). The floristics of Victorian salt marshes are relatively well served by the reports of Bridgewater (1975, 1982), Bridgewater and Kaeshagen (1979), Bridgewater *et al.* (1981) and Bridgewater and Cresswell (2003), and those of New South Wales' salt marshes by the detailed research by Adam (1981a, 1981b, 2009), Adam *et al.* (1988) and Saintilan (2009b, 2009c). Tasmanian salt marshes were described by Kirkpatrick and Glasby (1981). The recent book edited by Sainty *et al.* (2012) also provides a detailed overview and species-by-species descriptions of the plant taxa found in estuaries in south-eastern Australia.

When seen together, these studies indicate that two plant families dominate salt marshes in south-eastern Australia: Chenopodiaceae and, to a lesser extent, Poaceae. The abundance (in both number of species and area covered) of chenopods is a fundamental difference between coastal salt marshes in south-eastern Australia and coastal salt marshes in the northern hemisphere. The latter are often dominated exclusively by grasses and/or herbs (e.g. *Spartina* and *Puccinellia* spp.), whereas southern Australian coastal salt marshes as a group are far more complex, both structurally and floristically. The structural complexity is due to the range of life forms present in the south-east Australian flora, and the fact that species with different architectures are often assembled into spatially and temporally complex patterns. Succulent shrubs such as *Tecticornia* spp. form one distinctive structural type; large tussock-forming monocots (e.g. *Austrostipa stipoides* and *Gahnia filum*) make up one graminoid form, and low rhizomatous or stoloniferous grasses another (e.g. *Distichlis distichophylla* and *Sporobolus virginicus*); succulent herbs (e.g. *Sarcocornia* spp., *Hemichroa pentandra, Disphyma clavellatum*) and prostrate shrubs (e.g. *Frankenia pauciflora*) constitute another form. Moreover, local patches dominated by annuals are not uncommon; submerged macrophytes (e.g. *Ruppia* spp.) in the shallow pools of low-lying depressions are yet another structural form. Many, but by no means all, of the taxa present in saline wetlands are clonal and can reproduce and spread by asexual means (Hatton *et al.* 2008; Morris *et al.* 2008). Under the environmental conditions that often prevail in coastal wetlands, clonality can be a valuable life history trait (Xiao *et al.* 2011).

The strong north–south gradient in species richness of mangroves along the eastern seaboard of Australia has been noted above. There is a similar, but inverse, gradient in the number of salt marsh species along the coast. Specht and Specht (1999: 327), for example, noted that there were ~30 species of vascular plants in coastal salt marshes at 40°S, compared with ~5 at 10°S. Similarly, Turner *et al.* (2004) provided a diagram showing the strong latitudinal gradient in salt marsh plant species, with <10 species in northern Queensland and 10–30 species in southern Australia.

A difficulty in quantifying the species gradient within salt marsh vegetation is that, compared with the relatively easily distinguished mangrove taxa, there is far less certainty about what constitutes a species of salt marsh plant. For example, species such as *Juncus kraussii*, *Phragmites australis* and *Typha domingensis* are considered core salt marsh plant species in New South Wales, but not in Victoria (Boon *et al.* 2015). Moreover, estimates of the number of vascular plants in coastal salt marshes given by Specht and Specht (1999) and Turner *et al.* (2004) are likely to be markedly too low, probably by an order of magnitude: more than 130 indigenous and 118 exotic vascular plant species have been identified in Victorian coastal wetlands (Boon *et al.* 2015; see also Carr 2012a, 2012b).

## Distribution

Coastal salt marsh occurs sporadically along most of the Tasmanian coastline, but is best developed along the highly indented south-eastern coast and the far north-western coast (Kirkpatrick and Glasby 1981). Although some of the earliest European reports on coastal salt marsh in Australia concern marshes in Tasmania (Potts *et al.* 2006), little else has been published on these systems with the exception of the studies by Kirkpatrick and his co-workers. The recent paper by Prahalad *et al.* (2011), on changes in Tasmanian salt marshes with sea-level rise over the past three decades, is a noteworthy exception.

**Figure 9.2:** Coastal salt marsh at Western Port, Victoria. Note the proximity of the housing to coastal marsh, with clear implications for the impacts of sea-level rise on residential land use. (Photo: Paul Boon.)

Salt marsh is found along many parts of the Victorian coast, although it is most extensive along the western coast of Port Phillip Bay, northern parts of Western Port, in the Corner Inlet-Nooramunga complex, and behind the sand dunes that line the Ninety Mile Beach in Gippsland, especially in Lake Reeve (Sinclair and Boon 2012; Boon *et al.* 2015). Figure 9.2 shows coastal salt marsh on the western shore of Western Port, Victoria as an example. Barson and Calder (1981) concluded that coastal salt marsh was best developed in the central region of Victoria, between Barwon Heads and Corner Inlet. To the west of Barwon Heads, notable patches occur at Breamlea, the mouths of the Anglesea River and Painkalac Creek, Aireys Inlet, Port Fairy, and in the estuary of the Glenelg River. In many of these cases in western Victoria, as in other parts of Australia, salt marshes occur in estuarine settings along the lower reaches of coastal rivers and their floodplains. The rivers may be quite substantial (e.g. Barwon River), but are mostly small (e.g. Painkalac Creek). To the east of Corner Inlet, salt marsh fringes large areas of the shoreline of the Gippsland Lakes and then extends as far east as the mouth of the Snowy River, then to Wingan Inlet and to Mallacoota, near the New South Wales border (Boon *et al.* 2015).

In much of eastern Victoria, salt marsh occurs along the shore of coastal embayments with little freshwater inflow and largely oceanic salinity regimes (e.g. Western Port, Corner Inlet-Nooramunga, Lake Reeve). The salt marshes of Lake Reeve are of particular note, as they are extensive, floristically diverse and often neglected in ecological analyses of the Gippsland Lakes (which tend to stress the lakes' marine or estuarine habitats, rather than its hypersaline habitats). Salt marshes associated with the Gippsland Lakes are particularly dynamic, and this makes it especially difficult to quantify changes that have occurred since European colonisation (Sinclair and Boon 2012).

There are two general forms of salt marsh on the Victorian coast (Barson and Calder 1981). The first, a 'dry' type, is present in the central-western parts of Victoria, where low summer rainfall and high temperatures lead to intensely hypersaline conditions in the more elevated sites over summer. The vegetation is dominated by *Tecticornia pergranulata* and *Tecticornia halocnemoides*. The second, a 'wet' type, is found in areas that experience higher rainfall; here, the vegetation is dominated by samphires such as *Sarcocornia* spp. and *Tecticornia arbuscula*. The geographic differentiation of 'dry' and 'wet' types is a direct consequence of patterns in rainfall and evaporation across the state. Annual rainfall ranges from more than 1000 mm in far south-west and parts of eastern Victoria and New South Wales, to 400–600 mm in orographic rain shadow areas of the western shores of Port Phillip Bay and Corio to Torquay, as well as in a relatively small area near Loch Sport, Gippsland (Boon *et al.* 2015). A small low-rainfall area (<600 mm) occurs also near Lake Reeve (Bureau of Meteorology and Walsh 1993). In these Mediterranean-climate regions, summers are typically hot and dry, which leads to hypersaline conditions developing over summer in the mid to upper salt marsh, relieved only by rainfall in the wetter seasons of autumn, winter and spring (Boon *et al.* 2015).

In addition to the mostly east–west differentiation into 'dry' and 'wet' salt marsh, Victorian authors have often distinguished between salt marshes on the basis of elevation and tidal inundation: an 'upper' salt marsh and a 'lower' salt marsh (Barson 1977; Barson and Calder 1981). In Victoria, the 'lower' salt marsh occurs along better-watered parts of the coast and is typically dominated by chenopods such as *Salicornia quinqueflora* and *Tecticornia arbuscula*. The upper, more elevated parts, even in well-watered sites, are often dominated by a tussock grassland of *Austrostipa stipoides*, *Gahnia filum*, and rush lands of *Juncus kraussii* and *Apodasmia brownii*. In the drier, western parts of the state, *Tecticornia* (previously *Halosarcia*) *pergranulata* is often dominant in these upper sites (Barson and Calder 1981).

### Current area

Until the recent publication of the statewide assessment undertaken by Boon *et al.* (2015), there were few estimates of the area of coastal salt marsh in Victoria and, as with mangroves, the available studies showed little agreement. Drawing from the work of Bucher and Saenger (1991), Kelleway *et al.* (2009) stated that there were 125 km$^2$ of salt marsh in Victoria. In contrast, Carr (1979) had earlier estimated the area of Victorian coastal salt marsh, based on mapping of orange-bellied parrot (*Neophema chrysogaster*) habitat, as <60 km$^2$. The OzCoasts (2009) database reported that there were 68.5 km$^2$ of coastal salt marsh in New South Wales, out of an Australian-wide area of 13 029 km$^2$, but no values were provided in the database for Victoria. According to vegetation mapping currently available from the Department of Sustainability and Environment, the area of Ecological Vegetation Class 9 Coastal Saltmarsh Aggregate in Victoria is ~132 km$^2$. The results of the mapping and inventory studies undertaken by Boon *et al.* (2015), undoubtedly the most accurate to date, indicate the current statewide extent of coastal salt marsh is 192.12 km$^2$.

The most recent salt marsh mapping in New South Wales is based on interpretation of 1:25 000 aerial photographs and extensive field reconnaissance by Williams *et al.* (2006), who updated earlier work by West *et al.* (1985). They estimated the total area of coastal salt marsh in New South Wales to be approximately 72 km$^2$.

In Tasmania, the total area of salt marsh is 77 km$^2$, accounted for by roughly equal areas of saline grassland, saline aquatic herbland and succulent saline herbland (F. Faulkner, Department of Primary Industries, Parks, Water and Environment, pers. comm.; see also http://dpipwe.tas.gov.au/conservation/development-planning-conservation-assessment/tools/tasmanian-reserve-estate-spatial-layer).

## Freshwater and brackish meadows, reed beds, rushlands, sedgelands, grasslands and aquatic herbfields

### Description and distribution

Although salt marsh dominates many of the more elevated intertidal habitats along Australia's south-eastern coastline, a wide range of other treeless vegetation also occurs along less saline parts of the coast. Adam *et al.* (1985), for example, identified a variety of tall reed beds, short grasslands and sedgelands, non-salt marsh sedgelands, wet meadows, semi-aquatic and aquatic herbfields and open freshwater swamps in their inventory of New South Wales coastal wetlands. In Victoria, there are at least 11 Ecological Vegetation Classes (EVCs), in addition to seagrass beds, mangrove woodlands and paperbark-dominated systems, recognised by the Department of Sustainability and Environment within coastal wetlands and floodplains (Department of Sustainability and Environment 2012; Boon 2012). Figure 9.3 shows one example of a non-mangrove and non-salt marsh wetland type: extensive beds of common reed (*Phragmites australis*) along the lower (estuarine) reaches of the Mitchell River, which discharges into the Gippsland Lakes in southern Victoria. In the Victorian typology, these reed beds would be classified as EVC 952 Estuarine Reedbed (with a halophytic ground layer, mostly in estuarine settings) or EVC 821 Tall Marsh (with a glycophytic ground layer, mostly in freshwater settings).

In estuarine systems, coastal salt marshes in particular may be juxtaposed with a complex suite of other non-woody wetland types and inundated by saline water and freshwater over highly variable spatial and temporal scales (Keith *et al.* 2007). Even nominally freshwater wetlands in estuaries may be influenced by seawater during periods of extremely high tides or after storm surges. A second context where coastal salt marsh can be difficult to distinguish from other estuarine wetlands is where freshwater flows emerge at the coast as estuaries or from groundwater seepage. These transitional mosaics are often characterised by dense stands of perennial monocots, including *Juncus kraussii* and *Phragmites australis* in the seaward reaches, and *Bolboschoenus caldwellii* and *Schoenoplectus pungens* in areas further from the influence of seawater (Adam 1994; Sinclair and Sutter 2008). Under these conditions, halophytic and glycophytic species are often intermixed to form complex patterns and mosaics that reflect the interplay of site elevation, saline water and freshwater inundation, and groundwater flows (see Clarke and Hannon 1970 for a classical analysis of Sydney salt marshes). The co-existence of such diverse plant taxa creates real difficulties in differentiating between various types of coastal wetlands, with obvious implications for mapping and the preparation of inventories.

**Figure 9.3:** Extensive beds of common reed (*Phragmites australis*) fringing the lower (estuarine) reaches of the Mitchell River, Victoria. (Photo: Paul Boon.)

In parts of the low-lying coastal floodplain, reed beds often give way to meadows dominated by shorter sedges and grasses, including several species of *Paspalum*, *Panicum*, *Pseudoraphis*, *Eleocharis* and *Isolepis*, of which *Paspalum distichum* and *Eleocharis equisetina* are the most common (Pressey 1989a, 1989b; Keith *et al.* 2007). These species tend to occupy the brackish backplains between the channels of larger rivers, where they receive periodic overbank flows or sheet flows from lithic slopes adjacent to the floodplain so that tidal influence is diluted, but water tables are retained close to the surface for extended periods. The physical conditions that segregate short meadows from tall reed beds are uncertain, but likely to be related to complex spatial and temporal variations in hydrology and salinity. Winning and Saintilan (2009), for example, found that mean water depth, soil salinity and surface water salinity were similar in meadows and reed beds within Hexham Swamp on the Hunter River floodplain north of Sydney, although the depth and salinity of surface water was slightly more variable in meadows than in reed beds.

Punctuating the mosaics of floodplain meadows and forested wetlands are various depressions, billabongs, intermittent stream channels, swales between terraces, and levees and depressions below the toeslopes of lithic terrain on the floodplain margins. These are inundated permanently or semi-permanently with freshwater and contain a variety of emergent, floating and submerged aquatic plants including *Phylidrum lanuginosum*, *Persicaria* spp., *Potamogeton tricarinatus*, *Ranunculus inundata*, *Alisma plantago-aquatica*, *Lemna* spp., *Azolla filiculoides*, *Myriophyllum* spp., *Vallisneria* spp. and *Triglochin* spp. (Keith 2004). The vegetation of these floodplain lagoons is highly dynamic (Yen and Myerscough 1989; Keith 2004), with large variations in species composition occurring over short time scales in relation to floods and extended drying, as well as to wind and wave action.

The large floodplains of north-eastern New South Wales, including those of the Richmond, Clarence, Macleay, Hastings, Manning and Hunter rivers, are strongholds of treeless reed beds, meadows and swamps of brackish and freshwater habitats (Pressey 1989a, 1989b; Winning and Saintilan 2009). Pressey (1989a, 1989b) surveyed almost 300 km$^2$ of wetlands on the Clarence and Macleay floodplain in north-east New South Wales, and found that two-thirds of the area was occupied by non-woody freshwater wetlands dominated by sedges, grasses, reeds and aquatic forbs. Smaller occurrences extend through the New South Wales south coast (Goodrick 1970; de Jong 1997; Keith *et al.* 2007), as well as around the Victorian coastline (Sinclair and Sutter 2008).

### Current area

Given their floristic diversity and complex spatial and temporal patterning, it is perhaps not surprising that there are few estimates of the area of freshwater and brackish non-woody wetlands for coastal south-eastern Australia. In Victoria, there is a reliable and up to date estimate of the area of EVC 10 Estuarine Wetland, at 32.27 km$^2$ (Boon *et al.* 2015). Ecological vegetation class 10 is described as rushland/sedgeland vegetation, variously with a component of small halophytic herbs, occurring in coastal areas where freshwater flows augment otherwise saline environments. The characteristic species is *Juncus kraussii*. There are no up to date and adequately ground-truthed estimates for the other non-mangrove and non-salt marsh wetland types in Victoria as a whole, although estimates are available for some discrete sections of the coast (e.g. Sinclair and Sutter 2008 for south-western Victoria; Yugovic 2011 for Koo Wee Rup swamp, on the northern shore of Western Port, Victoria).

Recent survey and mapping of coastal floodplains in New South Wales (D. Keith, S. Rodoreda and J. Elith, unpubl. data) suggests that the combined extent of treeless freshwater coastal wetlands is in the order of 150 km$^2$. This estimate supersedes earlier mapping of permanent and semi-permanent freshwater swamps and meadows by Goodrick (1970), which produced an estimate of 220 km$^2$. The difference partly reflects improvements in the resolution of mapping, although the overall extent of these wetlands is likely to have declined in recent decades (Winning and Saintilan 2009). In Tasmania, the total area of freshwater and brackish meadows, reed beds, sedgelands, grasslands and aquatic herbfields is estimated at 120 km$^2$ (F. Faulkner, pers. comm.; see also http://dpipwe.tas.gov.au/conservation/development-planning-conservation-assessment/tools/tasmanian-reserve-estate-spatial-layer).

## Brackish-water paperbark and swamp she-oak wetlands

### Description and distribution

Although mangroves are the most obvious and well-known type of estuarine wetland dominated by woody angiosperms, a number of other coastal wetland types in south-eastern Australia are similarly dominated by woody taxa. In this section we discuss wetlands dominated by brackish-water paperbark (*Melaleuca* spp.) or swamp she-oak (*Casuarina glauca*), and in the next section we discuss the diverse range of other types of largely salt-intolerant forested coastal wetlands, including those dominated by various *Eucalyptus* species.

*Melaleuca*-dominated swamps occur across much of the Australian coast in low-lying areas adjacent to estuaries and between terrestrial and tidally inundated vegetation (Turner *et al.* 2004). As a group, coastal *Melaleuca* species are commonly known as paperbarks, but there is considerable differentiation in common names among the different species. *Melaleuca ericifolia*, for example, is commonly known as swamp paperbark in eastern Victoria, whereas *Melaleuca quinquenervia* is known variously as broad-leaved paperbark or paperbark tea tree in northern New South Wales and southern Queensland; the species found in western Victoria and coastal South Australia, *Melaleuca halmaturorum*, is known there as salt paperbark. As a generic moniker, however, the term 'paperbark' is not inappropriate to describe these sorts of *Melaleuca*-dominated swamps.

In New South Wales the two major coastal *Melaleuca* species are *M. quinquenervia* (in the north of the state) and *M. ericifolia* (generally dominant towards the south). Forests and scrubs dominated by *M. quinquenervia* extend northwards from Jervis Bay. The canopy stratum often includes *Eucalyptus robusta*, *E. resinifera* subsp. *hemilampra* and other eucalypts, although *M. quinquenervia* is usually present in highest densities (Keith and Scott 2005). The understorey includes a range of sedges, shrubs and ferns, with *Gahnia clarkei* and *Blechnum indicum* notable characteristic species. These *Melaleuca*-dominated swamps generally occur on sandier alluvial soils and slightly more brackish hydrological conditions than forests and woodlands dominated by *Casuarina glauca*, although the two types do intergrade. A form of *M. quinquenervia* forest also occurs in poorly drained dune swales of coastal sandplains, although these forests tend to have a greater diversity and cover of sclerophyllous shrubs than those on the floodplains (Keith and Scott 2005). *Melaleuca quinquenervia* forests reach their greatest development on the lower Richmond River floodplain, where extensive areas of sandplain have been eroded, mixed with silt and redistributed by fluvial action of the river.

Swamp she-oak scrubs and forests extend north from Bermagui in New South Wales and traverse substantial local gradient in soil salinity from freshwater to brackish on lower floodplains and estuaries. *Casuarina glauca* is typically the only tree species, although *M. ericifolia*, *M. styphelioides* and *M. quinquenervia* may co-occur, as may eucalypts in less saline environments where these communities adjoin the forested wetlands described below. On more saline soils, *C. glauca* typically has an understorey of salt-tolerant species, including *Sarcocornia quinqueflora*, *Samolus repens*, *Sporobolus virginicus* and *Juncus kraussii*. On less saline soils, the understorey is dominated by grasses and forbs, including *Microlaena stipoides*, *Oplismenus* spp., *Commelina cyanea*, *Centella asiatica*, *Carex appressa*, *Persicaria* spp. and *Lomandra longifolia* (Keith and Scott 2005). The vine *Parsonsia straminea* is often conspicuous on the trunks of *Casuarina* or among the ground stratum.

*Melaleuca ericifolia* dominates or co-dominates closed scrub or closed forest on poorly drained coastal sites in southern New South Wales, Victoria and Tasmania (Kirkpatrick and Harwood 1983; Bowkett and Kirkpatrick 2003; Turner *et al.* 2004; Harris and Kitchener 2005). Figure 9.4 shows a swamp paperbark-dominated wetland, Dowd Morass, of the Gippsland Lakes in eastern Victoria. This large (1500 ha) wetland lies adjacent to the southern shore of the Latrobe River, which discharges into the lakes via the westernmost lagoon, Lake Wellington. The life history attributes of *M. ericifolia* in relation to wetting and drying cycles and salinity regimes are reasonably well understood (Bird 1962; Ladiges *et al.* 1981; Gibson *et. al.* 1987; Robinson *et al.* 2006, 2008, 2012; Salter *et al.* 2007, 2010a, 2010b; Raulings *et al.* 2007, 2010, 2011; Boon *et al.* 2008; Morris *et al.* 2008; Hamilton-Brown *et al.* 2009; Boon 2011). This species can tolerate moderate-to-high levels of salinity and can maintain growth, usually via clonal mechanisms, under chronically waterlogged conditions, sometimes for decades.

The floristics of swamp scrubs dominated by *M. ericifolia* vary according to light availability, water regime and soil fertility, and the understory may be dominated variously by bryophytes, sedges, submerged macrophytes or herbaceous species (Bowkett and Kirkpatrick 2003; Raulings

**Figure 9.4:** Brackish-water marsh dominated by swamp paperbark (*Melaleuca ericifolia*) at Dowd Morass, Gippsland Lakes, Victoria. Along the fringing Latrobe River and Lake Wellington, the understorey of this paperbark marsh is dominated by common reed *Phragmites australis*. (Photo: Paul Boon.)

*et al.* 2010, 2011). In Tasmania, brackish marsh species such as *Samolus repens* and *Selleria radicans* occur under tall dense thickets of *M. ericifolia* in estuarine areas that have high levels of available nutrients (Bowkett and Kirkpatrick 2003). In swales of coastal dune systems, *M. ericifolia* forms an open forest with an understorey of *Leucopogon australis*, *Viola hederacea* and *Leptospermum scoparium* (Bowkett and Kirkpatrick 2003). These swamp paperbark wetlands are frequently surrounded by eucalypts and banksias on gentle rises and by reed beds and coastal salt marsh at lower elevations (Bird 1962).

The other common *Melaleuca* species in southern Victoria is moonah, *Melaleuca lanceolata*. Moonah woodlands tend to occur on swales, lower slopes and upper south-facing slopes of coastal dune systems and mallee or open woodlands further inland. They are often adjacent to wetlands, but are not a coastal wetland community per se and so are not discussed further in this chapter.

### Current area

Recent survey and mapping of coastal floodplains in New South Wales (D. Keith, S. Rodoreda and J. Elith, unpubl. data) suggests that the combined extent of paperbark swamps dominated by *M. quinquenervia* is in the order of 600 km$^2$, while swamp she-oak forests occupy approximately a further 500 km$^2$.

In Victoria, *Melaleuca*-dominated coastal wetlands are described by two ecological vegetation classes: EVC 53 Swamp Scrub and EVC 953 Estuarine Scrub. *Melaleuca ericifolia* and/or *Leptospermum lanigerum* are the characteristic canopy species in swamp scrub, and the ground cover is mostly either a complex of moss, lichen and liverwort or is herbaceous (Department of Sustainability and Environment 2010). Estuarine scrub has a canopy layer of *M. ericifolia*, and is differentiated from swamp scrub by the presence of a halophytic ground layer dominated by graminoids and herbs. We have not been able to obtain accurate estimates of these two EVCs for Victorian coastal areas.

In Tasmania, the area of swamp forest dominated by *M. ericifolia* is 78 km$^2$ (F. Faulkner, pers. comm.; see also http:www.dpiw.tas.gov/tasreserveestate).

## Other types of forested wetlands on coastal floodplains

### Description

Although paperbark and swamp she-oak wetlands sometimes attain a forest structure, they are confined to the lower parts of coastal floodplains where brackish water inundates the surface or infiltrates the subsoil. Other forested wetlands, dominated by eucalypts or rainforest genera, occur on more elevated parts of floodplains influenced only by freshwater flows from the catchment. The most widespread form of forest is dominated by eucalypts that often exceed 40 m in the mature state (Keith and Scott 2005), but structurally complex rainforest also occurs, mainly at subtropical latitudes, as gallery forest along the banks of major rivers and on instream islands at the lower extremes of freshwater dominance (Floyd 1990; Williams 1990).

The open forests of floodplains (Fig. 9.5) are typically dominated by one or more species of red gum (*Eucalyptus tereticornis*, *E. amplifolia*, *E. seeana*) and rough-barked apples (*Angophora floribunda* or *A. subvelutina*). River red gum (*E. camaldulensis*), a dominant species of inland floodplain and riparian landscapes, becomes part of the coastal floodplain communities in south-western Victoria and South Australia, as well as on the upper floodplain of the Hunter River in the central New South Wales coast. In warm temperate and subtropical latitudes, open forests may include a wide range of eucalypt mahoganies (*E. resinifera*, *E. acmenioides*) and blue gums (*E. grandis*, *E. saligna*, *E. deanei*), while peppermints and ribbon gums (*E. elata*, *E. viminalis*) may occur at cool temperate latitudes (Keith and Scott 2005). The open forests commonly have a variable sub-canopy or shrub layer of *Acacia*, *Melaleuca*, *Casuarina* and mesophyllous woody species and vines. Their ground layer consists of a diversity of forbs and grasses. Keith and Scott (2005), for example, recorded more than 500 vascular plant taxa in a limited sample of quadrats in two eucalypt communities on floodplains of the New South Wales coast. The total flora is likely to be considerably larger, although most species are not unique to coastal floodplains.

Floodplain rainforests of the region are highly variable in species composition and dominance at local scales and also subject to a strong latitudinal gradient, with declining diversity to the south (Floyd 1990). *Ficus* species are often emergent, while common canopy species include *Elaeocarpus* spp., *Heritiera trifoliata*, *Sloanea* spp., *Syzygium* spp., with palms *Archrontophoenix cunninghamii* and *Livistona australis*. Some characteristic tree species of floodplains include *Waterhousia floribunda*, *Castanospermum australe* and *Grevillea robusta*, but these are not widespread throughout the distribution.

**Figure 9.5:** Remnant of eucalypt-dominated floodplain forest at Kings Bush, Camden, on the upper floodplain of the Hawkesbury-Nepean River, New South Wales. The three trees in the foreground are *Eucalyptus baueriana*, with *Angophora subvelutina* on the right. The understorey includes a stratum of shrubs, saplings and vines, while the ground layer includes a diversity of native grasses and forbs. This remnant survived in church grounds and has recently undergone major restoration works to control weeds. (Photo: Peter Cuneo.)

Widespread tree species of floodplain rainforests include *Acmena smithii*, *Alphitonia excelsa*, *Dendrocnide excelsa*, *Elaeocarpus obovatus*, *Ficus macrophylla*, *F. coronata*, *F. obliqua*, *Glochidion ferdinandi*, *Livistona australis*, *Synoum glandulosum*, *Toona ciliata*, *Tristaniopsis laurina* and *Waterhousia floribunda* (Floyd 1990; Williams 1990; Keith and Scott 2005). Species such as *Araucaria cunninghamii*, *Castanospermum australe* and *Grevillea robusta* are characteristic of floodplain rainforests in subtropical latitudes. Vines are diverse and abundant throughout, but the ground layer is sparse, except in canopy gaps, and includes a variety of ferns, forbs and sedges (Floyd 1990; Keith and Scott 2005).

Forested wetlands are often less clearly distinguishable from vegetation in surrounding landscapes than are other floodplain communities. Where floodplains adjoin lithic terrain, with soils derived in situ from rocks, the transitions in both soils and vegetation can be gradual and subtle. Vegetation mosaics may occur where floodplains adjoin low-relief lithic terrain. These mosaics may include slight depressions that are occasionally inundated by surface flows descending from surrounding slopes. The depressions support elements of floodplain biota, even though they have limited development of alluvial soil profiles and may include a variable mix of species from the surrounding lithic soils. Examples from northern New South Wales include *Eucalyptus saligna*, *E. siderophloia* and *Lomandra longifolia*, each of which is common on non-alluvial soils. Floodplain rainforests undergo similar transitions from lithic rainforest types as streams enter the floodplain from rich lithic substrates such as basalt. Transitions also occur with other floodplain communities. As the influence of saline groundwater increases, open eucalypt forests and rainforests are replaced by paperbark forests and thickets. The dominant genera of the latter, *Melaleuca* and *Casuarina*, are often components of the open forest sub-canopy, even at locations with little or no saline influence. There are also transitional stands between open eucalypt forests and rainforest, as some eucalypt forests have well-developed rainforest sub-canopies and some gallery rainforests include emergent eucalypts.

### Distribution

Although now largely transformed by anthropogenic forces, the greatest expanses of open forests were historically on the upper parts of floodplains, on elevated terraces and on levees of large coastal rivers, notably the Richmond, Clarence, Macleay, Hastings, Manning, Hunter and Hawkesbury-Nepean in New South Wales, with smaller occurrences along coastal lowland rivers between Sydney and Melbourne (Keith and Scott 2005). However, much of the remaining distribution occurs along flats associated with smaller streams in the coastal lowlands, where discrimination from forests of lithic substrates and coastal sand plains is difficult.

Floodplain rainforests are distributed primarily on floodplains of north-eastern New South Wales, particularly those such as the Tweed, Richmond and Hastings, with nutrient-rich substrata in their catchments. The best examples are found on instream islands, such as Stott Island in the Tweed River, although a few stands remain on stream banks, as at Wingham Brush on the Manning River.

### Current area

There are few estimates of the remaining area of forested wetlands on coastal floodplains. Unpublished mapping data (D. Keith, S. Rodoreda and J. Elith) for New South Wales suggest that there may be approximately 1500 km$^2$ of eucalypt-dominated forest and 200 km$^2$ of rainforest on the coastal floodplains, most of it north of Sydney.

## CHANGES TO COASTAL FLOODPLAIN AND WETLAND VEGETATION SINCE EUROPEAN COLONISATION

Because of their position along the coast, where land has been appropriated for human settlement, coastal wetlands and floodplains across the world have experienced a long history of reclamation (Sinclair and Boon 2012). Almost all estuarine wetlands in south-eastern Australia have experienced some degree of reclamation or habitat destruction. In the Koo Wee Rup area bordering northern Western Port in Victoria, for example, large areas of salt marsh, mangrove and coastal *M. ericifolia* wetlands were reclaimed for agriculture in the late 19th and early 20th centuries, as described enthusiastically by East (1935) and Roberts (1985). Yugovic and Mitchell (2006) and Yugovic (2011) have recently described historical vegetation changes at Koo Wee Rup swamp. Similar enthusiasm for the conversion of 'unproductive' coastal vegetation to 'productive' agricultural land was shown by Callaghan (1948) in South Australia. Bird (1993) noted that about one-third of the original area of coastal wetland, presumably a mix of salt marsh, reed beds and swamp paperbark, probably with some mangroves, had been reclaimed from the eastern and northern shores of Shallow Inlet in South Gippsland, Victoria.

Saintilan and Williams (2000) reviewed the available evidence on the loss of salt marsh in eastern Australia and concluded that, based on 28 surveys employing historical aerial photographs, there had been a widespread decline of salt marsh from estuaries along the entire coast since the end of World War II. At the time of Saintilan and Williams' (2000) review, information on salt marsh loss was available mainly for New South Wales and Queensland; few or no quantitative data were available for Victoria or Tasmania. The extent of loss reported by Saintilan and Williams (2000) ranged from 78% from 1943 to 1991 at Hinchinbrook Island (Queensland), 72% in the Tweed River in northern New South Wales from 1947 to 1986, and 'extensive' losses in Corner Inlet (Victoria). There is some evidence that the area of coastal salt marsh in New South Wales has increased over the past 20 years. In their analysis of 125 estuaries along the New South Wales coast, Williams *et al.* (2006) reported that 92 estuaries contained salt marsh, the total area of which had increased since 1985 from 48 $km^2$ to 60 $km^2$. The area of salt marsh had increased in 77 estuaries, but decreased in 22, and no change was discernible in another 22. The degree to which this apparent increase is due to errors in quantifying the true area of salt marsh in the 1985 baseline study is, however, not clear (e.g. see Williams and Thiebaud 2007).

The extent of loss of coastal salt marsh, mangroves and *Juncus kraussii*-dominated wetlands in Victoria has recently been quantified by Sinclair and Boon (2012) and Boon *et al.* (2015) on the basis of archival maps and field observations to reconstruct the pre-European (i.e. 'pre-1750') distribution of intertidal wetland vegetation. Losses of coastal marshes have been very uneven across the Victorian coast, with some regions having suffered massive loss (notably Corner Inlet and parts of the Nooramunga coast, Anderson Inlet, Shallow Inlet and Port Phillip Bay), but others having retained most of their pre-colonial area of intertidal wetland. In some regions (e.g. eastern shore of Western Port, possibly parts of the Gippsland Lakes) there is good evidence of expansion. Some of the greatest losses have occurred along the South Gippsland coast, where it was estimated that ~60% of coastal salt marsh had been lost around Shallow Inlet and in the Powlett-Kilcunda region. Statewide, the loss of coastal marsh vegetation (mangrove shrublands, coastal salt marsh and EVC 10 Estuarine Wetland combined) was estimated at ~5–20%, depending largely on assumptions about changes in the Gippsland Lakes region (Sinclair and Boon 2012). In the Ramsar-listed wetlands of the Gippsland Lakes there have also been significant shifts across plant species since the 1950s, as outlined for the largest of these wetlands, Dowd Morass, by Boon *et al.* (2008).

The impacts of hydrological modification and subsequent conversion of wetlands to agriculture has been especially severe on treeless floodplain wetlands. Goodrick (1970) estimated that 84% of 'fresh meadows' and 'seasonal fresh swamps' throughout coastal New South Wales had been drained and converted to agriculture by 1969. Subsequent surveys and mapping suggest a very substantial ongoing decline in the distribution of the remainder in subsequent decades. For the Tweed River floodplain, Pressey and Griffith (1992) estimated that only 3% of treeless and forested freshwater wetlands remained. Similarly, large expanses of treeless freshwater wetlands on the Clarence, Macleay and Manning river floodplains have been converted to cane fields or pastures (Keith 2004). Recent survey and mapping, with modelled reconstruction of historical distributions, suggest that less than 5% of the estimated pre-European area of *Casuarina*- or *Melaleuca*-dominated forests, eucalypt forests and rainforests remain on the coastal floodplains of New South Wales (D. Keith, S. Rodoreda and J. Elith, unpubl. data).

## THREATS TO COASTAL FLOODPLAINS AND WETLANDS

Coastal wetlands are subject to a wide range of threats (Duke 2006; Laegdsgaard 2006; Silliman *et al.* 2009; Thomsen *et al.* 2009; Weis and Butler 2009; Sinclair and Boon 2012; Boon *et al.* 2015). In New South Wales, several ecological communities of coastal floodplains are listed as endangered under the state's threatened-species legislation. These include rainforest, two eucalypt forest communities, a paperbark swamp community, a swamp she-oak community, non-woody freshwater wetlands and coastal salt marsh. Collectively, these cover all the remaining native vegetation on coastal floodplains in New South Wales. Four major threatening processes common to all coastal floodplain communities – 'reclamation' and habitat destruction, hydrological modification, weed infestations and climate change – are discussed below.

### Land claim and habitat destruction

'Reclamation' is the word often used to describe the filling-in of coastal wetlands for human use, but is clearly inappropriate as it implies the regaining of land that was originally non-tidal. Indeed, some authors (e.g. Strong and Ayres 2009: 11) enclose the word in quotation marks to indicate its inappropriateness to describe the loss of coastal floodplains and wetlands. In the work from 2008 to 2011 that provided the first statewide inventory of Victorian coastal wetlands (Boon *et al.* 2015), the term 'land claim' was judged to be better suited to what had been euphemistically termed 'reclamation' in the past. Interestingly, and quite independently, the same word was coined by Thomsen *et al.* (2009: 367) to describe losses in New Zealand coastal marshes. In many ways the word 'land claim' overlaps with what is often called 'habitat destruction', but is preferable in many cases because it indicates that the land has not been merely modified but, in fact, completely cleared and converted to other uses.

Land claim and the related phenomenon of generalised habitat destruction take many forms in coastal floodplains and wetlands, but in almost all cases are prompted by the desire to create land for agricultural, industrial or urban uses. Figure 9.2 showed how close urban housing can penetrate to coastal wetlands. Indeed in this case it is possible that areas of former wetland were filled in to facilitate the housing estate. The consequences for natural systems vary according to which aim is pursued. In the case of agricultural uses, it may be possible for the modified wetland to retain some ecological values (Adam 1990). In-filling for industrial or urban uses, however, usually involves works that raise the soil profile to prevent flooding. The modified landscape then usually retains few or none of its original ecological values. Moreover, the degradation that occurs following in-filling for industrial or urban use is usually compounded by erection of sea walls to further limit tidal inundation, the use of poor-quality or contaminated materials for fill (e.g. rubble and building waste) with the possibility of their leaching of noxious substances into tidal waterways, and the imposition of polluting industries (e.g. petrochemical, steel-making) on the newly formed land. In many cases, the in-filling of coastal wetlands involves the creation of roads, and road-building can have a direct impact in terms of lost land as well as an indirect impact in terms of fragmentation and altered hydrology. Figure 9.6 shows an extreme example of habitat destruction, in fact better described as a land claim, of a coastal wetland; the area, currently a sports field, was previously used as a council waste dump and, before that, was a complex mangrove salt marsh wetland of high ecological value. The remnants of the once-extensive wetland are visible on the right-hand side of the photograph.

Although complete in-filling is the most severe form, other types of habitat destruction also take place in coastal wetlands (Sinclair and Boon 2012; Boon *et al.* 2015). These areas are commonly used as dumping grounds, both legally and illegally, and for inappropriate forms of recreation (e.g. vehicle access) that leave long-lasting impacts. Regardless of the causal process, almost all types of habitat destruction result in the fragmentation of the remaining wetland. Studies of other ecosystem types have shown the role of fragmentation in adversely affecting the ecological values of the remaining system (Hobbs and Yates 2003). These principles are likely to apply to coastal wetlands as well (Laegdsgaard 2006). Large salt marshes, for example, provide habitat for substantial populations

**Figure 9.6:** An extreme example of habitat destruction in a coastal wetland: the conversion of former mangroves and coastal salt marsh to a playing field at Hastings, Victoria. (Photo: Paul Boon.)

of crabs and molluscs absent from smaller salt marshes (Ross *et al.* 2009). As the patch size is further reduced, food webs may become increasingly constrained and might eventually be limited to few, small predators. The migratory birds attracted to large salt marshes probably choose these sites because they offer, among other advantages, good visual protection from predators. Furthermore, fragmentation is unlikely to take place in the absence of other disturbances: a fragmented salt marsh is likely to be one transgressed by roads and other developments that modify hydrological conditions, and thus are afflicted by the other stressors that such developments place on coastal wetlands.

## Hydrological modification

A wide range of developments can affect the hydrology of coastal wetlands, and vegetation in these environments is almost always highly susceptible to altered hydrology (Hickey and Bruce 2010). Impacts can arise from changes in the extent of oceanic (i.e. tidal) influence or of freshwater (often riverine) influence. In tidal portions of floodplains, the most severe outcomes often arise from the construction of floodgates (in New South Wales) or roads (in Victoria and Tasmania) through coastal wetlands, which limit tidal influence in the landward portions. Walking tracks can also obstruct tidal flows, especially if they are constructed of impervious materials such as concrete and lack suitable culverts or other mechanisms to maintain tidal inundation.

Many of the larger floodplains have undergone major transformation as a result of systematic hydrological engineering for development of agricultural land, especially in the early 20th century (East 1935; Pressey and Middleton 1982; Pressey and Griffith 1992; Bird 1993; Keith and Scott 2005). Construction of drain networks and tidal gates channelised much of the floodplain sheet flow, lowered the water table, reduced salinity and oxidised surface soils enabling extensive conversion of wetlands to agricultural land by the early decades of the 20th century (e.g. see Yugovic and Mitchell (2006) and Yugovic (2011) for Koo Wee Rup swamp, Victoria). Construction of floodgates and tidal gates has had profound impacts on coastal floodplain ecosystems, even where wetlands have not subsequently been converted to agriculture. There appears to be a close association between floodplain engineering and enhanced acidity, altered ionic ratios, increases in dissolved organic carbon, and sulfide oxidation in the soil profile (Keith and Scott 2005). In some cases, these changes have led to episodic release of acidic leachates into adjacent waterways, with associated major impacts on aquatic biota and ecosystem processes (White *et al.* 1997; Wilson *et al.* 1999; Johnston *et al.* 2003, 2005).

Where native vegetation remains, hydrological transformation has in some cases resulted in major shifts in biogeochemical processes (Pressey and Middleton 1982) and, more broadly, is likely to have a wide range of ecosystem-scale impacts (Boon 2006). In north-east New South Wales, for example, Johnston *et al.* (2003) attributed an expansion of *Melaleuca quinquenervia* forests into open freshwater wetlands to artificial drainage and shortening of the period over which the wetlands were inundated. Keith *et al.* (2007) inferred a similar change from historical information and aerial photographs on the lower Shoalhaven River floodplain, where *Casuarina glauca* apparently expanded across freshwater meadows and coastal salt marsh after construction of tidal gates. In many areas, livestock grazing may temporarily arrest the transition to woody vegetation. While this model of ecosystem dynamics remains to be fully tested, the legacy of reduced salinity, partial oxidation and acidification of soils appears to be an important pre-requisite for establishment of myrtaceous woody vegetation on abandoned agricultural land of the coastal floodplains.

In other cases, construction of tidal gates and drains appears to have both reduced salinity and increased waterlogging. On the lower Hunter River floodplain, Winning and Saintilan (2009) recorded a spectacular replacement of salt marsh and brackish swamp by reed beds and swamp she-oak woodlands after construction of tidal gates in 1971, coinciding with an increase in mean water depth and reduced salinity of soils and groundwater. A staged opening of the tidal gates is currently underway, with the aim of restoring a more variable saline water regime and increasing the area of salt marsh (Winning and Saintilan 2009). Increased disturbance and nutrient enrichment may also promote expansion of *Phragmites australis* into salt marsh and other wetland types (Minchinton and Bertness 2003).

Alterations can also occur in freshwater inputs to coastal wetlands. The most obvious impact arises from regulation and reduction in flows to rivers that discharge into the ocean. Almost all of the coastal rivers in Victoria and New South are affected by reduced flows caused by extraction for livestock and domestic supplies and, in some cases, also by more extensive extraction for industry and potable use (e.g. Yarra, Latrobe-Thomson-Macalister, Shoalhaven and Hawkesbury-Nepean river systems). Water extraction diminishes flows especially during dry times and reduces the frequency of small- and medium-sized flood events (Pressey and Middleton 1982; Boulton and Brock 1999). More subtle changes to freshwater inputs arise from the discharge of storm water into wetlands, often as a consequence of residential construction or the building of recreational facilities (e.g. golf courses) adjacent to intertidal and coastal wetlands (e.g. the Bellarine Peninsula, Victoria). Increased freshwater discharge in these cases can result in the replacement of halophytic plant communities (e.g. salt marsh) with glycophytic communities (e.g. reed beds). If water is ponded, mangroves can be lost as a result of the permanent inundation. The maintenance of hydrologic connectivity between marine and coastal habitats is therefore essential for the functioning of coastal wetlands.

Less common, but equally disruptive, are modifications to oceanic influences on coastal floodplains and wetlands. Perhaps the best example is provided by the Gippsland Lakes and its fringing wetlands (Bird 1966). Prior to the creation of a permanent opening to the Southern Ocean in 1889 at Lakes Entrance (to aid navigation for coastal shipping), the Gippsland Lakes was a largely freshwater, intermittently open-and-closed coastal lagoon. Bird (1966) argued that increased salinity in the Gippsland Lakes, resulting from the permanent connection with the ocean, would cause swamp paperbark woodlands to replace the existing common reed communities. There is good evidence that swamp paperbark has been progressively replacing common reed in at least some of the fringing wetlands (Boon *et al.* 2008), although the mechanism is as yet unclear (Morris *et al.* 2008). The appearance of mangroves in Cunninghame Arm, one of the most marine-influenced sections of the Gippsland Lakes, since the late 1980s or early 1990s is also notable. Impacts of saline intrusions on coastal floodplains, including on paperbark-dominated wetlands, have been reported also for northern Australia (Winn *et al.* 2006; Cobb *et al.* 2007; Finlayson *et al.* 2013), but are beyond the geographical scope of this review.

## Weed infestations

Wetlands are widely recognised as being particularly susceptible to weed invasions (Zedler and Kercher 2004) and those plant species capable of invading wetlands may pose a serious threat, even though it is often thought (incorrectly) that such areas are immune to weed infestations by virtue of their high salinity and/or periodic waterlogging. A number of authors (e.g. Adam 1990) have noted the susceptibility of the upper levels of salt marshes to invasion by exotic plant species. King *et al.* (1990) argued that while the upper salt marsh had provided habitat for many alien plant species, the potential for them to spread was limited. Recent analyses by Boon *et al.* (2015) suggest, however, that this was an overly optimistic assessment. A large number of aggressive weeds are found in coastal wetlands in south-eastern Australia, including the tussock-forming *Juncus acutus* and *Cortaderia selloana* as well as groundsel (*Baccharis halimifolia*). Exotic taxa in New South Wales coastal salt marsh include *Plantago coronopus*, *Limonium binervosum*, the daisy *Aster subulatus* and the grasses *Parapholis incurva* and *Polypogon monspeliensis* (Morrisey 1995). In Victoria, the most problematic weed species in upper salt marsh include *Lophopyrum ponticum*, *Parapholis incurva*, *Hordeum marinum* and *Juncus acutus*. Other common exotic species include *Atriplex prostrata*, *Parapholis strigosa* and *Hordeum marinum*.

The very lowest levels of Victorian coastal wetlands can be invaded by *Spartina anglica* and/or *Spartina* x *townsendii* (Bird and Boston 1968; Nightingale and Weiller 2005). The case of *Spartina* deserves particular attention, as the species is highly invasive to both mangroves and salt marshes and may, in fact, invade even into areas that were formerly seagrass beds. Boston (1981) provided a detailed history of the introduction of *Spartina* into Australia. *Spartina* is very seriously invasive in mangrove and lower salt marsh vegetation over, potentially, many areas of coastal southern Australia (e.g. see Kriwoken and Hedge 2000 for Tasmania). Presently no other weed species is invasive in these environments, and *Spartina*, with its tolerance to salinity and inundation, is probably unique in the exotic wetland flora. What also makes *Spartina* infestations different from many other coastal-verge weed problems is that the plant was almost always deliberately, not accidentally, introduced into Australian estuaries, at least in the beginning. It has since been able to spread, probably primarily via clonal means.

Forested wetlands are also particularly susceptible to weed invasion. Factors pre-disposing them to invasion include: their naturally high-nutrient status; propensity for further enrichment from agricultural, urban and industrial areas upslope; and exposure to recurring floods that disperse weed propagules and sediments and create gaps by displacing existing vegetation. One-third of all vascular plants recorded in floodplain eucalypt forest communities described by Keith and Scott (2005) were exotic species (D.A. Keith, unpubl. data). The majority of these were ground layer species, particularly asteraceous forbs and Poaceae (grasses), with potential to dominate the biomass and to exclude native species. Thicket-forming woody species, such as *Ligustrum* spp., *Lantana camara*, *Rubus* spp. and *Olea europea*, were among the most frequently recorded species and, similarly, have the potential to exclude native species from the community. *Ligustrum sinense* and *L. camara* were recorded at more than 40% of sampled sites (D.A. Keith, unpubl. data). Exotic vines, including *Anredera cordifolia*, *Asparagus* spp., *Cardiospermum grandiflorum*, *Ipomea* spp., *Macfadyena unguis-cati* and *Passiflora* spp. also pose a significant threat due to their ability to smother and exclude native species.

Floating exotic species such as *Salvinia molesta* and *Eichhornia crassipes* are capable of rapidly smothering open freshwater wetlands and precipitating top-down regime shifts in the aquatic ecosystem (Carpenter 2003). Such regime shifts have been reported in Australia, but as yet only for inland aquatic systems (Morris *et al.* 2003, 2004). It is not clear whether infestations with floating or semi-emergent taxa could commonly occur in estuarine and inter-tidal wetland systems. In these environments high salinity may impose a barrier to infestations by many of the potentially problematic freshwater taxa that have floating and semi-emergent growth forms (e.g. see Toomey and Boon 2007 in the case of *Myriophyllum aquaticum*).

## Climate change and sea-level rise

Estuarine wetlands are among the habitats most likely to be affected adversely by climate change, largely because of their coastal location and inherent susceptibility to sea-level rise, but also because mangrove, salt marsh and estuarine-wetland vegetation is particularly sensitive to subtle changes to salinity and hydrological regimes (Vanderzee 1990; Cahoon *et al.* 2006; Burley *et al.* 2012).

In pre-colonial times, salt marshes and other types of coastal wetland in Australia could respond to natural sea-level rise by retreating into the terrestrial hinterland or by increasing sediment elevation. Today, the first is a possibility only if the hinterland is sufficiently flat, is not alienated by sea walls or other structures, and is not already developed for human use. In very few parts of the south-eastern coast are these conditions met. The second mechanism, increased sediment elevation, remains a tantalising option for coastal wetlands to adapt to sea-level rise without the need for migration into the hinterland (Cahoon *et al.* 2006). If the combined effect of abiotic sediment deposition and biogenic increases in sediment height (e.g. via root production and peat formation) exceed rates of sea-level rise, coastal wetlands could keep track with climate-change induced impacts (for examples in south-eastern Australia see Rogers *et al.* 2006, 2012). If, however, the converse process is dominant and sediments compact rather than expand and become more elevated, coastal wetlands are likely to be doomed (for an example in the Mississippi Delta, USA, see Day *et al.* 2011). Several processes may change the ability of coastal wetlands to increase the elevation of their sediments. These include altered rates of sediment deposition, arising from altered patterns of freshwater discharge or altered rates of catchment erosion, and changes in the rate of below-ground biomass accumulation due to nutrient enrichment shifting the balance between above-ground and below-ground production (see Lovelock *et al.* 2009; Turner *et al.* 2009).

Using a sea level and marine marsh (SLAMM) model adapted for coastal environments in south-eastern Queensland, combined with a high-resolution digital elevation model, information on vegetation–salinity relationships and dispersal rates, Traill *et al.* (2011) predicted a substantial contraction in the distribution of salt marsh and freshwater wetlands, whereas mangroves were predicted to keep pace with sea-level rise over the 21st century. Along much of the south-east Australian coast, salt marsh and mangrove communities are constrained by development in the hinterland and have nowhere to retreat. In other situations, salt marsh may migrate inland, but only at the expense of existing freshwater wetlands, whose range will necessarily contract between the advancing saline wetlands and the topographic limits of flat land. A similar process has been posited to occur along the coastline of the south-eastern United States with increasing sea levels, where forested freshwater wetlands are at particular risk (Doyle *et al.* 2010).

The fourth assessment of global climate change (Intergovernmental Panel on Climate Change 2007) presented a number of projections for future sea-level rises. Under the high emissions scenario, a rise in mean sea levels of between 0.26 and 0.59 m by 2100 was projected. These projections do not take into account consequences of possible ice-sheet collapse on Greenland and western Antarctica, and even higher sea levels are anticipated by other researchers. Rahmstorf *et al.* (2007), for example, argued that the sea-level rise by 2100 would likely be of the order of 0.5 to 1.4 m, and if an exponential relationship between average warming and sea-level rise is adopted, the mean ocean level may rise by up to 5 m by the end of the century (Pittock 2009). The most recent assessment of likely sea-level rise for south-eastern Australia is the report by CSIRO-BOM (2015). It concluded that there was a very high confidence that sea levels would continue to rise. The projected range of sea-level rise by 2030 was 0.07 to 0.19 m above the 1986–2005 level, with only minor differences among emission scenarios. As the century progressed, however, projections became increasingly sensitive to concentration pathways. By 2090, for example, the intermediate emissions case was associated with a sea-level rise of 0.27 to 0.66 m and the high emissions case with a rise of 0.38 to 0.89 m. The report noted that under certain circumstances sea-level rises higher than these may occur.

The implications of such rises in mean sea level for coastal wetlands, especially the inter-tidal ones, are severe. A simple application of the Bruun rule (Walsh 2004) suggests that, under Rahmstorf *et al.*'s (2007) estimates of mean sea-level rise, the coast will retreat by 50–140 m by the end of the century. Such rises are likely to submerge much of the existing area of mangrove and salt marsh, and cause significant saline incursion into the remaining freshwater wetlands of the coastal floodplains. In their review of the sensitivity of Australian wetlands to climate change, Finlayson *et al.* (2013) concluded that sea-level rise threatened wetlands along the whole of the nation's coastline.

In addition to its impact on mean and extreme sea-level rise, climate change will have significant effects on other aspects of the coastal environment, including temperature, rainfall and evaporation.

Changes in these three weather-related variables are likely to affect not only the floristic composition of estuarine wetlands, but also the spatial relationships between coastal freshwater and brackish-water wetlands, salt marshes and mangroves. Gradual warming, for example, will decrease the incidence and severity of frosts in coastal southern Australia. Frosts probably limit the vigour and distribution of the mangrove *Avicennia marina* subsp. *australasica* along the southern Victorian coast (Oliver 1982) and it may be expected that mangroves could become more aggressive colonisers of areas currently vegetated with salt marshes with further climate change. Already there is strong evidence that mangroves have transgressed into many salt marshes along the New South Wales coast (Saintilan and Williams 1999); similar cases of mangrove expansion have been reported also in the US (Comeaux *et al.* 2012). The strong latitudinal trend in mangrove diversity in Australia suggests that northern mangrove species will expand southwards as air and water temperatures increase, particularly as propagule dispersal may not be a limiting factor for many taxa.

It is often assumed that different types of coastal wetlands will respond in the same way to climate change and to sea-level rise, but Friess *et al.* (2012) have shown that this is an unwarranted assumption. It is likely that mangroves, for example, will respond in different ways than salt marshes. Moreover, sea-level rise may result in mangroves expanding in area at the expense of salt-intolerant woody taxa of coastal wetlands, which are likely to be compromised by saline intrusions up coastal floodplain rivers and into estuaries (Doyle *et al.* 2010). It is highly likely, therefore, that climate change will affect differently the spatial and temporal patterning of different coastal wetlands types, and their resident taxa. A good example is provided by the saline wetlands that fringe the Victorian coast. As noted earlier, those parts of coastal Victoria in orographic rain shadows often support extensive areas of hypersaline salt marsh, generally vegetated with *Tecticornia pergranulata*, whereas the wetter parts of the coast support other salt marsh taxa, typically including *Sarcocornia* spp. A shift to a drier, hotter climate may well favour the more desiccation-tolerant and salt-tolerant taxa over others. Woo and Takekawa (2012) have reported how one North American salt marsh species, *Sarcocornia pacifica*, might respond to altered inundation and salinity regimes.

Climate change projections for southern Australia generally indicate that wetlands are likely to dry more frequently and for longer periods than in the recent past. Shallow coastal wetlands will be particularly at risk, not only from desiccation but from hypersalinisation. Conversely, rainfall is projected to increase in subtropical latitudes of coastal New South Wales (Pitman and Perkins 2008), and consequent increases in catchment runoff may alter proportions of freshwater and saltwater inputs, initiating changes in the distributions of different floodplain communities (Minchinton 2002). This may facilitate the increased migration of exotic species between hydrologically connected wetlands.

Changes in $CO_2$ concentrations are also likely to affect competitive interactions among salt marsh plants, especially among taxa with different photosynthetic pathways (Hughes 2003). Plants possessing the $C_3$ photosynthetic pathway (e.g. most salt marsh and mangrove taxa) have high rates of photorespiration and a low photosynthetic capacity (Bonan 2002). In contrast, plants with $C_4$ photosynthetic pathways (e.g. *Spartina* and many salt-tolerant grasses, such as *Distichlis*) show little light saturation and, at full sunlight, can be twice as productive as comparable $C_3$ plants. Moreover, because of their more efficient use of $CO_2$, $C_4$ plants use less water to achieve the same rate of primary production as $C_3$ plants. The optimal temperature for $C_4$ plants is usually greater than that of $C_3$ plants. Thus for a number of physiological reasons, a warmer, drier climate would be expected in general to favour $C_4$ plants over $C_3$ plants. The complication is that the $CO_2$ compensation point is lower for $C_3$ than $C_4$ plants. We can therefore expect a complex interaction, and likely shifts between taxa (and plant life forms) with different photosynthetic pathways, in wetlands subject to climate change. Adam (2009) concluded that it was almost certain that there would be changes in the balance between $C_3$ and $C_4$

plants in coastal salt marshes as a result of climate change.

## CONCLUSIONS

Temperate coastal floodplain wetlands are among the most depleted and degraded ecosystems in Australia. They are also among the least well studied (Boon 2012). However, they are adaptable and perhaps resilient systems. Recurrent flushing by seawater and freshwater offers a mechanism by which some of the essential wetland functions and environmental properties can be restored. In addition, much of the native plant and animal biota is widely dispersed, providing a means by which their populations may be sustained, once an appropriate level of wetland function is restored to degraded sites. Although complete restoration of wetland structure and function is largely beyond our current capacity (Zedler and Callaway 1999), the resilience of coastal wetlands suggests that projects aiming to restore inundation regimes, such as opening of tidal gates on Hexham swamp (Winning and Saintilan 2009), have a good chance of restoring many of the indigenous vegetation patterns and processes.

Even where rehabilitation is only partially successful, the resulting improved wetlands can contribute to the viability of other remnant wetlands that have so far survived anthropogenic transformation of the Australian temperate coastal landscape. Although they are typically small fragments that have lost some of their component species, these remnants include some outstanding examples of unique ecosystems. The persistence of remnant coastal floodplain wetlands is contingent on effective zoning and sustainable design of coastal development, management of freshwater flows, storm water and tides, control of invasive species and anticipatory adaptation to pending climate change and sea-level rise.

## ACKNOWLEDGEMENTS

Portions of this chapter were based on the more detailed work undertaken by Sinclair and Boon (2012) and Boon *et al.* (2015). Components of these earlier studies were reviewed by Paul Adam and Neil Saintilan, and we thank them for their insightful comments on those early drafts. PIB was supported by the Gippsland Lakes Environment Fund project # EG 1314.30.297. DK was supported by Environmental Trust Grant # RD/2007/0014.

## REFERENCES

Adam P (1981a) Australian saltmarshes. *Wetlands (Australia)* **1**, 8–10.

Adam P (1981b) Saltmarsh plants of New South Wales. *Wetlands (Australia)* **1**, 11–19.

Adam P (1990) *Saltmarsh Ecology*. Cambridge University Press, Cambridge, UK.

Adam P (1994) Saltmarsh and mangrove. In *Australian Vegetation*. 2nd edn. (Ed. RH Groves) pp. 395–435. Cambridge University Press, Cambridge, UK.

Adam P (2002) Saltmarshes in a time of change. *Environmental Conservation* **29**, 39–61. doi:10.1017/S0376892902000048

Adam P (2009) Australian saltmarshes in global context. In *Australian Saltmarsh Ecology*. (Ed. N Saintilan) pp. 1–21. CSIRO Publishing, Melbourne.

Adam P, Unwin N, Weiner P, Sim I (1985) *Coastal Wetlands of New South Wales*. Coastal Council of New South Wales, Sydney.

Adam P, Wilson NC, Huntley B (1988) The phytosociology of coastal saltmarsh vegetation in New South Wales. *Wetlands (Australia)* **7**, 35–57.

Ahern KS, Udy JW, Pointon SM (2006) Investigating the potential for groundwater from different vegetation, soil and landuses to stimulate blooms of the cyanobacterium, *Lyngbya majuscula*, in coastal waters. *Marine and Freshwater Research* **57**, 177–186. doi:10.1071/MF05115

Barson MM (1977) Tidal salt marshes in Victoria. *Victoria's Resources* **18**, 11–14.

Barson MM, Calder DM (1981) The vegetation of the Victorian coast. *Proceedings of the Royal Society of Victoria* **92**, 55–65.

Bateman IJ, Abson D, Beaumont N, Darnell A, Fezzi C, Hanley N, Kontoleon A, Maddison D, Morling P, Morris J, Mourato S, Pascual U, Perino G,

Sen A, Tinch D, Turner RK, Valatin G (2011) Economic values from ecosystems. In *The UK National Ecosystem Assessment. Technical Report.* pp. 1067–1152. United Nations Environment Program, Cambridge, UK.

Beeftink WG (1977) The coastal salt marshes of western and northern Europe: an ecological and phytosociological approach. In *Ecosystems of the World. 1. Wet Coastal Ecosystems.* (Ed. VJ Chapman) pp. 109–155. Elsevier, Amsterdam.

Bird ECF (1962) The swamp paper-bark. *The Victorian Naturalist* **79**, 72–81.

Bird ECF (1966) The impact of man on the Gippsland Lakes, Australia. In *Geography as Human Ecology. Methodology by Example.* (Eds SR Eyre and GRJ Jones) pp. 55–73. Edward Arnold, London.

Bird ECF (1993) *The Coast of Victoria. The Shaping of Scenery.* Melbourne University Press, Melbourne.

Bird ECF, Boston KG (1968) *Spartina* in Victoria. *The Victorian Naturalist* **85**, 11–18.

Bleakney JS (1972) Ecological implications of annual variation in tidal extremes. *Ecology* **53**, 933–938. doi:10.2307/1934311

Bonan GB (2002) *Ecological Climatology. Concepts and Applications.* Cambridge University Press, Cambridge, UK.

Boon PI (2006) Biogeochemistry, ecology and management of hydrologically dynamic wetlands. In *The Ecology of Freshwater and Estuarine Wetlands.* (Eds DP Batzer and RR Sharitz) pp. 115–176. University of California Press, Berkeley, CA.

Boon PI (2011) The rehabilitation of coastal wetlands: why small-scale variations in topography are critical to success. *Australasian Plant Conservation* **20**, 9–10.

Boon PI (2012) Coastal wetlands of temperate eastern Australia: will Cinderella ever go to the ball? *Marine and Freshwater Research* **63**, 845–855. doi:10.1071/MF12205

Boon PI, Raulings E, Roache M, Morris K (2008) Vegetation changes over a four-decade period in Dowd Morass, a brackish-water wetland of the Gippsland Lakes, south-eastern Australia. *Proceedings of the Royal Society of Victoria* **120**, 403–418.

Boon PI, Allen T, Carr G, Frood D, Harty C, McMahon A, Mathews S, Rosengren N, Sinclair S, White M, Yugovic J (2015) Coastal wetlands of Victoria, south-eastern Australia: providing the inventory and condition information needed for their effective management and conservation. *Aquatic Conservation: Marine and Freshwater Ecosystems* **25**, 454–479. doi:10.1002/aqc.2442

Boorman LA (2009) The role of freshwater flows on salt marsh growth and development. In *Coastal Wetlands. An Integrated Ecosystem Approach.* (Eds GME Perillo, E Wolanski, DR Cahoon and MM Brinson) pp. 493–514. Elsevier, Amsterdam.

Boston KG (1981) 'The introduction of *Spartina townsendii* (s.l.) to Australia.' Occasional Paper N.6. Melbourne State College, Melbourne.

Boulton MA, Brock MA (1999) *Australian Freshwater Ecology: Processes and Management.* Geneagles Publishing, Glen Osmond, SA.

Bowkett LA, Kirkpatrick JB (2003) Ecology and conservation of remnant *Melaleuca ericifolia* stands in the Tamar Valley, Tasmania. *Australian Journal of Botany* **51**, 405–413. doi:10.1071/BT02071

Bridgewater PB (1975) Peripheral vegetation of Westernport Bay. *Proceedings of the Royal Society of Victoria* **87**, 69–78.

Bridgewater PB (1982) Phytosociology of coastal saltmarshes in the Mediterranean climatic region of Australia. *Phytocoenologia* **10**, 257–296. doi:10.1127/phyto/10/1982/257

Bridgewater PB (1999) Mangroves. In *Flora of Australia. Volume 1. Introduction.* 2nd edn. (Ed. AE Orchard) pp. 416–420. Department of the Environment and Heritage, Canberra and CSIRO Publishing, Melbourne.

Bridgewater PB, Cresswell ID (2003) Identifying biogeographic patterns in Australian saltmarsh and mangal systems: a phytogeographic analysis. *Phytocoenologia* **33**, 231–250. doi:10.1127/0340-269X/2003/0033-0231

Bridgewater PB, Kaeshagen D (1979) Changes induced by adventive species in Australian plant communities. In *Werden und Vergehen von Plflanzengesellschaften.* (Eds O Wilmanns and R Tüxen) pp. 561–579. Cramer, Lehre, Germany.

Bridgewater PB, Rosser C, de Corona A (1981) *The Saltmarsh Plants of Southern Australia.* Botany Department, Monash University, Clayton.

Bucher D, Saenger P (1991) An inventory of Australian estuaries and enclosed marine waters: an overview of results. *Australian Geographical Studies* **29**, 370–381. doi:10.1111/j.1467-8470.1991.tb00726.x

Bureau of Meteorology, Walsh NG (1993) Climate of Victoria. In *Flora of Victoria. Volume 1. Introduction.* (Eds DB Foreman and NG Walsh) pp. 47–60. Inkata Press, Melbourne.

Burley JG, McAllister RRJ, Collins KA, Lovelock CE (2012) Integration, synthesis and climate change adaptation: a narrative based on coastal wetlands at the regional scale. *Regional Environmental Change* **12**, 581–593. doi:10.1007/s10113-011-0271-4

Cahoon DR, Hensel PF, Spencer T, Reed DJ, McKee KL, Saintilan N (2006) Coastal wetland vulnerability to relative sea-level rise: wetland elevation trends and process controls. *Ecological Studies* **190**, 271–292. doi:10.1007/978-3-540-33187-2_12

Callaghan AR (1948) From tea-tree swamp to pasture. *Walkabout* **49**, 29–32.

Carpenter S (2003) *Regime Shifts in Lake Ecosystems: Pattern and Variation.* International Ecology Institute, Oldendorf/Luhe, Germany.

Carr GW (1979) 'Survey of Victorian coastal saltmarsh distribution in relation to the habitat of the orange-bellied parrot.' Report to Kinhill Planners and ICI Australia. Ecology Australia, Fairfield, Victoria.

Carr G (2012a) Overview of coastal saltmarsh and mangrove vegetation in Victoria. In *Estuary Plants and What's Happening to Them in South-eastern Australia.* (Eds G Sainty, J Hosking, G Carr and P Adam) pp. 376–379. Sainty and Associates, Potts Point, NSW.

Carr G (2012b) Inventory of Victorian marine, estuarine and saltmarsh vascular plant species. In *Estuary Plants and What's Happening to Them in South-eastern Australia.* (Eds G Sainty, J Hosking, G Carr and P Adam) pp. 379–421. Sainty and Associates, Potts Point, NSW.

Carter ES, White SM, Wilson AM (2008) Variation in groundwater salinity in a tidal salt marsh basin, North Inlet estuary, South Carolina. *Estuarine, Coastal and Shelf Science* **76**, 543–552. doi:10.1016/j.ecss.2007.07.049

Chapman MG, Underwood AJ (1995) Mangrove forests. In *Coastal Marine Ecology of Temperate Australia.* (Eds AJ Underwood and MG Chapman) pp. 187–204. UNSW Press, Sydney.

Claridge D, Burnett J (1993) *Mangroves in Focus.* Wet Paper, Ashmore, QLD.

Clarke LD, Hannon NJ (1967) The mangrove swamp and salt marsh communities of the Sydney district I. Vegetation, soils and climate. *Journal of Ecology* **55**, 753–771. doi:10.2307/2258423

Clarke LD, Hannon NJ (1969) The mangrove swamp and saltmarsh communities of the Sydney district. II. The holocoenotic complex with particular reference to physiography. *Journal of Ecology* **57**, 213–234. doi:10.2307/2258216

Clarke LD, Hannon NJ (1970) The mangrove swamp and salt marsh communities of the Sydney district III. Plant growth in relation to salinity and waterlogging. *Journal of Ecology* **58**, 351–369. doi:10.2307/2258276

Clough BF (Ed.) (1982) *Mangrove Ecosystems in Australia. Structure, Function and Management.* Australian Institute of Marine Science, Townsville and Australian National University Press, Canberra.

Cobb SM, Saynor MJ, Eliot M, Eliot I, Hall R (2007) 'Saltwater intrusion and mangrove encroachment of coastal wetlands in the Alligator rivers region, Northern Territory, Australia.' Supervising Scientist Report 191, Darwin.

Coles R, McKenzie L, Campbell S (2003) The seagrasses of eastern Australia. In *World Atlas of Seagrasses.* (Eds EP Green and FT Short) pp. 119–133. University of California Press, Berkeley, CA.

Collins MI (1921) On the mangrove and saltmarsh vegetation near Sydney, New South Wales, with special reference to Cabbage Tree Creek, Port Hacking. *Proceedings of the Linnean Society of New South Wales* **46**, 376–392.

Comeaux RS, Allison MA, Bianchi TS (2012) Mangrove expansion in the Gulf of Mexico with climate change: implications for wetland health and resistance to rising sea levels. *Estuarine,*

*Coastal and Shelf Science* **96**, 81–95. doi:10.1016/j.ecss.2011.10.003

Creighton C, Boon PI, Sheaves M, Brookes J (2015) Revitalising Australia's estuaries – what benefits at what cost? *Marine and Freshwater Research* **66**, 493–507. doi:org/10.1071/MF14041

Crowe AS, Shikaze SG (2004) Linkages between groundwater and coastal wetlands of the Laurentian Great Lakes. *Aquatic Ecosystem Health & Management* **7**, 199–213. doi:10.1080/14634980490461560

CSIRO-BOM (2015) *Southern Slopes Cluster Report. Climate Change in Australia Projections for Australia's Natural Resource Management Regions.* CSIRO and Bureau of Meteorology, Canberra, <http://www.climatechangeinaustralia.gov.au/en/impacts-and-adaptation/nrm-regions/>.

Curtis WM, Somerville J (1948) Boomer marsh – a preliminary botanical and historical survey. Papers and Proceedings of the Royal Society of Tasmania 151–157.

Day JW, Kemp GP, Reed DJ, Choon DR, Boumans RM, Suhayda JM, Gambrell R (2011) Vegetation death and rapid loss of surface elevation in two contrasting Mississippi delta marshes: the role of sedimentation, autocompaction and sea-level rise. *Ecological Engineering* **37**, 229–240. doi:10.1016/j.ecoleng.2010.11.021

de Jong N (1997) An analysis of plant communities at Coomonderry swamp with comparisons to other wetlands on the south coast of New South Wales. *Cunninghamia* **5**, 81–128.

Department of Sustainability and Environment (2010). *EVC/Bioregion Benchmark for Vegetation Quality Assessment. Gippsland Plains Bioregion.* Government of Victoria, Melbourne.

Department of Sustainability and Environment (2012) *A Field Guide to Victorian Wetland Ecological Vegetation Classes for the Index of Wetland Condition.* 2nd edn. Government of Victoria, Melbourne.

Doyle TW, Krauss KW, Conner WH, From AS (2010) Predicting the retreat and migration of tidal forests along the northern Gulf of Mexico under sea-level rise. *Forest Ecology and Management* **259**, 770–777. doi:10.1016/j.foreco.2009.10.023

Duke N (2006) *Australia's Mangroves. The Authoritative Guide to Australia's Mangrove Plants.* University of Queensland Press, St Lucia, QLD.

East LR (1935) Swamp reclamation in Victoria. *Journal of the Institution of Engineers, Australia* **7**, 77–91.

Environment Australia (2001) *A Directory of Important Wetlands in Australia.* 3rd edn. Environment Australia, Canberra, ACT.

Finlayson CM, Davis JA, Gell PA, Kingsford RT, Parton KA (2013) The status of wetlands and the predicted effects of climate global climate change: the situation in Australia. *Aquatic Sciences* **75**, 73–93. doi:10.1007/s00027-011-0232-5

Floyd AG (1990) *Australian Rainforests in New South Wales.* National Parks and Wildlife Service of New South Wales, Sydney.

Friess DA, Krauss KW, Horstman EM, Balke T, Bouma TJ, Galli D, Webb EL (2012) Are all intertidal wetlands naturally created equal? Bottlenecks, thresholds and knowledge gaps to mangrove and saltmarsh ecosystems. *Biological Reviews of the Cambridge Philosophical Society* **87**, 346–366. doi:10.1111/j.1469-185X.2011.00198.x

Galloway RW (1981) An inventory of Australia's coastal lands. *Australian Geographical Studies* **19**, 107–116. doi:10.1111/j.1467-8470.1981.tb00379.x

Galloway RW (1982) Distribution and physiographic patterns of Australian mangroves. In *Mangrove Ecosystems in Australia. Structure, Function and Management.* (Ed. BF Clough) pp. 31–54. Australian Institute of Marine Science, Townsville and Australian National University, Canberra, ACT.

Gibson N, Williams K, Marsden-Smedley J, Brown MJ (1987) Regeneration characteristics of a swamp forest in northwestern Tasmania. *Proceedings of the Royal Society of Tasmania* **121**, 93–99.

Goodrick GN (1970) 'A survey of wetlands of coastal New South Wales.' Technical memorandum no. 5. CSIRO, Canberra.

Hamilton AA (1919) An ecological study of the saltmarsh vegetation in the Port Jackson district. *Proceedings of the Linnean Society of New South Wales* **44**, 463–513.

Hamilton-Brown S, Boon PI, Robinson R, Morris K, Raulings EJ (2009) Aerial seed storage in *Melaleuca ericifolia* Sm. (swamp paperbark): environmental triggers for seed release. *Hydrobiologia* **620**, 121–133. doi:10.1007/s10750-008-9623-6

Harris S, Kitchener A (2005) *From Forest to Fjaeldmark: Descriptions of Tasmania's Vegetation*. Department of Primary Industries, Parks, Water and Environment, Hobart.

Harty C (1997) *Mangroves in New South Wales and Victoria*. Vista, Melbourne.

Hatton M, Boon PI, Robinson RR (2008) Distribution of clonal and non-clonal wetlands plants at Clydebank Morass, Gippsland Lakes, in relation to elevation and salinity gradients. *The Victorian Naturalist* **125**, 11–18.

Hickey D, Bruce E (2010) Examining tidal inundation and salt marsh vegetation distribution patterns using spatial analysis (Botany Bay, Australia). *Journal of Coastal Research* **26**, 94–102. doi:10.2112/08-1089.1

Hobbs RJ, Yates CJ (2003) Turner review no 7: impacts of ecosystem fragmentation on plant populations: generalizing the idiosyncratic. *Australian Journal of Botany* **51**, 471–488. doi:10.1071/BT03037

Hughes L (2003) Climate change and Australia: trends, projections and impacts. *Austral Ecology* **28**, 423–443. doi:10.1046/j.1442-9993.2003.01300.x

Hutchings PA, Recher HF (1982) The fauna of Australian mangroves. *Proceedings of the Linnean Society of New South Wales* **106**, 83–121.

Hutchings PA, Saenger P (1987) *The Ecology of Mangroves*. University of Queensland Press, Brisbane.

Intergovernmental Panel on Climate Change (2007) *Climate Change 2007 – Synthesis Report, a Report of the Intergovernmental Panel on Climate Change, IPCC*. Cambridge University Press, Cambridge, UK.

Johnston SG, Slavich PG, Sullivan LA, Hirst P (2003) Artificial drainage of floodwaters from sulfidic backswamps: effects on deoxygenation in an Australian estuary. *Marine and Freshwater Research* **54**, 781–795. doi:10.1071/MF02016

Johnston SG, Slavich PG, Hirst P (2005) Changes in surface water quality after inundation of acid sulfate soils of different vegetation cover. *Australian Journal of Soil Research* **43**, 1–12. doi:10.1071/SR04073

Keith D (2004) *Ocean Shores to Desert Dunes. The Native Vegetation of New South Wales and the ACT.* Department of Environment and Conservation, Sydney.

Keith DA, Scott J (2005) Native vegetation of coastal floodplains – a diagnosis of the major plant communities in New South Wales. *Pacific Conservation Biology* **11**, 81–104.

Keith DA, Simpson C, Tozer MG, Rodoreda S (2007) Contemporary and historical descriptions of the vegetation of Brundee and saltwater swamps on the lower Shoalhaven River floodplain, southeastern Australia. *Proceedings of the Linnean Society of New South Wales* **128**, 123–153.

Kelleway J, Williams RJ, Laegdsgaard P (2009) Mapping, assessment and monitoring of saltmarshes. In *Australian Saltmarsh Ecology.* (Ed. N Saintilan) pp. 211–229. CSIRO Publishing, Melbourne.

King RJ, Adam P, Kuo J (1990) Seagrasses, mangroves and saltmarsh plants. In *Biology of Marine Plants.* (Eds MN Clayton and RJ King). pp. 213–239. Longman Cheshire, Melbourne.

Kirkpatrick JB, Glasby J (1981) *Salt Marshes in Tasmania – Distribution, Community Composition and Conservation*. Occasional Paper 8, Department of Geography, University of Tasmania, Hobart.

Kirkpatrick JB, Harwood CE (1983) Plant communities of Tasmanian wetlands. *Australian Journal of Botany* **31**, 437–451. doi:10.1071/BT9830437

Krest JM, Moore WS, Gardner LR, Morris JT (2000) Marsh nutrient export supplied by groundwater discharge: evidence from radium measurements. *Global Biogeochemical Cycles* **14**, 167–176. doi:10.1029/1999GB001197

Kriwoken LK, Hedge P (2000) Exotic species and estuaries: managing *Spartina anglica* in Tasmania, Australia. *Ocean and Coastal Management* **43**, 573–584. doi:10.1016/S0964-5691(00)00047-8

Ladiges PY, Foord PC, Willis RJ (1981) Salinity and waterlogging tolerance of some populations of *Melaleuca ericifolia* Smith. *Australian Journal of Ecology* **6**, 203–215. doi:10.1111/j.1442-9993.1981.tb01291.x

Laegdsgaard P (2006) Ecology, disturbance and restoration of coastal saltmarsh in Australia: a review. *Wetlands Ecology and Management* **14**, 379–399. doi:10.1007/s11273-005-8827-z

Lear R and Turner T (1977) *Mangroves of Australia.* University of Queensland Press, St Lucia, QLD.

Lovelock CE, Ball MC, Martin KC, Feller IC (2009) Nutrient enrichment increases mortality of mangroves. *PLoS ONE* **4**, e5600. doi:10.1371/journal.pone.0005600

Macnae W (1966) Mangroves in eastern and southern Australia. *Australian Journal of Botany* **14**, 67–104. doi:10.1071/BT9660067

Minchinton T (2002) Precipitation during El Niño correlates with increasing spread of *Phragmites australis* in New England, USA, coastal marshes. *Marine Ecology Progress Series* **242**, 305–309. doi:10.3354/meps242305

Minchinton T, Bertness M (2003) Disturbance-mediated competition and the spread of *Phragmites australis* in a coastal marsh. *Ecological Applications* **13**, 1400–1416. doi:10.1890/02-5136

Morgan GJ (1986) *A Survey of Macrobenthos in the Waters of Corner Inlet and the Nooramunga, Southern Victoria, with an Assessment of the Extent of* Posidonia *Seagrass.* Fisheries and Wildlife Paper 31. Department of Conservation, Forests and Lands, Melbourne.

Morris KL, Bailey PC, Boon PI, Hughes L (2003) Stable states in the aquatic vegetation of shallow lakes. II. A novel mechanism for the catastrophic loss of macrophytes in shallow lakes. *Marine and Freshwater Research* **54**, 185–200. doi:10.1071/MF02002

Morris KL, Harrison KA, Bailey PC, Boon PI (2004) Domain shifts in the aquatic vegetation of shallow urban lakes: the relative roles of light limitation and anoxia in the catastrophic loss of the submerged angiosperm *Vallisneria americana. Marine and Freshwater Research* **55**, 749–758. doi:10.1071/MF03193

Morris K, Boon PI, Raulings EJ, White SE (2008) Floristic shifts in wetlands: the effects of environmental variables on the interaction between *Phragmites australis* (common reed) and *Melaleuca ericifolia* (swamp paperbark). *Marine and Freshwater Research* **59**, 187–204. doi:10.1071/MF07072

Morrisey D (1995) Saltmarshes. In *Coastal Marine Ecology of Temperate Australia.* (Eds AJ Underwood and MG Chapman) pp. 205–220. University of New South Wales Press, Sydney.

Morrisey DJ, Swales A, Dittman S, Morrison MA, Lovelock CE, Beard CM (2010) The ecology and management of temperate wetlands. *Oceanography and Marine Biology – an Annual Review* **48**, 43–160.

Nightingale ME, Weiller CM (2005) *Spartina.* In *Flora of Australia. Volume 44B. Poaceae 3.* (Ed. K Mallet) pp. 310–311. Department of the Environment and Heritage, Canberra.

Oliver J (1982) The geographic and environmental aspects of mangrove communities: climate. In *Mangrove Ecosystems in Australia.* (Ed. BF Clough) pp. 19–30. Australian Institute of Marine Science, Townsville and Australian National University, Canberra.

Orchard AE (Ed.) (1999) *Flora of Australia. Volume 1. Introduction.* 2nd edn. Department of the Environment and Heritage, Canberra and CSIRO Publishing, Melbourne.

OzCoasts (2009) *National Intertidal/Subtidal Benthic Habitat.* Geoscience Australia, Australian Government, <http://www.ozcoasts.org.au/nrm_rpt/habitat_extent.jsp>.

Patton RT (1942) Ecological studies in Victoria. IV. Salt marsh. *Proceedings of the Royal Society of Victoria* **54**, 131–144.

Perillo GME (2009) Tidal courses: classification, origin and functionality. In *Coastal Wetlands: an Integrated Ecosystem Approach.* (Eds GME Perillo, E Wolanski, DR Cahoon and MM Brinson) pp. 185–209. Elsevier, Amsterdam.

Perillo GME, Wolanski E, Cahoon DR, Brinson MM (Eds) (2009) *Coastal Wetlands: an Integrated Ecosystem Approach.* Elsevier, Amsterdam.

Pidgeon IM (1940) The ecology of the central coast of New South Wales. III. Types of primary succession. *Proceedings of the Linnean Society of New South Wales* **65**, 221–249.

Pitman AJ, Perkins SE (2008) Regional projections of future seasonal and annual changes in rainfall and temperature over Australia based on skill-selected AR4 models. *Earth Interactions* **12**, 1–50. doi:10.1175/2008EI260.1

Pittock AB (2009) *Climate Change*. 2nd edn. CSIRO Publishing, Melbourne.

Potts BM, Kantvilas G, Jarman SJ (2006) *Janet Somerville's Botanical History of Tasmania 1642–1820*. University of Tasmania and Tasmanian Museum and Art Gallery, Hobart.

Prahalad VN, Kirkpatrick JB, Mount RE (2011) Tasmanian coastal saltmarsh community transitions associated with climate change and relative sea level rise 1975–2009. *Australian Journal of Botany* **59**, 741–748.

Pressey RL (1989a) Wetlands of the lower Clarence floodplain, northern coastal New South Wales. *Proceedings of the Linnean Society of New South Wales* **111**, 143–155.

Pressey RL (1989b) Wetlands of the lower Macleay floodplain, northern coastal New South Wales. *Proceedings of the Linnean Society of New South Wales* **111**, 157–168.

Pressey RL, Griffith SJ (1992) Vegetation of the coastal lowlands of Tweed shire, northern New South Wales, species and conservation. *Proceedings of the Linnean Society of New South Wales* **113**, 203–243.

Pressey RL, Middleton M (1982) Impacts of flood mitigation works on coastal wetlands in New South Wales. *Wetlands (Australia)* **2**, 27–44.

Rahmstorf S, Cazenave A, Church JA, Hansen JE, Keeling RF, Parker DE, Somerville RCJ (2007) Recent climate observations compared to projections. *Science* **316**, 709. doi:10.1126/science.1136843

Ranwell DS (1972) *Ecology of Salt Marshes and Sand Dunes*. Chapman & Hall, London.

Raulings E, Boon PI, Bailey PC, Morris K, Roache MC, Robinson RR (2007) Rehabilitation of swamp paperbark (*Melaleuca ericifolia*) wetlands in south-eastern Australia: effects of hydrology, microtopography, plant age and planting technique on the success of community-based revegetation trials. *Wetlands Ecology and Management* **15**, 175–188. doi:10.1007/s11273-006-9022-6

Raulings E, Morris K, Roache M, Boon PI (2010) The importance of water regimes operating at small spatial scales for the diversity and structure of wetland vegetation. *Freshwater Biology* **55**, 701–715. doi:10.1111/j.1365-2427.2009.02311.x

Raulings E, Morris K, Roache M, Boon PI (2011) Is hydrological manipulation an effective management tool for rehabilitating chronically flooded, brackish-water wetlands? *Freshwater Biology* **56**, 2347–2369. doi:10.1111/j.1365-2427.2011.02650.x

Roberts D (1985) *From Swampland to Farmland. History of the Koo-Wee-Rup Flood Protection District*. Rural Water Commission of Victoria, Melbourne.

Robertson AI, Alongi DM (Eds) (1992) *Tropical Mangrove Ecosystems*. American Geophysical Union, Washington, DC.

Robinson RR, Boon PI, Bailey PC (2006) Germination characteristics of *Melaleuca ericifolia* Sm. (swamp paperbark) and their implications for the rehabilitation of coastal wetlands. *Marine and Freshwater Research* **57**, 703–711. doi:10.1071/MF06006

Robinson RR, Boon PI, Sawtell N, James L, Cross R (2008) Effects of environmental conditions on the production of hypocotyl hairs in seedlings of *Melaleuca ericifolia* (swamp paperbark). *Australian Journal of Botany* **56**, 564–573. doi:10.1071/BT06186

Robinson RW, James EA, Boon PI (2012) Population structure in the woody wetland plant *Melaleuca ericifolia* Sm. (Myrtaceae): an analysis using historical aerial photographs and molecular techniques. *Australian Journal of Botany* **60**, 9–19. doi:10.1071/BT11292

Rogers K, Wilton KM, Saintilan N (2006) Vegetation dynamics and surface elevation dynamics in estuarine wetlands of southeast Australia. *Estuarine, Coastal and Shelf Science* **66**, 559–569. doi:10.1016/j.ecss.2005.11.004

Rogers K, Saintilan N, Copeland C (2012) Modelling wetland surface elevation dynamics and its application to forecasting the effects of sea-level rise on estuarine wetlands. *Ecological Modelling* **244**, 148–157. doi:10.1016/j.ecolmodel.2012.06.014

Ross P, Minchinton T, Ponder W (2009). The ecology of molluscs in Australian saltmarshes. In *Australian Saltmarsh Ecology*. (Ed. N Saintilan) pp. 75–114. CSIRO Publishing, Melbourne.

Saenger P, Specht MM, Specht RL, Chapman VJ (1977) Mangrove and coastal salt-marsh communities in Australasia. In *Ecosystems of the World. 1. Wet Coastal Ecosystems*. (Ed. VJ Chapman) pp. 293–345. Elsevier, Amsterdam.

Saintilan N (Ed.) (2009a) *Australian Saltmarsh Ecology*. CSIRO Publishing, Melbourne.

Saintilan N (2009b) Biogeography of Australian saltmarsh plants. *Austral Ecology* **34**, 929–937.

Saintilan N (2009c) Distribution of Australian saltmarsh plants. In *Australian Saltmarsh Ecology*. (Ed. N Saintilan) pp. 23–52. CSIRO Publishing, Melbourne.

Saintilan N, Williams R (1999) Mangrove transgression into saltmarsh environments in New South Wales, Australia. *Global Ecology and Biogeography* **8**, 117–124. doi:10.1046/j.1365-2699.1999.00133.x

Saintilan N, Williams RJ (2000) Short note: the decline of saltmarsh in southeast Australia: results of recent surveys. *Wetlands (Australia)* **18**, 49–59.

Sainty G, Hosking J, Carr G, Adam P (Eds) (2012) *Estuary Plants and What's Happening to Them in South-eastern Australia*. Sainty and Associates, Potts Point, NSW.

Salter J, Morris K, Bailey PCB, Boon PI (2007) Interactive effects of salinity and water depth on the growth of seedling swamp paperbark (*Melaleuca ericifolia* Sm). *Aquatic Botany* **86**, 213–222. doi:10.1016/j.aquabot.2006.10.002

Salter J, Morris K, Read J, Boon PI (2010a) Impact of long-term, saline flooding on condition and reproduction of the clonal wetland tree, *Melaleuca ericifolia* (Myrtaceae). *Plant Ecology* **206**, 41–57. doi:10.1007/s11258-009-9623-2

Salter J, Morris K, Read J, Boon PI (2010b) Understanding the potential effects of water regime and salinity on recruitment of *Melaleuca ericifolia* Sm. *Aquatic Botany* **92**, 200–206. doi:10.1016/j.aquabot.2009.11.008

Silliman BR, Grosholz ED, Bertness MD (Eds) (2009) *Human Impacts on Salt Marshes: a Global Perspective*. University of California Press, Berkeley, CA.

Sinclair S, Boon PI (2012) Changes in the area of coastal marsh in Victoria since the mid 19th century. *Cunninghamia* **12**, 153–176.

Sinclair SJ, Sutter GR (2008) 'Estuarine wetland vegetation mapping, Glenelg Hopkins Catchment Management Authority.' Technical Report Series 178. Arthur Rylah Institute for Environmental Research, Melbourne.

Specht RL, Specht A (1999) *Australian Plant Communities*. Oxford University Press, South Melbourne.

Strong DR, Ayres DR (2009) *Spartina* introductions and consequences in salt marshes. In *Human Impacts on Salt Marshes: a Global Perspective*. (Eds BR Silliman, ED Grosholz and MD Bertness) pp. 3–22. University of California Press, Berkeley, CA.

Swinbanks DD (1982) Intertidal exposure zones: a way to subdivide the shore. *Journal of Experimental Marine Biology and Ecology* **62**, 69–86. doi:10.1016/0022-0981(82)90217-9

Tagliapietra D, Sigovini M, Ghirardini AV (2009) A review of terms and definitions to categorise estuaries, lagoons and associated environments. *Marine and Freshwater Research* **60**, 497–509. doi:10.1071/MF08088

Thomsen MS, Adam P, Silliman BR (2009) Anthropogenic threats to Australasian coastal salt marshes. In *Human Impacts on Salt Marshes: a Global Perspective*. (Eds BR Silliman, ED Grosholz and MD Bertness) pp. 361–390. University of California Press, Berkeley, CA.

Toomey M, Boon PI (2007) Response of Brazilian milfoil (*Myriophyllum aquaticum*) to salinity and water-level fluctuations and its potential to invade wetlands of the Gippsland Lakes, south-east Victoria. In *Proceedings of the 3rd Victoria Weed Conference*. 3–4 October 2007, Bendigo. (Ed. S Shepherd) pp. 22–27. The Weed Society of Victoria, Frankston.

Traill LW, Perhans K, Lovelock CE, Prohaska A, McFallan S, Rhodes JR, Wilson KA (2011) Managing for change: wetland transitions under sea-level rise and outcomes for threatened species. *Diversity & Distributions* **17**, 1225–1233. doi:10.1111/j.1472-4642.2011.00807.x

Turner L, Tracey D, Tilden J, Dennison WC (2004) *Where River Meets Sea: Exploring Australia's Estuaries*. CRC for Coastal Zone, Estuary and Waterway Management, Brisbane.

Turner RE, Howes BL, Teal JM, Milan CS, Swenson EM, Goehringer-Toner DD (2009) Salt marshes and eutrophication: an unsustainable outcome. *Limnology and Oceanography* **54**, 1634–1642. doi:10.4319/lo.2009.54.5.1634

Ungar IA (1962) Influence of salinity on seed germination in succulent halophytes. *Ecology* **43**, 763–764. doi:10.2307/1933476

Ungar IA (1978) Halophyte seed germination. *Botanical Review* **44**, 233–264. doi:10.1007/BF02919080

Vanderzee MP (1990) Changes in saltmarsh vegetation as an early indicator of sea-level rise. In *Greenhouse: Planning for Climate Change.* (Ed. GI Pearman) pp. 147–160. CSIRO Publishing, Melbourne.

Walker DI (1999) The marine angiosperms. In *Flora of Australia. Volume 1. Introduction.* 2nd edn. (Ed. AE Orchard) pp. 420–426. Department of the Environment and Heritage, Canberra and CSIRO Publishing, Melbourne.

Walsh K (2004) *Climate Change and Coastal Response.* CRC for Coastal Zone, Estuary and Waterway Management, Indooroopilly, QLD.

Weis JS, Butler CA (2009) *Salt Marshes: a Natural and Unnatural History.* Rutgers University Press, New Brunswick, NJ.

Werner AD, Lockington DA (2006) Tidal impacts on riparian salinities near estuaries. *Journal of Hydrology* **328**, 511–522. doi:10.1016/j.jhydrol.2005.12.011

West RJ, Thorogood C, Walford T, Williams RJ (1985) *An Estuarine Inventory for New South Wales, Australia.* Department of Agriculture, Sydney.

White I, Melville MD, Wilson BP, Sammut J (1997) Reducing acidic discharges from coastal wetlands in eastern Australia. *Wetlands Ecology and Management* **5**, 55–72. doi:10.1023/A:1008227421258

Williams G (1990) Riverine rainforest remnants in the Manning Valley. *Wetlands (Australia)* **9**, 68–76.

Williams RJ, Thiebaud I (2007) *An Analysis of Changes to Aquatic Habitats and Adjacent Land-use in the Downstream Portion of the Hawkesbury Nepean River Over the Past Sixty Years.* Department of Primary Industries, Port Stephens, NSW.

Williams RJ, West G, Morrison D, Creese RG (2006) *Estuarine Resources of New South Wales.* Department of Primary Industries, Port Stephens, NSW.

Wilson BP, White I, Melville MD (1999) Floodplain hydrology, acid discharge and change in water quality associated with a drained acid sulfate soil. *Marine and Freshwater Research* **50**, 149–157. doi:10.1071/MF98034

Wilton KM, Saintilan N (2000) *Protocols for Mangrove and Saltmarsh Habitat Mapping.* Australian Catholic University, Sydney.

Winn KO, Saynor MJ, Eliot MJ, Eliot I (2006) Saltwater intrusion and morphological change at the mouth of the East Alligator River, Northern Territory. *Journal of Coastal Research* **22**, 137–149. doi:10.2112/05A-0011.1

Winning G (1991) Some problems in determining the boundaries of SEPP 14 wetlands. *Wetlands (Australia)* **11**, 10–20.

Winning G, Saintilan N (2009) Vegetation changes in Hexham Swamp, Hunter River, New South Wales, since the construction of floodgates in 1971. *Cunninghamia* **11**, 185–194.

Woo I, Takekawa JY (2012) Will inundation and salinity levels associated with projected sea level rise reduce the survival, growth, and reproductive capacity of *Sarcocornia pacifica* (pickleweed)? *Aquatic Botany* **102**, 8–14. doi:10.1016/j.aquabot.2012.03.014

Woodroffe CD, Davies G (2009) The morphology and development of tropical coastal wetlands. In *Coastal Wetlands: an Integrated Ecosystem Approach.* (Eds GME Perillo, E Wolanski, DR Cahoon and MM Brinson) pp. 65–88. Elsevier, Amsterdam.

Woodward RT, Yui Y-S (2001) The economic value of wetland services: a meta-analysis. *Ecological Economics* **37**, 257–270. doi:10.1016/S0921-8009(00)00276-7

Xiao K, Yu D, Wang L, Han Y (2011) Physiological integration helps a clonal macrophyte spread into competitive environments and coexist with other species. *Aquatic Botany* **95**, 249–253. doi:10.1016/j.aquabot.2011.07.002

Yen S, Myerscough PJ (1989) Co-existence of three species of amphibious plants in relation to spatial and temporal variation: field evidence. *Australian Journal of Ecology* **14**, 291–303. doi:10.1111/j.1442-9993.1989.tb01438.x

Yugovic J (2011) Ecology of the Kooweerup swamp and associated grasslands. *Proceedings of the Royal Society of Victoria* **123**, 172–188.

Yugovic J, Mitchell S (2006) Ecological review of the Koo-Wee-Rup swamp and associated grasslands. *The Victorian Naturalist* **123**, 223–234.

Zedler JB, Callaway JC (1999) Tracking wetland restoration: do mitigation sites follow desired trajectories? *Restoration Ecology* **7**, 69–73. doi:10.1046/j.1526-100X.1999.07108.x

Zedler JB, Kercher S (2004) Causes and consequences of invasive plants in wetlands: opportunities, opportunists, and outcomes. *Critical Reviews in Plant Sciences* **23**, 431–452. doi:10.1080/07352680490514673

# 10

# Riverine vegetation of inland south-eastern Australia

*Jane Roberts, Matthew Colloff and Tanya Doody*

## INTRODUCTION

Inland south-eastern Australia is virtually synonymous with the Murray–Darling Basin, one of Australia's largest drainage basins, and a highly significant area for agricultural production. Agricultural enterprises in this region are diverse, ranging from dryland cropping and pastoralism to irrigated agriculture, and there is a heavy reliance on rivers to supply water for a range of purposes. As a result of agricultural activities, much of the riverine vegetation in inland south-eastern Australia has been completely cleared, or considerably modified, and the condition of the remainder is a matter of national concern. The importance of riverine vegetation for river health has been studied extensively, including for the Murray–Darling Basin, but the spatial ecology of riverine vegetation of inland south-eastern Australia is poorly known. The environmental controls on species and communities have been little considered, other than for flow regime, which has been a major focus, particularly in regulated systems. As a result, the surface water requirements, life history, and management considerations of common floodplain plants are relatively well known (Roberts and Marston 2011; Rogers and Ralph 2011) whereas understanding of upland species, and the role of disturbance, climate and edaphic factors, is rudimentary.

There are remarkably few accounts of the riverine vegetation and communities of inland south-eastern Australia. General descriptive overviews are given by Beadle (1981), Smith and Smith (1990), Roberts (2001), Westbrooke *et al.* (2004) and Keith (2004). Although coarse-scale vegetation mapping has been undertaken (e.g. Beadle 1948, 1981; Neldner 1984), mapping that targets riverine vegetation is relatively recent and available for only a few parts of inland south-eastern Australia (e.g. MPPL 1990; Eco Logical Australia 2009; Johnstone *et al.* 2011). More detailed, comprehensive accounts of vegetation communities as ecological units, including their classification, distribution and mapping, represent long-term undertakings and are still in progress, including those provided by Benson *et al.* (2006, 2010), Benson (2008) and DSE (2011). These accounts reveal that a relatively high number of riverine communities in this region are characterised by woody species, as indicated by the prevalence of forests, woodlands and shrublands. The New South Wales vegetation classification and assessment database lists over 70 classes with woody species for floodplains of inland New South Wales (Benson *et al.* 2010), even though tree species diversity itself is very low. Other characteristic features are the patchy, regional or local distribution of any particular vegetation class, and that community diversity is driven by understorey plants. These three characteristics – low tree diversity, high understorey diversity and mosaic distribution – are the end result of a multiplicity of adaptive evolutionary responses to environmental gradients in south-eastern Australia, and are inextricably linked to the effects of European settlement, such as clearing, altered fire and disturbance regimes and different herbivory pressures, as well as the far longer history of land management and modification by Aboriginal people (Colloff 2014).

This chapter provides an overview of riverine vegetation of inland south-eastern Australia, focusing on the woody component and its environment. Woody plant adaptations are discussed in relation to types of riverine habitat, using flood frequency and flow energy to construct a habitat framework, the whole-of-river framework, suitable for overviewing rivers. The adaptations discussed focus on growth and persistence, and cover only some aspects of regeneration. A full consideration of adaptation to flow regime and habitat requires more extensive knowledge of propagule longevity, dispersal and species phenology than is currently available for individual species of riverine trees and shrubs for this part of Australia.

## RIVERINE LANDSCAPES OF INLAND SOUTH-EASTERN AUSTRALIA

Inland south-eastern Australia is characterised by strong environmental gradients (Fig. 10.1), notably an east–west gradient in temperature and a north–south weather pattern that brings summer rainfall in the north and winter rainfall in the south-east. Runoff is generally low, and rivers flow in response to rainfall in the upper catchments. Much of the area is a low-elevation flat-to-undulating Quaternary depositional landscape, with a semi-arid climate, characterised by low and variable rainfall, high rates of evaporation, and warm to hot temperatures. Inland south-eastern Australia is a large natural drainage depression, the Murray–Darling Basin, that is ringed to the east and south by the Great Dividing Range where temperatures are much cooler and rainfall higher. The Great Dividing Range is highest in the south-east corner of the region, gradually becoming lower as it tapers to the north and west, and hence rivers in these areas,

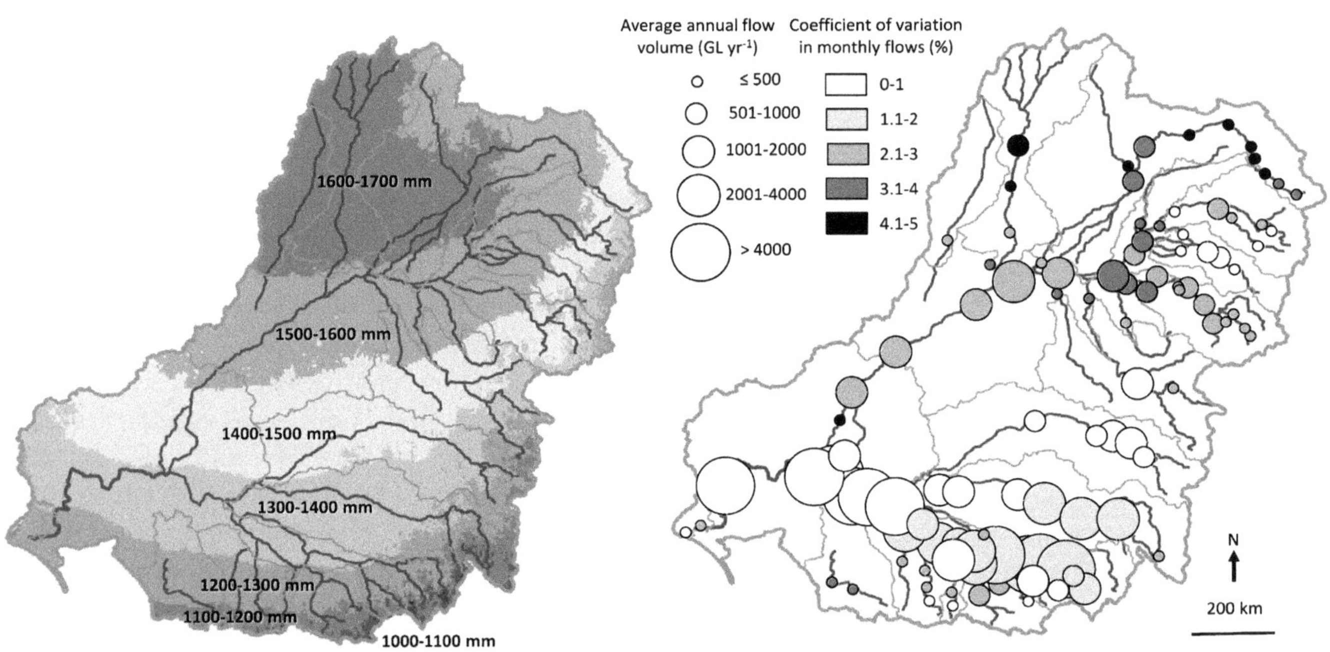


**Figure 10.1:** Environmental gradients in the Murray–Darling Basin, south-eastern Australia. Left: annual potential aerial evapotranspiration (modified from Chiew *et al.* 2008). Right: flow variation in the major rivers. Circles represent flow gauges. Rivers with high flow volume and low variability tend to be downstream of areas with a strongly positive net water balance in the figure on the left, whereas high flow variability and low volume correspond with areas with negative net water balance (modified from Kirby *et al.* 2008). Grey lines are catchment boundaries used in hydrological modelling (CSIRO 2008).

such as the Wimmera, are of particularly low gradient. The highest parts of the Great Dividing Range are snow-bound in winter, and rivers draining from these parts, such as the Murrumbidgee, Murray, Mitta Mitta, Ovens and Kiewa, receive snow melt in spring, producing a marked seasonal pattern of flow, described as 'stable winter baseflow' (Kennard *et al.* 2010). However, the reliability of these high spring flows (relative to rivers further inland) means the lower reaches of these rivers are (with the exception of the Ovens) intensively regulated due to water being used for irrigation, domestic consumption and hydro-electricity generation. At the other extreme, rivers rising in the low ranges of the north and north-west, such as the Warrego, Moonie and Paroo, have high flow variability, are categorised as 'unpredictable summer highly intermittent' (Kennard *et al.* 2010), and have little or no regulation.

There are about twenty major tributaries to the Murray and Darling rivers, the two rivers that give the Murray–Darling Basin its name. The number is imprecise because there is no standard for how river valleys are defined: adjacent and interconnected systems are sometimes treated as one, such as the Goulburn-Broken in northern Victoria. However defined, these 20 or so rivers and their tributaries and distributaries collectively total over 93 million km of channels, and hence a similar length of riverine habitat. Most of this habitat is at low elevations: 41.2% lies below 200 m Australian height datum (AHD), 35.4% between 200 and 400 m AHD, and only 7.6% lies above 700 m AHD (from data provided by Murray–Darling Basin Authority, Canberra). Geomorphic classification suggests that, as well as being a disproportionate part of the total length of riverine habitat, low elevations show considerable geomorphic diversity as indicated by the number of functional process zones occurring there (Fig. 10.2). A functional process zone is a length of river with similar discharge and sediment regime (Chapter 2, this volume; Thoms *et al.* 2004), so is a stretch of river channel with distinctive valley dimensions and cross-sectional morphology, and a particular slope, stream power, sediment characteristics and channel planform features. Collectively, these provide particular combinations of habitat and so influence the distribution of ecological communities of the channel and floodplain (Chapter 2, this volume; Thoms *et al.* 2004). Of the eight functional process zones recognised for rivers in the Murray–Darling Basin, only two are given a high or very high rating for stream power, 'armoured' and 'confined', and these account for only 11.5 and <5% of total river length, respectively (Thoms *et al.* 2007). The other six are rated low to moderate for stream power and two, 'anabranch' and 'distributary', collectively account for 64% of total stream length or 39% and 25%, respectively (Thoms *et al.* 2007), hence their characteristics dominate riverine habitats in inland south-eastern Australia (Fig. 10.2).

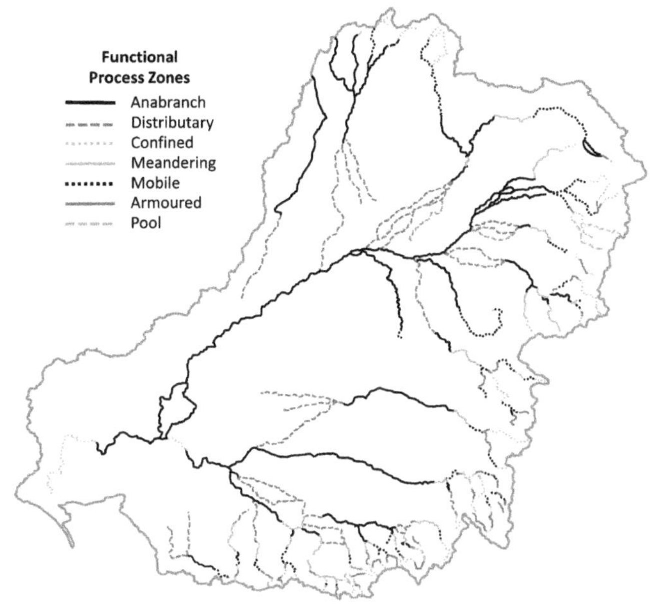


**Figure 10.2:** Changing geomorphic character and functional process zones along river valleys in the Murray–Darling Basin (redrawn and adapted from Thoms *et al.* 2007).

The anabranch zone is characterised by channels that flow across very broad floodplains, have relatively high sinuosity, low stream power, very low slope (i.e. ≤0.0003) and high relative bed stability with dominant sediments being sand, silt and clay. The distributary zone consists of bifurcating channels, either persisting independently of the main channel for many kilometres before re-joining it or as a secondary channel, or terminating in a floodplain. This zone is characterised by low stream power and slope (i.e. ≤0.0002) and high relative bed stability, with sediments of silt and clay. In contrast, the two high-energy functional process zones, confined and armoured, have relatively high slope (i.e. 0.01–0.004) with high or very high stream power, low relative bed stability and the dominant sediments are boulder, cobble or gravel.

These functional process zones are not randomly distributed. The confined and armoured zones occur in upland areas of high rainfall and low evaporation, which are in the eastern and southern upper catchments, whereas the anabranch and distributary zones occur in lowland areas with low rainfall and high evaporation, which are in the lower parts of central, northern and western catchments.

## RIVERINE PLANTS OF INLAND SOUTH-EASTERN AUSTRALIA

### Floristic characteristics

From the general descriptive overviews and maps of contemporary riverine vegetation available for south-eastern Australia (MPPL 1990; Keith 2004; Benson *et al.* 2006; Benson 2008; DSE 2009; Eco Logical Australia 2009; Johnstone *et al.* 2011) it is clear that the trees and shrubs that dominate and structurally characterise riverine vegetation of the region are mostly in the family Myrtaceae (genera such as *Eucalyptus*, *Melaleuca*, *Callistemon*, *Leptospermum*), with lesser but still significant (i.e. extensive) representation from Chenopodiaceae (*Atriplex*, *Chenopodium*), Mimosaceae (*Acacia*), Myoporaceae (*Eremophila*), Casuarinaceae (*Casuarina*) and Polygonaceae (*Duma*). For non-native trees and shrubs, the significant families are Salicaceae (*Salix* and *Populus*) and Rosaceae (*Rubus* and *Rosa*).

These native families are part of the autochthonous sub-element in the Australian flora (Barlow 1994), families that generally show remarkable diversification across Australia. In inland south-eastern Australia, this diversification has been in response to edaphic developments and increasing aridity. Successful adaptation to aridity is indicated for those trees and shrubs whose geographic range extends into the semi-arid and arid areas (Fig. 10.3). The wide distribution of these species is in marked contrast to species occurring along the cooler, mesic, upland rim, or to species with southerly extensions from the tropics, and to the very restricted *Callistemon wimmerensis* (Wimmera bottlebrush). The number of riverine species, and the identity of characteristic and dominant species in the Murray–Darling Basin thus differs from other parts of Australia, such as Tasmania and coastal New South Wales (Daley and Kirkpatrick 2004; Williams and Roberts 2005). In the Myrtaceae, for example, just three eucalypt tree taxa, river red gum (*Eucalyptus camaldulensis* subsp. *camaldulensis*), black box (*Eucalyptus largiflorens*) and coolibah (*Eucalyptus coolabah* subsp. *coolabah*) dominate and characterise the lowland riverine forests and most woodlands of inland south-eastern Australia (Fig. 10.3), whereas in Tasmania, black gum (*Eucalyptus ovata*) is a significant riparian species in lowland areas, but eucalypt richness in the riparian zone is boosted by the presence of several non-riparian species (Daley and Kirkpatrick 2004).

### Adaptations of woody plants to riverine habitats

Woody plants growing in riverine habitats alternate between being flooded and being dry. Both these conditions can adversely affect plant growth and, if sustained, can threaten plant survival. If flooding results in the soil becoming waterlogged or preventing gas exchange to the atmosphere, then soil oxygen is depleted to low (i.e. hypoxia) or zero (i.e. anoxia) levels, which is stressful or even lethal to roots. If flooding is deep enough to submerge some (or all, as can happen with seedlings and juveniles) of the above-ground parts, then plant metabolism (i.e. photosynthesis, respiration) is negatively affected and growth arrested. Trees can survive periods of hypoxia and anoxia in the rhizosphere by increasing oxygen supply to the roots (Kozlowski 1997), and this can be achieved by increasing oxygen uptake through the

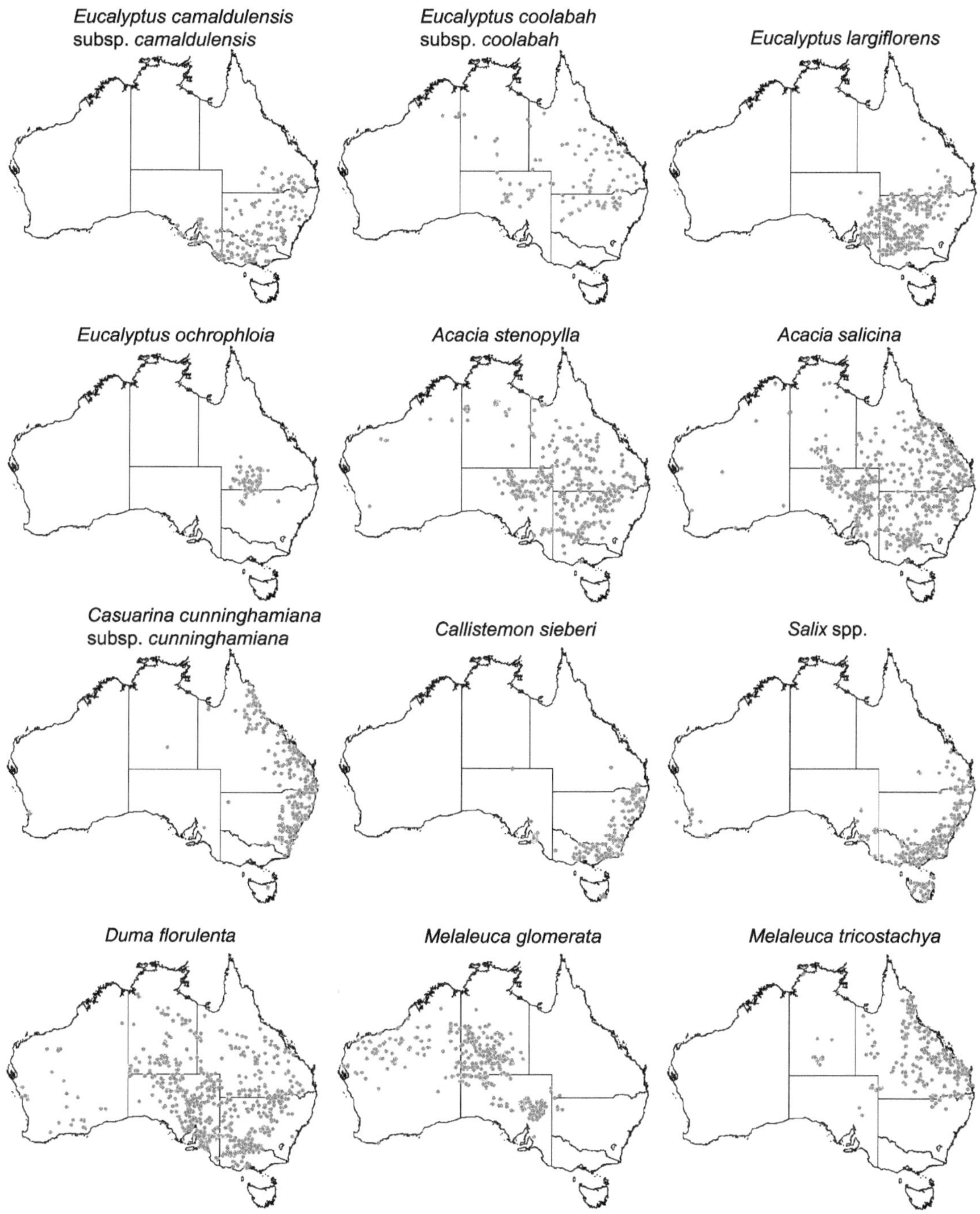


**Figure 10.3:** Contrasting distributions for 12 tree and shrub species important in the riverine vegetation of inland south-eastern Australia, plotted from herbarium records. Specimen data obtained from Australia's Virtual Herbarium, Council of Heads of Australasian Herbaria, http://avh.chah.org.au in 2011. These data are available under a Creative Commons Attribution 3.0 Australia licence.

development of adventitious (sometimes called 'aerial') roots from a trunk or branch while submerged, through hypertrophic growth from lenticels, or by increasing internal gas transport capacity through the development of aerenchymatous tissue with large intercellular spaces. Other than in certain wetland shrubs (Carter *et al.* 2006), such adaptations have been little studied in woody species in

Australia. Extensive adventitious root development in response to flooding is currently known for at least two riverine trees of inland south-eastern Australia, river red gum (*E. camaldulensis*) (Sena Gomes and Kozlowski 1980; Heinrich 1990) and river paperbark (*Melaleuca trichostachya*) (Fig. 10.4). The root mass looks small relative to the tree but its surface area, which is the critical feature for gas exchange, is extensive. The survival advantage of these roots is still to be definitively established for mature trees but is accepted for seedlings (Sena Gomes and Kozlowski 1980). For established trees, survival under prolonged inundation is more likely to be determined by soil characteristics, specifically time for soil oxygen to become depleted, than by adaptations per se.

High-energy flows are a particular type of flooding. Woody plants can only persist in habitats subject to periodic high-energy flows if they have, in addition to inundation tolerance, adaptations that reduce the potentially damaging effects of abrasion, shearing and scour, if they can readily recover from loss of plant parts, and if they can regenerate promptly once flows cease to be high energy. Abrasion occurs

**Figure 10.4:** Adventitious roots on mature riverine trees induced by several weeks of flooding. (a) *Eucalyptus camaldulensis* (river red gum) showing a bunch of adventitious roots emerging from the bole, waterhole on Paroo River. (b) *Melaleuca trichostachya* (river paperbark) on a bench on the Culgoa River showing adventitious roots hanging from low branch. (c) River paperbark on Culgoa River showing bunches of roots developed from under the bark of the bole, and matted with sediment. (d) Adventitious roots developed around the base of the bole of a river paperbark on the Culgoa River. (Photos: Jane Roberts.)

when floating woody debris is forcibly rafted against boles and branches; certain types of bark may confer a degree of protection against such damage. Shearing occurs when leaves, stems or even branches are broken off; morphological traits that reduce plant resistance to flow include fine or trailing foliage, small elongate leaves, and pliant rather than rigid stems and branches. Post-flood recovery traits include sprouting from a damaged stump or toppled tree (i.e. layering), and root development from toppled stems. The capacity of a species to readily colonise newly formed fluvial features such as benches and bars depends on phenology, and seed and propagule attributes such as longevity, storage, dispersal and germination, as well as seedling tolerances.

The value of many of these morphological traits is accepted, but rarely demonstrated. Pliancy, for example, is either an observed or an inferred adaptive trait, based on post-flood observations of boles broken off close to the base (e.g. Fielding *et al.* 1997) or through reach-scale studies correlating species and their attributes with flow vectors through modelling (e.g. Bendix 1999; Evans 2003). Energy, as unit stream power, is rarely reported in field studies or experiments. Velocity tends to be treated as a surrogate for energy, but as it is just one of the terms used to calculate energy (e.g. unit stream power; Bendix 1999) the relationship is not perfect (e.g. Sandercock and Hooke 2010). The effects of high-energy flows on Australian riparian species are poorly established. For introduced taxa such as willows (*Salix* spp.), the challenge is in translating studies done overseas to the taxa and hybrids found in inland south-eastern Australia.

Trees adapted to high-energy habitats are potentially significant as geomorphic agents and ecosystem engineers; the principal Australian examples known are *Casuarina cunninghamiana* (river she-oak) and *Melaleuca argentea* (silver paperbark), both from coastal systems. The model proposed for river she-oak, which is for a highly dynamic sand bed stream (Erskine *et al.* 2009), makes this engineering role clear. The scouring initiated by major or 'catastrophic' floods (*sensu* Erskine *et al.* 2009), and the depositional processes following scour result in new unvegetated bars and benches. Effectively these are competition-free habitats where plants can establish, provided they can survive being submerged and exposed, and can grow on such substrates. Once established, both seedlings and saplings provide localised flow conditions suitable for sediment deposition and accretion, and their root mats help to consolidate these transient fluvial features, at least until the next catastrophic flood. The density and spacing of individual trees and shrubs, so-called 'group strategies', determine an intricate feedback between plant growth, flow and sediment dynamics (Fielding *et al.* 1997) as exemplified by the coastal riverine tree, silver paperbark. Similar examples in inland south-eastern Australia are yet to be demonstrated.

The importance of adaptations to dry periods depends on regional context. Water stress, for example, is generally less or shorter in temperate uplands than in semi-arid lowlands. Given that aridity has been a principal driving force in the evolution of the flora of inland Australia, and that multi-year droughts are a feature of the climate affecting lowland riverine habitats, adaptations to dry and to hot conditions are likely to be more strongly developed in riverine woody plants than adaptations to inundation in lowland habitats, but relatively less important in upland habitats. Tolerance and avoidance are the two strategies for coping with drought, and each can be achieved by mechanisms ranging from physiological to morphological and behavioural. Among eucalypts, for example, these drought-coping mechanisms include: tolerance of moderately low water potentials; capacity for osmotic adjustment; leaf orientation that reduces heat loading in the canopy; lowering water demand by reducing canopy size through leaf shedding or even branch shedding; strong stomatal control over transpiration; and root systems that allow access to soil water deep within the profile (Bell and Williams 1997). These adaptive traits occur across many species and are not necessarily present in all species. Riparian trees switch from using water in the soil surface as seedlings to the use of more constant sources of deep water as they mature (Dawson and Ehleringer 1991). As a means of avoiding drought and competition, groundwater uptake is likely to be relatively common among riparian trees.

## SPATIAL PATTERNS IN RIPARIAN VEGETATION

### Longitudinal and elevational patterns

Given the strong correlation between flow energy, climate and elevation, and between habitat diversity and elevation apparent in the riverine landscapes of the Murray–Darling Basin, equivalent patterns might be expected in the distribution, abundance and diversity of riverine communities and species. Indeed, such patterns are evident in riverine communities of the Namoi valley, one of very few catchments where riverine plant communities have been surveyed from the headwaters to the end of the valley using a uniform method and vegetation typology (Eco Logical Australia 2009). Consistent with the patterns across the basin, most of the riverine habitat in the Namoi valley, when measured as total stream length (for Strahler stream orders 3 to 8, elevations 100 to 1200 m AHD inclusive), occurs at lower elevations, with five to six times more kilometres of stream below 500 m AHD than above 1000 m (Fig. 10.5). This distribution is largely related to topography as drainage density does not trend with altitude, but remains in the range of 16.4 to 20.8 km/km$^2$ for each 100 m vertical elevation band. Riverine community richness in the Namoi valley tracks habitat abundance (Fig. 10.5). Of the 15 riverine communities recognised in the valley, 14 occur below 600 m AHD and only two above. Most of the communities have a narrow elevation range (i.e. 100–300 m) and only two, river red gum woodlands and forests and river she-oak riparian woodland, have a wide elevation range (i.e. ~500 m and 1000 m, respectively).

The distribution of riverine communities in the Namoi valley shows a notable shift in community richness around 500 m AHD, which coincides with a change in functional process zones. Below 500 m AHD the streams are categorised as anabranch and distributary functional process zones (Fig. 10.2) and woody vegetation types in these habitats are more diverse, comprising shrublands and woodlands of river red gum, black box, river cooba (*Acacia stenophylla*) and coolibah (Fig. 10.5). Above 500 m AHD the streams are categorised as being in armoured, meandering, confined and mobile zones (Thoms *et al.* 2007: Table 28.4), with relatively high

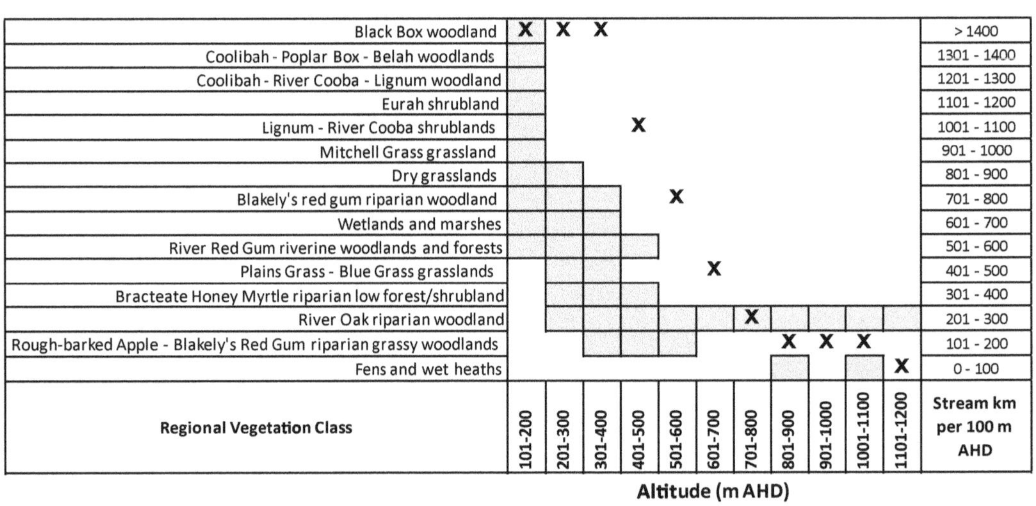


**Figure 10.5:** Elevation distribution of riverine communities, mapped as regional vegetation classes (boxes) and length (km) of streams that are order 3 to 8 (Strahler) in the Namoi valley per 100-m elevation band. Shaded boxes indicate presence of a community per 100-m altitude band; crosses indicate stream length per 100-m altitude band. Figure produced from data provided by Eco Logical Australia in 2011.

slopes, moderate to high stream power, active river channels and coarse, unstable sediments. Woody vegetation in these habitats comprises only a single vegetation type, river she-oak woodland.

Similar patterns are evident in the distribution of individual species, with species turnover within a narrow elevation band suggesting that species share similar responses and boundaries. Along the Murrumbidgee River valley, for example, there is a hinge point around 500 m: river red gums are absent above 400 m, river she-oaks are absent above 570 m, whereas manna gum (*E. viminalis*) does not occur below 590 m, and three riparian shrubs, river bottlebrush (*Callistemon sieberi*), river tea tree (*Leptospermum obovatum*) and silver wattle (*Acacia dealbata*), do not occur below 550 m (Evans 2003). The explorer Charles Sturt (1833: vol. 1, xxxii) also noted this longitudinal zonation of river she-oak and river red gum (which he called 'blue-gum') along the Murrumbidgee and the Macquarie valleys:

*It was observed both on the Macquarie River and the Morumbidgee [sic.], that the casuarinæ ceased at a particular point. On the Macquarie particularly, these trees which had often excited our admiration from Wellington Valley downwards, ceased to occupy its banks below the cataract, nor were they again noticed until we arrived on the banks of the Castlereagh. The blue-gum trees, again, were never observed to extend beyond the secondary embankments of the rivers, occupying that ground alone which was subject to flood and covered with reeds.*

The cataract Sturt mentions was near present-day Gin Gin Weir (~225 m AHD), between Narromine and Warren, indicating pre-regulation rivers had greater stream power extending to lower elevations than today.

## Lateral patterns

Woody riverine vegetation is typically organised laterally in response to flow-related gradients in stream power (Bendix 1999; Sandercock and Hooke 2010) or flooding frequency (van Coller *et al.* 2000). In inland south-eastern Australia, this is well known for lowland rivers and particularly those with large floodplains that are so characteristic of the Murray–Darling Basin, for example, Barmah Forest (Bren and Gibbs 1986) and Yanga National Park (Wen *et al.* 2009). Here, the riverine vegetation forms distinct zones relative to the channel. Because these zones can be related to flood characteristics such as frequency, duration and sometimes variability, vegetation condition can be related to recent changes in flood and flow characteristics such as those caused by river regulation and drought (e.g. Wen *et al.* 2009; Thomas *et al.* 2010). Riverine vegetation zones can be distinguished as much by their structural characteristics (e.g. density, height, structure and number of strata in the understorey) as by their dominant species (Fig. 10.6), reflecting the importance of water availability in combination with species adaptations.

Figure 10.6 is a simple case of lateral organisation, with a single gradient away from the

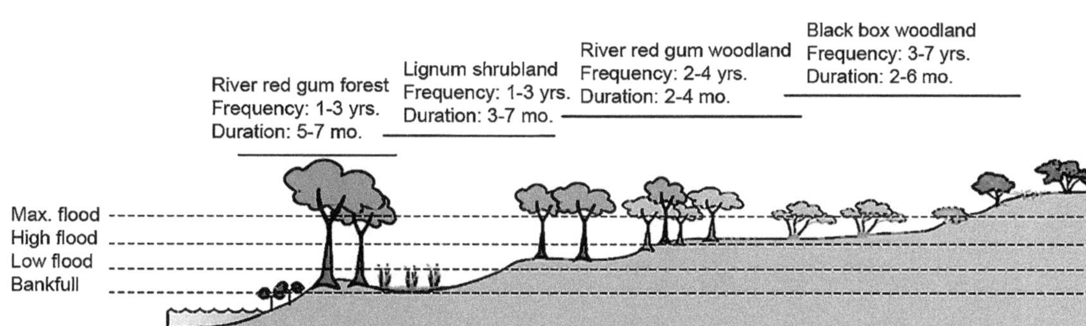


**Figure 10.6:** Diagram showing the ecotonal nature of floodplain vegetation communities in relation to the frequency and duration of flooding required to maintain vigorous growth. Flood frequency and duration data from Roberts and Marston (2011).

river channel. In lowland areas with multiple channels, such as the Macquarie Marshes or the Condamine-Balonne (e.g. Thoms and Parsons 2011), these lateral patterns may become broken and dissected, but are still present although less obvious. In contrast, lateral organisation of riverine habitats and vegetation for upland areas is currently not well known. This knowledge gap results partly from the scale used for mapping vegetation (typically 1:50 000), which is too coarse to show different vegetation types in a relatively narrow riparian strip. There is also inadequate knowledge of relevant environmental gradients, such as velocity and shear stress in these areas as well as a computational challenge in objectively identifying transition points along a transect (e.g. Mac Nally *et al.* 2008).

## RIVERINE HABITATS AND PLANT TRAITS: A WHOLE-OF-RIVER FRAMEWORK

Because flows and floods have strong influences on the structure, distribution and extent of riverine plant communities (Amoros and Bornette 2002; van Eck *et al.* 2005), their characteristics can be used to describe riverine habitat in terms of the traits and adaptations of riverine plants. The conceptual model of Bornette *et al.* (2008), henceforward 'the habitat heterogeneity model', emphasises the importance of physical processes in forming plant habitats. This model is a synthesis of how variations in flood frequency affect the intensity of erosional and depositional processes, and allows for prediction of plant life history traits, recruitment, establishment and growth along gradients of erosion-deposition, disturbance intensity and river slope. The model presents flows in terms of their disturbance intensity, and the spatial and temporal scales at which disturbance operates. These variables are used to circumscribe a three-dimensional space within which plant strategies and adaptations can be placed. River reaches dominated by one process, whether erosion or deposition, have low habitat diversity; reaches with both processes are more heterogeneous, and reaches affected equally have the highest habitat diversity and the lowest habitat specialisation, and are thus likely to have highest plant species diversity.

Although derived for floodplain rivers, the habitat heterogeneity model is based on fundamental physical processes so can be applied to upland reaches and whole river valleys. At the valley scale, however, resolution becomes an issue and reach-to-reach variability is lost: either erosion or deposition appears as the dominant process over long stretches of river, which are broadly equivalent to uplands and lowlands. Thus, at the whole-of-river scale, the model reduces to just two dimensions, flow energy and flood frequency, and is valuable in its own right as a means of exploring plant diversity and adaptations across inland south-eastern Australia. This reduced two-dimensional model is referred to here as the whole-of-river framework (Table 10.1).

In the whole-of-river framework, each of the two dimensions is a gradient, with flow energy ranging from high to low, and flood frequency from frequent to infrequent. For exploring plant habitats, however, each gradient is divided into two categories, high and low, with the combinations giving four types of habitat (Table 10.1). The characteristics of each habitat as applied to inland south-eastern Australia are here assigned drawing on the regional-scale correlations (noted above) between river slope, elevation and climate.

### Low-energy, frequently flooded habitat

This type of habitat floods frequently, notionally every one to five years, has hot, dry summers with high evaporation (Table 10.1). It typically occurs along seasonal watercourses and on lowland floodplains, where the soils tend to be dense or cracking clays, interspersed at depth with sandy palaeochannel aquifers. For trees and shrubs to persist in this habitat, they must endure several months of flooding then several months of hot and dry conditions, and also regenerate.

**Table 10.1.** Characteristics of riverine habitats in relation to flow energy and flood frequency in inland south-eastern Australia, including temperature, mean annual rainfall and evapotranspiration: the whole-of-river framework

| | Low-energy flow | High-energy flow |
|---|---|---|
| Infrequently flooded | High variation in flood frequency: floods irregular, unpredictable | High variation in flood frequency: floods irregular, unpredictable |
| | Lowland floodplains, high connectivity between channel and broad floodplain | Moderate elevation |
| | Hot dry summer climate | Hot dry summer climate |
| | Low rainfall (200–400 mm) | Low rainfall (200–40 mm) |
| | High evapotranspiration | High evapotranspiration |
| Frequently flooded | Regular, predictable flooding: annual-to-bi-annual | Annual sustained or more frequent flashy floods |
| | Lowland floodplains | Uplands; low hydraulic connectivity between channel and narrow floodplain |
| | Warm to hot dry summer climate | Warm temperate climate |
| | Moderate rainfall (300–700 mm) | High rainfall (700–1600 mm) |
| | High evapotranspiration | Low evapotranspiration |

## Growth traits

Throughout inland south-eastern Australia, just one riverine tree, a subspecies of river red gum (*Eucalyptus camaldulensis* subsp. *camaldulensis*), characterises and dominates this habitat. Adaptations of river red gums to this habitat are the architecture of its root system, its water-conducting vessels, and seedling growth rates and seedling flood tolerance. It also has some of the common eucalypt adaptations to water stress and mechanisms for recovery, such as leaf shedding and epicormic buds (but unlike other subspecies, has no lignotuber).

Well-watered river red gums are productive, and can have high transpiration rates. Conventionally, their water needs are described in terms of the flooding regime required to sustain vigorous growth (e.g. Roberts and Marston 2011), but the value of surface flooding is mostly because it recharges soil and groundwater. Significantly, depth of groundwater is predictive of stand condition, dependent on salinity (Cunningham *et al.* 2011). In the mid-Murray forests, surface flooding can percolate as much as 6 m (Bacon *et al.* 1993). River red gums take up groundwater or soil water that is saline in preference to surface water that is fresh (Mensforth *et al.* 1994; Thorburn and Walker 1994), presumably because the sandy palaeochannel aquifers are a more reliable supply.

River red gums have an extensive root system, with tap roots, sinker roots and lateral roots. This not only allows mature trees to take up water from deep within the soil profile, but means that most of the root system is well removed from the parts of the profile potentially saturated by floods, which presumably helps to confer resistance to flooding. The roots of mature river red gums may extend to depths of 9 m or so (Davies 1953) and the effective root zone may be as deep as 10–30 m (Heinrich 1990). Sinker roots, which drop vertically at intervals from the lateral roots, have very high hydraulic conductivity and their secondary xylem vessels, at around 0.2 mm diameter, are among the largest recorded (Heinrich 1990). These characteristics enable river red gums to take up water rapidly when available, which is part of its capacity for recovery following a period of drought and water stress. Roots of river red gums show 'hydraulic redistribution', a passive process driven by differences in water potential of wet soil, dry soil and roots, whereby water is moved from wet to drier soil layers, typically at night when transpiration is low, and is available for re-uptake during the day (Eamus *et al.* 2006).

Mechanisms used by river red gums to postpone dehydration when water stressed are as for other eucalypts, and include extensive leaf shedding to reduce water demand and heat load (Briggs and Maher 1983; Gibson and Bachelard 1994), and branch shedding. Under severe water stress, river red gums gradually reduce their water-conducting sapwood area. Following an increase in water availability, the trees can then rapidly produce new canopy (Doody and Overton 2009). Recovery from being acutely stressed or defoliated is achieved through epicormic growth, which re-establishes the canopy.

### Reproductive traits

The particular subspecies that occurs in the Murray–Darling Basin flowers in December–January over summer (McDonald *et al.* 2009), and flowering is not triggered by flooding. Although river red gums are not dependent on flooding for regeneration, floods provide conditions favourable for seedling establishment and certainly promote it. Mass recruitment events occur after recession of large floods (Jacobs 1955). Seedling survival is greatest and growth most rapid when seed germinates on moist soil during spring and there is a relatively short window of opportunity for establishment before onset of summer conditions when the clay soil dries and hardens (Dexter 1978). River red gums are able to take advantage of this opportunity because of the flood tolerance and growth characteristics of seedlings. This flood tolerance increases with age and height, for example, 6-week-old seedlings can tolerate being partly submerged for 40 days before showing signs of stress (Sena Gomes and Kozlowski 1980) whereas most seedlings 50–60 cm tall can tolerate inundation for 4–6 months (Dexter 1978). Root growth is rapid, much more rapid than shoots, and the root to shoot length ratios are around 4.5 (Dexter 1978). These tap roots can rapidly penetrate the sub-soil and so access the soil moisture there over the following summer.

## Low-energy, infrequently flooded habitat

This type of habitat occurs mostly along lowland rivers, on floodplains at higher elevation or greater distance from the main channel than the low-energy frequently flooded habitats, and along intermittently flowing channels, mostly in semi-arid regions (Table 10.1). Trees and shrubs are dependent on flooding, rainfall and groundwater, which is often saline, but may be refreshed when recharged by infrequent floods or rainfall events.

### Growth traits

Adaptations of woody vegetation to this habitat include: drought-hardiness; flexible and opportunistic use of water from different sources, including the capacity to take up saline groundwater; and a capacity to recover from poor condition, re-establish vigour and reproduce relatively rapidly after flooding. These characteristics are likely to manifest as slow growth and low transpiration for prolonged periods. Regeneration is linked to flooding as it provides the soil moisture needed for seedlings to establish in a climate that is otherwise hot and dry. These traits are known to be present in black box, and some are known for coolibah, river cooba and tangled lignum (*Duma florulenta*, previously known as *Muehlenbeckia florulenta*) (Fig. 10.5).

Black box is distributed throughout the western and northern Murray–Darling Basin, occurring as open, monospecific woodlands, particularly along the Barwon-Darling, Paroo, Warrego and Condamine-Balonne rivers (Keith 2004: 250). Black box is able to use saline groundwater while coping with negative water potentials of ~3.6 MPa (compared with 1.5 MPa in the flushed zone or after flooding) (Holland *et al.* 2006; Doody *et al.* 2009). The uptake rate for saline water is about six times lower than for freshwater (Streeter *et al.* 1996; Doody *et al.* 2009), indicating an ability to lower stomatal conductance (Jolly and Walker 1995). Black box is both flexible and opportunistic in its water use, as documented at a saline site 90 m away from the main river channel on the Chowilla floodplain, South Australia. Here, black box used surface soil water in early summer, soil water and groundwater during summer, and groundwater thereafter (Doody *et al.* 2009), with the most energetically favourable water source being generally taken up first (Jolly and Walker 1995). Black box tends to be very slow-growing, in part because transpiration is low (Akeroyd *et al.* 1998): their height is typically 10–20 m with diameter at breast height (DBH) >10 cm only after 100 years (George *et al.* 2005). Like most eucalypts, black box trees shed leaves in response to water stress, reducing their water requirements.

River cooba uses water at rates similar to river red gum and black box when located in the flushed zones of a watercourse (~0.27 mm/day), but shows a 40-fold reduction in daily water use under saline conditions: it can tolerate saline groundwater of ~18 000 mg/L (Doody *et al.* 2009). Leaf water potentials of stressed river cooba, recorded as around –2 MPa, indicate that it exerts extreme stomatal control over water losses from the long narrow phyllodes that function as leaves.

Coolibah is even more salt-tolerant than black box and river cooba, and is able to continue water uptake at chloride concentrations of at least 20 000–40 000 mg/L (seawater is ~35 000 mg/L). When soil water reserves are depleted, it persists 'in a delicate three-way balance' through transpiration, frequency of flows that replenish that reserve, and capillary rise of hypersaline groundwater (Costelloe *et al.* 2008). Like black box and river cooba, coolibah copes with extreme salinity by keeping transpiration low, by growing at sites where soil water is periodically replenished and flushed with freshwater from stream flows, and by taking up water at very low osmotic potentials. Transpiration rates of coolibah are very low, <0.11 m/year. Although similar to black box (Akeroyd *et al.* 1998; Costelloe *et al.* 2008), these rates are considerably lower than riparian trees with access to freshwater. The capacity to use highly saline groundwater makes coolibah uniquely adapted to low-energy infrequently flooded habitats, as occur in the arid zone.

The multi-stemmed shrub tangled lignum (*Duma florulenta*) is renowned for its observed capacity to endure prolonged drought, surviving on rainfall for up to 10 years and then regenerating via its rootstock after being flooded (Roberts and Marston 2011). The mechanisms for this have not been formally established, but a root system penetrating 2–3 m into heavy clay (Craig *et al.* 1991) and drought deciduousness must contribute. Tangled lignum has small leaves that are produced as a post-flooding growth flush but shed within a few months, and this presumably helps to reduce water usage as the habitat moves from being flooded to unflooded. The root system remains alive after the above-ground parts have died. Tangled lignum is a divaricating woody plant. It has small leaves with very high density of tough, wiry stems with multiple growing tips and a large number of branch orders. These characteristics are considered to have evolved as adaptations to arid conditions during periods of glaciation (McGlone and Webb 1981) and as co-adaptations to limit or deter browsing pressure from large ratite birds (Greenwood and Atkinson 1977). Browsing pressure from now extinct megafauna, and subsequently from emu and kangaroo, would likely have been widespread on floodplains in south-eastern Australia. Tangled lignum's New Zealand relative *Muehlenbeckia astonii* is highly palatable to grazing animals (Gibbs 2006: 204). Its capacity to resist arid conditions, deter browsing and regenerate from rootstock after prolonged drought makes tangled lignum uniquely adapted to a low-energy infrequently flooded habitat.

### Reproductive traits

Various stages of reproduction and regeneration, although not strictly dependent on flooding, are strongly favoured by it. Black box, for example, can flower more than once a year, in response to flooding and rainfall (Parsons and Zubrinich 2010) while river cooba can flower at any time of the year (Whibley 1986; Cowan 1996), which may indicate an adaptation to unseasonal flooding. Tangled lignum flowers within four weeks of flooding (Roberts and Marston 2011). Flooding increases the likelihood of regeneration being successful as it provides soil moisture conditions needed for seedling establishment in semi-arid areas (as does heavy rainfall). It is instrumental in dispersal of seeds, which means establishment occurs away from the parent tree, thereby eliminating direct competition between seedling and mature tree at a critical life history stage (Treloar 1959). In black box, germination and seedling survival require soil moisture of 10–25%, while seeds that germinate in floodwater need to be stranded at the water line within 10 days if they are to continue growing (Jensen 2008). Germination and establishment conditions are little known for river cooba and coolibah, although field observations (Cunningham *et al.* 1981; Roberts 1993) of high seedling abundance along the high water mark after a major flood suggest dispersal and germination are synchronised with flooding. Tangled lignum is distinctive in having a clear dependency on flooding: sequential or sustained flooding is required for flowering, seed set and dispersal, as well as floodwater recession to elicit germination (Roberts and Marston 2011).

Unlike seedlings in the low-energy frequently flooded habitat, seedlings in the low-energy infrequently flooded habitat are more suited to drying conditions. Black box seedlings show stress after a month of flooding (Heinrich 1990), and after two months they have lost their original root system but have not developed a replacement. Tangled lignum seedlings, although they can survive being partly submerged for up to six months, grow very little despite developing adventitious roots within a month of being flooded (Capon *et al.* 2009). They grow more in drying conditions, and most under damp conditions, with shoots 70 cm tall and roots nearly 300 mm deep after 180 days (Capon *et al.* 2009).

## High-energy, frequently flooded habitat

This habitat occurs mainly in upland areas where the climate is generally temperate, and rainfall is high enough to produce high-energy flows at least once and possibly more than once a year (Table 10.1). Pockets of this habitat may occur elsewhere, for example, in warmer and drier areas, on in-channel benches of some lowland rivers, or along steep but low-elevation streams.

### Growth traits

The adaptive characteristics that differentiate woody vegetation in the upland habitat from that of the low-energy frequently flooded habitat are morphological features and growth strategies that reduce the otherwise potentially damaging effects of high-energy flows (i.e. abrasion, shearing); capacity to establish under high and variable flows; capacity to tolerate submergence and recover from mechanical damage, with a range of post-flood recovery mechanisms. Native trees and shrubs known to have such adaptations are river bottlebrush, river tea tree and river she-oak. Willows, which are non-native trees widespread in inland south-eastern Australia (Fig. 10.3), are renowned for having many such adaptations (Karrenberg *et al.* 2002).

Spongy bark, present in paperbarks, and thick rugose bark present in river she-oaks may provide protection against abrasion from woody debris being rafted downstream in floods. Canopy resistance to flow can be reduced in different ways. River she-oak has needle-like photosynthetic stems known as cladodes with miniscule leaves, resulting in a canopy that is fine and pendulous, and trails when submerged in flowing water. River tea tree has leaves that are longer than they are wide and quite small, only 5–20 mm long, as well as pliant branches and multiple stems, which flex with flow. Weeping willow (*Salix babylonica*) has thin leaves on highly flexible stems and branchlets, and crack willow (*Salix fragilis*) has breaking points at the twig base, effectively controlling where damage occurs (Karrenberg *et al.* 2002).

### Reproductive traits

Seedlings of river she-oak, river bottlebrush and river tea tree grow more (relative growth rate based on height increments) after having been partly submerged for up to four days in relatively fast-flowing (0.7 m/s) water than in slow- (0.2 m/s) or non-flowing water, and retained all their leaves (Evans 2003). River she-oak seedlings that are cut to simulate mechanical damage have faster height relative growth rate than uncut seedlings (Evans 2003).

When completely submerged (in non-flowing water), river she-oak, river bottlebrush and river tea tree continue to grow at least initially, but eventually die; however, river she-oak and river bottlebrush have an astonishing capacity to survive under (non-flowing) water, taking ~200 days to die whereas river tea tree takes only ~50 days (Evans 2003). Despite their capacity to withstand inundation, river she-oak seedlings do not grow as well in saturated soils as when the water table is 5–10 m below the surface (Woolfrey and Ladd 2001).

Post-flood recovery is exemplified by willows. Fragments and branches that are broken off by floods and transported downstream become new individuals if deposited in wet ground or shallow water. Willows also show layering, which is the capacity to develop roots from a branch of a tree pushed sideways but not dislodged by floods (Roberts and Marston 2011).

## High-energy, infrequently flooded habitat

This habitat occurs where average rainfall is low, but may be delivered as infrequent but intense or prolonged rainfall events, of variable duration, and is most common in the arid zone of inland south-eastern Australia. The woody riverine vegetation is generally reliant on rainfall and groundwater. Pockets of this habitat occur in parts of the riparian zone that are set back from, or perched above, streams in upland areas.

### Growth traits

Adaptive characteristics for woody riverine vegetation of this habitat are drought and heat tolerance; capacity to utilise groundwater, of uncertain quality; capacity to withstand high-energy floods; capacity to recover from flood disturbance by specific types of re-growth, such as layering; capacity to regenerate opportunistically in response to temporary but unpredictable availability of water. A subspecies of river red gum (*Eucalyptus camaldulensis* subsp. *arida*) that occurs in channel beds and among rocks in this habitat along streams in semi-arid central and western areas (Benson *et al.* 2006; Dunkerley 2008) is a different subspecies from that in the low-energy, frequently flooded habitat. Distribution of river red gum along these arid zone creeks is discontinuous, notably it is absent from the fastest sections or chutes (Dunkerley 2008). Other riverine species are the inland paperbark (*Melaleuca glomerata*) (Tooth and Nanson 2000) and the invasive tree tamarisk (*Tamarix aphylla*) (Griffin *et al.* 1989).

## Synthesis

The adaptations of woody riverine species can be summarised and compared by mapping them within the two-dimensional space defined by the whole-of-river framework (Fig. 10.7). Black box, for example, occurs over most of the flood frequency gradient, but is restricted to low-energy (depositional) environments, and willows occur in both erosional and depositional environments, but have a high water requirement during their growing season, so are restricted to channel-side or very frequently flooded sites.

Generic adaptations of woody riverine vegetation to high-energy habitats relate to surviving

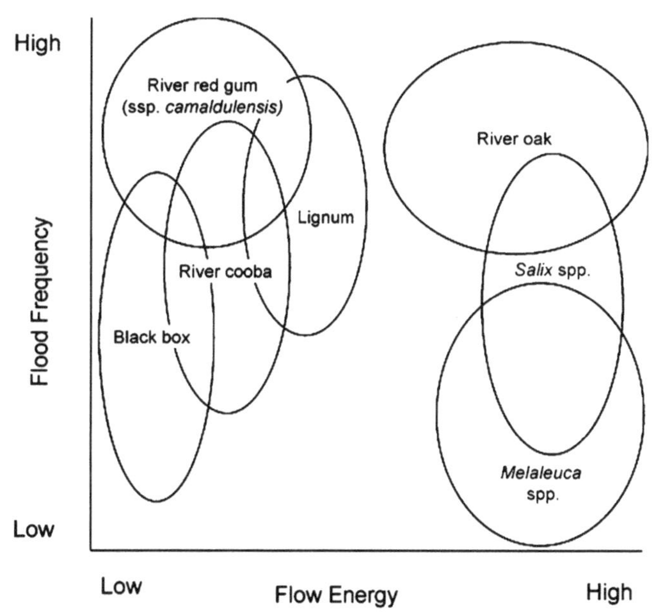


**Figure 10.7:** Distribution of woody species along gradients of flow energy and flood frequency according to the whole-of-river framework. The circles and ellipses are indicative, rather than precise. Coolibah (*Eucalyptus coolabah*) overlaps with black box (*E. largiflorens*) and is omitted for clarity.

physical damage and rapid recolonisation after disturbance; generic adaptations to low-energy habitats, whether frequently or infrequently flooded, relate to ensuring constancy of water supply by accessing groundwater. In infrequently flooded habitats, trees adapted to take up highly saline groundwater over long periods incur an energy penalty, with the cumulative effect of slow growth. Trees in frequently flooded habitats can take up saline water, but at low concentrations, and do not have to sustain this for as long.

## VEGETATION RESPONSES TO CHANGING FLOOD REGIMES

Water resources development and a history of over-allocation has led to changes in flood regimes in most river valleys in inland south-eastern Australia (CSIRO 2008), generally resulting in floods of shorter duration, reduced or altered frequency, over less area, and a shift in timing. The effect of these hydrological changes on riverine vegetation has been a series of shifts in the distribution and abundance of communities as altered flow regimes favour some species at the expense of others. These shifts are often structural, involving expansion–contraction between non-woody and woody dominants. For example, at Macquarie Marshes there has been a dramatic decline in areas of water couch (*Paspalum distichum*) and an increase in the extent of chenopod shrubland (Thomas *et al.* 2010). The Barmah Forest has seen a decline in areas of Moira grass (*Pseudoraphis spinescens*) (Chesterfield 1986; Colloff *et al.* 2014) with a concomitant increase in giant rush (*Juncus ingens*) (Mayence *et al.* 2010) and river red gum (Bren 1992). Where a change in flooding regime results in water requirements for particular woody species not being met, the result is widespread water stress and decline in condition, as for river red gums along the Murray River (Cunningham *et al.* 2007), and dramatically altered regeneration patterns. Lack of regeneration has been apparent in many black box woodlands in the southern Murray–Darling Basin for some time (Roberts and Marston 2011).

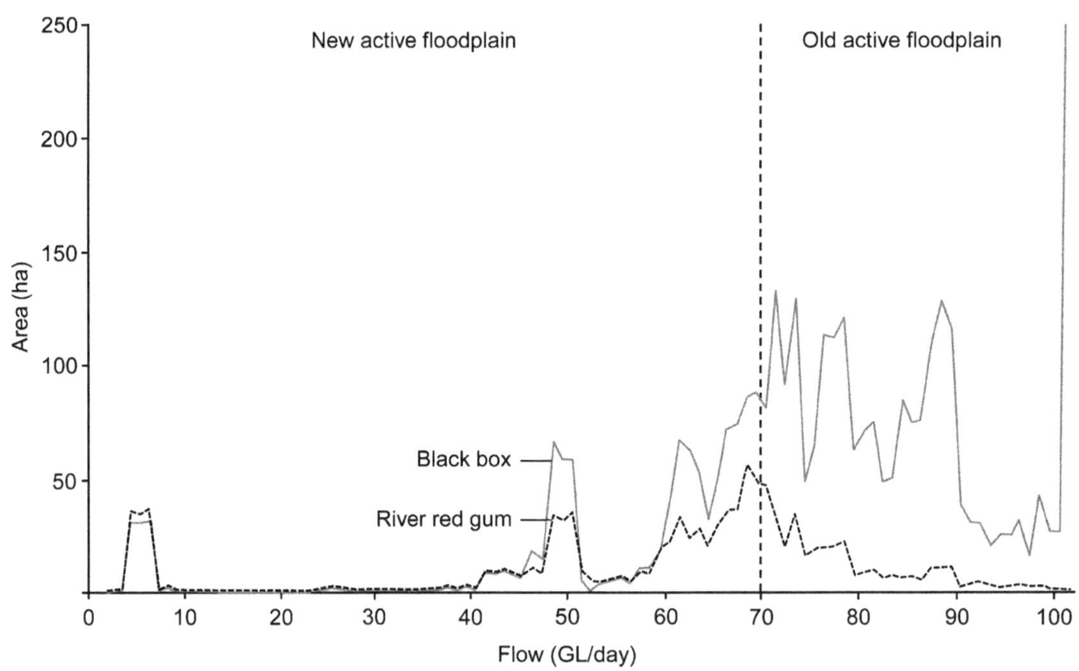


**Figure 10.8:** Abundance (area) of river red gum and black box dominated vegetation on the Chowilla floodplain, South Australia, in relation to commence-to-flood flows (GL/day). The 'active floodplain' is the area inundated by a flood with an average return interval of 1-in-13 years. The lines from 'Black box' and 'River red gum' indicate regeneration (from Overton and Doody 2008).

The long-term and cumulative ecological consequences are a reduction in extent of riverine communities and a transition towards terrestrial vegetation as established trees and shrubs die, and as regeneration opportunities shift in time and space. This process is already underway on several rivers, and in both lowland and upland reaches. On the Chowilla floodplain, for example, the 'active floodplain' (Overton and Doody 2008), which can be considered as the area inundated by Murray River floods with a 1-in-13-year average return interval, has shifted from flows of over 100 GL/day to just over 70 GL/day (Fig. 10.8), effectively stranding the higher parts of the floodplain. Regeneration (indicated by arrows) now occurs only next to the main river channel and on the lowest parts of the floodplain. Similar changes are occurring on other floodplains. Those with low stream power, extremely low slope, and thus a disproportionate relationship between reduction in flow volume and area inundated, are likely to be most affected (Sims *et al.* 2012). On the middle Murrumbidgee River, for example, a recent survey found that non-riparian and fire-tolerant eucalypts were establishing within the river corridor in stands of black cypress pine (*Callitris endlicheri*) and river she-oak affected by severe wildfire, decadal drought and river regulation (Johnstone *et al.* 2011).

## CONCLUSIONS

Revisiting the habitat heterogeneity model of Bornette *et al.* (2008) shows some important differences between it and the characteristics of riverine vegetation communities in south-eastern Australia, as synthesised here within the whole-of-river framework. We contend this is in part due to the context in which the habitat heterogeneity model was developed, namely, the geomorphologically relatively young (late Pleistocene to early Holocene) landscapes and river valleys, structured and recolonised afresh following recent extensive glaciation in the northern hemisphere during the last ice age. In contrast, the fluvial landscapes of inland south-eastern Australia were not extensively glaciated and were instead shaped by long-term cycles of arid and pluvial conditions (Page *et al.* 2009; Pietsch and Nanson 2011). Whereas post-glacial landscapes are rarely water-limited, inland floodplains of south-eastern Australia generally are water-limited but with periodic floods. In a landscape of such physiological extremes, woody plant strategies can be expected to emphasise survival, as those that cannot survive dry spells between regeneration events would soon be eliminated, leaving a landscape populated by short-lived species only.

In the context of riverine ecosystems in inland south-eastern Australia, disturbance and stressors equate not just with the frequency and magnitude of flood events, but with the vast range of environmental variation between flood and drought conditions. Of the major woody plant species that occur on lowlands and floodplains, only river red gum has multiple adaptations for maximising water use for longer periods. By so doing, it optimises its growth and productivity compared with black box and coolibah, which are adapted to minimising water use under dry conditions and whose growth is slow. Thus, river red gums 'thrive to survive'.

Disturbance for south-eastern Australian riverine vegetation is more complex than habitat reconfiguration by floods, as it includes climatic extremes such as drought, and potentially fire. In this context, reducing the frequency of habitat disturbance will not necessarily result in a prevalence of biotic interactions (i.e. competition and herbivory), or favour more competitive species as predicted by Bornette *et al.* (2008). Instead, reducing the frequency of one type of disturbance (e.g. flooding) may result in the onset of another (e.g. drought), which would tend to select for species with adaptive traits to multiple stressors, and limit the likelihood for biotic interactions.

Riverine ecosystems of inland south-eastern Australia are adapted to natural patterns of wetting and drying brought about by the long-term aridification of the Australian continent, commencing 15 million years ago and intensifying in the past 1–4 million years (Byrne *et al.* 2008; Cohen *et al.* 2011). Regional scale gradients of evapotranspiration, rainfall, flow energy and flood frequency have resulted in a series of vegetation communities that are structured by their adaptive responses to these environmental drivers. These adaptations are characterised by an ability to withstand both flooding and drought, relatively low rates of water uptake and transpiration, salinity tolerance, plasticity in accessing different sources of water, vegetative regeneration and the capacity to optimise reproduction and recruitment to coincide with the occurrence of favourable conditions. This formidable suite of adaptations is associated with only a few woody species and these now form the dominant elements of the riverine vegetation in south-eastern Australia.

The whole-of-river framework is effective in describing landscape-scale patterns in the distribution of riverine plant communities. This trait-based approach should also be effective predictively, by giving insight into the effects of regional-scale shifts such as climate change.

## ACKNOWLEDGEMENTS

We thank the Murray–Darling Basin Authority for providing data on stream length and altitude. Tanya Doody and Matthew Colloff acknowledge the support of CSIRO Water for a Healthy Country National Research Flagship.

## REFERENCES

Akeroyd MD, Tyerman SD, Walker GR, Jolly ID (1998) Impact of flooding on the water use of semi-arid riparian eucalypts. *Journal of Hydrology* **206**, 104–117. doi:10.1016/S0022-1694(98)00092-4

Amoros C, Bornette G (2002) Connectivity and biocomplexity in waterbodies of riverine floodplains. *Freshwater Biology* **47**, 761–776. doi:10.1046/j.1365-2427.2002.00905.x

Bacon PE, Stone C, Binns DL, Leslie DJ, Edwards DW (1993) Relationship between water availability and *Eucalyptus camaldulensis* growth in a riparian forest. *Journal of Hydrology* **150**, 541–561. doi:10.1016/0022-1694(93)90125-S

Barlow BA (1994) Phytogeography of the Australian region. In *Australian Vegetation*. 2nd edn. (Ed. RH Groves) pp. 3–35. Cambridge University Press, Cambridge, UK.

Beadle NCW (1948) *The Vegetation and Pastures of Western New South Wales With Special Reference to Soil Erosion*. Government Printer, Sydney.

Beadle NCW (1981) *The Vegetation of Australia*. Cambridge University Press, Cambridge, UK.

Bell DT, Williams JE (1997) Eucalypt ecophysiology. In *Eucalypt Ecology: Individuals to Ecosystems*. (Eds HJ Williams and J Woinarski) pp. 168–195. Cambridge University Press, Cambridge, UK.

Bendix J (1999) Stream power influence on southern Californian riparian vegetation. *Journal of Vegetation Science* **10**, 243–252. doi:10.2307/3237145

Benson JS (2008) New South Wales vegetation classification and assessment: part 2 plant communities of the NSW south-western slopes bioregion and update of NSW western plains plant communities, version 2 of the NSWVCA database. *Cunninghamia* **10**, 599–673.

Benson JS, Allen C, Togher C, Lemmon J (2006) New South Wales vegetation classification and assessment: part 1 plant communities of the NSW western plains. *Cunninghamia* **9**, 383–451.

Benson JS, Richards PG, Waller S, Allen CB (2010) New South Wales vegetation classification and assessment: part 3 plant communities of the NSW Brigalow belt south, Nandewar and west New England bioregions and update of NSW western plains and south-western slopes plant communities, version 3 of the NSWVCA database. *Cunninghamia* **11**, 457–579.

Bornette G, Tabacchi E, Hupp C, Puijalon S, Rostan CJ (2008) A model of plant strategies in fluvial hydrosystems. *Freshwater Biology* **53**, 1692–1705. doi:10.1111/j.1365-2427.2008.01994.x

Bren LJ (1992) Tree invasion of an intermittent wetland in relation to changes in the flooding frequency of the River Murray, Australia. *Australian Journal of Ecology* **17**, 395–408. doi:10.1111/j.1442-9993.1992.tb00822.x

Bren LJ, Gibbs NL (1986) Relationship between flood frequency, vegetation and topography in a river red gum forest. *Australian Forest Research* **16**, 357–370.

Briggs SV, Maher MT (1983) Litter fall and leaf decomposition in a river red gum (*Eucalyptus camaldulensis*) swamp. *Australian Journal of Botany* **31**, 307–316. doi:10.1071/BT9830307

Byrne M, Yeates DK, Joseph L, Kearney M, Bowler J, Williams MAJ, Cooper S, Donnelan C, Keogh S, Leys R, Melville J, Murphy DJ, Porch N, Wyrwoll K-H (2008) Birth of a biome: insights into the assembly and maintenance of the Australian arid zone biota. *Molecular Ecology* **17**, 4398–4417. doi:10.1111/j.1365-294X.2008.03899.x

Capon SJ, James C, Williams L, Quinn G (2009) Response to flooding and drying in seedlings of a common Australian desert floodplain shrub: *Muehlenbeckia florulenta* Meisn. (tangled lignum). *Environmental and Experimental Botany* **66**, 178–185. doi:10.1016/j.envexpbot.2009.02.012

Carter JL, Colmer TD, Veneklaas EJ (2006) Variable tolerance of wetland tree species to combined salinity and waterlogging is related to regulation of ion uptake and production of organic solutes. *New Phytologist* **169**, 123–134. doi:10.1111/j.1469-8137.2005.01552.x

Chesterfield E (1986) Changes in the vegetation of the river red gum forest at Barmah, Victoria. *Australian Forestry* **49**, 4–15. doi:10.1080/00049158.1986.10674458

Chiew FHS, Teng J, Kirono D, Frost A, Bathols J, Vaze J, Viney N, Hennesy K, Cai W (2008) 'Climate data for hydrologic scenario modelling across the Murray–Darling Basin.' A report to the Australian Government from the CSIRO Murray-Darling Basin Sustainable Yields Project. CSIRO, Canberra, <http://www.csiro.au/science/MDBSY.html>.

Cohen TJ, Nanson GC, Jansen JD, Jones BG, Jacobs Z, Treble P, Price DM, May J-H, Smith AM, Ayliffe LK, Hellstrom JC (2011) Continental aridification and the vanishing of Australia's megalakes. *Geology* **39**, 167–170. doi:10.1130/G31518.1

Colloff MJ (2014) *Flooded Forest and Desert Creek: Ecology and History of the River Red Gum*. CSIRO Publishing, Melbourne.

Colloff MJ, Ward KA, Roberts J (2014) Ecology and conservation of grassy wetlands dominated by spiny mud grass *Pseudoraphis spinescens* in the southern Murray–Darling Basin, Australia. *Aquatic Conservation: Marine and Freshwater Ecosystems* **24**, 238–255. doi:10.1002/aqc.2390

Costelloe J, Payne E, Woodrow I, Irvine E, Western A (2008) Water sources accessed by arid zone riparian trees in highly saline environments, Australia. *Oecologia* **156**, 43–52. doi:10.1007/s00442-008-0975-4

Cowan RS (1996) *Acacia* group 2. In *Flora of Victoria Vol. 3: Dicotyledons, Winteraceae to Myrtaceae.* (Eds NG Walsh and TJ Entwisle) pp. 602–607. Inkata Press, Melbourne.

Craig AE, Walker KF, Boulton AJ (1991) Effects of edaphic factors and flood frequency on the abundance of lignum (*Muehlenbeckia florulenta* Meissner) (Polygonaceae) on the River Murray floodplain, South Australia. *Australian Journal of Botany* **39**, 431–433. doi:10.1071/BT9910431

CSIRO (2008) 'Water availability in the Murray–Darling Basin.' A report to the Australian Government from the CSIRO Murray–Darling Basin Sustainable Yields Project. CSIRO, Canberra, <http://www.csiro.au/partnerships/MDBSY>.

Cunningham GM, Mulham WE, Milthorpe PE, Leigh JH (1981) *Plants of Western New South Wales*. Inkata Press, Melbourne.

Cunningham SC, Read J, Baker PJ, Mac Nally R (2007) Quantitative assessment of stand condition and its relationship to physiological stress in stands of *Eucalyptus camaldulensis* (Myrtaceae). *Australian Journal of Botany* **55**, 692–699. doi:10.1071/BT07031

Cunningham SC, Read J, Baker PJ, Mac Nally R (2011) Groundwater change forecasts widespread forest dieback across an extensive floodplain system. *Freshwater Biology* **56**, 1494–1508. doi:10.1111/j.1365-2427.2011.02585.x

Daley EA, Kirkpatrick JB (2004) Native riparian vegetation in Tasmania. *Cunninghamia* **8**, 409–430.

Davies N (1953) 'Investigations on the soil and water relations of the river red gum forests.'

Resources branch report no. R124. Forestry Commission of New South Wales, Mathoura.

Dawson TE, Ehleringer JR (1991) Streamside trees that do not use stream water. *Nature* **350**, 335–337. doi:10.1038/350335a0

Dexter BD (1978) Silviculture of the river red gum forests of the central Murray floodplain. *Proceedings of the Royal Society of Victoria* **90**, 175–194.

Doody TM, Overton IC (2009) Environmental management of riparian tree health in the Murray–Darling Basin, Australia. In *River Basin Management V*. (Ed. C Brebbia) pp. 197–206. WIT Press, Southampton, UK.

Doody TM, Holland K, Benyon R, Jolly I (2009) Effect of groundwater freshening on riparian vegetation water balance. *Hydrological Processes* **23**, 3485–3499. doi:10.1002/hyp.7460

DSE (2009) *A Field Guide to Victorian Wetland Ecological Vegetation Classes*. Department of Sustainability and Environment, Melbourne.

DSE (2011) *Ecological Vegetation Class (EVC) Benchmarks by Bioregion*. Department of Sustainability and Environment, Melbourne, <http://www.dse.vic.gov.au>.

Dunkerley D (2008) Flow chutes in Fowlers Creek, arid western new South Wales, Australia: evidence for the diversity in influence of trees on ephemeral channel form and process. *Geomorphology* **102**, 232–241. doi:10.1016/j.geomorph.2008.05.004

Eamus D, Hatton T, Cook P, Colvin C (2006) *Ecohydrology: Vegetation Function, Water and Resource Management*. CSIRO Publishing, Melbourne.

Eco Logical Australia (2009) 'Riverine vegetation in the Namoi catchment. An assessment of type and condition.' Final report prepared for Cotton Catchment Communities CRC and Namoi Catchment Management Authority. Eco Logical Australia, Coffs Harbour, NSW.

Erskine W, Chalmers A, Keene A, Cheetham M, Bush R (2009) Role of a rheophyte in bench development on a sand-bed river in southeast Australia. *Earth Surface Processes and Landforms* **34**, 941–953.

Evans LJ (2003) The influence of fluvial geomorphology on riparian vegetation in upland river valleys: south-eastern Australia. PhD thesis. University of Canberra, ACT.

Fielding CR, Alexander J, Newman-Sutherland E (1997) Preservation of in situ, arborescent vegetation and fluvial bar construction in the Burdekin River of north Queensland, Australia. *Palaeogeography, Palaeoclimatology, Palaeoecology* **135**, 123–144. doi:10.1016/S0031-0182(97)00022-9

George AK, Walker KF, Lewis MM (2005) Population status of eucalypt trees on the River Murray floodplain, South Australia. *River Research and Applications* **21**, 271–282. doi:10.1002/rra.846

Gibbs G (2006) *Ghosts of Gondwana: the History of Life in New Zealand*. Craig Potton Publishing, Nelson, NZ.

Gibson A, Bachelard AP (1994) Relationships between site characteristics and survival strategies of *Eucalyptus camaldulensis* seedlings. In *Australian Tree Species Research in China*. (Ed. AG Brown) pp. 91–95. ACIAR proceedings series no. 48, Australian Centre for International Agricultural Research, Canberra.

Greenwood RM, Atkinson IAE (1977) Evolution of divaricating plants in New Zealand in relation to Moa browsing. *Proceedings of the New Zealand Ecological Society* **24**, 21–33.

Griffin GF, Stafford Smith DM, Morton SR, Allan GE, Masters KA (1989) Status and implications of the invasion of tamarisk (*Tamarix aphylla*) on the Finke River, Northern Territory, Australia. *Journal of Environmental Management* **29**, 297–315.

Heinrich P (1990) 'The eco-physiology of riparian river red gum (*Eucalyptus camaldulensis*).' Final report to the Australian Water Resources Advisory Council. Forestry Section, University of Melbourne, Creswick, Victoria.

Holland K, Tyreman S, Mensforth L, Walker G (2006) Tree water sources over shallow, saline groundwater in the lower River Murray, south eastern Australia: implications for groundwater recharge mechanisms. *Australian Journal of Botany* **54**, 193–205.

Jacobs MR (1955) *Growth Habits of the Eucalypts*. Forestry and Timber Bureau, Canberra.

Jensen A (2008) The roles of seed banks and soil moisture in recruitment of semi-arid floodplain plants: the River Murray, Australia. PhD thesis, University of Adelaide.

Johnstone L, Skinner S, Ishiyama L, Sharp S (2011) 'Survey of vegetation and habitat in key riparian zones: Murrumbidgee River ACT.' Technical report 22. Environment and Sustainable Development Directorate, Canberra.

Jolly ID, Walker GR (1995) *A Sketch of Salt and Water Movement in the Chowilla Floodplain*. CSIRO Division of Water Resources, Adelaide.

Karrenberg S, Edwards PJ, Kollman J (2002) The life history of Salicaceae living in the active zone of floodplains. *Freshwater Biology* **47**, 733–748. doi:10.1046/j.1365-2427.2002.00894.x

Keith DA (2004) *Ocean Shores to Desert Dunes: the Native Vegetation of New South Wales and the ACT*. New South Wales Department of Environment and Conservation, Sydney.

Kennard MJ, Pusey BJ, Olden JD, Mackay SJ, Stein JL, Marsh N (2010) Classification of natural flow regimes in Australia to support environmental flow management. *Freshwater Biology* **55**, 171–193. doi:10.1111/j.1365-2427.2009.02307.x

Kirby JM, van Dijk AIJM, Mainuddin M, Peña-Arancibia U, Guerschman J-P, Liu Y, Marvanek S, McJannet DL, Paydar Z, McVicar TR, van Niel TG, Li LT (2008) 'River water balance accounts across the Murray–Darling Basin, 1990–2006.' A report to the Australian Government from the CSIRO Murray–Darling Basin Sustainable Yields Project. CSIRO, Canberra.

Kozlowski TT (1997) *Responses of Woody Plants to Flooding and Salinity*. Tree physiology monograph No. 1. Heron Publishing, Victoria, Canada.

Mac Nally R, Molyneux G, Thomson JR, Lake PS, Read J (2008) Variation in widths of riparian-zone vegetation of higher elevation streams and implications for conservation management. *Plant Ecology* **198**, 89–100. doi:10.1007/s11258-007-9387-5

Mayence CE, Marshall DJ, Godfree R (2010) Hydrological and mechanical control for an invasive wetland plant, *Juncus ingens*, and implications for rehabilitating and managing Murray River floodplain wetlands, Australia. *Wetlands Ecology and Management* **18**, 717–730. doi:10.1007/s11273-010-9191-1

McDonald MW, Brooker MIH, Butcher PA (2009) A taxonomic revision of *Eucalyptus camaldulensis* (Myrtaceae). *Australian Systematic Botany* **22**, 257–285. doi:10.1071/SB09005

McGlone M, Webb CJ (1981) Selective forces influencing the evolution of divaricating plants. *New Zealand Journal of Ecology* **4**, 20–28.

Mensforth L, Thorburn P, Tyerman S, Walker G (1994) Sources of water used by riparian *Eucalyptus camaldulensis* overlying highly saline groundwater. *Oecologia* **100**, 21–28. doi:10.1007/BF00317126

MPPL (1990) 'Riparian vegetation of the River Murray.' Report prepared for the Murray–Darling Basin Commission by Margules and Partners, P. and J. Smith Ecological Consultants, and Department of Conservation Forests and Lands, Victoria.

Neldner VJ (1984) *Vegetation Survey of Queensland: South West Queensland*. Queensland botany bulletin no. 3. Queensland Department of Primary Industries, Brisbane.

Overton IC, Doody TM (2008) 'Groundwater, surface water, salinity and responses to a proposed regulator on Chowilla Creek.' Water for a Healthy Country report prepared for the South Australian Murray–Darling Basin Natural Resource Management Board.

Page KJ, Kemp J, Nanson GC (2009) Late Quaternary evolution of riverine plain palaeochannels, south-eastern Australia. *Australian Journal of Earth Sciences* **56**, S19–S33. doi:10.1080/08120090902870772

Parsons R, Zubrinich TM (2010) The green-leaved variant of *Eucalyptus largiflorens*: a story involving hybridization and observant local people. *Cunninghamia* **11**, 415–418.

Pietsch T, Nanson G (2011) Bankfull hydraulic geometry: the role of in-channel vegetation and downstream declining discharges in the anabranching and distributary channels of the Gwydir distributive fluvial system, southeastern

Australia. *Geomorphology* **129**, 152–165. doi:10.1016/j.geomorph.2011.01.021

Roberts J (1993) Regeneration and growth of coolibah, *Eucalyptus coolabah* subsp. *arida*, a riparian tree, in the Cooper Creek region of South Australia. *Australian Journal of Ecology* **18**, 345–350. doi:10.1111/j.1442-9993.1993.tb00461.x

Roberts J (2001) Large plants. In *Rivers as Ecological Systems: the Murray–Darling Basin*. (Ed. WJ Young) pp. 187–221. Murray-Darling Basin Commission, Canberra.

Roberts J, Marston F (2011) *Water Regime for Wetland and Floodplain Plants. A Source Book for the Murray–Darling Basin.* National Water Commission, Canberra.

Rogers K, Ralph T (Eds) (2011) *Floodplain Wetland Biota in the Murray–Darling Basin: Water and Habitat Requirements.* CSIRO Publishing, Melbourne.

Sandercock PJ, Hooke JM (2010) Assessment of vegetation effects on hydraulics and of feedbacks on plant survival and zonation in ephemeral channels. *Hydrological Processes* **24**, 695–713. doi:10.1002/hyp.7508

Sena Gomes AR, Kozlowski TT (1980) Effects of flooding on *Eucalyptus camaldulensis* and *Eucalyptus globulus* seedlings. *Oecologia* **46**, 139–142. doi:10.1007/BF00540117

Sims NC, Chariton A, Jin H, Colloff MJ (2012) A classification of floodplain ecosystems in the Murray–Darling Basin based on changes in flooding between pre-regulation and current conditions. *Wetlands* **32**, 239–248. doi:10.1007/s13157-011-0231-2

Smith P, Smith J (1990) Floodplain vegetation. In *The Murray.* (Eds N Mackay and D Eastburn) pp. 215–228. Murray–Darling Basin Commission, Canberra.

Streeter TC, Tyerman SD, Walker GR (1996) Water-use strategies of *Eucalyptus largiflorens.* In *Salt and Water Movement in the Chowilla Floodplain.* (Eds GR Walker, ID Jolly and SD Jarwal) pp. 78–89. Water resources series no. 15. CSIRO Division of Water Resources, Canberra.

Sturt C (1833) *Two Expeditions into the Interior of Southern Australia During the Years 1828, 1829, 1830 and 1831; with Observations on the Soil, Climate, and General Resources of the Colony of New South Wales.* Smith Elder, London. Facsimile edition (1963) reprinted in two volumes by the Public Library of South Australia, Adelaide.

Thomas RF, Bowen S, Simpson SL, Cox SJ, Sims NC, Hunter SJ, Lu Y (2010) Inundation response of vegetation communities of the Macquarie Marshes in semi-arid Australia. In *Ecosystem Response Modelling in the Murray–Darling Basin.* (Eds N Saintilan and I Overton) pp. 137–150. CSIRO Publishing, Melbourne.

Thoms M, Parsons M (2011) Patterns of vegetation community distribution in a large, semi-arid floodplain landscape. *River Systems* **19**, 271–282. doi:10.1127/1868-5749/2011/019-0050

Thoms MC, Hill SM, Spry MJ, Chen XJ, Mount TJ, Sheldon F (2004) The geomorphology of the Darling River. In *The Darling.* (Eds R Breckwoldt, R Boden and J Andrew) pp. 68–105. Murray–Darling Basin Commission, Canberra.

Thoms MC, Rayburg SC, Neave MR (2007) Physical diversity and assessment of a large river system: the Murray–Darling Basin, Australia. In *Large Rivers: Geomorphology and Management.* (Ed. A Gupta) pp. 587–607. John Wiley & Sons, Chichester, UK.

Thorburn P, Walker G (1994) Variations in stream water-uptake by *Eucalyptus camaldulensis* with differing access to stream water. *Oecologia* **100**, 293–301. doi:10.1007/BF00316957

Tooth S, Nanson GC (2000) The role of vegetation in the formation of anabranching channels in an ephemeral river, northern plains, arid central Australia. *Hydrological Processes* **14**, 3099–3117. doi:10.1002/1099-1085(200011/12)14:16/17<3099::AID-HYP136>3.0.CO;2-4

Treloar GK (1959) Some factors affecting seedling survival of *Eucalyptus largiflorens* F. Muell. *Australian Forestry* **23**, 46–48. doi:10.1080/00049158.1959.10675863 van Coller AL, Rogers KH, Heritage GL (2000) Riparian vegetation-environment relationships: complementarity of gradients versus patch hierarchy approaches. *Journal of Vegetation Science* **11**, 337–350. doi:10.2307/3236626

van Eck WHJM, van de Steeg HM, Blom CWPM, de Kroon H (2005) Recruitment limitation along disturbance gradients in river floodplains. *Journal of Vegetation Science* **16**, 103–110.

Wen L, Ling J, Saintilan N, Rogers K (2009) An investigation of the hydrological requirements of river red gum (Eucalyptus camaldulensis) forest, using classification and regression tree modelling. *Ecohydrology* **2**, 143–155.

Westbrooke M, Leversha J, Kerr M (2004) The vegetation of the Darling Basin. In *The Darling*. (Eds R Breckwoldt, R Boden and J Andrew) pp. 142–167. Murray–Darling Basin Commission, Canberra.

Whibley DJE (1986) Mimosoideae. In *Flora of South Australia Volume 2. Leguminoseae-Rubiaceae*. 4th edn. (Eds JP Jessop and HR Toelken) pp. 511–569. South Australian Government Printing Division, Adelaide.

Williams DG, Roberts J (2005) *Riparian Vegetation Diversity in the Sydney Catchment Authority's Area of Operation*. Cooperative Centre for Freshwater Ecology, Canberra.

Woolfrey AR, Ladd PG (2001) Habitat preference and reproductive traits of a major Australian riparian tree species (*Casuarina cunninghamiana*). *Australian Journal of Botany* **49**, 705–715. doi:10.1071/BT01009

# 11

# Pattern and process in riparian vegetation of the treeless high country of Australia

*Jamie Kirkpatrick*

## INTRODUCTION

Australia's high mountains tend more to elevated vegetated plains than the steep, bare crumbling mountains copiously decorated by snow and ice characteristic of the plate collision alpine archetype. Consequently, these landscapes are rich in gentle streams in the depositional mode, as well as those that rush and cut. Unlike most of Australia, the high mountains receive more moisture from precipitation than they lose through evaporation, and a substantial part of the moisture input arrives as snow, hail or rime, to cover the landscape on time scales that vary spatially and from days to seasons.

The riparian vegetation that is the subject of this chapter occurs in the alpine and subalpine zones of Australia, not including the subantarctic islands. The alpine zone is defined as land above the limit of trees, which roughly corresponds to a mean daily temperature of the warmest month of 10°C. Treeless vegetation of similar species composition to alpine vegetation also occurs extensively in the subalpine zone, defined roughly by the lower altitudinal limit of *Eucalyptus niphophilus* on mainland Australia (Costin *et al.* 2000), by the upper limit of the hummock sedge *Gymnoschoenus sphaerocephalus* in western Tasmania (Kirkpatrick and Brown 1987) and by the lower altitudinal limit of *Eucalyptus coccifera* in central and eastern Tasmania (Kirkpatrick 1999). Treeless vegetation in the subalpine zone and alpine vegetation are henceforth collectively called 'alpine vegetation' and the area they occupy 'alpine' following Kirkpatrick (1982, 1997).

In alpine areas, lotic and lentic wetlands tend to be less distinct than at lower altitudes, with snow patches, springs, lakes, ponds, bogs, fens and streams being frequently interacting parts of the same surface hydrological system and sharing many vascular plant species. All of these elements of the hydrological system are regarded in the present chapter as providing riparian habitat.

### Variation in the environmental context of Australian alpine riparian vegetation

Australian alpine vegetation has been divided into four intergrading floristic and environmental regions: mainland, eastern Tasmania, central Tasmania and western Tasmania (Kirkpatrick 1982, 1986, 1997). The environmental conditions in each of these regions, and their different species pools, have potentially strong influences on the nature of riparian vegetation.

Alpine climates in Australia vary from continental to maritime. Because summer temperatures control the distribution of alpine vegetation, maritime alpine areas below the nival zone, which does not exist in any of the four alpine regions, have infrequent and unpredictable snow cover. Thus, the most alpine (i.e. summer cool) of the mountains of the islands of Australia and Tasmania, Frenchmans Cap in maritime western Tasmania (Fig. 11.1), is not covered with persistent snow in winter, except for one small snow patch on its lee slopes, while most of the continental Snowy Mountains and Bogong High Plains alpine areas are deeply covered with snow for all of winter and some of spring, with snow that can persist a month at a time extending to the lower altitudinal limit of the subalpine zone, and many snow patches that typically persist well into summer.

The nature of riparian vegetation is strongly influenced by variability in flow (Hughes 1987), and, in particular, extreme high flows (e.g. Wintle and Kirkpatrick 2007). Extensive and persistent areas of snow cover result in alpine streams with very different hydrological and geomorphological characteristics to those found where snow cover is both temporally and spatially limited. Flow is likely to be much more constant where snow is not a strong seasonal event, resulting in less scouring of beds and banks and a greater tendency for stream vegetation cover with the same amount of annual flow.

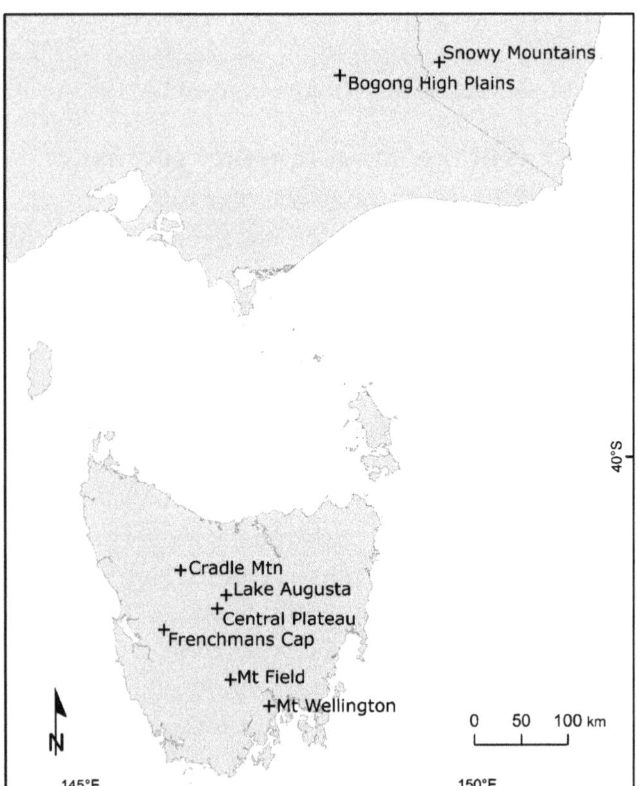


**Figure 11.1:** Locations mentioned in the text.

Snow cover and much colder winters than are experienced in the more maritime Tasmanian alpine areas limit the growing season in the mainland alpine areas, which can be extremely short where snow melts in late spring to early summer. In response to a very limited growing season, some short alpine herbfield species associated with melt water below snow patches, such as *Caltha intraloba*, use carbohydrate reserves in their root stocks to commence flowering before the snow melts above them (Costin *et al.* 2000). In contrast, Gibson (1990) demonstrated that bolster plants (i.e. hard cushion plants) grow in winter on Mount Field in Tasmania.

The heavy and prolonged snow cover on the higher peaks of the mainland mountains is associated with a lack of native vertebrate herbivores, whereas in Tasmania Bennett's wallabies (*Macropus rufogriseus*) and wombats (*Vombatus ursinus*) are found in all parts of the alpine zone (Kirkpatrick 1989).

Precipitation is also highly variable within and between Australia's alpine regions. Using a model based on cloud cover sensed from satellites, calibrated with temporary and permanent ground stations, Nunez *et al.* (1996) estimated that annual precipitation was almost 5000 mm per annum in the far southern mountains of Tasmania. At the lower altitudes of the subalpine zone on the Central Plateau, mean annual rainfall approximates 1000 mm. In Tasmania, annual precipitation in the alpine zone increases from the north-east to the south-west, as does the degree of concentration of precipitation in the cooler months. The mainland alpine areas have annual precipitation that varies from that received in the eastern alpine areas of Tasmania to that received in the central alpine areas of Tasmania (Kirkpatrick 1982). There is a gradient in the seasonal distribution of precipitation from the south, where there is a strong winter maximum, to the north, where precipitation becomes more even between the seasons.

Palynological evidence suggests that the north-east to south-west precipitation gradient in Tasmania was more subdued at the peak of the last glacial event than at present (Kirkpatrick and Fowler 1998), but the gradient in altitude of glacial cirque basins

(Peterson 1968) leaves little doubt that it existed, as the amount of precipitation is a major control on the formation and extent of glaciers. Most alpine areas of Tasmania lie within the limits of ice during the last glacial event (Colhoun and Fitzsimons 1990). Their landforms are largely a product of glacial scraping, deposition of moraines during glacial retreat and the frost-shattering of nunataks. Glacial erosion over several cold periods during the Quaternary created a large number of deep waterbodies in extensive loch and lochan landscapes, numerous cirque lakes and very shallow soils, which largely consist of organic material accumulated since the ice retreated (Colhoun and Fitzsimons 1990). Scree deposits are widespread around nunatak peaks.

Outside the limits of ice, extensive boulder fields have resulted from the removal of fine material by water erosion from previously deeply weathered mantles. These so-called block streams are usually underlain by clay, protected by the lag deposit of dolerite boulders. Water flows beneath the boulders, emerging to be captured by cascades of peat-dammed ponds. On the eastern Central Plateau, there are extensive glacio-fluvial outwash plains, where wind deflation has created shallow waterbodies and associated leeward sand dunes composed of sand-sized particles of glacially ground dolerite (Pharo and Kirkpatrick 1994). Fluvial deposits of silts and clays are also found inside and outside the limits of glaciation, wherever the siltstones and mudstones of the Parmeener supergroup have been exposed adjacent to gently sloping valleys. A contrast in texture between the coarser surface deposits and underlying finer material has resulted in extensive piping, its existence evident from occasional surface collapses and subterranean gurgling. The larger streams that flow through these plains have vertical, well-vegetated margins and contain many soil islands.

In contrast to Tasmania, the signs of Quaternary glaciation on the mainland are confined to some of the higher peaks in the Snowy Mountains (Barrows *et al.* 2001). Evidence of periglacial activity is more widespread, although some of the features, such as contour terraces, that have been suggested to be relicts of solifluction, do not require soil freezing to form (Kirkpatrick 1997). Screes and boulder streams similar to those found so extensively in eastern and central Tasmania are found on many mainland alpine mountains.

The soils of the mainland alpine areas are deeper and more fertile than those of the Tasmanian alpine zone, largely because of top dressing with finely textured wind-transported sediments from the interior of Australia (Walker and Costin 1971; Johnston 2001). The eastern and central Tasmanian alpine soils, largely formed on dolerite and sedimentary rocks, are more fertile than those of western Tasmania, largely formed on quartzite. Average soil depth, pH, extractable phosphorus and extractable potassium values are highest on the mainland, next highest in eastern Tasmania and lowest in western Tasmania (Kirkpatrick 1997). Extractable phosphorus is the variable most associated with floristic variation as a whole in Australian alpine vegetation (Kirkpatrick and Bridle 1998, 1999).

## RIPARIAN FLORA

Because vegetation influenced by flooding, although usually distinct in its overall composition, can be, and usually is, dominated by species found in the adjacent and upstream non-riparian vegetation, the identification of riparian species is difficult. This is especially the case in alpine riparian vegetation, where water is usually available in abundance in the whole of the landscape.

To qualify for inclusion in the list of alpine riparian species (Table 11.1), a taxon had to either be noted to occur in at least one of stream sides, pools, bogs, fens, short alpine herbfield or bolster heath by Costin *et al.* (2000) or Kirkpatrick (1997), or have more than a 30% frequency in at least one community known to be associated with moving or fluctuating water (Kirkpatrick 1997; Daley and Kirkpatrick 2004; McDougall and Walsh 2007). Two hundred and thirty-five species satisfy one or both of these criteria, 175 in Tasmania and 143 on the mainland, with 89 species in common (Table 11.1). Of the 235 species, 111 are associated with streams,

**Table 11.1.** Species recorded from Australian alpine riparian vegetation, by vegetation type and island

Nomenclature follows McDougall and Walsh (2007) for species listed for mainland Australia and Buchanan (2005) for other species. Aust, mainland Australia; BH, bolster heath; Riv, river; SAH, short alpine herbfield; Tas, Tasmania.

| Species | Family | Tas | Aust | Riv | Pool | Fen | Bog | SAH | BH |
|---|---|---|---|---|---|---|---|---|---|
| *Abrotanella forsteroides* | Asteraceae | 1 | – | – | – | – | – | – | 1 |
| *Abrotanella nivigena* | Asteraceae | – | 1 | 1 | – | – | – | 1 | – |
| *Acaena montana* | Rosaceae | 1 | – | 1 | – | – | – | – | 1 |
| *Acaena* sp.[1] | Rosaceae | – | 1 | 1 | – | – | – | – | – |
| *Aciphylla simplicifolia* | Apiaceae | – | 1 | 1 | – | – | – | – | – |
| *Actinotus suffocatus* | Apiaceae | 1 | – | – | – | – | – | – | 1 |
| *Agrostis muelleriana* | Poaceae | 1 | 1 | – | – | 1 | 1 | – | – |
| *Agrostis* aff. *parviflora* | Poaceae | 1 | 1 | 1 | – | 1 | 1 | – | – |
| *Agrostis parviflora* | Poaceae | 1 | 1 | 1 | – | – | – | – | – |
| *Almaleea subumbellata* | Fabaceae | 1 | – | 1 | – | – | – | – | – |
| *Amphibromus recurvatus* | Poaceae | 1 | – | – | 1 | – | – | – | – |
| *Anemone crassifolia* | Ranunculaceae | 1 | – | – | – | – | – | – | 1 |
| *Argyrostegium fordianum* | Asteraceae | 1 | 1 | 1 | – | – | – | – | – |
| *Argyrotegium mackayi* | Asteraceae | – | 1 | – | – | – | – | 1 | – |
| *Argyrostegium nitidulum* | Asteraceae | – | 1 | 1 | – | – | 1 | – | – |
| *Argyrostegium poliochlorum* | Asteraceae | – | 1 | 1 | – | – | 1 | – | – |
| *Asperula gunnii* | Rubiaceae | 1 | 1 | – | – | – | 1 | – | – |
| *Astelia alpina* | Liliaceae | 1 | 1 | – | – | – | 1 | – | 1 |
| *Astelia psychrocharis* | Liliaceae | – | 1 | – | – | – | 1 | – | 1 |
| *Baeckea gunniana* | Myrtaceae | 1 | 1 | 1 | – | – | 1 | – | – |
| *Baeckea utilis* | Myrtaceae | – | 1 | 1 | – | – | – | – | – |
| *Baloskion australe* | Restionaceae | 1 | 1 | – | – | – | 1 | – | 1 |
| *Bauera rubioides* | Cunoniaceae | 1 | – | – | – | – | – | – | 1 |
| *Bellendena montana* | Proteaceae | 1 | – | 1 | – | – | – | – | – |
| *Blechnum penna-marina* | Blechnaceae | 1 | 1 | 1 | – | – | 1 | 1 | – |
| *Brachyscome* aff. *tadgellii*[1] | Asteraceae | – | 1 | 1 | – | 1 | 1 | 1 | – |
| *Brachyscome nivalis* | Asteraceae | 1 | 1 | – | – | – | – | 1 | – |
| *Brachyscome obovata* | Asteraceae | – | 1 | 1 | – | 1 | 1 | – | – |
| *Brachyscome radicans* | Asteraceae | 1 | 1 | 1 | – | – | – | – | – |
| *Brachyscome scapigera* | Asteraceae | – | 1 | – | – | – | – | 1 | – |
| *Brachyscome stolonifera* | Asteraceae | – | 1 | – | – | – | – | 1 | – |
| *Caladenia alpina* | Orchidaceae | 1 | 1 | – | – | – | 1 | – | – |
| *Callistemon pityoides* | Myrtaceae | – | 1 | – | – | – | 1 | – | – |
| *Caltha intraloba* | Ranunculaceae | – | 1 | 1 | – | – | – | 1 | – |
| *Caltha phylloptera* | Ranunculaceae | 1 | – | 1 | – | – | – | 1 | – |
| *Cardamine astoniae* | Brassicaceae | – | 1 | 1 | – | – | – | – | – |
| *Cardamine robusta* | Brassicaceae | 1 | 1 | 1 | – | – | – | – | – |
| *Carex appressa* | Cyperaceae | 1 | 1 | – | – | – | 1 | – | – |

**Table 11.1.** (Continued)

| Species | Family | Tas | Aust | Riv | Pool | Fen | Bog | SAH | BH |
|---|---|---|---|---|---|---|---|---|---|
| *Carex* aff. *diandra*[2] | Cyperaceae | 1 | – | – | – | – | 1 | – | – |
| *Carex breviculmis* | Cyperaceae | 1 | 1 | – | – | – | – | 1 | – |
| *Carex cephalotes* | Cyperaceae | 1 | 1 | – | – | 1 | 1 | 1 | – |
| *Carex curta* | Cyperaceae | 1 | 1 | – | – | 1 | 1 | – | – |
| *Carex echinata* | Cyperaceae | – | 1 | – | – | 1 | – | 1 | – |
| *Carex gaudichaudiana* | Cyperaceae | 1 | 1 | 1 | 1 | 1 | 1 | – | – |
| *Carex hypandra* | Cyperaceae | 1 | 1 | – | – | 1 | 1 | 1 | – |
| *Carex polyantha* | Cyperaceae | 1 | – | – | – | 1 | 1 | 1 | – |
| *Carpha alpina* | Cyperaceae | 1 | 1 | – | – | 1 | 1 | – | – |
| *Carpha nivicola* | Cyperaceae | – | 1 | – | – | 1 | 1 | – | – |
| *Carpha rodwayi* | Cyperaceae | 1 | – | – | – | – | – | – | 1 |
| *Celmisia asteliifolia* | Asteraceae | 1 | – | – | – | – | 1 | – | 1 |
| *Celmisia saxifraga* | Asteraceae | 1 | – | – | – | – | – | – | 1 |
| *Celmisia sericophylla* | Asteraceae | – | 1 | 1 | – | – | – | – | – |
| *Celmisia costiniana* | Asteraceae | – | 1 | – | – | – | – | 1 | – |
| *Celmisia* aff. *pugioformis* | Asteraceae | – | 1 | 1 | 1 | – | – | – | – |
| *Celmisia tomentella* | Asteraceae | – | 1 | – | – | – | 1 | – | – |
| *Centrolepis monogyna* | Centrolepidaceae | 1 | – | – | 1 | – | – | – | 1 |
| *Centrolepis muscoides* | Centrolepidaceae | 1 | – | – | 1 | – | – | 1 | 1 |
| *Chionogentias demissa* | Gentianaceae | 1 | – | – | – | – | – | – | 1 |
| *Colobanthus nivicola* | Caryophyllaceae | – | 1 | – | – | – | – | 1 | – |
| *Coprosma moorei* | Rubiaceae | 1 | 1 | – | – | – | – | – | 1 |
| *Coprosma perpusilla* | Rubiaceae | 1 | 1 | – | – | – | 1 | – | 1 |
| *Coprosma pumila* | Rubiaceae | 1 | 1 | 1 | 1 | | | | |
| *Cotula alpina* | Asteraceae | 1 | 1 | – | – | – | 1 | 1 | – |
| *Craspedia alba* | Asteraceae | – | 1 | 1 | 1 | 1 | – | – | – |
| *Craspedia jamesii* | Asteraceae | – | 1 | 1 | – | – | – | – | – |
| *Craspedia leucantha* | Asteraceae | – | 1 | 1 | – | – | – | – | – |
| *Craspedia paludicola* | Asteraceae | 1 | – | 1 | – | – | – | – | – |
| *Cyathodes dealbata* | Ericaceae | 1 | – | – | – | – | – | – | 1 |
| *Deschampsia caespitosa* | Poaceae | 1 | 1 | 1 | 1 | 1 | – | – | – |
| *Deyeuxia affinis* | Poaceae | – | 1 | 1 | – | – | – | 1 | – |
| *Deyeuxia carinata* | Poaceae | 1 | 1 | 1 | 1 | 1 | 1 | – | – |
| *Dichondra repens* | Convolvulaceae | – | 1 | – | 1 | – | – | 1 | – |
| *Dichosciadeum ranunculaceum* | Apiaceae | 1 | 1 | – | – | – | 1 | 1 | 1 |
| *Diplaspis hydrocotyle* | Apiaceae | 1 | – | – | – | – | – | 1 | 1 |
| *Diplaspis nivis* | Apiaceae | – | 1 | – | – | – | 1 | 1 | – |
| *Donatia novae-zelandiae* | Donatiaceae | 1 | – | – | – | – | – | – | 1 |
| *Dracophyllum milliganii* | Ericaceae | 1 | – | – | – | – | – | – | 1 |
| *Dracophyllum minimum* | Ericaceae | 1 | – | – | – | – | – | – | 1 |

(Continued)

**Table 11.1.** (Continued)

| Species | Family | Tas | Aust | Riv | Pool | Fen | Bog | SAH | BH |
|---|---|---|---|---|---|---|---|---|---|
| *Drosera arcturi* | Droseraceae | 1 | 1 | – | – | – | 1 | 1 | 1 |
| *Drosera pygmaea* | Droseraceae | 1 | – | 1 | 1 | – | – | 1 | – |
| *Ehrharta oreophila* | Poaceae | 1 | – | – | 1 | – | – | – | – |
| *Eleocharis acuta* | Cyperaceae | 1 | 1 | 1 | 1 | – | – | – | – |
| *Empodisma minus* | Restionaceae | 1 | 1 | 1 | – | – | 1 | – | – |
| *Epacris breviflora* | Ericaceae | – | 1 | – | – | – | 1 | – | – |
| *Epacris glacialis* | Ericaceae | – | 1 | 1 | – | 1 | 1 | 1 | – |
| *Epacris microphylla* | Ericaceae | – | 1 | – | – | – | 1 | – | – |
| *Epacris navicularis* | Ericaceae | 1 | – | – | – | – | – | – | 1 |
| *Epacris paludosa* | Ericaceae | – | 1 | – | – | – | 1 | – | – |
| *Epacris petrophila* | Ericaceae | 1 | 1 | – | – | – | 1 | – | – |
| *Epacris serpyllifolia* | Ericaceae | 1 | – | – | – | – | 1 | – | 1 |
| *Epilobium curtisiae* | Onagraceae | 1 | 1 | – | – | – | – | 1 | – |
| *Epilobium gunnianum* | Onagraceae | 1 | 1 | – | – | 1 | 1 | – | – |
| *Epilobium tasmanicum* | Onagraceae | 1 | 1 | – | – | – | – | 1 | – |
| *Erigeron* aff. *stellatus*[2] | Asteraceae | 1 | – | – | – | – | – | 1 | 1 |
| *Erigeron gunnii* | Asteraceae | 1 | – | – | – | – | – | – | 1 |
| *Erigeron nitidus* | Asteraceae | – | 1 | – | – | – | – | 1 | – |
| *Erigeron paludicola* | Asteraceae | – | 1 | – | – | 1 | 1 | – | – |
| *Erigeron setosus* | Asteraceae | – | 1 | 1 | – | – | – | 1 | – |
| *Erigeron stellatus* | Asteraceae | 1 | – | – | – | – | – | – | 1 |
| *Erigeron tasmanicus* | Asteraceae | 1 | – | – | – | – | 1 | – | – |
| *Euchiton collinus* | Asteraceae | 1 | 1 | – | – | – | – | 1 | – |
| *Euphrasia collina* | Scrophulariaceae | 1 | 1 | – | – | – | – | 1 | – |
| *Euphrasia gibbsiae* | Scrophulariaceae | 1 | 1 | – | – | – | 1 | – | 1 |
| *Euphrasia hookeri* | Scrophulariaceae | 1 | – | – | – | – | – | – | 1 |
| *Ewartia meredithiae* | Asteraceae | 1 | – | – | – | – | – | – | 1 |
| *Forstera bellidifolia* | Stylidiaceae | 1 | – | – | – | – | – | – | 1 |
| *Gaimardia fitzgeraldii* | Centrolepidaceae | 1 | – | – | – | – | – | – | 1 |
| *Gaimardia setacea* | Centrolepidaceae | 1 | – | – | – | – | – | – | 1 |
| *Gaultheria tasmanica* | Ericaceae | 1 | – | – | – | – | – | – | 1 |
| *Gentianella muelleriana* | Gentianaceae | – | 1 | 1 | – | – | – | – | – |
| *Geranium sessiliflorum* | Geraniaceae | 1 | 1 | 1 | – | – | – | – | – |
| *Gleichenia alpina* | Gleicheniaceae | 1 | – | – | – | – | 1 | – | – |
| *Gonocarpus micranthus* | Haloragaceae | 1 | 1 | 1 | 1 | 1 | 1 | 1 | – |
| *Gratiola nana* | Scrophulariaceae | 1 | 1 | 1 | 1 | – | – | 1 | – |
| *Gunnera cordifolia* | Gunneraceae | 1 | – | 1 | – | – | – | 1 | – |
| *Helichrysum milliganii* | Asteraceae | 1 | – | – | – | – | – | – | 1 |
| *Helichrysum scorpioides* | Asteraceae | 1 | 1 | 1 | – | – | – | – | – |
| *Hierochloe redolens* | Poaceae | 1 | 1 | 1 | – | – | – | – | – |

**Table 11.1.** (Continued)

| Species | Family | Tas | Aust | Riv | Pool | Fen | Bog | SAH | BH |
|---|---|---|---|---|---|---|---|---|---|
| *Hydatella filamentosa* | Hydatellaceae | 1 | 1 | 1 | – | – | – | – | – |
| *Hydrocotyle algida* | Apiaceae | – | 1 | – | – | – | 1 | – | – |
| *Hydrocotyle muscosa* | Apiaceae | 1 | – | 1 | – | – | – | 1 | – |
| *Hydrocotyle sibthorpioides* | Apiaceae | 1 | 1 | – | 1 | – | – | 1 | – |
| *Hydrocotyle tripartita* | Apiaceae | – | 1 | 1 | 1 | – | – | – | – |
| *Hypericum japonicum* | Hypericaceae | 1 | 1 | – | – | – | 1 | 1 | – |
| *Isoetes gunnii* | Isoetaceae | 1 | – | 1 | 1 | – | – | – | – |
| *Isoetes humilior* | Isoetaceae | 1 | – | 1 | 1 | – | – | – | – |
| *Isoetes muelleri* | Isoetaceae | 1 | 1 | 1 | – | – | – | – | – |
| *Isolepis alpina* | Cyperaceae | 1 | 1 | – | 1 | – | – | – | – |
| *Isolepis aucklandica* | Cyperaceae | 1 | 1 | 1 | 1 | 1 | 1 | 1 | – |
| *Isolepis crussiuscula* | Cyperaceae | 1 | 1 | 1 | 1 | 1 | – | 1 | – |
| *Isolepis habra* | Cyperaceae | 1 | 1 | 1 | 1 | – | 1 | 1 | – |
| *Isolepis montivaga* | Cyperaceae | 1 | 1 | 1 | 1 | – | – | 1 | – |
| *Isolepis subtillissima* | Cyperaceae | 1 | 1 | 1 | 1 | – | 1 | 1 | – |
| *Isotoma fluviatilis* | Campanulaceae | 1 | – | 1 | – | – | – | 1 | – |
| *Juncus antarcticus* | Juncaceae | 1 | 1 | 1 | – | – | 1 | 1 | – |
| *Juncus curtisiae* | Juncaceae | 1 | – | – | – | – | 1 | – | – |
| *Juncus falcatus* | Juncaceae | 1 | 1 | – | – | 1 | 1 | 1 | – |
| *Juncus ratkowskyanus* | Juncaceae | 1 | – | – | – | – | 1 | – | – |
| *Juncus sandwithii* | Juncaceae | 1 | – | 1 | – | – | 1 | – | 1 |
| *Juncus thompsonianus* | Juncaceae | – | 1 | 1 | – | – | 1 | 1 | – |
| *Kelleria laxa* | Thymeleaceae | – | 1 | – | 1 | – | – | – | – |
| *Lachnagrostis lacunarum* | Poaceae | 1 | – | – | 1 | – | – | – | – |
| *Lachnagrostis meionectes* | Poaceae | 1 | 1 | 1 | 1 | 1 | 1 | 1 | – |
| *Leptospermum rupestre* | Myrtaceae | 1 | – | 1 | – | – | – | – | – |
| *Lilaeopsis polyantha* | Apiaceae | 1 | 1 | 1 | 1 | – | – | – | – |
| *Liparophyllum gunnii* | Menyanthaceae | 1 | – | 1 | 1 | – | – | 1 | – |
| *Lobelia gelida* | Campanulaceae | – | 1 | – | 1 | – | – | – | – |
| *Lobelia peduculata* | Campanulaceae | 1 | 1 | 1 | – | – | – | – | – |
| *Lobelia surrepens* | Campanulaceae | 1 | 1 | – | 1 | 1 | 1 | 1 | – |
| *Luzula acutifolia* | Juncaceae | – | 1 | 1 | – | – | – | 1 | – |
| *Luzula atrata* | Juncaceae | 1 | 1 | 1 | – | 1 | 1 | – | – |
| *Luzula modesta* | Juncaceae | 1 | 1 | – | – | – | 1 | – | – |
| *Lycopodium fastigiatum* | Lycopodiaceae | 1 | 1 | 1 | – | – | 1 | – | – |
| *Microcachrys tetragona* | Podocarpaceae | 1 | – | – | – | – | – | – | 1 |
| *Microseris* sp. 2[3] | Asteraceae | – | 1 | 1 | – | – | – | – | – |
| *Miligania densiflora* | Liliaceae | 1 | – | 1 | – | – | – | – | 1 |
| *Milligania lindoniana* | Liliaceae | 1 | – | 1 | – | – | – | 1 | – |
| *Milligania stylosa* | Liliaceae | 1 | – | 1 | – | – | – | – | 1 |

(Continued)

**Table 11.1.** (Continued)

| Species | Family | Tas | Aust | Riv | Pool | Fen | Bog | SAH | BH |
|---|---|---|---|---|---|---|---|---|---|
| *Mimulus repens* | Scrophulariaceae | 1 | – | 1 | – | – | – | – | – |
| *Mitrasacme archeri* | Loganiaceae | 1 | – | – | – | – | – | 1 | – |
| *Myriophyllum austropygmaeum* | Haloragaceae | 1 | – | 1 | 1 | – | – | – | – |
| *Myriophyllum pedunculatum* | Haloragaceae | 1 | 1 | 1 | 1 | 1 | 1 | 1 | – |
| *Myriophyllum simulans* | Haloragaceae | 1 | – | 1 | 1 | – | – | – | – |
| *Myriophyllum variifolium* | Haloragaceae | – | 1 | 1 | 1 | – | – | – | – |
| *Neopaxia australasica* | Portulacaceae | 1 | 1 | 1 | 1 | – | – | 1 | – |
| *Nertera depressa* | Rubiaceae | 1 | – | – | – | – | – | 1 | – |
| *Nertera granadensis* | Rubiaceae | – | 1 | – | – | – | 1 | – | – |
| *Nertera* sp.[1] | Rubiaceae | – | 1 | 1 | – | 1 | 1 | 1 | – |
| *Nymphoides exigua* | Menyanthaceae | 1 | – | 1 | 1 | – | – | – | – |
| *Nymphoides montana* | Menyanthaceae | – | 1 | 1 | 1 | – | – | – | – |
| *Olearia algida* | Asteraceae | 1 | 1 | – | – | – | 1 | – | – |
| *Olearia obcordata* | Asteraceae | 1 | – | 1 | – | – | – | – | – |
| *Oreobolus distichus* | Cyperaceae | 1 | 1 | – | – | – | 1 | 1 | – |
| *Oreobolus oligocephalus* | Cyperaceae | 1 | – | – | – | – | – | – | 1 |
| *Oreobolus pumilio* | Cyperaceae | 1 | 1 | 1 | 1 | 1 | – | 1 | 1 |
| *Oreomyrrhis ciliata* | Apiaceae | 1 | 1 | – | – | 1 | 1 | 1 | 1 |
| *Oreomyrrhis eriopoda* | Apiaceae | 1 | 1 | 1 | – | – | – | – | – |
| *Oreomyrrhis pulvinifica* | Apiaceae | – | 1 | – | – | – | – | 1 | 1 |
| *Orites revoluta* | Proteaceae | 1 | – | 1 | – | – | – | – | 1 |
| *Oschatzia cuneifolia* | Apiaceae | – | 1 | – | – | – | 1 | – | – |
| *Ourisia integrifolia* | Scrophulariaceae | 1 | – | 1 | – | – | – | – | – |
| *Ozothamnus* aff. *hookeri*[1] | Asteraceae | – | 1 | – | – | – | 1 | – | – |
| *Ozothamnus alpinus* | Asteraceae | – | 1 | – | – | – | 1 | – | – |
| *Ozothamnus rodwayi* | Asteraceae | 1 | – | – | – | – | – | – | 1 |
| *Parantennaria uniceps* | Asteraceae | – | 1 | 1 | – | – | 1 | 1 | – |
| *Plantago alpestris* | Plantaginaceae | – | 1 | 1 | – | – | 1 | 1 | – |
| *Plantago daltonii* | Plantaginaceae | 1 | – | 1 | – | – | – | – | – |
| *Plantago glacialis* | Plantaginaceae | 1 | 1 | – | – | – | – | 1 | – |
| *Plantago gunnii* | Plantaginaceae | 1 | – | – | – | – | – | – | 1 |
| *Plantago muelleri* | Plantaginaceae | – | 1 | 1 | – | – | 1 | 1 | – |
| *Plantago paradoxa* | Plantaginaceae | 1 | – | 1 | – | – | – | 1 | – |
| *Poa fawcettiae* | Poaceae | 1 | 1 | – | – | – | – | 1 | – |
| *Poa costiniana* | Poaceae | 1 | 1 | 1 | – | 1 | 1 | – | – |
| *Prasophyllum alpinum* | Orchidaceae | 1 | – | – | – | – | – | – | 1 |
| *Prasophyllum alpestre* | Orchidaceae | 1 | 1 | – | – | – | 1 | – | – |
| *Prasophyllum tadgellianum* | Orchidaceae | 1 | 1 | – | – | – | 1 | – | – |
| *Pterygopappus lawrencei* | Asteraceae | 1 | – | – | – | – | – | – | 1 |
| *Ranunculus amphitrichus* | Ranunculaceae | 1 | 1 | 1 | 1 | – | – | – | – |

**Table 11.1.** (Continued)

| Species | Family | Tas | Aust | Riv | Pool | Fen | Bog | SAH | BH |
|---|---|---|---|---|---|---|---|---|---|
| *Ranunculus anemoneus* | Ranunculaceae | – | 1 | 1 | – | – | – | 1 | – |
| *Ranunculus collicolus* | Ranunculaceae | 1 | – | – | 1 | – | – | 1 | – |
| *Ranunculus collinus* | Ranunculaceae | 1 | 1 | 1 | 1 | – | – | 1 | – |
| *Ranunculus decurvus* | Ranunculaceae | 1 | – | – | – | – | 1 | – | – |
| *Ranunculus dissectifolius* | Ranunculaceae | – | 1 | 1 | – | – | – | – | – |
| *Ranunculus glabrifolius* | Ranunculaceae | 1 | – | 1 | – | – | – | 1 | – |
| *Ranunculus gunnianus* | Ranunculaceae | 1 | 1 | 1 | – | – | – | – | – |
| *Ranunculus jugosus* | Ranunculaceae | 1 | – | 1 | – | – | – | – | – |
| *Ranunculus millanii* | Ranunculaceae | – | 1 | 1 | 1 | 1 | 1 | – | – |
| *Ranunculus nanus* | Ranunculaceae | 1 | – | 1 | 1 | – | – | 1 | – |
| *Ranunculus niphophilus* | Ranunculaceae | – | 1 | 1 | – | – | – | 1 | – |
| *Ranunculus pimpinellifolius* | Ranunculaceae | 1 | 1 | 1 | 1 | – | 1 | 1 | – |
| *Ranunculus setaceus* | Ranunculaceae | 1 | – | 1 | 1 | – | – | 1 | – |
| *Ranunculus triplodontus* | Ranunculaceae | 1 | – | 1 | – | – | – | 1 | – |
| *Richea* aff. *curtisiae*[2] | Ericaceae | 1 | – | – | – | – | – | – | 1 |
| *Richea continentis* | Ericaceae | – | 1 | – | – | – | 1 | – | – |
| *Richea gunnii* | Ericaceae | 1 | – | 1 | – | – | 1 | – | – |
| *Richea scoparia* | Ericaceae | 1 | – | – | – | – | 1 | – | 1 |
| *Rubus gunnianus* | Rosaceae | 1 | – | – | – | – | 1 | – | 1 |
| *Rytidosperma australe* | Poaceae | 1 | 1 | 1 | – | – | – | 1 | – |
| *Rytidosperma nudiflorum* | Poaceae | 1 | 1 | – | – | 1 | – | 1 | – |
| *Rytidosperma nivicola* | Poaceae | 1 | 1 | – | – | 1 | – | – | – |
| *Rytidosperma pauciflorum* | Poaceae | 1 | – | – | – | – | – | – | 1 |
| *Schoenus calyptratus* | Cyperaceae | 1 | 1 | – | – | – | – | 1 | 1 |
| *Schoenus pygmaeus* | Cyperaceae | 1 | – | – | – | – | – | 1 | – |
| *Senecio gunnii* | Asteraceae | 1 | 1 | 1 | – | – | – | – | – |
| *Sprengelia montana* | Ericaceae | 1 | – | – | – | – | – | – | 1 |
| *Stackhousia pulvinaris* | Stackhousiaceae | 1 | 1 | – | 1 | – | – | – | – |
| *Stylidium graminifolium*[1] | Stylidiaceae | 1 | 1 | – | – | 1 | 1 | – | – |
| *Thelymitra cyanea* | Orchidaceae | – | 1 | – | – | – | 1 | – | – |
| *Uncinia compacta* | Cyperaceae | 1 | – | – | – | – | – | – | 1 |
| *Utricularia dichotoma* | Lentibulariaceae | 1 | 1 | – | – | – | – | 1 | 1 |
| *Utricularia monanthos* | Lentibulariaceae | 1 | 1 | 1 | 1 | – | – | 1 | – |
| *Velleia montana* | Goodeniaceae | 1 | 1 | 1 | – | – | – | 1 | – |
| *Veronica gracilis* | Scrophulariaceae | 1 | 1 | 1 | – | – | – | – | – |
| *Viola cunninghamii* | Violaceae | 1 | – | 1 | – | – | – | 1 | – |
| *Wahlenbergia ceracea* | Campanulaceae | 1 | 1 | 1 | – | 1 | 1 | – | – |

[1] *sensu* Costin *et al.* (2000).
[2] *sensu* Kirkpatrick (1997).
[3] *sensu* Ross and Walsh (2003).

48 are associated with ephemeral pools or lakes, 35 are associated with fens, 86 with bogs, 84 with short alpine herbfield and 59 with bolster heath.

Seventy-three per cent of the pool and lake species also occur in vegetation associated with streams. The equivalent figures for other vegetation types are 54% for fen, 54% for short alpine herbfield, 36% for bog and 10% for bolster heath.

The most speciose family is the Asteraceae (43), followed by the Cyperaceae (25), Ranunculaceae (18), Poaceae (17), Ericaceae (16) and Apiaceae (14). No other family has more than nine species (Table 11.1). The most speciose genus is *Ranunculus* (15). The many intergrading alpine species in this genus have evolved reproductive strategies that enable it to dominate much of the alpine riparian environment (Pickering 1997). The equally taxonomically difficult *Carex* (9) and *Epacris* (7) are the next most speciose genera (Table 11.1).

## RIPARIAN VEGETATION TYPES

There have been many detailed descriptive studies of all (Kirkpatrick 1989; Kirkpatrick and Bridle 1998, 1999), or substantial parts of (Costin 1954; McVean 1969; McDougall 1982; Kirkpatrick 1983, 1989, 1997; Walsh *et al.* 1984; Costin *et al.* 2000; McDougall and Walsh 2007), Australia's alpine vegetation. At a broad level, the vegetation formation typology developed by Costin (1954) has proven to be robust for the mainland Australian alpine areas when tested by quadrat floristic data (Kirkpatrick and Bridle 1999). The typology has been extended to account for formations that characterise Tasmanian alpine vegetation, but are rare, or absent from, the mainland mountains (Kirkpatrick 1983, 1997). None of these formations are confined to the riparian zone, as defined by the area of influence of streams, although several are strongly associated with water and do occur widely, and sometimes exclusively, in the riparian zone as defined herein.

At the finer taxonomic level of plant community, defined by floristic composition, several alpine riparian communities have been described in the mainland classification of McDougall and Walsh (2007), and one from the Tasmania-wide riparian vegetation classification of Daley and Kirkpatrick (2004). This section provides some details of our existing knowledge of the riparian vegetation types, organised by formation.

### Aquatic vegetation

Aquatic species are those that can complete their life cycles on inundated ground. Many of these species can also survive and flourish when waters recede, and many other species will survive prolonged inundation but require its lack for regeneration. The alpine aquatic vegetation of Australia is extremely poorly known. McDougall and Walsh (2007), on the basis of one quadrat and their personal observations, give a list of common taxa in aquatic vegetation on the Australian mainland, among which *Myriophyllum pedunculatum*, *Nymphoides montana* and *Ranunculus amphitrichus* are widely regarded to be true aquatics. Kirkpatrick and Harwood (1983a, 1983b) surveyed some of the many lentic wetlands of the Central Plateau as part of a much wider study, but did not discriminate any distinctly alpine communities. The closest to an obligate alpine community is the open fern tussockland dominated by *Isoetes gunnii*, which occurs widely on the floors of tarns and lakes. The quill-like fronds of *I. gunnii* form drifts on alpine beaches and shores. Its congenor, *I. humilior* is found in lakes and slow-flowing alpine streams. In such situations in fertile country, such as on the Central Plateau, *Myriophyllum simulans* can be prominent in deeper water and *Eleocharis acuta* in the edge shallows. *Myriophyllum pedunculatum* and *M. austropygmaeum* are prominent in the zone of fluctuating water levels, as are *Nymphoides exigua* and several immersion-resistant species of *Ranunculus*, *Schoenus* and *Isolepis*. It is clear that more work is needed on Tasmanian alpine lotic aquatic vegetation, which has greater concentrations of Tasmanian endemic aquatic species than Tasmanian lotic aquatic vegetation in general.

### Short alpine herbfield

Short alpine herbfield was defined by Kirkpatrick (1997) as vegetation with more than 50% cover that is not dominated by woody plants or cushion species, and in which the dominant stratum is less than 4 cm tall. This definition fits the snow patch herbfield, *Lobelia surrepens–Ranunculus millanii* herbfield, *Hypericum japonicum–Ranunculus pimpinellifolius* herbfield and short turf snow patch grassland communities of McDougall and Walsh (2007), but not all of their short alpine herbfield, which includes some vegetation dominated by the cushion species *Oreobolus pumilio*. However, the *Plantago muelleri–Montia* (= *Neopaxia*) *australasica* alliance of Costin (1954), which occurs in both of the snow patch herbfield and short alpine herbfield of McDougall and Walsh (2007), does fit this definition.

Snow patches are the origin of many alpine streams. In Tasmania, the melt water from these snow patches often contains fine sediment which deposits in lobes downslope (Fig. 11.2; Gibson and Kirkpatrick 1985), with water flowing above or close to the surface when the lobes are not snow-covered. These lobes are usually covered by short alpine herbfield of highly variable species composition between localities. In Tasmania, short alpine herbfield is also found adjacent to alpine streams, where it is associated with deposits of fine sediments in places accessible to vertebrate grazing, on the margins of lakes and covering ephemeral ponds, where it has been called marginal herbfield in a Tasmanian classification of lentic wetland vegetation (Kirkpatrick and Harwood 1983a), and on wind-exposed shallow soils on gentle slopes on the highest peaks (Kirkpatrick 1997; Taylor 2006). In all but the latter situation it is subject to occasional inundation.

On the mainland, short alpine herbfield dominated by *Neopaxia australasica* and characterised by *Ranunculus niphophilus* and *R. anemoneus*, occurs in areas with short growing seasons, subject to strong irrigation, below snow patches (Costin *et al.* 2000). Short alpine herbfield also occurs in small patches along streams, in shallow ephemeral ponds, and on the margins of lakes (McDougall and Walsh 2007), in places in which mat-forming species, such as *Plantago muelleri*, *P. glacialis*, *Caltha intraloba*, *Lobelia surrepens*, *Ranunculus millanii* and *R. pimpinellifolius* avoid being overwhelmed by upright sedges, cushions and shrubs. On the Bogong High Plains a short turf snow patch grassland characterises snow patches below the climatic treeline (McDougall 1982; McDougall and Walsh 2007: Fig. 12). A structurally similar community, also dominated by *Poa*, also occurs in similar situations in the central and eastern Tasmanian mountains (see Kirkpatrick 1997: plates 2 and 3). In mainland subalpine grassy plains, occasional ephemeral ponds and soaks also support short alpine herbfield, as is also the case in Tasmania, with many of the frequent species in common between Tasmania and mainland Australia.

**Figure 11.2:** Alpine stream in short alpine herbfield on snow patch lobe at Mt Field, Tasmania.

## Fen

In Australia, but not elsewhere in the world, fen has been defined as vegetation dominated by the upright rhizomatous sedges *Carex gaudichaudiana* or *C. hypandra* (Costin 1954; Kirkpatrick 1997; McDougall and Walsh 2007). This species-poor formation occurs in fertile places (Kirkpatrick and Bridle 1999) that are inundated for much of the year, including stream margins, lakes and ponds. It does not occur in the infertile western Tasmanian mountains (Kirkpatrick and Bridle 1999), but is widespread elsewhere.

## Bog

The way that 'bog' is used in Australia is unfortunate in many ways, the most important of which is the confusion that it causes among northern hemisphere mire specialists, who associate the term with ombrotrophic wetlands, which are defined as receiving water only from precipitation. There appear to be no mires in alpine Australia that are ombrotrophic; raised mires in Tasmania are associated with springs (di Folco 2007). The only commonality between a northern hemisphere bog and an Australian bog lies in the presence of the moss genus *Sphagnum*, although the *Sphagnum* species are not shared. The presence of *Sphagnum* is the sole defining feature of 'bog' as used herein and by Costin (1954) and Kirkpatrick (1997). McDougall and Walsh (2007) prefer the more accurate term 'wet heathlands', but their wet heathland includes communities in which *Sphagnum* is not prominent.

A substantial proportion of the alpine bogs of Australia are not riparian in any sense, occurring in places that are constantly moist, but not associated with steams, ponds or lakes (Whinam *et al.* 1989, 2003; Wahren *et al.* 2001; Whinam and Chilcott 2002; Shannon and Morgan 2007). However, bogs are found widely along streams (Fig. 11.3), damming ponds (Whinam and Kirkpatrick 1994), around hillside springs (Costin *et al.* 2000) and on the margins of lakes. The form of bogs is controlled by *Sphagnum cristatum* mounds, which are used as substrate by sedges, rushes and shrubs, the latter often constituting an apparent superstructure for the moss. In most cases the bright yellow to orange surface of *S. cristatum* mounds is largely obscured by vascular plants. Shrub species that dominate bogs in mainland Australian alpine vegetation include *Richea continentis*, *R. victoriana*, *Baeckea gunniana*, *B. utilis*, *Callistemon pityoides*, *Epacris paludosa*, *E. breviflora*, *E. petrophila* and *E. glacialis*. Of these species, only *B. gunniana* and *E. petrophila* occur in Tasmania, and only the former is a common shrub in alpine bogs. *Richea gunnii* is almost exclusively associated with alpine bogs in Tasmania, and the more widespread *R. scoparia* is also a widespread dominant. Bogs are absent from the infertile mountains of western Tasmania (Whinam *et al.* 1989).

## Bolster heath

Bolsters are hard cushion plants, some of which are shrubs, some herbs and some monocotyledonous plants. They are well adapted to strong winds in winter (Gibson 1990), which is probably why they

**Figure 11.3:** Bog along the upper Snowy River, Snowy Mountains, New South Wales.

are more diverse and widespread in maritime alpine Tasmania than in the more continental alpine areas of mainland Australia. Most of Australia's many bolster species are found in constantly wet places, although none appear to be able to survive long-term inundation. They can dominate in mixed species stands, called mosaic bolster heath, or can occur as monospecific dominants (Kirkpatrick *et al.* 1985; Gibson 1990).

Bolster heath dominated by the cushion sedge *Oreobolus pumilio* occurs alongside and in high altitude intermittent streams and ephemeral ponds in mainland alpine vegetation. In Tasmania, this species is rarely a single dominant of bolster heath, but rather is found in combination with one or more other Tasmanian bolster species. In saddles and on plateaus at or near the sources of streams, bolster species commonly form the walls of dams that contain small staircase ponds (Kirkpatrick and Gibson 1984; Gibson and Kirkpatrick 1992). The bolster species that are common in these dams in Tasmania include the sedges *Carpha rodwayi* and *Oreobolus pumilio*, the daisies *Abrotanella forsteroides* and *Pterygopappus lawrencei* and the heath *Dracophyllum minimum*. Two dicotyledonous bolster herbs, *Donatia novae-zelandiae* and *Mitrasacme archeri*, are also frequent.

## Alpine sedgeland

This formation is dominated by sedges (not including *Carex gaudichaudiana* and *C. hypandra*), rushes, lilies, irids, restiads or ferns more than 4 cm high (Kirkpatrick 1997). The only mainland alpine vegetation type described by McDougall and Walsh (2007) that fits this definition is *Carex appressa* sedgeland, but this community was confined to the moist sections of block streams in alpine areas.

On the flatter parts of the quartzite mountains of western Tasmania, alpine sedgeland dominated by *Isophysis tasmanica* grows adjacent to small streams and pools, as does alpine sedgeland dominated by various combinations of *Empodisma minus*, *Baloskion australe*, *Milligania* spp., *Astelia alpina* and *Gleichenia alpina* in parts of the central and eastern mountains. As with bog and bolster heath, alpine sedgeland can form peat dams that block drainage to form small ponds (see Kirkpatrick 1997: plate 20). However, the alpine sedgeland vegetation next to streams and pools differs little from alpine sedgeland growing away from water.

## Alpine tussock grassland

This formation of Kirkpatrick (1997), defined by the dominance of Poaceae species, is equivalent to the 'sod tussock grassland' of Costin *et al.* (2000) and includes several of the grassland floristic communities of McDougall and Walsh (2007), all except one of which, 'alpine valley grassland', are riparian only by topographic accident. Alpine valley grassland is associated with waterways in the high subalpine zone of the Snowy Mountains, is dominated by *Poa costiniana* and has many alpine wetland plant species in its understorey (McDougall and Walsh 2007). In Tasmania, the cosmopolitan tussock grass *Deschampsia caespitosa* dominates some periodically inundated flats on the subalpine parts of the Central Plateau and does not extend to better-drained ground.

## Alpine heath

This formation constitutes most of the alpine vegetation of Tasmania and a substantial proportion of the alpine vegetation of the Australian mainland. It is dominated by woody plants that are neither deciduous nor gymnosperms and are more than 4 cm tall in vegetation in which *Sphagnum* does not occur (Kirkpatrick 1997). Only one of the many heathland communities of McDougall and Walsh (2007), *Epacris* moist heathland, fits this definition and is associated with water. *Epacris glacialis* and *E. gunnii* are the most frequent dominants. The understorey is usually constituted of mat-forming plants. The *Orites acicularis–Baeckea gunniana–Richea acerosa–Hierochloe redolens–Poa costiniana* grassy heath in the riparian vegetation classification of Daley and Kirkpatrick (2004) was recorded from the subalpine treeless zone of the Central Plateau of Tasmania.

## Tall alpine herbfield

Tall alpine herbfield has been suggested to be the formation occupying 'normal', therefore non-riparian, sites in the high alpine areas of mainland Australia (Costin 1954; Costin *et al.* 2000), although McVean (1969) thought it might be a product of disturbance. Only three of the floristic communities of McDougall and Walsh (2007) had tall herbs listed as major dominants. One of these, *Themeda triandra–Leucochrysum albicans* grassland, lies outside my definition of alpine, and is not riparian.

The *Poa fawcettiae–Celmisia costiniana* snow patch grassland is associated with steep south-east-facing slopes that receive abundant flowing water from snow melt (McDougall and Walsh 2007). The dominants of this floristic community co-occur more widely than this situation (Costin *et al.* 2000), relevant quadrats having been subsumed in wider communities, as was the case with earlier nationwide floristic classifications of Australian alpine vegetation (Kirkpatrick 1989; Kirkpatrick and Bridle 1998). *Celmisia sericophylla* herbfield typically occurs where water flows among rocky ground. The community, and dominant species, is confined to the State of Victoria.

Tall alpine herbfield is extremely rare in Tasmania for reasons that will be discussed later in this chapter. However, small areas of the formation have been recorded on the upstream extremities of islands in deep streams on the Central Plateau. The dominant herb is *Craspedia paludicola* (Bridle and Kirkpatrick 1998).

### Other formations

The fjaeldmark, mat heath, coniferous heath and deciduous heath formations of Kirkpatrick (1997) are atypically associated with surface water, although deciduous beech (*Nothofagus gunnii*) and the gymnosperms are highly fire-sensitive (Kirkpatrick and Dickinson 1984), so may only survive in many recently burnt areas where protected by water, giving the appearance of a propensity for a riparian habitat.

## RIPARIAN ECOSYSTEM DYNAMICS

The riparian environment has a deserved reputation for dynamism. Fluvial geomorphic processes provide regeneration niches for a wide variety of plant species in situations in which water is seldom a limiting factor, and where nutrients are concentrated by downslope movement of water and sediment. These geomorphic processes are, in turn, affected by the vegetation of the riparian zone, which can capture sediment and detritus by slowing water movement. Palatable vegetation and water in the riparian zone attract animals, which can affect vegetation dynamics and be agents in geomorphic processes. These interactions are likely to be particularly strong in the alpine headwaters of streams, where the limited size of catchments and more constant rainfall than in the lowlands mitigate extreme flood events that could potentially overwhelm the effects of biota.

This section of the chapter summarises our present knowledge of some ecosystem interactions that result in particular expressions of geomorphic features and riparian vegetation in the Australian alpine zone.

### Interactions involving vertebrate herbivores

Above Lake Augusta, the Ouse River on the Central Plateau of Tasmania meanders through alpine plains riddled with subterranean pipes. In this situation, the stream, which may have formed from the collapse of such pipes, has vertical, fully vegetated banks ~2 m tall, and numerous islands with similarly steep and well-vegetated banks. Some of these islands can be accessed by Bennet's wallabies or, if very close, wombats. Others are too far from shore for a safe hop, as indicated by a total lack of animal scats (Bridle and Kirkpatrick 1998). The animals also find it impossible to reach the vegetation on most of the vertical walls. The islands and the walls are some of the few places in Tasmania where tall herbs, in this case *Craspedia paludicola*, dominate alpine vegetation (Kirkpatrick 1989). However, the dominance of tall herbs in these ungrazed riparian situations is confined to places where flood action could create disturbance, such as the upstream extremities of the islands. Elsewhere, highly palatable wide-bladed grasses, most notably *Hierochloe redolens*, take advantage of the lack of grazers to the exclusion of tall herbs (Bridle and Kirkpatrick 1998).

Below Lake Augusta, the Ouse River has gentler banks suitable for animal access. Where fine sediments have been deposited in the lower parts of the zone of fluctuating water levels, short alpine herbfield forms bright green mats. Scat numbers, chewed-off leaves of *Carex gaudichaudiana* and chewed-off stems of shrub seedlings indicate that this is marsupial lawn, created by the interaction of heavy grazing pressure from Bennet's wallabies and wombats on constantly moist sites. A two-year exclosure experiment in floristically similar lawns in a swale in alpine sand dunes has shown a potential for alpine shrubs and taller grasses and sedges to replace the lawns in the absence of grazing (Roberts *et al.* 2011). The marsupial lawns on the Central Plateau are not confined to wetlands, but also develop on well-drained road embankments and car parks (Taylor 2006). Short alpine herbfield that is associated with neither water disturbance nor human disturbance also occurs on flat surfaces on the higher peaks of the plateau, where shallow soils and exposure to strong winds may exclude taller species.

The short alpine herbfield of the high alpine peaks of the mainland is definitely not marsupial lawn, native vertebrate grazers being absent (McDougall and Walsh 2007). A short growing season and the abrasive powers of firn (compressed snow) may account for its diminutive stature where associated with snow patches. A similarly short growing season and a different form of abrasion may occur in the short alpine herbfield along streams. The short vegetation in ephemeral pools and ponds may result from a lack of inundation constant enough to favour fen.

## The role of vegetation in forming pools and ponds

In the Snowy Mountains, on gentle slopes between rocky ridges and flats, there are features designated as contour terraces, which consist of pools and dams orientated along the contours (Costin *et al.* 2000: 64, 65) in the form of the string bogs of the northern hemisphere (Galloway 2004). Morphologically similar features have been documented in Tasmania (Kirkpatrick and Gibson 1984; Whinam and Kirkpatrick 1994).

Costin *et al.* (2000) suggested that the contour terraces may be relict solifluction features. Terraces may indeed have an ultimate origin in the rearrangement of sediments by solifluctions and overland flow as ice and year-long snow disappeared from most of the Snowy Mountains ~15 000 years ago and all of it by 9000–10 000 years ago (Costin *et al.* 2000). However, it is unlikely that their present surface form is that ancient. Kirkpatrick and Gibson (1984) documented only a weak association between the surface and underlying topography in the features at Newdegate Pass, Mount Field, Tasmania. Here, as at Mt Wellington (Whinam and Kirkpatrick 1994), the dams and ponds have formed on a gently sloping and slightly terraced dolerite block stream, within which there is some interstitial clay. An examination of a core from the tallest dam showed that bolster plants had formed all of it except the very base, where the macrofossils were from aquatic plants (Gibson and Kirkpatrick 1992). Several of the ponds had in situ drowned shrubs and inundated peat, while most of them had rocky floors. Some ponds had drained after the water level got high enough and the basal peat amorphous enough to force basal tunnels. These dams and pools are obviously dynamic features of the landscape, not relicts of past climatic conditions. Growth of the fibric horizon formed by the cushions is ~6 cm per decade, judging from growth against pins, while losses can be rapid, as the dams oxidise after drainage of the ponds (Gibson and Kirkpatrick 1992). In the case of the Mt Wellington features, the 1967 fire caused breaching of some dams, presumed to be because of peat loss resulting from the burning of *Richea* bushes (Whinam and Kirkpatrick 1994).

The dams that form the ponds and spread the upper catchment drainage are the result of a positive feedback process, with the water trapped in the ponds inhibiting growth within them and promoting growth on the dams. If one part of the dam were to grow faster than another, there would be a negative feedback from less moisture supply. The dams do not have to be totally composed of organic material for these feedback processes to occur. Air- or water-borne sediment is more likely to be trapped by the sod tussock grassland that grows on the dams of the Snowy Mountain pools than by the pools themselves.

Tree fall has been observed as a mechanism for forming ponds (pers. obs.). A subalpine plain to the north-west of Cradle Mountain in Tasmania, which has been steadily expanded by fire at the expense of subalpine rainforest, exhibits all stages in this development. Fire kills the trees that previously dominated the area. The dead stags are eventually blown over, leaving a hole previously occupied by the root system, which fills with water for part of the year. Eventually, the water works the hole into a circular shallow, flat-bottomed wetland occupied by short alpine herbfield, with steep sides defined by the caespitose root systems of the dominant *Poa* tussocks on the plain. By this stage there is no sign of the wood of the fallen tree, only a subdued mound of earth that marks the past location of the upended root system with its attached plate of soil. The ponds on this plain are likely to persist through the same positive feedback mechanisms that create natural dammed ponds. It seems likely that any process that creates a hole on flat ground, in a situation in which that hole can fill with water for much of the year, will result in the type of persistent circular shallow depression regarded as mysterious in origin by McDougall and Walsh (2007: Fig. 6). The contour trenches of McElroy (1952) may be the ultimate response to a hole on sloping ground, with the form of the shallow depression becoming more elongate along the contour as the slope steepens.

Some alluvial plains in alpine areas exhibit, in miniature, the full range of landforms associated with mighty rivers in their depositional phase, including levee banks, levee bank blocked wetlands and the ox bow lakes that result from cut-off meanders. Levee banks are also self-perpetuating features, as their vegetation is not only more productive than the back swamps, but also effectively traps sediment when streams overflow.

Lentic wetlands are often regarded as ephemeral features of the landscape, destined for ultimate infilling. This will not be the case where sediment and organic matter accumulation are equal or greater on the land surrounding the wetland than in the wetland itself. The data of Gibson and Kirkpatrick (1992) and Clarke and Martin (1999) indicate that particular alpine wetland types can persist for thousands of years while actively accreting, although Clarke and Martin (1999) did find some more recent cases of replacement of fen by bog.

## CONCLUSIONS

Despite a limited knowledge base, it seems that Australian alpine riparian vegetation, in its broadest sense, is highly distinctive in Australia and globally. This is partly the product of the extreme global distinctiveness of the Tasmanian alpine vegetation as a whole (Kirkpatrick *et al.* 1993; Balmer *et al.* 2004). The subdued topography, lack of a widespread history of glaciation and deep soils of the mainland mountains are also globally unusual (Good 1989, 1992; Kirkpatrick 1993; Costin *et al.* 2004). Soils do not freeze to depth anywhere in the Australian alpine zone (Gibson 1990; Costin *et al.* 2000), making it distinct from most continental alpine areas, where freeze–thaw processes have a strong influence on vegetation patterns.

The alpine riparian vegetation of Australia has a highly distinct flora. The mainland alpine areas have almost all of the persistent winter snow cover in Australia, creating distinct riparian environments, a substantial number of local endemic plant species being associated with snow patches and snow melt flushes and streams. More than half of the species that occur in the western Tasmanian bolster heath and alpine sedgeland communities that abut flowing water are Tasmanian endemics. In the eastern mountains, Tasmanian endemism is also high, even among the more aquatic of the riparian species. In Australian alpine riparian vegetation as a whole there are very few species that occur naturally outside Australia.

Riparian vegetation differs between the mainland and Tasmania in both species composition and process. There are few species in common between the riparian vegetation of western Tasmania and that of the alpine parts of the Snowy Mountains. In Tasmania, herbivores create riparian marsupial lawns and largely exclude tall herbs from riparian vegetation. On the mainland, native vertebrate herbivores do not occur in country with prolonged snow cover. The greater constancy of stream flow in Tasmania than on the mainland results in well-vegetated banks, often little different in their vegetation to land upslope. The gravelly waterways below mainland snow patches, with their distinct suite of species, are largely absent, replaced by well-vegetated and well-grazed lobes of fine sediment. Peat-forming plants create dams that spread drainage parts of the upper catchments. Similar patterning of dams and ponds can be found on some of the mainland mountains, but the dams are typically composed of largely mineral soils covered by sod tussock grasslands.

The future of Australian alpine riparian vegetation is as uncertain as the future of snow in Australia (Green 1998). Will climate changes associated with global warming lead to massive changes in riparian vegetation patterns and fluvial processes in our high mountains? The climatic systems are too complex and the predictive models too simple and coarse to be able to predict specific outcomes in specific localities, despite some pessimism on the prospects for both snow and alpine vegetation (Green 1998; Pickering and Armstrong 2003). For example, there are parts of the Australian alpine zone where it appears to have become colder (Kirkpatrick *et al.* 2002) while the continent as a whole warmed. Desperation for water in drying lowlands, expressed through engineering works in the alpine zone, may prove to be as harmful to alpine riparian systems, and more imminent, than the effects of global warming itself, as orographically induced precipitation is unlikely to completely disappear. The drainage systems of areas formerly grazed by stock are still in the slow process of recovery documented by Wimbush and Costin (1983) and Wahren *et al.* (1994) and some riparian vegetation is also recovering slowly from recent extensive fires (Walsh and McDougall 2004; Kirkpatrick and Bridle 2013), which could be more frequent with global warming. Partly because of the impacts of fire on bogs, alpine fens and bogs have been recently listed as threatened at the national level. On the bright side, stock grazing has now been excluded from all alpine public land, and most of the area of alpine vegetation in Australia is within the Western Tasmanian Wilderness World Heritage Area or national parks listed as nationally significant under the *Environment Protection and Biodiversity Conservation Act 1999* (Commonwealth).

## REFERENCES


Balmer J, Whinam J, Kelman J, Kirkpatrick JB, Lazarus E (2004) 'A review of the floristic values of the Tasmanian Wilderness World Heritage Area.' Nature conservation report 2004–03, Department of Primary Industries, Water and Environment, Hobart.

Barrows TT, Stone JO, Fifield LK, Cresswell RG (2001) Late Pleistocene glaciation of the Kosciuszko Massif, Snowy Mountains. *Quaternary Research* **55**, 179–189. doi:10.1006/qres.2001.2216

Bridle KL, Kirkpatrick JB (1998) Why do tall herbs rarely dominate Tasmanian alpine vegetation? Evidence from islands in the Ouse River system. *Papers and Proceedings of the Royal Society of Tasmania* **132**, 9–14.

Buchanan AM (2005) *A Census of the Vascular Plants of Tasmania.* 4th edn. Tasmanian Museum and Art Gallery, Hobart.

Clarke PJ, Martin ARH (1999) *Sphagnum* peatlands of Kosciuszko National Park in relation to altitude, time and disturbance. *Australian Journal of Botany* **47**, 519–536. doi:10.1071/BT97129

Colhoun EA, Fitzsimons SJ (1990) Late Cainozoic glaciation in western Tasmania, Australia. *Quaternary Science Reviews* **9**, 199–216. doi:10.1016/0277-3791(90)90018-6

Costin AB (1954) *A Study of the Ecosystems of the Monaro Region of New South Wales*. Government Printer, Sydney.

Costin AB, Gray M, Totterdell C, Wimbush D (2000) *Kosciuszko Alpine Flora*. CSIRO Publishing, Melbourne.

Costin AB, Wimbush D, Kirkpatrick JB (2004) Flora values. In *An Assessment of the Values of Kosciuszko National Park*. (Ed. Independent Scientific Committee) pp. 55–72. New South Wales National Parks and Wildlife Service, Sydney.

Daley EA, Kirkpatrick JB (2004) Native riparian vegetation in Tasmania. *Cunninghamia* **8**, 409–430.

di Folco M-B (2007) The organosols of Tasmania. PhD thesis, University of Tasmania.

Galloway R (2004) Earth science values. In *An Assessment of the Values of Kosciuszko National Park*. (Ed. Independent Scientific Committee) pp. 21–26. New South Wales National Parks and Wildlife Service, Sydney.

Gibson N (1990) The environments and primary productivity of cushion species at Mt Field and Mt Wellington, Tasmania. *Australian Journal of Botany* **38**, 229–243. doi:10.1071/BT9900229

Gibson N, Kirkpatrick JB (1985) Vegetation associated with localised snow accumulation at Mount Field West, Tasmania. *Australian Journal of Ecology* **10**, 91–99. doi:10.1111/j.1442-9993.1985.tb00869.x

Gibson N, Kirkpatrick JB (1992) Dynamics of a Tasmanian cushion heath community. *Journal of Vegetation Science* **3**, 647–654. doi:10.2307/3235832

Good RB (Ed.) (1989) *The Scientific Significance of the Australian Alps*. Australian Alps Liaison Committee, Canberra.

Good RB (1992) *Kosciusko Heritage*. New South Wales National Parks and Wildlife Service, Sydney.

Green K (Ed.) (1998) *Snow: a Natural History; an Uncertain Future*. Australian Alps Liaison Committee, Canberra.

Hughes JMR (1987) A study of riverine plant communities in Tasmania, with especial reference to central east coast rivers. PhD thesis, University of Tasmania.

Johnston SW (2001) The influence of aeolian dust deposits on alpine soils in south-east Australia. *Australian Journal of Soil Research* **39**, 81–88. doi:10.1071/SR99121

Kirkpatrick JB (1982) Phytogeographical analysis of Tasmanian alpine floras. *Journal of Biogeography* **9**, 255–271. doi:10.2307/2844668

Kirkpatrick JB (1983) Treeless plant communities of the Tasmanian high country. *Proceedings of the Ecological Society of Australia* **12**, 61–77.

Kirkpatrick JB (1986) Conservation of plant species, alliances and associations of the treeless high country of Tasmania. *Biological Conservation* **37**, 43–57. doi:10.1016/0006-3207(86)90033-9

Kirkpatrick JB (1989) The comparative ecology of mainland Australian and Tasmanian alpine vegetation. In *The Scientific Significance of the Australian Alps*. (Ed. RB Good) pp. 127–142. Australian Academy of Science, Canberra.

Kirkpatrick JB (1993) *The International Significance of the Natural Values of the Australian Alps*. Australian Alps Liaison Committee, Canberra.

Kirkpatrick JB (1997) *Alpine Tasmania: an Illustrated Guide to the Flora and Vegetation*. Oxford University Press, Melbourne.

Kirkpatrick JB (1999) Grassy vegetation and subalpine eucalypt communities. In *Vegetation of Tasmania*. (Eds JB Reid, RS Hill, MJ Brown, M Hovendon) pp. 265–285. Australian Government Publishing Service, Canberra.

Kirkpatrick JB, Bridle KL (1998) Environmental relationships of floristic variation in the alpine vegetation of south-east Australia. *Journal of Vegetation Science* **9**, 251–260. doi:10.2307/3237124

Kirkpatrick JB, Bridle KL (1999) Environment and floristics of ten Australian alpine vegetation formations. *Australian Journal of Botany* **47**, 1–21. doi:10.1071/BT97058

Kirkpatrick JB, Bridle KL (2013) Natural and cultural histories of fire differ between Tasmanian and mainland alpine vegetation. *Australian Journal of Botany* **61**, 465–474. doi:10.1071/BT13128

Kirkpatrick JB, Brown MJ (1987) The nature of the transition from sedgeland to alpine vegetation in south-west Tasmania. I. Altitudinal vegetation change on four mountains. *Journal of Biogeography* **14**, 539–550. doi:10.2307/2844879

Kirkpatrick JB, Dickinson KJM (1984) The impact of fire on Tasmanian alpine vegetation and soils. *Australian Journal of Botany* **32**, 613–629. doi:10.1071/BT9840613

Kirkpatrick JB, Fowler M (1998) Locating likely glacial refugia in Tasmania using palynological and ecological information to test alternative climatic models. *Biological Conservation* **85**, 171–182. doi:10.1016/S0006-3207(97)00144-4

Kirkpatrick JB, Gibson N (1984) Dynamics of a Tasmanian bolster heath string fen. *Vegetatio* **58**, 71–78. doi:10.1007/BF00044930

Kirkpatrick JB, Harwood CE (1983a) Tasmanian wetland plant communities. *Australian Journal of Botany* **31**, 437–451. doi:10.1071/BT9830437

Kirkpatrick JB, Harwood CE (1983b) The conservation of Tasmanian wetland macrophytic species and communities. *Papers and Proceedings of the Royal Society of Tasmania* **117**, 5–20.

Kirkpatrick JB, Minchin P, Davies J (1985) Floristic composition and macroenvironmental relationships of Tasmanian vegetation containing bolster plants. *Vegetatio* **63**, 89–96.

Kirkpatrick JB, Hutchinson MN, MacQuillan PB (1993) Alpine ecosystems. In *Tasmanian Wilderness – World Heritage Values*. (Eds S Smith and MR Banks) pp. 73–79. Royal Society of Tasmania, Hobart.

Kirkpatrick JB, Bridle KL, Lynch AJJ (2002) Changes in vegetation and landforms at Hill One, Tasmania. *Australian Journal of Botany* **50**, 753–759. doi:10.1071/BT02039

McDougall KL (1982) The alpine vegetation of the Bogong High Plains. Environmental studies publication no. 357. Ministry for Conservation, Melbourne.

McDougall KL, Walsh NG (2007) Treeless vegetation of the Australian Alps. *Cunninghamia* **10**, 1–58.

McElroy CT (1952) Contour trench formations in uplands plains of New South Wales. *Journal of the Royal Society of New South Wales* **85**, 53–63.

McVean DN (1969) Alpine vegetation of the central Snowy Mountains, New South Wales. *Journal of Ecology* **57**, 67–86. doi:10.2307/2258208

Nunez M, Kirkpatrick JB, Nilsson C (1996) Rainfall estimation in south west Tasmania using satellite images and phytosociological calibration. *International Journal of Remote Sensing* **17**, 1583–1600. doi:10.1080/01431169608948724

Peterson JA (1968) Cirque morphology and Pleistocene ice formation conditions in southeastern Australia. *Australian Geographical Studies* **6**, 67–83. doi:10.1111/j.1467-8470.1968.tb00178.x

Pharo E, Kirkpatrick JB (1994) Vegetation of the alpine sand dunes at Lake Augusta, Tasmania. *Australian Journal of Ecology* **19**, 319–327. doi:10.1111/j.1442-9993.1994.tb00495.x

Pickering CM (1997) Reproductive strategies and constraints of alpine plants as illustrated by five species of Australian alpine *Ranunculus*. *Opera Botanica* **132**, 101–108.

Pickering CM, Armstrong T (2003) The potential impact of climatic change on plant communities in the Kosciuszko alpine zone. *The Victorian Naturalist* **120**, 15–24.

Roberts C, Kirkpatrick JB, McQuillan PB (2011) Tasmanian lentic wetland lawns are maintained by grazing rather than inundation. *Austral Ecology* **36**, 303–309. doi:10.1111/j.1442-9993.2010.02168.x

Ross JH, Walsh NG (2003) *Census of the Vascular Plants of Victoria*. 7th edn. Royal Botanic Gardens, South Yarra, Victoria.

Shannon JM, Morgan JW (2007) Floristic variation in *Sphagnum*-dominated peatland communities of the central highlands, Victoria. *Cunninghamia* **10**, 59–76.

Taylor S (2006) The ecology of short alpine herbfield on the Central Plateau, Tasmania. Honours thesis. School of Geography and Environmental Studies, University of Tasmania.

Wahren C-HA, Papst WA, Williams RJ (1994) Long-term vegetation change in relation to cattle grazing in sub-alpine grassland and heathland on the Bogong High-Plains: an analysis of vegetation records from 1945 to 1994. *Australian Journal of Botany* **42**, 607–639. doi:10.1071/BT9940607

Wahren C-HA, Williams RJ, Papst WA (2001) Alpine and subalpine snow patch vegetation on the Bogong High Plains, SE Australia. *Journal of*

*Vegetation Science* **12**, 779–790. doi:10.2307/3236865

Walker PH, Costin AB (1971) Atmospheric dust accession in south-eastern Australia. *Australian Journal of Soil Research* **9**, 1–5. doi:10.1071/SR9710001

Walsh NG, McDougall KL (2004) Progress in the recovery of the flora of treeless subalpine vegetation in Kosciuszko National Park after the 2003 fires. *Cuninghamia* **8**, 439–452.

Walsh NG, Barley RH, Gullan PK (1984) 'The alpine vegetation of Victoria, excluding the Bogong High Plains.' Environmental studies publication no. 376. Department of Conservation, Forests and Lands, Melbourne.

Whinam J, Chilcott N (2002) Floristic description and environmental relationships of *Sphagnum* communities in NSW and the ACT and their conservation management. *Cunninghamia* **7**, 463–500.

Whinam J, Kirkpatrick JB (1994) The Mt Wellington string bogs, Tasmania. *Papers and Proceedings of the Royal Society of Tasmania* **128**, 63–68.

Whinam J, Eberhard S, Kirkpatrick JB, Moscal A (1989) *Ecology and Conservation of Tasmanian* Sphagnum *Peatlands*. Tasmanian Conservation Trust, Hobart.

Whinam J, Chilcott N, Morgan JW (2003) Floristic composition and environmental relationships of *Sphagnum*-dominated communities in Victoria. *Cunninghamia* **8**, 162–174.

Wimbush DJ, Costin AB (1983) Trends in drainage characteristics in the subalpine zone at Kosciusko. *Proceedings of the Ecological Society of Australia* **12**, 143–154.

Wintle B, Kirkpatrick JB (2007) The response of riparian vegetation to flood-maintained habitat heterogeneity. *Austral Ecology* **32**, 592–599. doi:10.1111/j.1442-9993.2007.01753.x

# 12

# Riparian vegetation of tropical northern Australia

*Neil Pettit, John Dowe and Ian Dixon*

## INTRODUCTION

Riparian zones, as linear systems penetrating the landscape, are ecologically influential in regulating the movement of species and propagules as well as water and nutrients. In tropical northern Australia – here defined as the area stretching from the Fitzroy River in Western Australia, across near-coastal northern Australia to the Burdekin River in Queensland – riparian zones can represent areas of high productivity, particularly in savannas and lower rainfall regions, where available moisture can be a major factor limiting plant productivity. For the majority of river systems, this effect is visually obvious with taller, greener and denser riparian forest in linear arrangements within a matrix of drier, sparser woodlands (Fig. 12.1). Overall, the floristic composition of most riparian systems in northern Australia is rather homogeneous, with many species having wide-ranging distribution. However, the large differences in rainfall and stream flow that exist between the wet and dry seasons in northern Australia strongly influence plant community structure and species distribution (Pettit *et al.* 2011). The environmental factors that are significant contributors to riparian vegetation community structure at waterholes and perennial and intermittent rivers in northern Australia include flow regime, channel width (as a proxy for other influences such as position in the catchment or ecosystem size) and groundwater influence (Warfe *et al.* 2013).

Riparian zones in high rainfall areas, such as the wet tropics region of north Queensland and some rainforest locations on Cape York Peninsula, have less of a dramatic interface with adjacent vegetation than do those in savannas and lower rainfall regions (Fig. 12.2). In areas such as Bellenden Ker Ranges, where annual average rainfall is up to 5000 mm, riparian vegetation is often composed of trees and shrubs that also occur in nearby forest, and floristic distinctions between the riparian zone and adjacent forests occur primarily in the pteridophyte and bryophyte floras (Lawson *et al.* 2007; Dowe 2008). Nevertheless, the floristic composition is relatively heterogeneous compared with river systems in drier areas, supporting many local endemics and species with narrow distribution ranges. The riparian zones and streambeds of steeply graded watercourses may host rheophytes and suites of bryophytes and ferns that are associated with flowing water (Webb 1968; Clifford and Specht 1986). Indeed adaptation to fast-flowing and/or permanent flowing waters and the presence of rheophytes are among the basic floristic distinctions between savanna and rainforest riparian plant assemblages (Meave *et al.* 1991; Dowe 2008).

The relatively greater productivity of riparian zones in savanna and low-rainfall regions compared

**Figure 12.1:** Contrast between the higher productivity of the vegetation in the riparian zone (left) and the surrounding savanna woodland on the Ord River in the Kimberley region of north-western Australia.

**Figure 12.2:** The riparian vegetation in many dynamic wet tropics rivers is floristically similar to that in adjacent forests, Barron River, wet tropics bioregion of Queensland.

with adjacent woodlands and environments, contrasts with a relatively lower productivity in the riparian zones of high rainfall streams when compared with adjacent forest. This results primarily from erosion and debris removal from the riparian zone during times of high flow causing plant death and/or reductions in plant biomass. Consequently, we suggest that there is an ecological continuum with regards to riparian zone productivity from streams in low-rainfall to high-rainfall regions (Naiman *et al.* 2005) that has scarcely been researched in Australia to date (Dowe 2008; Pettit *et al.* 2011).

Riparian zones can also provide refugia for tropical rainforest species as they line watercourses in areas that are too dry in the contemporary climate to support rainforest in the surrounding terrestrial landscape. These riparian areas can provide some protection to the gallery rainforest from fire (Kellman and Tackaberry 1993). In tropical regions, general determinants of the more mesic or xeric components of riparian communities include seasonal predictability, amount of rainfall and the effects of fire. Where rainfall is more or less consistent, but highly seasonal, riparian vegetation must be adapted to large hydrological variations (Dodds 1997). In more arid areas, where rainfall is not consistent, tolerance to drought and dependence on surface and subsurface hydrology are features of the riparian vegetation (Bowman and McDonough 1991). The degree of isolation from fire will also determine the development of riparian communities (Kellman *et al.* 1998). Consequently, inundation and disturbance are key ecological processes in these riparian environments. A clonal growth habit with underground rhizomes or roots represents a general adaptation among riparian trees to a range of disturbances, including fire and flooding (Pettit and Naiman 2007), both of which are typical disturbances in most northern Australian riparian landscapes (see Chapter 16).

To date, studies of riparian vegetation in northern Australia have generally been either highly localised and confined to single catchments, or broad in scale and lacking detail and local relevance (Thurgate 1994; Pettit 2002; Dowe 2008). For example, although most of the vegetation of northern Australia has been mapped using aerial photographs and Landsat imagery, the scale is insufficient to confidently identify specific elements, and so riparian vegetation has been generally incorporated into broader vegetation types (Fox *et al.* 2001). The most comprehensive classification of riparian vegetation communities in northern Australia is provided by Beadle (1981). This classification included the broad tropical riparian category and detailed several communities based on various dominant species including the *Eucalyptus camaldulensis* (river red gum) alliance, *Melaleuca leucadendra* (weeping paper bark) alliance, *Casuarina cunninghamiana* (river she-oak) alliance and, the *Eucalyptus microtheca* alliance, among others. Recently, the focus on riparian vegetation communities in this region has seen them recognised and treated as distinct vegetation units primarily related to land management and biodiversity issues rather than as distinct floristic or functional units.

This chapter discusses aspects of riparian plants and vegetation that occur within the Timor, Gulf of Carpentaria and north-east drainage divisions of the northern third of Australia that lies within the tropic zone. We include areas of riparian communities within the wet tropics and the wet–dry tropical savannah, as well as tropical floodplains as a special type of riparian environment. We summarise the characteristics of riparian communities within each of these climatic zones and provide examples from representative rivers within each of these regions. We also briefly discuss some aspects of the ecology of riparian vegetation as they apply to tropical regions, and examine the various impacts and management issues that confront riparian communities in the tropics.

## BIOGEOGRAPHY OF RIVERINE PLANT COMMUNITIES IN NORTHERN AUSTRALIA

In tropical Australia there are generally two distinct seasons that are broadly described as the 'wet' and the 'dry'. The wet season lasts from approximately November through to April and is generally hot and humid with almost all of the year's rain falling during this period. Rainfall is usually the result of heavy bursts from thunderstorms, widespread monsoon depressions or from the passage of associated tropical cyclones (Gentilli 1972). The dry season, from May through to October, is generally dry and cooler (although temperatures are warm throughout the year), with little rain and low humidity. The amount of rainfall varies across the region and in relation to latitude, due mainly to topography and distance from the coast. Annual average rainfall gradually diminishes with distance inland and to the south, from more than 2400 mm to less than 800 mm (Bureau of Meteorology). The majority of this area of northern Australia is described as tropical savanna or grassy woodlands, with some areas of rainforest and tall sclerophyll forest in the wet tropics of north-east Queensland.

A comprehensive study of riparian vegetation in northern Australia was undertaken as part of a larger project on Australian tropical rivers (Tropical Rivers Inventory and Assessment Project). In this study Dowe (2008) identified a total of 264 plant species that were recognised as occurring in the riparian zone across northern Australia. Of these, 23 species were identified as obligate riparian species (Table 12.1), with 126 facultative riparian species; 84 species were identified as freshwater aquatic plants and 28 as mangrove species. The report by Dowe (2008) provides a detailed description of the taxonomy, ecology and distribution of each of these species in riparian zones throughout northern Australia.

### Wet tropics

In contrast to the highly seasonal rainfall and high average temperatures that are associated with most of the river systems in northern Australia, the systems in the wet tropics of north-east Queensland and Cape York Peninsula generally have perennial flow, though with a significant flow peak in the wet season, and comparatively lower temperatures in mid to high elevation areas (Balston 2008). The high rainfall areas, defined by the 1200 mm or greater isohyet, are most often associated with mountain ranges and upland areas, and most of the river systems in these areas are highly dynamic and perennial. There is a marked floristic difference in the riparian vegetation between the wet tropics river systems and those highly seasonal systems in drier environments that are considered more 'typical' for northern Australia (Metcalfe and Ford 2008). In the wet tropics, however, there is much less distinction in productivity and species composition between the riparian zone and upland areas, which is such a distinctive feature of riparian areas in the dry regions of the tropics.

**Table 12.1.** Obligate riparian species recorded for northern Australian rivers (from Dowe 2008)

| Species | Family | Life form |
|---|---|---|
| *Aeschynomene indica* | Fabaceae | Perennial herb |
| *Asteromyrtus symphyocarpa* | Myrtaceae | Tall shrub |
| *Calycopeplus casuarinoides* | Euphorbiaceae | Small tree |
| *Casuarina cunninghamiana* | Casuarinaceae | Large tree |
| *Cathormion umbellatum* | Mimosaceae | Small tree |
| *Chrysopogon oblongatus* | Poaceae | Perennial grass |
| *Corymbia bella* | Myrtaceae | Medium tree |
| *Cyperus difformis* | Cyperaceae | Perennial herb |
| *Eucalyptus camaldulensis* | Myrtaceae | Tall tree |
| *Flueggea virosa* | Euphorbiaceae | Shrub |
| *Goodenia strangfordii* | Goodeniaceae | Shrub |
| *Livistona rigida* | Arecaceae | Medium tree |
| *Lophostemon grandiflorus* | Myrtaceae | Tall tree |
| *Melaleuca argentea* | Myrtaceae | Medium to tall tree |
| *Melaleuca bracteata* | Myrtaceae | Small tree |
| *Melaleuca fluviatilis* | Myrtaceae | Medium to tall tree |
| *Melaleuca leucadendra* | Myrtaceae | Medium to tall tree |
| *Melaleuca stenostachya* | Myrtaceae | Small tree |
| *Melaleuca trichostachya* | Myrtaceae | Small tree |
| *Persicaria attenuata* | Polygonaceae | Perennial herb |
| *Sesbania cannabina* | Fabaceae | Shrub |
| *Sesbania erubescens* | Fabaceae | Shrub |
| *Vitex acuminata* | Lamiaceae | Small tree |

*Melaleuca leucadendra* riparian forest is distributed throughout the wet tropics region. It occurs on well-drained alluvial soils fringing streams and may include vine forest elements and species such as *Eucalyptus tereticornis, Tristaniopsis exiliflora* and *Xanthostemon chrysanthus*. On alluvial soils along lowland creeks and river levees, semi-deciduous vine forest can occur, commonly with *Nauclea orientalis, Cryptocarya hypospodia* and *Castanospermum australe*. On outwash plains of lowlands of the wetter rainfall areas, a complex notophyll vine forest may be found with emergent *Argyrodendron polyandrum, Corymbia torelliana, Acacia celsa* and *A. mangium*, and canopy species such as *Elaeocarpus grandis, Antidesma bunius, Grevillea baileyana, Flindersia pimenteliana* and *Alstonia muelleriana*. This community has generally been reduced to discontinuous strips along stream banks throughout its range and is subject to human disturbances, particularly alteration of natural fire regimes.

## Savanna riparian areas

Riparian vegetation throughout much of semi-arid and monsoonal savanna regions of northern Australia is dominated by a small number of tree species (Cole 1986; Dowe 2008). Prominent among these species are *Eucalyptus camaldulensis*, broad-leaf (*Melaleuca argentea, M. fluviatilis, M. leucadendra, M. viridiflora*) and narrow-leaf *Melaleuca* spp. (*M. bracteata, M. trichostachya*), *Casuarina cunninghamiana, Terminalia* spp. and *Lophostemon grandiflorus*, among others. Both *M. argentea* and *M. leucadendra* are broadly distributed across northern Australian streams, although *M. leucadendra* is generally more widely distributed and frequently occurs on poorly drained sites isolated from streams (Franklin *et al.* 2007). *Melaleuca leucadendra* appears to be more tolerant of higher-energy floods in the higher stream gradients or better able to capitalise on recruitment opportunities provided by flood scouring (Pettit and Froend 2001a). Alternatively, *M. argentea* may need greater access to groundwater, which supplies up to 50% of its dry season water requirement (Lamontagne *et al.* 2005). Overall, it has been demonstrated that riparian trees lower down on stream slopes and closer to the river are more reliant on groundwater than trees upslope (O'Grady *et al.* 2006).

On the Burdekin River in the savanna region of north Queensland, species diversity is generally low but with locally high individual species abundance and with a flood-tolerant flora dominated by the *Melaleuca* species (Dowe 2004). The morphology of these *Melaleuca* species suggests adaptation to flood disturbance, including total submergence, as well as mitigation of impact damage during dynamic wet season runoff events (Fielding *et al.* 1997). Adaptations include a reclined downstream-trailing habit, multiple-stemmed form, modified crown with weeping foliage, thick spongy bark and clonal regeneration. Re-sprouting of seedlings, which are exposed to otherwise catastrophic breakage during flood events, is also a feature of some of the dominant riparian trees (Chong *et al.* 2007). For most of the seasonal streams, riparian species display large-scale distribution patterns that are linear to stream flow, and dependant on changes in stream morphology, sediment regimes and bedrock characteristics (Dowe 2008). For small-scale distribution patterns, species are primarily distributed relative to average high and low water flow limits and stream flow characteristics, as well as light availability and soil nutrient levels (Capon and Dowe 2007).

The Daly River in the savanna region of the Northern Territory is an example of a perennial flowing river and demonstrates some unique features of the riparian vegetation. The banks of this perennially flowing river generally consist of a series of benches below the floodplain created by high wet season flows as well as areas of groundwater seepage (Erskine *et al.* 2003). This habitat structure results in distinct vertical zonation of riparian vegetation from the river bed across the benches to the levee crest of the floodplain and into the terrestrial vegetation of the savanna. Bank slopes are generally dominated by six species (*M. argentea*, *M. leucadendra*, *Corymbia bella*, *Cathormion umbellatum*, *Nauclea orientalis* and *Casuarina cunninghamiana*) with *M. argentea* and *M. leucadendra* trees dominating along the lower bank. Closed monsoon forests of *C. cunninghamiana*, *N. orientalis*, *Barringtonia acutangula*, *Ficus racemosa*, *C. umbellatum*, *Terminalia microcarpa* and *Strychnos lucida* occur behind this fringing vegetation on benches on the mid to upper slopes, with a eucalypt woodland present on the levees with *Corymbia bella*, *E. tectifica*, *Erythrophloem chlorostachys*, *Planchonia careya* and *Terminalia ferdinandiana* generally common. The large riparian trees *Acacia auriculiformis* and *C. cunninghamiana* are common throughout all zones. Patches of monsoon rainforest containing *N. orientalis* and *Lophostemon grandiflorus* with dense *Pandanus* species fringes also occur (Price *et al.* 2003).

Intermittently flowing savanna rivers of the gulf country contrast in productivity with the surrounding savanna and are generally subject to high usage by cattle. Consequently, these riparian zones are often infested with weeds such as rubber vine (*Cryptostegia grandiflora*), which is becoming a major threat to riparian communities in this area. Riparian communities on alluvial soils of the major watercourses are typically dominated by *M. argentea* and less frequently *M. leucadendra* in open forest, together with *Acacia auriculiformis*, *Syzygium forte* and *N. orientalis*, as well as occasional *Ficus racemosa* and *Terminalia platyphylla*. A sub-canopy layer is usually present including *Diospyros* spp., *Barringtonia acutangula*, *Acacia torulosa*, *Strychnos lucida* and *Leptospermum madidum*. Stream bank trees can also include *Casuarina cunninghamiana* and *Eucalyptus tereticornis*, and scattered vines, including *Flagellaria indica*, *Austrosteenisia blackii*, *Lygodium flexuosum* and *Smilax australis*, may also be present.

## Floodplains

On flat alluvial floodplains of savanna region rivers such as the Mitchell River, *Eucalyptus microtheca* with some *E. chlorophylla* occur as low open woodlands. Other species present can include *Corymbia papuana*, *C. bella* and *Grevillea striata*. The ground layer can be dense and dominated by grasses such as *Themeda arguens*, *Brachyachne convergens*, *Sarga plumosum*, *Aristida* spp. and *Eriachne* spp. On major channels on the fine-textured alluvial plains *E. microtheca* and *E. camaldulensis* woodland occur, generally with a ground layer of *Chrysopogon fallax* (golden beard grass) and *Bothriochloa decipiens*. In lower lying areas of the floodplains, *E. microtheca* can occur with *Excoecaria parvifolia* and a ground storey of *Dichanthium* spp. In floodplain waterholes that retain water for longer periods, aquatic plants such as *Nelumbo nucifera* and *Nymphaea* spp. can occur, and sedges (*Eleocharis* spp.), together with the semi-aquatic grasses *Pseudoraphis spinescens* and *Leersia hexandra*, occupy the shallower margins.

On floodplains in higher rainfall regions such as the South Alligator River floodplain, three riparian community types are recognised (Petty and Douglas 2010), including: (1) floodplains where the degree of inundation varies and consists of a closed paperbark (*Melaleuca* spp.) forest with areas of emergent herblands (Finlayson 2005) and shrublands of *Cathormion umbellatum*, *Barringtonia acutangula* and *Strychnos lucida* among grass and sedgeland; (2) evergreen to semi-deciduous 'rainforest' communities of diverse species assemblages fringing floodplains or associated with springs, including closed forests fringing perennial streams (Russell-Smith 1991); and (3) seasonally flooded open woodlands dominated by the trees *Corymbia bella*, *C. polycarpa* and *Eucalyptus alba* (Bowman and Dunlop 1986; Bowman and Minchin 1987), with *Erythrophleum chlorostachys*, *Syzygium suborbiculare* and *Lophostemon lactifluus* at slightly higher elevations (Bowman and McDonough 1991).

Vegetative communities on the floodplains of the Alligator Rivers region of the Northern Territory are related to microtopography and water depth during the wet season (Bowman and Wilson 1986; Finlayson *et al.* 1990; Finlayson 2005). Common vegetation includes broad-leaf paperbark (*Melaleuca* spp.) forests, open perennial and annual swamps, billabongs and grass and sedge herbfields (Williams 1979). The widespread grassland communities on the floodplains are dominated by three species: *Hymenachne acutigluma* (15% of the floodplain), *Pseudoraphis spinescens* (14%) and *Oryza meridionalis* (12%) (Finlayson *et al.* 1993). In recent years, large areas of *O. meridionalis* (wild rice) and other wetland plants on the Magela floodplain have been replaced by exotic para grass (*Urochloa mutica*) (Bayliss *et al.* 2006), introduced during the early 1980s as a pasture plant before Kakadu was established as a national park.

High rainfall floodplain rivers, such as those in the Top End, operate with predictable seasonal rains in the wet season which provide replenishing floodwaters and boost plant production (Pettit *et al.* 2011). In drier regions, such as in the southern Gulf of Carpentaria and the Kimberley region of Western Australia, flooding is much reduced in space and time (Ward *et al.* 2013), so that growth of aquatic plants on the floodplains during the inundation phase is much reduced and terrestrial grasses generally dominate floodplains. Recent work indicates that flooding as a disturbance exerts selective pressures on river biota while at the same time enhances productivity and diversity (Lake *et al.* 2006). The predictability of the timing and magnitude of flooding is a key factor determining whether floods act to restore carbon and nutrients, enhancing productivity and diversity, or act as a disturbance, altering physical templates, damaging biota and setting new recovery trajectories. The balance between these two conflicting effects is accomplished through the yearly predictability of flood events (Jardine *et al.* 2015).

## TEMPORAL AND SEASONAL DYNAMICS OF RIPARIAN VEGETATION

Riparian vegetation dynamics across northern Australia are closely related to the natural flow regime. Patterns in reproductive phenology and seedling establishment, as well as plant community composition and structure, tend to reflect the duration and frequency of flooding (Pettit *et al.*

2001; Pettit and Froend 2001b). The Ord River in the Kimberley region of tropical Western Australia is a large floodplain river, and the catchment above the Ord River dam is fairly typical of those found in northern Australia. For this type of river, flows are strongly seasonal as a consequence of climatic patterns driven by monsoonal rainfall in the wet season (Gentilli 1972), and flow variability is high for all months. Flows are usually of short duration, even in the wet season, and result from high-intensity rainfall events. There is typically limited groundwater storage to sustain higher base flows (Pettit *et al.* 2001). On the Ord River, Pettit *et al.* (2001) found there was a good correlation between riparian community structure and both the duration and frequency of flooding, demonstrating the importance of the flow regime in driving the zonation of the vegetation along a gradient of distance from the low-flow water level. On these types of tropical rivers, fluvial processes are dominated by intermittent, large, high-energy floods that are capable of destroying the existing riparian vegetation community and preventing the establishment of stable mature plant community states. Except in areas protected from flooding, riparian vegetation dynamics are characterised by long periods of transition between short-lived stable states (Pettit and Froend 2001a). This is evidenced by the existence of large mature trees on these river reaches only at high elevations, in protected river bends or far from the river where only extreme floods can reach. High magnitude wet season floods are, however, highly episodic and, in intervening years, smaller floods allow the development of existing trees and recruitment of younger trees. This type of riparian ecosystem, common in northern Australia, is thus driven at mid- to long-term time scales by allogenic or physical processes rather than vegetation successional (autogenic) processes (Baker and Walford 1995).

Regeneration from seed in these habitats can occur through several processes (Fig. 12.3). Reproductive phenology of many dominant riparian tree species in tropical rivers reflects their adaptation to their environment. Maturation of flowers and seed takes place within a couple of months in many tropical eucalypts (Brooker and Kleinig 1994; Eldridge *et al.* 1993) as opposed to more than two years in some southern eucalypts (Ashton 1975). For common northern Australian riparian trees, such as *E. camaldulensis* and *M. leucadendra*, the period of greatest seedfall coincides with floodwater recession, when seed germination and survival are favoured by the exposure to moist sediments (Pettit and Froend 2001a). Seed is released over a short period of time at the end of the wet season, which provides the only opportunity for recruitment in a season. Other northern Australian *Eucalyptus* species have been reported to display a similar reproductive phenology (Bowman *et al.* 1991; Fensham 1992; Setterfield and Williams 1996). For tropical northern Australia, many species of eucalypts have short reproductive cycles with flowering occurring at the end of the dry season (Brooker and Kleinig 1994) and all seed released by the end of the wet season. In contrast to *E. camaldulensis* in

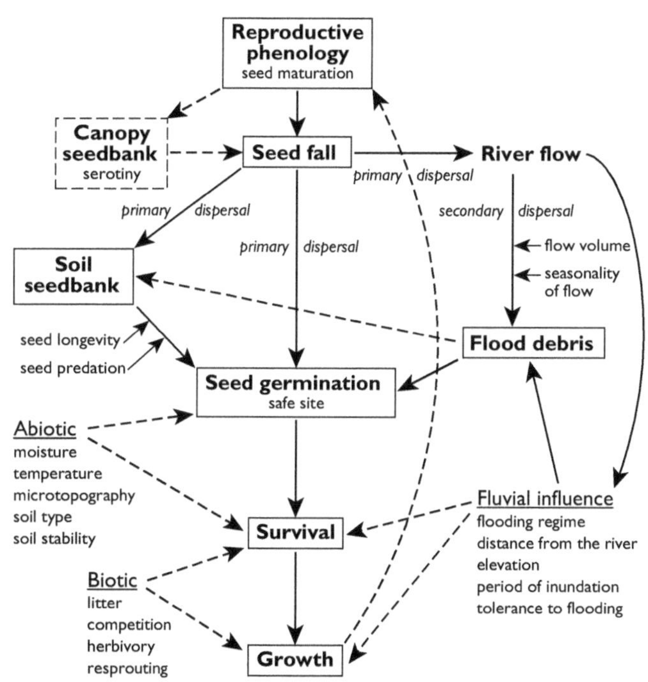


**Figure 12.3:** A general model of regeneration from seed of riparian vegetation. Many trees of tropical riparian areas tend not to have much capacity for seed storage either on the plant (serotiny) or in the soil and a greater reliance on consistent wet season rains or secondary dispersal by water.

northern Australia, *E. camaldulensis* on the Murray River in south-eastern Australia usually retains mature seed in the canopy for at least 12 months (Dexter *et al.* 1986; Pettit and Froend 2001b). Although there is usually a time of peak seed release, some seed is released all year round, so there is potential for germination of seed whenever conditions are favourable (Dexter *et al.* 1986).

After the production of seed and primary seed dispersal, important secondary dispersal by water in river flows occurs for many species, usually at the time of high river flows. At this time, seed can disperse rapidly downstream away from competition from mature trees, as well as laterally to higher elevations which may afford protection to germinating seedlings from subsequent inundation. *Cathormiom umbellatum* is a small to medium-sized riparian tree which occurs fringing tropical rivers and floodplains, and is an example of a plant adapted to dispersal by water. It has buoyant pods that dehisce, usually into or near the water, with large non-buoyant seeds that remain intact and are only released after decomposition of the pod. The occurrence of seed in flood debris also indicates the importance of secondary dispersal of seed by water. This has been reported for tropical riparian habitats in Australia and southern Africa (Pettit and Froend 2001b; Pettit and Naiman 2006).

A large proportion of riparian plants regenerate after flooding via vegetative re-sprouting. Vegetative re-sprouting is particularly common in communities receiving frequent periodic disturbance (James 1984). Re-sprouting from uprooted trees after a large flood is a significant form of recovery of vegetation biomass by obligate riparian tree species on a subtropical African river (Pettit *et al.* 2006). By vegetative re-sprouting, these species are able to quickly dominate the canopy of the new riparian forest, an important process for facilitating the rapid regeneration of riparian vegetation. Clonal regeneration, via roots or lateral branches, as an adaptation to flood disturbance is also commonly seen among *Melaleuca* species in tropical Australian rivers (Fielding *et al.* 1997; Pettit and Froend 2001a). Another tropical riparian plant that regenerates largely through the development of clonal tillers, but also from seed, is the arborescent bamboo *Bambusa arnhemica*. This species occurs in riparian zones in north-western parts of the Northern Territory. Flowering occurs gregariously and plants then die synchronously across large areas. Regeneration occurs from seed with the first substantial rains of the wet season. Seedlings are generally tolerant of inundation, although prolonged flooding and soil scouring can prevent seedlings establishing in the lowest-lying areas of the riparian zone (Franklin *et al.* 2007).

*Casuarina cunninghamiana* (river she-oak) is one of the few truly riparian trees in Australia and is common along rivers east of the Dividing Range, from southern New South Wales through Queensland to the Northern Territory. As with *E. camaldulensis* and *Melaleuca* species, *C. cunninghamiana* in the tropics is not serotinous, with samaras released from cones when they are mature, to coincide with relatively predictable monsoonal rainfall favourable for seedling establishment. Recruitment from seed usually occurs yearly, although episodic events (floods, drought) may enhance or decrease establishment (Woolfrey and Ladd 2001). This species also spreads clonally through the reiteration of lateral branches, which develop into upright stems when the parent axis is laid horizontally by flood (Ladd 1989). This allows the plants to have much longer site tenure despite infrequent establishment opportunities from seed due to the high disturbance regime in lower sections of the riparian zone (Woolfrey and Ladd 2001).

## STATUS, THREATS AND MANAGEMENT OF TROPICAL RIPARIAN LANDSCAPES

Much of the large-scale degradation of catchments and their streams in Australia is concentrated in southern temperate regions. In contrast, northern Australia's tropical rivers and wetlands are generally in relatively good condition, except where European colonisation has been more intense, such as parts of north-eastern Queensland (Junk 2002).

Due to the remoteness and relatively undeveloped landscape of the wet–dry tropics, there has been little attempt to study the impacts of various pressures on its riparian vegetation, and much of our current knowledge is based on local observations and point-source disturbance studies (Douglas and Pouliot 1997; van Dam *et al.* 2008). Furthermore, multiple pressures on riparian zones often occur simultaneously resulting in complex management issues with management that targets one pressure sometimes exacerbating another.

Tropical rivers feature large flood events during the annual wet season that can drastically modify the structural community of riparian vegetation and cause severe bank erosion. Erosion potential is greatest in smaller streams, but is also high in all streams with steep profile banks (Saynor and Erskine 2006). Although riparian vegetation may recover from bank erosion under natural conditions, other threats, including weed invasion, feral animals, fire and overgrazing (Bartolo *et al.* 2008) may inhibit native plant recruitment and stabilisation of riverbanks. Furthermore, riparian areas are often perceived as being at risk of disproportionately high levels of degradation due to the combined effects of other pressures and natural flood disturbance, as well as the intensity of riparian land use. The small but highly productive area occupied by riparian vegetation in the savanna landscape, for example, is subject to particularly intense use for cattle watering.

## Grazing

Pastoralism is the dominant land use across northern Australia. Although most stock grazing occurs on 'unimproved' pastures of native vegetation at low stocking rates, intensive cattle grazing in riparian zones can have several degrading impacts such as loss of native plant cover and diversity, particularly groundcover species, soil compaction and bank erosion, and loss of riparian habitat and food resources, resulting in changed faunal composition and the introduction and spread of weeds (Sattler 1993; Douglas and Pouliot 1997). In recent history, poor grazing practices and unregulated cattle access to streams in the Ord, Fitzroy and Victoria river catchments have led to major losses of riparian vegetation, especially understorey species, and severe bank erosion that is exacerbated by seasonal floods (Winter 1990; Anon 1992). As surface water diminishes throughout the dry season, many seasonal waterholes become dry season water refuges, thus putting additional pressure on riparian vegetation and bank stability (Burrows 2001).

In parts of the Burdekin River catchment, off-stream fences adjacent to alluvial flats have been installed to control access to riparian areas during vulnerable times (Burrows 2001). Exclusion fences are difficult on large properties across northern Australia, due to the costs of construction and maintenance, and additional expenses for the installation of off-stream watering points, and repair of flood-damaged fence lines. Once implemented, managers have reported reduced trampling of riparian vegetation, reduced soil compaction and erosion, intact habitat for native fauna, easier stock access to watering points, more even grazing pressure across the landscape, more efficient mustering, and fewer stock losses due to bogging, drowning and crocodile attack (Douglas and Pouliot 1997; Panton 2001). However, excluding cattle completely from riparian zones may lead to management neglect of that area, resulting in expansion of weeds such as *Urochloa mutica* (para grass), *Hymenachne amplexicaulis* (olive hymenachne), *Cryptostegia grandiflora* (rubber vine), *Zizyphus mauritiana* (chinee apple) and *Jatropha gossypifolia* (bellyache bush), harbouring of feral animals such as *Sus scrofa* (pigs) and *Bubalus bubalis* (buffalo), and increased fire fuel loads (Burrows 2001; Panton 2001). Alternatively, with regulated access to the stream, cattle could be used as a management tool to suppress weed invasion in riparian zones. In this case, the benefits of cattle (i.e. controlling weeds) may outweigh the threats if appropriate stocking rates are maintained (Dowe 2005). Determining sustainable stocking rates specific to biodiversity and ecosystem conservation in riparian zones requires further investigation (Douglas and Pouliot 1997).

In parts of Queensland, determining a balance in cattle numbers, weeds and fires poses a major challenge to land managers. Grazing in the riparian zone has kept fuel loads low, but has simultaneously promoted the growth of rubber vine, which is one of Queensland's most destructive weeds (Sattler 1993). Following removal of stock, fuel loads have increased allowing for hotter fires that help suppress rubber vine (Grice 1997). Based on experimental work, Radford *et al.* (2008) recommend that low-intensity, early wet-season burning, or early dry-season burning, is best for control of rubber vine in order to minimise the fire intensity and loss of large trees in riparian habitats.

## Weeds

Several grass species introduced for pasture improvement are considered environmental weeds in tropical northern riparian zones as they have detrimental effects on native flora and fauna, especially when left ungrazed or uncontained within pastoral lands. Grassy weeds, such as *Pennisetum ciliare* (buffelgrass), *Andropogon gayanus* (gamba), *Megathyrsus maximus* (Guinea grass) and *Pennisetum polystachion* (mission grass), have been observed to grow to the channel edge and dominate the riparian zone of numerous streams across northern Australia (Dixon and Douglas 2007; Dowe and Dixon 2007; Dixon *et al.* 2010). These species have the potential to alter the fire regime, as they have greater fuel loads and create hotter and taller fires than native grasses (Grice 2004; NTG 2009). Riparian vegetation can be sensitive to fire, especially in the late dry season when fires can reduce the abundance and diversity of riparian vegetation by increasing mortality of established trees (see Chapter 16; Douglas *et al.* 2003). The effects of fire on water quality and riparian zone erosion can otherwise be highly variable, dependent on stream morphology, soil type and time-lag between burning and commencement of wet-season flow and runoff (Townsend and Douglas 2000). Fringing riparian vegetation is most vulnerable to late dry season fires as they are the first point of impact from a savanna fire. The potential destructive forces of hotter and taller fires fuelled by grassy weeds may threaten the establishment of native riparian trees resulting in an encroaching grassy weed population. Loss of riparian trees has further effects by reducing bank stability and stream shading.

As a result of their typically large seed production, weeds are often the first species to invade disturbed areas. Given the annual flood disturbance, higher soil moisture and greater soil fertility in riparian zones, weeds favour these areas compared with adjacent uplands (Grice 2004). Controlling their spread is difficult, as cattle and feral animals create disturbed areas in riparian zones and transport seeds via their faeces or attached to their hides (Cowie and Werner 1993; Lonsdale 1993; Douglas and Pouliot 1997). Weed invasion is exacerbated in the dry season when animals focus their activity around water refuges, and in the wet season when flood waters transport seeds downstream. Spread of weeds also occurs across catchments via birds and human activities.

Managers face several challenges when implementing weed control programs. Often a 'weed complex' of trees, shrubs, vines, herbs and grasses occur together, making it harder to manage as not all weeds will respond to a particular control technique (Grice 2001). Legislative constraints also arise when attempting large-scale weed control, such as herbicide restrictions near water or clearing restrictions that may prevent mechanical removal (chaining) of larger weeds near streams (Rea and Storrs 1999; Grice 2001). Use of fire for weed control may also damage habitat for native fauna, such as birds and reptiles, requiring consideration of season and frequency of burns in fire management strategies (Valentine *et al.* 2007; Valentine and Schwarzkopf 2009). Furthermore, eradication of weeds may result in invasion by other weed species (e.g. para grass replacing mimosa after control; Cook and Setterfield 1995), or establishment of stress-tolerant native plants not typical of uninfested areas (e.g. Lane *et al.* 1997). Weed control needs to consider causes of weed invasion and spread to prevent re-establishment, rather than just localised treatment (Rea and Storrs 1999).

### Feral animals

Of the many introduced animals present in tropical Australia, pigs and buffalo are infamous for being highly destructive to wetland and riparian habitats. Pigs are renowned for their uprooting of vegetation, wallowing and transportation of weed propagules, such as mimosa seeds in Kakadu National Park (Lindner 1995). Buffalo impacts are similar to those caused by cattle, but due to the unique way they use the landscape they are also capable of changing the surface hydrology. In Kakadu National Park, high densities of buffalo in the past have led to the breakdown of natural levee banks between freshwater wetlands and tidal channels. This has decreased wetland retention and led to saltwater intrusion into the wetlands and riparian vegetation, resulting in the destruction of large forests of *Melaleuca* trees fringing the swamps (Finlayson *et al.* 1997). Artificial levee banks (i.e. barrages) have helped promote successful recovery of areas following saltwater intrusion (e.g. Mary River wetlands east of Darwin; Fulton 1995). To ensure long-term success of management programs, such as weed eradication and levee bank rehabilitation, feral animal control should be considered a fundamental action.

### Water resources development

As pressures for agricultural, urban and industrial development grow across northern Australia, greater exploitation of water resources is increasingly likely. Tropical riparian vegetation has evolved with contrasting high flow in the wet season and low to no flow in the dry season. Change to this natural water regime may impact on riparian communities if not managed carefully. Evidence from the few regulated rivers in the north, as described by Doupé and Pettit (2002) and Start (2000), suggests that riparian vegetation undergoes structural and recruitment changes in response to flow regime modification. In regulated systems that provide a constant release of flow downstream of dams, the abundant supply of water supports a greater density of some riparian plants and other invading species that would not usually survive there. With dam walls reducing the frequency and intensity of floods downstream, dense stands of riparian vegetation are not thinned by flood surges. Long-standing dense riparian vegetation can prevent development of larger-sized trees, with communities then stabilising as saplings and recruitment opportunities diminish. Reduction of powerful floods prevents the erosion of instream and point bars and allows for the encroachment of riparian vegetation. Also, a lack of large floods can limit the potential for seed dispersal into the riparian zone and surrounding floodplain, preventing the seed drop and seedling emergence that is typically triggered by floodwaters. The challenge for water managers is to mimic the dynamic nature of tropical wet–dry season flows while supplying other agricultural, ecological and socio-economic requirements. However, this is not always possible where, for example, dry season flows are required to dilute agricultural effluent or natural stream channels act as conduits for irrigation.

Water extraction from unregulated rivers for agricultural irrigation also has significant consequences. Petty and Douglas (2010) propose that groundwater availability and fire are key factors determining the riparian forest boundary. It follows that if water extraction lowers the water table, a combination of water stress and fire during the dry season could be particularly detrimental to this vegetation. Under such circumstances, water extraction could lead to a decrease in riparian width and native habitat, the terrestrialisation of the riparian zone (Naiman *et al.* 2005) and the encroachment of savanna vegetation.

## CONCLUSIONS

Riparian vegetation communities in northern Australia are as diverse as the environments in which they occur. In many cases, riparian zones are refugial habitats for rare and poorly known species and plant communities. In the drier regions, they provide corridors of high productivity in the generally low-productivity savanna landscape. Although the floristics and biogeography of plants that make up riparian communities in the tropics of northern Australia are becoming better known, there is still much that is poorly understood, especially ecological processes related to riparian vegetation. This is particularly the case with regards to studies linking the biology of important riparian plants to river flow regimes and other environmental factors. For example, management for environmental water provisions is based on the effects of altered flows on the aquatic environment, but also needs to take into account water requirements of riparian vegetation. In addition, the functional role of riparian vegetation in aquatic ecosystems is not well understood in northern Australia. For example, riparian subsidies in the form of leaf litter, flowers and fruit as well as insects dropping into the stream from overhanging vegetation can be an important source of energy for aquatic food webs (Douglas *et al.* 2005; Lau *et al.* 2009). Shading by riparian vegetation can also reduce primary productivity in streams and increase the importance of riparian inputs (Davies *et al.* 2008). Information about these ecological processes is critical to the proper management of northern river systems.

Ecological studies should complement and improve monitoring programs that report on the condition of riparian communities. New studies looking at the classification of Australian tropical rivers (Kennard *et al.* 2010) based on ecological characteristics can provide a starting point to classify riparian vegetation in northern Australia. Despite the existence of some exogenous disturbance on tropical rivers, such as cattle grazing, riparian ecosystems in northern Australia are generally much less altered at present by anthropogenic developments than those in temperate regions of Australia (e.g. much of the Murray–Darling Basin). There is, therefore, an opportunity to use these northern riparian systems as a benchmark for riparian ecosystem condition, not only to inform any future agricultural or industrial developments in northern Australia, but also to provide a blueprint for the requirements of restoration of degraded riparian habitats throughout Australia.

The maintenance and restoration of riparian habitats requires strategies and knowledge of the environmental requirements and functions of key riparian plant species. In addition, we need to develop methods for minimising the impacts of the many continuing threats to riparian vegetation in tropical northern Australia. It is clear, therefore, that an ecosystem-function level of understanding is critical for the effective management of these unique and important habitats.

## REFERENCES

Anon (1992) *State of the Environment Report.* Government of Western Australia, Perth.

Ashton DH (1975) Studies of flowering behaviour in *Eucalyptus regnans* F. Muell. *Australian Journal of Botany* **23**, 399–411. doi:10.1071/BT9750399

Baker WL, Walford GM (1995) Multiple stable states and models of riparian vegetation succession on the Animas River, Colorado. *Annals of the Association of American Geographers. Association of American Geographers* **85**, 320–338. doi:10.1111/j.1467-8306.1995.tb01797.x

Balston J (2008) Impacts of climate variability and climate change on the wet tropics of north-eastern Australia. In *Living in a Dynamic Tropical Forest Landscape.* (Eds NE Stork and SM Turton) pp. 267–281. Blackwell Publishing, Carlton, Victoria.

Bartolo R, van Dam R, Bayliss P (2008) Semi-quantitative risk assessments – the relative risk model. In *Ecological Risk Assessment for Australia's Northern Tropical Rivers. Sub-project 2 of Australia's Tropical Rivers – an Integrated Data Assessment and Analysis (DET18).* (Eds R Bartolo, P Bayliss and R van Dam) pp. 162–270. A report to Land and Water Australia, Canberra. Environmental Research Institute of the Supervising Scientist, National Centre for Tropical Wetland Research, Darwin.

Bayliss P, van Dam R, Boyden J, Walden D (2006) Ecological risk assessment of Magela floodplain to differentiate mining and non-mining impacts. In *ERISS Research Summary 2004–2005.* (Eds KG

Evans, J Rovis-Hermann, A Webb and DR Jones) pp. 172–185. Supervising scientist report 189, Supervising Scientist, Darwin.

Beadle NCW (1981) *The Vegetation of Australia*. Gustav Fischer Verlag, Stuttgart, Germany.

Bowman DMJS, Dunlop CR (1986) Vegetation pattern and environmental correlates in coastal forests of the Australian monsoon tropics. *Vegetatio* **65**, 99–104. doi:10.1007/BF00044880

Bowman DMJS, McDonough L (1991) Tree species distribution across a seasonally flooded elevation gradient in the Australian monsoon tropics. *Journal of Biogeography* **18**, 203–212. doi:10.2307/2845293

Bowman D, Minchin P (1987) Environmental relationships of woody vegetation patterns in the Australian monsoon tropics. *Australian Journal of Botany* **35**, 151–169. doi:10.1071/BT9870151

Bowman DMJS, Wilson BA (1986) Wetland vegetation pattern on the Adelaide River flood plain, Northern Territory, Australia. *Proceedings of the Royal Society of Queensland* **97**, 69–77.

Bowman DMJS, Wilson BA, Woinarski JCZ (1991) Floristic and phenological variation in a northern Australian rocky savanna. *Proceedings of the Royal Society of Queensland* **101**, 79–90.

Brooker MIH, Kleinig DA (1994) *Field Guide to the Eucalypts: Northern Australia*. Inkata Press, Sydney.

Burrows D (2001) Livestock management in streams and wetlands: examples from the Burdekin catchment, north Queensland. In *Wise Use of Wetlands in Northern Australia: Grazing Management in Wetlands and Riparian Habitats*. (Eds B Myers, M McKaige, P Whitehead and M Douglas) pp. 17–19. Centre for Tropical Wetlands Management, Northern Territory University, Darwin.

Capon SJ, Dowe JL (2007) Diversity and dynamics of riparian vegetation. In *Principles for Riparian Lands Management*. (Eds S Lovett and P Price) pp. 13–32. Land and Water Australia, Canberra.

Chong C, Edwards W, Waycott M (2007) Differences in resprouting ability are not related to seed size or seedling growth in four riparian woody species. *Journal of Ecology* **95**, 840–850. doi:10.1111/j.1365-2745.2007.01241.x

Clifford HT, Specht RL (1986) *Tropical Plant Communities: Their Resilience, Functioning and Management in Northern Australia*. University of Queensland, St Lucia, QLD.

Cole MM (1986) *The Savannas: Biogeography and Geobotany*. Academic Press, London.

Cook G, Setterfield S (1995) Ecosystem dynamics and the management of environmental weeds in wetlands. In *Wetland Research in the Wet–Dry Tropics of Australia*. (Ed. CM Finlayson) pp. 200–205. Supervising scientist report 101. Commonwealth of Australia, Canberra.

Cowie ID, Werner PA (1993) Alien plant species invasive in Kakadu National Park, tropical northern Australia. *Biological Conservation* **63**, 127–135. doi:10.1016/0006-3207(93)90500-Z

Davies PM, Bunn SE, Hamilton SK (2008) Primary production in tropical streams and rivers. In *Tropical Stream Ecology*. (Ed. D Dudgeon) pp. 23–42. Elsevier, London.

Dexter BD, Rose HJ, Davies N (1986) River regulation and associated forest management problems in the River Murray red gum forests. *Australian Forestry* **49**, 16–27. doi:10.1080/00049158.1986.10674459

Dixon I, Douglas M (2007) *Riparian Condition in the Ord River Catchment, Western Australia: Application of a Rapid Assessment Tool*. Tropical Savannas Cooperative Research Centre, Darwin.

Dixon IH, Douglas MM, Townsend SA (2010) *Riparian Condition of the Darwin-Litchfield-Bynoe Regions, Northern Territory*. Tropical Rivers and Coastal Knowledge, Charles Darwin University, Darwin.

Dodds WK (1997) Distribution of runoff and rivers related to vegetative characteristics, latitude and slope: a global perspective. *Journal of the North American Benthological Society* **16**, 162–168. doi:10.2307/1468248

Douglas M, Pouliot A (1997) A review of the impacts of the northern Australian grazing industry on wetlands and riparian habitats. In *The Northern Australian Program: Catchment Management, Water Quality and Nutrient Flows in Northern Australia and the Northern Australian Beef Industry*. (Ed. R Hook) pp. R2-1–R2-19. NAP occasional publication no. 3. Meat and Livestock Australia, Melbourne.

Douglas MM, Townsend SA, Lake PS (2003) Streams. In *Fire in Tropical Savannas: the Kapalga Experiment*. (Eds AN Anderson, GD Cook and RJ Williams) pp. 59–78. Springer-Verlag, New York, NY.

Douglas MM, Bunn SE, Davies PM (2005) River and wetland food webs in Australia's wet–dry tropics: general principles and implications for management. *Marine and Freshwater Research* **56**, 329–342.

Doupé RG, Pettit NE (2002) Ecological perspectives on regulation and water allocation for the Ord River, Western Australia. *River Research and Applications* **18**, 307–320. doi:10.1002/rra.676

Dowe JL (2004) 'A survey of dominant riparian vegetation at selected sites in the Georgetown area, Northern Gulf region.' Australian Centre for Tropical Freshwater Research report no. 04/13., ACTFR, Townsville, QLD.

Dowe JL (2005) 'Changes in riparian vegetation after the establishment of stock exclosure plots at three sites in the Burdekin River catchment, Queensland.' Australian Centre for Tropical Freshwater Research report no. 05/08. ACTFR, Townsville, QLD.

Dowe JL (2008) Distribution and ecological preferences of riparian vegetation in northern Australia. In *A Compendium of Ecological Information on Australia's Northern Tropical Rivers. Sub-project 1 of Australia's Tropical Rivers – an Integrated Data Assessment and Analysis (DET18)*. (Eds GP Lukacs and CM Finlayson) pp. 1–55. A report to Land and Water Australia, Canberra. Australian Centre for Tropical Freshwater Research, National Centre for Tropical Wetland Research, Townsville, QLD.

Dowe JL, Dixon I (2007) *Riparian Condition in the Gilbert River and Walsh River Catchments Using the Tropical Rapid Appraisal of Riparian Condition (TRARC) Method*. Australian Centre for Tropical Freshwater Research, Townsville, QLD.

Eldridge K, Davidson J, Harwood C, van Wyk G (1993) *Eucalypt Domestication and Breeding*. Oxford University Press, Oxford, UK.

Erskine WD, Begg GW, Jolly P, Georges A, O'Grady A, Eamus D, Rea N, Dostine P, Townsend S, Padovan A (2003) 'Recommended environmental water requirements for the Daly River, Northern Territory, based on ecological, hydrological and biological principles.' Supervising scientist report 175, Darwin.

Fensham RJ (1992) The establishment of *Eucalyptus* seedlings in tropical savanna forest. *Northern Territory Naturalist* **13**, 30–36.

Fielding CR, Alexander J, Newman-Sutherland E (1997) Preservation of in situ, arborescent vegetation and fluvial bar construction in the Burdekin River of north Queensland, Australia. *Palaeogeography, Palaeoclimatology, Palaeoecology* **135**, 123–144. doi:10.1016/S0031-0182(97)00022-9

Finlayson CM (2005) Plant ecology of Australia's tropical floodplain wetlands: a review. *Annals of Botany* **96**, 541–555. doi:10.1093/aob/mci209

Finlayson CM, Cowie ID, Bailey BJ (1990) Characteristics of a seasonally flooded freshwater system in monsoonal Australia. In *Wetland Ecology and Management: Case Studies*. (Eds DF Whigham, RE Good and J Kvet) pp. 141–162. Kluwer Academic Publishers, Dordrecht, Netherlands.

Finlayson CM, Cowie ID, Bailey BJ (1993) Biomass and litter dynamics in a *Melaleuca* forest on a seasonally inundated floodplain in tropical, northern Australia. *Wetlands Ecology and Management* **2**, 177–188. doi:10.1007/BF00188152

Finlayson CM, Storrs MJ, Lindner G (1997) Degradation and rehabilitation of wetlands in the Alligator Rivers region of northern Australia. *Wetlands Ecology and Management* **5**, 19–36. doi:10.1023/A:1008271219441

Fox ID, Neldner VJ, Wilson GW, Bannink PJ (2001) *Vegetation of Northern Australia*. Queensland Herbarium, Environmental Protection Agency, Brisbane.

Franklin DC, Brocklehurst PS, Lynch D, Bowman DMJS (2007) Niche differentiation and regeneration in the seasonally flooded *Melaleuca* forests of northern Australia. *Journal of Tropical Ecology* **23**, 457–467. doi:10.1017/S0266467407004130

Fulton I (1995) Multiple use on the lower Mary River wetlands. In *Wetland Research in the Wet–Dry Tropics of Australia*. (Ed. CM Finlayson) pp. 80–85. Supervising scientist report 101, Commonwealth of Australia, Canberra.

Gentilli J (1972) *Australian Climate Patterns*. Thomas Nelson, Melbourne.

Grice AC (1997) Fire for the strategic management of rubbervine *Cryptostegia grandiflora*, an invasive shrub. In *Bushfire 97, Proceedings of the Australian Bushfires Conference*. 8–10 July, 1997, Darwin. CSIRO Tropical Ecosystem Research Centre, Darwin.

Grice A (2001) Weeds in wetlands: threats and opportunities. In *Wise Use of Wetlands in Northern Australia: Grazing Management in Wetlands and Riparian Habitats*. (Eds B Myers, M McKaige, P Whitehead M and Douglas) pp. 47–51. Centre for Tropical Wetlands Management, Northern Territory University, Darwin.

Grice AC (2004) Weeds and the monitoring of biodiversity in Australian rangelands. *Austral Ecology* **29**, 51–58. doi:10.1111/j.1442-9993.2004.01364.x

James S (1984) Lignotubers and burls – their structure, function and ecological significance in Mediterranean ecosystems. *Botanical Review* **50**, 225–266.

Jardine TJ, Bond NR, Burford MA, Ward DP, Bayliss P, Davies PM, Douglas MM, Hamilton SK, Kennard MJ, Melack JM, Naiman RJ, Olley JM, Pettit NE, Pusey BJ, Warfe DM, Bunn SE (2015) Does flood rhythm drive ecosystem responses in tropical riverscapes? *Ecology* **96**, 684–692.

Junk WJ (2002) Long-term environmental trends and the future of tropical wetlands. *Environmental Conservation* **29**, 414–435. doi:10.1017/S0376892902000310

Kellman M, Tackaberry R (1993) Disturbance and tree species coexistence in tropical riparian forest fragments. *Global Ecology and Biogeography Letters* **3**, 1–9. doi:10.2307/2997557

Kellman M, Tackaberry R, Rigg L (1998) Structure and function in two tropical gallery forest communities: implications for forest conservation in fragmented systems. *Journal of Applied Ecology* **35**, 195–206. doi:10.1046/j.1365-2664.1998.00300.x

Kennard MJ, Pusey BJ, Olden JD, Mackay S, Stein J, Marsh N (2010) Classification of natural flow regimes in Australia to support environmental flow management. *Freshwater Biology* **55**(1), 171–193. doi:10.1111/j.1365-2427.2009.02307.x

Ladd PG (1989) The status of Casuarinaceae in Australian forests. In *Australia's Ever-Changing Forests: Proceedings of the First National Conference on Australian Forest History*. (Eds KJ Frawley, NM Semple) pp. 63–85. Publication No. 1. Department Geography and Oceanography, Australian Defence Forces Academy, Campbell, Canberra.

Lake S, Bond N, Reich P (2006) Floods down rivers: from damaging to replenishing forces. *Advances in Ecological Research* **39**, 41–62. doi:10.1016/S0065-2504(06)39003-4

Lamontagne S, Cook PG, O'Grady A, Eamus D (2005) Groundwater use by vegetation in a tropical savanna riparian zone (Daly River, Australia). *Journal of Hydrology* **310**, 280–293. doi:10.1016/j.jhydrol.2005.01.009

Lane AM, Williams RJ, Muller WJ, Lonsdale WM (1997) The effects of the herbicide tebuthiuron on seedlings of *Mimosa pigra* and native floodplain vegetation in northern Australia. *Australian Journal of Ecology* **22**, 439–447. doi:10.1111/j.1442-9993.1997.tb00695.x

Lau DCP, Leunig KMY, Dudgeon D (2009) What does stable isotope analysis reveal about trophic relationships and the relative importance of allochthonous and autochthonous resources in tropical streams? A synthetic study from Hong Kong. *Freshwater Biology* **54**, 127–141. doi:10.1111/j.1365-2427.2008.02099.x

Lawson T, Gillieson D, Goosem M (2007) Assessment of riparian rainforest vegetation change in tropical north Queensland for management and restoration purposes. *Geographical Research* **45**(4), 387–397. doi:10.1111/j.1745-5871.2007.00477.x

Lindner G (1995) Wetland management issues in Kakadu National Park. In *Wetland Research in the Wet–Dry Tropics of Australia*. (Ed. CM Finlayson)

pp. 23–34. Supervising scientist report 101, Commonwealth of Australia, Canberra.

Lonsdale WM (1993) Rates of spread of an invading species: *Mimosa pigra* in northern Australia. *Journal of Ecology* **81**, 513–521. doi:10.2307/2261529

Meave J, Kellman M, MacDougall A, Rosales J (1991) Riparian habitats as tropical forest refugia. *Global Ecology and Biogeography Letters* **1**, 69–76. doi:10.2307/2997492

Metcalfe DJ, Ford AJ (2008) Floristics and plant biodiversity of the rainforests of the wet tropics. In *Living in a Dynamic Tropical Forest Landscape.* (Eds NE Stork and SM Turton) pp. 123–132. Blackwell Publishing, Carlton, Vic.

Naiman RJ, Décamps H, McClain ME (2005) *Riparia: Ecology, Conservation and Management of Streamside Communities.* Elsevier/Academic Press, San Diego, CA.

NTG (2009) *Weed Risk Assessment.* Northern Territory Government, Darwin.

O'Grady AP, Eamus D, Cook PG, Lamontagne S (2006) Groundwater use by riparian vegetation in the wet–dry tropics of northern Australia. *Australian Journal of Botany* **54**, 145–154. doi:10.1071/BT04164

Panton B (2001) Strategic stock exclusion fencing – who's responsible? In *Wise Use of Wetlands in Northern Australia: Grazing Management in Wetlands and Riparian Habitats.* (Eds B Myers, M McKaige, P Whitehead and M Douglas) pp. 69–73. Centre for Tropical Wetlands Management, Northern Territory University, Darwin.

Pettit NE (2002) 'Survey of riparian vegetation status on the Cape and Campaspe rivers, Burdekin Catchment, North Queensland.' Technical report no. 02/22. Australian Centre for Tropical Freshwater Research, Townsville, QLD.

Pettit NE, Froend RH (2001a) Variability in flood disturbance and the impact on riparian tree recruitment in two contrasting river systems. *Wetlands Ecology and Management* **9**, 13–25. doi:10.1023/A:1008471100136

Pettit NE, Froend RH (2001b) Availability of seed for recruitment of riparian vegetation: a comparison of a tropical and temperate river ecosystem in Australia. *Australian Journal of Botany* **49**, 515–528. doi:10.1071/BT00059

Pettit NE, Froend RH, Davies PM (2001) Identifying the natural flow regime and the relationship with riparian vegetation for two contrasting Western Australian rivers. *Regulated Rivers: Research and Management* **17**, 201–215. doi:10.1002/rrr.624

Pettit NE, Latterell JJ, Naiman RJ (2006) Formation, distribution and ecological consequences of flood-related wood debris piles in a bedrock confined river in semi-arid South Africa. *River Research and Applications* **22**, 1097–1110. doi:10.1002/rra.959

Pettit NE, Naiman RJ (2006) Flood-deposited wood creates regeneration niches for riparian vegetation on a semi-arid South African river. *Journal of Vegetation Science* **17**, 615–624. doi:10.1111/j.1654-1103.2006.tb02485.x

Pettit NE, Naiman RJ (2007) Fire in the riparian zone: characteristics and ecological consequences. *Ecosystems* **10**, 673–687. doi:10.1007/s10021-007-9048-5

Pettit NE, Townsend S, Dixon IH, Wilson D (2011) Plant communities of aquatic and riverine habitats. In *Aquatic Biodiversity in Northern Australia.* (Ed. BJ Pusey) pp. 37–50. Charles Darwin University Press, Darwin.

Petty A, Douglas M (2010) Scale relationships and linkages between woody vegetation communities along a large topical floodplain river, north Australia. *Journal of Tropical Ecology* **26**, 79–92. doi:10.1017/S0266467409990319

Price O, Milne D, Connors G, Harwood B, Woinarski JCZ, Butler M (2003) *A Conservation Plan for the Daly Basin Bioregion.* Draft for public consultation. Parks and Wildlife Commission of the Northern Territory, Darwin.

Radford IJ, Grice AC, Abbott BN, Nicholas DM, Whiteman L (2008) Impacts of changed fire regimes on tropical riparian vegetation invaded by an exotic vine. *Austral Ecology* **33**(2), 151–167. doi:10.1111/j.1442-9993.2007.01803.x

Rea N, Storrs MJ (1999) Weed invasion in wetlands of Australia's Top End: reasons and solutions.

*Wetlands Ecology and Management* **7**, 47–62. doi:10.1023/A:1008425109473

Russell-Smith J (1991) Classification, species richness, and environmental relations of monsoon rainforest vegetation in the northern territory. *Journal of Vegetation Science* **2**, 259–278. doi:10.2307/3235959

Sattler P (1993) Riparian zone management in Queensland and the Northern Territory: policy and practice. In *Ecology and Management of Riparian Zones in Australia*. (Eds SE Bunn, BJ Pusey and P Price) Land and Water Resources Research and Development Corporation occasional paper series no. 05/93. LWRRDC, Canberra and the Centre for Catchment and In-Stream Research, Griffith University, Queensland.

Saynor MJ, Erskine WD (2006) Spatial and temporal variations in bank erosion on sand-bed streams in the seasonally wet tropics of northern Australia. *Earth Surface Processes and Landforms* **31**, 1080–1099. doi:10.1002/esp.1310

Setterfield SA, Williams RJ (1996) Patterns of flowering and seed production in *Eucalyptus miniata* and *E. tetrodonta* in a tropical savanna woodland, northern Australia. *Australian Journal of Botany* **44**, 107–122. doi:10.1071/BT9960107

Start AN (2000) Riparian and aquatic vegetation. In *Recommendations for Estimation of Interim Ecological Water Requirements for the Ord River*. (Ed. Anon) Water and Rivers Commission, Perth.

Thurgate M (1994) *Natural Assessment of the Gulf Plains Biogeographic Region*. James Cook University, Townsville, QLD.

Townsend SA, Douglas MM (2000) The effect of three fire regimes on stream water quality, water yield and export coefficients in a tropical savanna (northern Australia). *Journal of Hydrology* **229**, 118–137. doi:10.1016/S0022-1694(00)00165-7

Valentine L, Schwarzkopf L (2009) Effects of weed-management burning on reptile assemblages in Australian tropical savannas. *Conservation Biology* **23**(1), 103–113. doi:10.1111/j.1523-1739.2008.01074.x

Valentine LE, Schwarzkopf L, Johnson C, Grice AC (2007) Burning season influences the response of bird assemblages to fire in tropical savannas. *Biological Conservation* **137**, 90–101. doi:10.1016/j.biocon.2007.01.018

van Dam R, Bartolo R, Bayliss P (2008) Identification of ecological assets, pressures and threats. In *Ecological Risk Assessment for Australia's Northern Tropical Rivers. Sub-project 2 of Australia's Tropical Rivers – an Integrated Data Assessment and Analysis (DET18)*. (Eds R Bartolo, P Bayliss and R van Dam) pp. 14–161. A report to Land and Water Australia, Canberra. Environmental Research Institute of the Supervising Scientist, National Centre for Tropical Wetland Research, Darwin.

Ward DP, Hamilton SK, Jardine TD, Pettit NE, Tews EK, Olley JM, Bunn SE (2013) Assessing the seasonal dynamics of inundation, turbidity and aquatic vegetation in the Australian wet–dry tropics using optical remote sensing. *Ecohydrology* **6**, 312–323.

Warfe DM, Pettit NE, Magierowski RH, Pusey BJ, Davies PM, Douglas MM, Bunn SE (2013) Hydrological connectivity structures concordant animal and plant assemblages according to niche rather than dispersal processes. *Freshwater Biology* **58**, 292–305. doi:10.1111/fwb.12058

Webb LJ (1968) Environmental relationships of the structural types of Australian rainforest vegetation. *Ecology* **49**(2), 296–311. doi:10.2307/1934459

Williams AR (1979) Vegetation and stream pattern as indicators of water movement on the Magela floodplain, Northern Territory. *Australian Journal of Ecology* **4**, 239–247.

Winter WH (1990) Australia's northern savanna: a time for change in management philosophy. *Journal of Biogeography* **17**, 525–529. doi:10.2307/2845386

Woolfrey AR, Ladd PG (2001) Habitat preference and reproductive traits of a major Australian riparian tree species (*Casuarina cunninghamiana*). *Australian Journal of Botany* **49**, 705–715. doi:10.1071/BT01009

# 13

# Vegetation of Australia's desert river landscapes

*Samantha Capon, John Porter and Cassandra James*

## INTRODUCTION

Australia is renowned for being the driest inhabited continent on Earth and 'a land of droughts and flooding rains' (Dorothea Mackellar). Over one-third of Australia's land area can be considered 'arid' and a further third 'semi-arid' (Martin 2006). Importantly, these areas are characterised not just by extended periods of low rainfall but also extremely patchy and unpredictable rainfall that sometimes produces enormous floods, driving some of the most expansive and dynamic riverine landscapes in Australia and the world.

Inland Australia hasn't always been arid, and for much of the Palaeocene supported lush rainforests and swamps of both angiosperms and gymnosperms. Gradual drying, with some wetter periods, has since occurred over a period of around 15 million years. A constantly changing climate throughout the Pleistocene, with glacial and interglacial cycling tending towards current levels of aridity, led to fluctuations in the distribution and abundance of many biological populations between periods of expansions and phases of contraction into refuges, which gradually favoured more adaptable taxa (Martin 2006). Palaeo-botanical studies indicate that many of Australia's arid zone species developed well before aridity took hold in central Australia and may have grown in these landscapes when still occupied by rainforest (Martin 2006). The current flora therefore largely comprises relict taxa able to cope with these climatic changes together with a few species that have evolved specific adaptations in response to arid conditions (Carolin 1982; Byrne *et al.* 2008).

Today, riverine vegetation in Australia's arid zone is often the tallest vegetation in the surrounding landscape where it also tends to be among the most taxonomically diverse and complex (Capon 2003, 2005; Brock *et al.* 2006). While dominated by few woody species in most places, desert river habitats can support a high diversity of herbs, forbs and algae, and a wide range of vegetation communities differing in composition, structure and dynamics. Desert riverine vegetation provides many important ecosystem functions, goods and services: shaping the physical landscape, providing significant habitat to wildlife, and supporting profitable grazing and tourism industries (Brock *et al.* 2006).

In this chapter, we consider the habitats, plants and vegetation of riverine landscapes in Australia's inland deserts. We begin with a description of rivers and the plant habitats they support then consider floristics and traits of common plants that enable them to persist under conditions of extreme hydrological variability and unpredictability. Next, we discuss major vegetation community types of this region and their temporal and spatial dynamics. Finally, we comment on management issues concerning desert riverine vegetation communities including the ecosystem goods and services they provide, major threats to which they are exposed and key approaches to their management.

## DESERT RIVER LANDSCAPES OF AUSTRALIA

Deserts are typically defined in relation to their mean annual precipitation and include both arid areas that receive below 200 mm mean annual rainfall, and semi-arid areas that receive between 200 and 500 mm mean annual rainfall (Davies *et al.* 1994). In Australia, deserts, thus defined, encompass an area of around 5.5 million km$^2$, accounting for almost three-quarters of the continent and spanning all of the mainland states and territories (Fig. 13.1; Stafford-Smith and Cribb 2009). Under the Australian (modified Köppen) climate classification, this area includes both deserts and grasslands that are mostly hot and persistently dry, especially in central Australia (Stern and de Hoedt 2000), with summer rainfall dominating in the north and winter rainfall in the south (Stafford-Smith and Cribb 2009). Across both the arid and semi-arid zones, however, rainfall is characterised by high inter-annual variability and spatial patchiness whereby most places experience extended dry periods interrupted by sporadic heavy rains (Morton *et al.* 2011). Rainfall is significantly more variable in Australia's northern desert region, which is comparable, in terms of precipitation, to the Indian Thar, Namib-Kalahari and Somali deserts, while the southern, winter rainfall arid zone is less variable and compares with deserts in North America, the Sahel, northern Sahara and Karoo (van Etten 2009).

Australia's desert region intersects substantial areas of seven major drainage divisions: the Lake Eyre Basin, the Murray–Darling Basin, the Tanami-Timor Sea coast, the north-western plateau, the south-western plateau, south-west coast and the

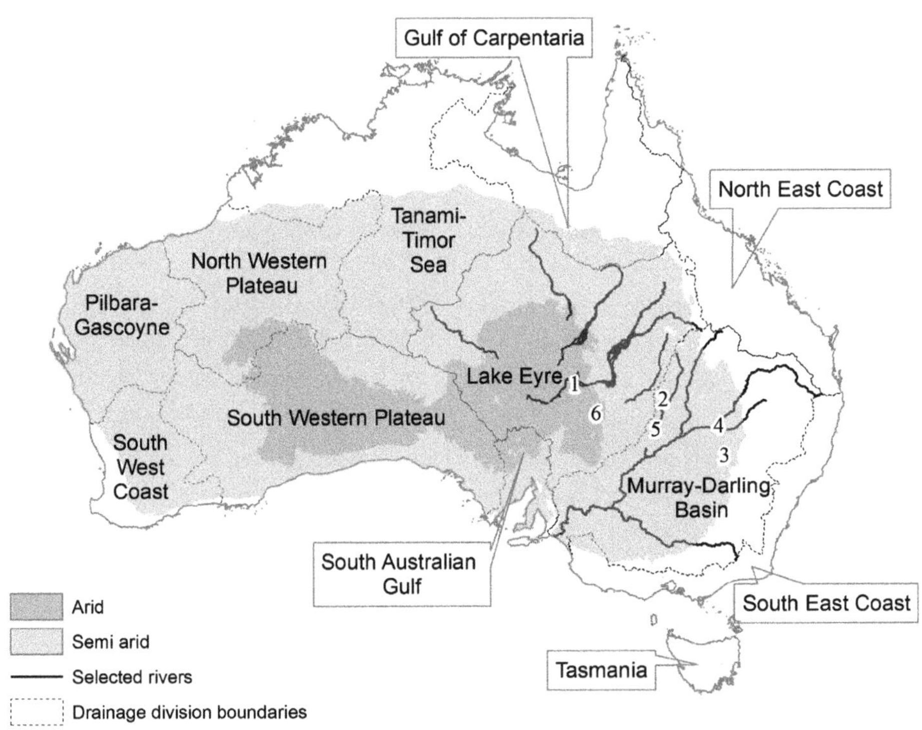


**Figure 13.1:** Map showing major drainage divisions intersected by Australia's desert region. Numbers indicate key locations referred to in the text: (1) Coongie Lakes, (2) Currawinya Lakes, (3) Macquarie Marshes, (4) Narran Lakes Nature Reserve, (5) Paroo River wetlands and (6) Lake Pinaroo. (N.B. Boundaries for arid and semi-arid regions are based on annual rainfall means averaged across 30 years from 1960 to 1990.)

Pilbara-Gascoyne (Fig. 13.1). In the east, these basins are characterised by the large, complex floodplain river systems of the Lake Eyre Basin (e.g. Cooper, Diamantina and Georgina rivers) and the northern and western Murray–Darling Basin (e.g. Warrego, Paroo and Darling rivers) as well as the internally draining Bulloo-Bancannia system, which straddles the Queensland and New South Wales borders between these larger basins. Predominantly, these rivers originate in upland regions beyond the arid zone, where higher rainfall drives downstream flows that feed channels and their associated floodplains and wetlands and that ultimately drain into large, shallow terminal wetland systems (e.g. Lake Eyre; Fig. 13.2a). In contrast, Australia's central and western desert regions contain few major rivers, with drainage lines tending to be relatively indistinct and discontinuous. In much of the arid Northern Territory, South Australia and Western Australia, the driest part of the continent, drainage is limited to ephemeral rivers that dissect sandy or stony plains and dune fields, and ultimately flow into inland 'floodouts' (playas) and temporary lakes (Duguid *et al.* 2005; Box *et al.* 2008; Bureau of Meteorology 2012). In the central Australian region of the Northern Territory, there are many low relief ranges (100–800 m) with numerous springs from which many such ephemeral rivers arise (Box *et al.* 2008). A few small arid rivers in the north-western plateau and south-western plateau discharge into the Southern and Indian oceans (Bureau of Meteorology 2012).

Runoff and streamflow throughout Australia's desert river systems are characterised by high levels of variability and unpredictability (Walker *et al.* 1995). For much of the time, surface water is limited to deep waterholes or short sections of spring-fed creeks (e.g. Halse *et al.* 1998; Puckridge *et al.* 1998). Groundwater discharge is likely to be important in most places where water persists for long durations, including riverine waterholes (Fig. 13.2b; Box *et al.* 2008). Some temporary wetlands, such as swamps, may also be influenced by groundwater (Box *et al.* 2008), although in other arid and semi-arid areas surface water is driven almost completely by inflows and evaporation, with little contribution from groundwaters (Fig. 13.2c; Bunn *et al.* 2006; Costelloe *et al.* 2007, 2009). Groundwater affecting riparian areas in semi-arid southern Australia tends to be highly saline, reflecting the accumulation of marine sediments during Tertiary marine incursions, in contrast to northern Australian groundwater which typically has low salinity (Costelloe *et al.* 2008).

When wet periods occur, the low relief, unconstrained topography and complex networks of major (in the east) and minor braided channels, which typify many of Australia's desert river landscapes, result in large areas being inundated. In the ill-defined and chaotic drainage systems of the central and western deserts, runoff events can result in days or weeks of flow in usually dry rivers and creeks, leading to significant overland flows in areas that lack distinct channels (Fig. 13.2d; Duguid *et al.* 2005). In the large eastern desert rivers of the Lake Eyre and Murray–Darling basins, high flows are typically associated with tropical cyclones that produce rainfall in the non-arid, upland regions (Roshier *et al.* 2001). Depending on the size of the discharge events that follow, such flows connect and replenish waterholes and, when sufficiently large, inundate vast areas of floodplain, filling wetlands and terminal lake systems, some of which then retain water for up to several years. Overall, the rivers of Australia's eastern desert region are among the world's most hydrologically variable and unpredictable (Davies *et al.* 1994; Puckridge *et al.* 1998). The Lake Eyre Basin, in particular, is remarkable for having the lowest mean annual runoff of any major drainage basin in the world (Kotwicki and Allan 1998) while still exhibiting huge flood events in which tens of thousands of square kilometres are inundated (Capon and Brock 2006).

The variable and unpredictable nature of rainfall and streamflow in Australia's desert river landscapes has an overriding influence on their physical, chemical and biological structure and function (Walker *et al.* 1995; Lake 2000; Kingsford *et al.* 2004). Dramatic responses among flora and fauna to

**Figure 13.2:** A selection of desert rivers and associated wetland habitats: (a) Lake Eyre, (b) semi-permanent waterhole on the Diamantina River, (c) Strzelecki desert dune swamps, (d) temporary freshwater swamp with islands near the Mulligan River, (e) Yantabulla Basin (background) and adjacent temporary claypan wetlands (foreground) showing complex hydrology, (f) Thomson River fringed with coolibah and black box, (g) Menindee Lakes fill from the Darling River, (h) Lake Numalla (freshwater, semi-permanent and fills predominantly from the Paroo River) and Lake Wyara (saline, fills mainly from local tributaries and runoff) form the Currawinya Lakes system.

changing hydrological conditions are frequently described as a 'boom and bust' ecology (Walker *et al.* 1997; Bunn *et al.* 2006; Leigh *et al.* 2010). With respect to plant habitats and vegetation dynamics, surface water hydrology tends to be the chief determinant of habitat availability and quality as well as the outcome of life history processes and plant interactions (Brock *et al.* 2006). Because of the spatial heterogeneity that results from variable flooding patterns, desert riverine landscapes can be described as 'dynamic patch mosaics' (Fig. 13.2e; Thoms *et al.* 2005) within which patches differ in physical and chemical character, resource availability (e.g. soil moisture and nutrients), stressors (e.g. soil anoxia) and other disturbance regimes such as fire (Puckridge *et al.* 2000; Stromberg 2001) and flow modification (Kingsford 2000; Poff and Zimmerman 2010). Floods tend to 'reset' these ecosystems, homogenising riverine landscapes, while droughts promote divergence among patches as local influences (e.g. soil character, grazing pressure, canopy cover) become more influential (Walker *et al.* 1995; Capon 2003).

## Plant habitats

Due to the high degree of temporal and spatial heterogeneity that characterises desert river landscapes, typical distinctions between riverine habitats that apply elsewhere (e.g. riverine versus non-riverine, lotic versus lentic) are often inappropriate (Porter *et al.* 2007). While some major habitat types available to plants throughout Australia's desert riverine landscapes can be identified, there is considerable variation within these broad types in terms of size and character (e.g. persistence, depth, salinity, turbidity). Furthermore, particular habitats can change considerably over time in relation to in situ physical and chemical changes (e.g. salinisation of freshwater lakes with drying; Timms 2001a, 2001b; Porter *et al.* 2007), landscape-scale changes (e.g. altered flow regimes; Sheldon *et al.* 2010) or climate change.

Instream channel habitats include near-permanent, semi-permanent and temporary waterholes that are intermittently connected during periods of river flow and tend to be highly turbid due to the suspension of fine sediments (Davies *et al.* 1994). Dry river beds of sand (e.g. Todd River) and clay (e.g. Cooper Creek) also provide habitat for plants in lowland areas (Steward *et al.* 2012). In many places, especially the channel country of south-west Queensland, complex distributary networks of smaller channels spread across the landscape in braided or anastomosing patterns (Fig. 13.2f). In-channel benches and other geomorphic features (e.g. islands) also provide plant habitats in some desert rivers (e.g. the Darling River in western New South Wales). In upland areas (e.g. gorges), channel habitats can include small, rocky creeks which contain temporary and permanent rock pools that experience flash flooding (Halse *et al.* 1998; Sheldon *et al.* 2010).

Watercourses are bordered by a variety of fringing or channel bank habitats, including naturally raised levees along many of the major rivers. These habitats typically grade into floodplains, which are extensive in the eastern desert rivers of the Lake Eyre and Murray–Darling basins. The Cooper Creek floodplain, for example, is greater than 60 km wide in places (Knighton and Nanson 1994). Floodplains, while present in deserts of Western Australia and the Northern Territory, are fewer and much smaller in the western region (Duguid *et al.* 2005; Bureau of Meteorology 2012).

A wide range of wetlands also occur within Australia's desert river catchments, including swamps, marshes, billabongs, claypans, lakes and terminal wetland systems of varying size and character (Fig. 13.2g; Kingsford and Porter 1999). These differ considerably with respect to their hydrological connectivity to river channels and some may never receive river water except in the most extreme of flood events. Local hydrological regimes of wetlands can therefore vary greatly between and within wetland types, with some only rarely holding water for short periods and others holding water for many years. Lake Gregory on the edge of the Great Sandy Desert in north-west Western Australia, for example, was reported as being wet almost continuously between 1969 and 1990 with only two possible drying events in 1974 and 1979 (Halse *et al.* 1998). Desert wetlands also exhibit a range of salinities and turbidities, with saline lakes particularly common in the drier central and western desert regions (Fig. 13.2h). An inverse relationship exists between salinity and turbidity, both of which fluctuate in response to flooding, drying and wave action (Williams 1998; Timms and Boulton 2001).

## PLANT DIVERSITY

### Floristics

In general, Australia's desert flora is not considered particularly distinctive when compared with other arid regions, comprising a similar mixture of grasses, chenopods, Asteraceae, Brassicaceae and Malvaceae (Barlow 1981; Morton *et al.* 2011). With respect to plants in riverine habitats, most species are widespread and common and many are cosmopolitan, also occurring well beyond the arid zone (Box *et al.* 2008). A few endemics are present in central Australia, often in association with relict habitats and usually considered to be remnants of a wetter past, e.g. numerous ferns that grow in shaded springs and seepage areas (Box *et al.* 2008). *Livistona mariae* subsp. *mariae* (red cabbage palm), for example, only grows in a small area near the Finke River and its tributaries around Palm Valley (Duguid *et al.* 2005). Recent genetic evidence suggests, however, that this population is likely to have originated from related species in northern tropical coastal rivers and may have been transported to desert habitats by Aboriginal people relatively recently (Kondo *et al.* 2012). Some specialised herbaceous species occur only on isolated groundwater-fed mound or artesian springs, including *Myriophyllum artesium* (Halford and Fensham 2001), the endangered salt pipewort *Eriocaulon carsonii* and the carnivorous bladderworts *Utricularia ameliae* and *U. fenshamii* (Fensham *et al.* 2004; Jobson 2013).

Aquatic habitats of Australia's desert regions support a moderately diverse algal flora (e.g. 188 genera identified from the Neales, lower Diamantina and lower Cooper; Costelloe *et al.* 2005) and often have a high abundance of charophytes, especially *Lamprothamnium* spp., *Chara* spp. and *Nitella* spp. (Fig. 13.3a, b; Sim *et al.* 2006; Porter *et al.* 2007). Submerged and floating-leaved macrophytes can be present with common species including *Myriophyllum* spp. (water milfoils), *Ruppia* spp. (sea tassels), *Nymphaea* species (waterlilies), *Nymphoides* spp. (marshworts), *Ottelia ovalifolia* (swamp lily), *Potamogeton* spp. (pond weeds), *Najas* spp. (water nymphs), and aquatic ferns *Azolla filliculoides* and *Marsilea* spp. (nardoo) (Fig. 13.3; Halse *et al.* 1998; Sainty and Jacobs 2003; Duguid *et al.* 2005; Porter *et al.* 2007). Common emergent macrophytes of Australia's desert watercourses and wetlands, particularly in littoral habitats, include a range of rushes (e.g. *Juncus aridicola* and *Typha domingensis*), sedges (e.g. *Cyperus* spp. and *Eleocharis* spp.) and forbs including *Lythrum* spp., *Persicaria* spp. (knotweeds), *Ranunculus* spp. (buttercups) and *Rumex* spp. (docks) (Fig. 13.3d; Kingsford and Porter 1999; Sainty and Jacobs 2003). Semi-aquatic and flood-tolerant sedges and grasses can also be abundant in wet and drying habitats, such as *Fimbristylis dichotoma* (summer fringe-rush), *Lipocarpha microcephala* (button rush), *Pseudoraphis spinescens* (spiny mud grass), *Eragrostis australasica* (swamp canegrass), *Echinochloa inundata* and *E. turneriana* (channel millet), *Leptochloa digitata* (umbrella canegrass), *Paspalidium jubiflorum* (Warrego summer grass) and *Sporobolus mitchellii* (rat's tail couch). Many other terrestrial grasses occur in riparian habitats or dominate vast swathes of floodplains or drying lake beds, for example, *Astrebla* spp. (Mitchell grasses), *Aristida* spp. (wiregrasses), *Eragrostis* spp. (lovegrasses), *Eriachne* spp. (Wanderrie grasses), *Iseilema* spp. (Flinders grasses) (Capon 2003, 2005; Duguid *et al.* 2005; Brock *et al.* 2006). Damp and drying riverine habitats in Australia's desert region can also support a high diversity of riparian forbs representing a wide range of plant habits, groups and families (see Chapter 7), many of which also occur in non-wetland habitats (Duguid *et al.* 2005). Common families represented include Asteraceae, Amaranthaceae, Chenopodiaceae, Euphorbiaceae, Fabaceae, Goodeniaceae, Malvaceae, Nyctaginaceae and Polygonaceae (Fig. 13.3g, h; Capon 2005; James *et al.* 2007).

**Figure 13.3:** A selection of aquatic plants of Australian desert wetlands and rivers: (a) *Chara* sp.aff. *preissii* is a submergent macroalgae (charophyte) commonly found in shallow temporary wetlands; (b) *Lamprothamnium macropogon* is a charophyte algae that grows prolifically in saline wetlands where it provides shelter and habitat for invertebrates and food for waterbirds; (c) *Ottelia ovalifolia* (floating leaves) and *Myriophyllum verrucosum* (submerged) are widespread in shallow temporary wetlands; (d) emergent grasses and sedges including *Eleocharis acuta* are commonly found at the margins of shallow water habitats; (e) *Duma florulenta* is a widespread woody shrub that can form extensive thickets along river channels, drainage lines and floodplains, and provides critical shelter and nesting habitat for fauna, especially waterbirds; (f) *Eucalyptus ochrophloia* (Yapunyah) is a riparian tree species that occurs mainly in the Bulloo, Paroo and Warrego river systems; (g) *Goodenia nocoleche* showing the emergent and (h) floating-leaved forms.

Shrubs and sub-shrubs are a dominant component of the vegetation in many drier riverine habitats (e.g. floodplains) and drying waterbodies (e.g. dry lake beds) of the Australia desert, with common and widespread species including *Duma florulenta* (tangled lignum; Fig. 13.3e) and a wide range of chenopods such as bluebushes (e.g. *Chenopodium auricomum*, Queensland bluebush, and *Maireana* spp.), saltbushes (e.g. *Atriplex nummularia*, old man saltbush), and copperburrs (e.g. *Sclerolaena* spp.). Small shrubs such as *Frankenia* spp., *Halosarcia* spp. and *Tecticornia* spp., variously referred to as samphires and glassworts, also occur in more saline and rarely flooded habitats, particularly in western regions (Read 1994; Kingsford and Porter 1999; Capon 2005). Taller shrubs and small trees grow in a range of riparian habitats including along small drainage lines, fringing lakes and other wetlands, or as part of the canopy on the banks of major watercourses. Common and widely distributed species include *Dodonaea viscosa* subsp. *mucronata* (sticky hop bush), *Eremophila* spp. (turkey bushes), *Myoporum montanum* (boobialla) and *Senna phyllodinea*. Acacias ranging from small shrubs to trees also occur frequently in desert riparian and floodplain habitats, for example, *Acacia stenophylla* (river cooba), *A. cambagei* (gidgee), *A. georgninae* (Georgina gidgee) and *A. aneura* (mulga) (Keith 2004).

The tree flora of Australia's desert river habitats, like elsewhere on the continent, is dominated by eucalypts (see Chapter 8), with *Eucalyptus camaldulensis* (river red gum) and *E. coolabah* (coolibah) particularly widespread throughout the region. Other locally abundant eucalypts include *E. ochrophloia* (Yapunyah) in the east (Fig. 13.3f), *E. microtheca* in the north and *E. victrix* (smooth-barked coolibah) in the west. Melaleucas are also widespread, with common species including *Melaleuca glomerata* (desert paperbark), *M. trichostachya* (river paperbark) and *M. bracteata* (black tea tree). Other relatively common and widely distributed trees occurring in riparian habitats of desert Australia include *Atalaya hemiglauca* (whitewood), *Bauhinia gilva* (bauhinia), *Erythrina vespertilio* (bat's wing coral tree) and *Geijera parviflora* (wilga).

## Adaptations to hydrological unpredictability

Plants of Australia's desert riverine habitats exhibit a wide range of adaptations to aridity, many of which involve traits that minimise water loss such as a waxy cuticle, succulence, crassulacean acid metabolism (CAM), sunken stomata, densely hairy leaves, steep leaf angles, the capacity to roll leaves to reduce light absorbance, or the ability to reduce leaf area (e.g. by leaf shedding) (Chaves *et al.* 2003; Harris 2003; Larcher 2003; Capon *et al.* 2009). Plants adapted to dry conditions may also demonstrate an ability to increase water uptake through the accumulation of greater root biomass or exhibit traits that facilitate increased water storage, for example, succulence or underground storage organs such as tubers, rhizomes and bulbs (Larcher 2003). Some common dryland riparian trees (e.g. *Eucalyptus coolabah* and *E. largiflorens*) minimise water use by having very low transpiration rates and can also extract water at very low osmotic potentials (Jolly and Walker 1996; Costelloe *et al.* 2008). Arid floodplain trees can additionally be capable of utilising a variety of water sources including rain and flood-derived soil moisture as well as groundwater (Jolly and Walker 1996). Some riparian plants, however, may depend largely on groundwater for persistence in these landscapes, for example, *Melaleuca faucicola* (desert bottlebrush), which is endemic to the MacDonnell Ranges, where it grows in shaded gullies and gorges fed by springs or seepages (Duguid *et al.* 2005; Box *et al.* 2008).

In general, Australia's desert river habitats support only a small proportion of plants that are truly aquatic, although many might be considered amphibious or at least partially flood tolerant. Species in the latter categories possess traits that enable their growth or survival during periods of inundation. The development of adventitious roots, for instance, appears to be particularly prevalent among common perennial species (e.g. Capon *et al.* 2009). However, most dominant perennial plants in these landscapes tend to demonstrate greater tolerance of drought than flooding (e.g. Capon 2003; Capon *et al.* 2009). More importantly, species that persist in these habitats must be able to cope with the variable and unpredictable availability of surface water and soil moisture that characterises these environments. Plant strategies therefore need to contend with the 'pulse-reserve' character of the arid zone, enabling survival during dry periods but also allowing plants to capitalise on unpredictable favourable periods for growth and reproduction (Morton *et al.* 2011).

The capacity of plants to cope with the environmental variability of Australia's arid zone is often attributed to phenotypic plasticity, that is, the capacity to respond adaptively to environmental fluctuations (Morton *et al.* 2011). In habitats that are as unpredictable and extreme as those of Australia's desert river landscapes, however, plasticity may be maladaptive as plant responses to one set of conditions (e.g. stem elongation in response to flooding) may not be advantageous under other conditions (i.e. drought) (Alpert and Simms 2002; Valladares *et al.* 2007; Capon *et al.* 2009). A few taxa may circumvent these problems by changing growth forms from submergent to emergent or terrestrial. For example, a recently described plant, *Goodenia nocoleche*, from the Paroo and Bulloo river region produces floating leaves on elongated petioles when submerged, but as waters recede responds by shedding aquatic leaves and producing a terrestrial radical herb form to match the drying conditions (Fig. 13.3g, h; Pellow and Porter 2005). For the majority of species, however, plant growth under environmental unpredictability may be favoured by 'adaptive fixity' (Alpert and Simms 2002; Capon *et al.* 2009). The emergent sedge *Cyperus gymnocaulos* (spiny sedge), for example, copes with fluctuating water levels by producing rhizomes that widely traverse a range of channel bank elevations, thereby enabling the establishment of plants at the water's edge (Costelloe *et al.* 2004).

Many plants in Australia's desert river habitats cope with the inherent variability and unpredictability of their environments via growth and life history patterns that alternate between periods of dormancy, during unfavourable conditions, and periods of rapid, opportunistic growth and reproduction in response to wetting. With respect to mature plants, such strategies are likely to involve the ability to store energy in stems and roots as well as a capacity to quickly mobilise energy reserves for the production of new growth and fruiting (Chaes *et al.* 2003). *Duma florulenta*, for example, persists during dry phases as a leafless tangle of drying stems which rapidly produce leaves and flowers after rainfall or flooding (Capon *et al.* 2009). Seedlings of some common trees and shrubs (e.g. *D. florulenta* and *Acacia stenophylla*) also exhibit stasis in growth during submergence, with rapid growth ensuing after recession of floodwaters and the cessation of the stresses these impose (e.g. anoxia) (Capon *et al.* 2009).

A high proportion of herbaceous plants in Australia's desert river landscapes persist for much of the time as dormant seeds stored in soil seed banks, which germinate primarily in response to sufficient water and light availability (Capon and Brock 2006; Capon 2007; James *et al.* 2007; Porter *et al.* 2007). Many graminoids and forbs that rely on long-lived soil seed banks for persistence during dry periods also exhibit extremely rapid life cycles that enable reproduction and replenishment of the soil seed bank before further drought (Brock *et al.* 2006). Furthermore, in many species, only a small proportion of seeds stored in the soil seed bank germinate in response to any particular wetting event, which 'bet-hedges' against the possibility of reproductively unsuccessful germination (Capon and Brock 2006; Brock *et al.* 2006; Capon 2007). A few species, including submerged and floating-leaved aquatic plants, can germinate, grow and reproduce during periods of inundation while others (e.g. *Glossostigma* spp., *Bergia trimera*, *Pratia* spp.) may grow rapidly when submerged but require the recession of floodwaters for reproduction to occur (Brock *et al.* 2006). Similarly, while some emergent plants in these habitats (e.g. *Echinochloa turneriana*, channel millet) are capable of germinating during periods of inundation, germination in the majority of desert wetland soil seed bank species is favoured by the waterlogged or damp soil conditions that follow flood drawdown (Brock *et al.* 2006; Capon 2007; James *et al.* 2007). Additionally, dominant herbaceous species in these environments tend to germinate from soil seed banks opportunistically, without the temperature cues required to break dormancy in many other plants with long-lived seeds (Baskin and Baskin 1998; Capon 2007).

# VEGETATION ECOLOGY

## Vegetation diversity

A significant proportion of vegetation types recognised in Australia's deserts are associated with riverine landscapes (Morton *et al.* 2011). Riparian and wetland vegetation are often the most diverse and structurally complex in these landscapes, with woody vegetation in many places largely limited to the margins of channels and other waterbodies (Brock *et al.* 2006).

Vegetation only tends to be absent from open water habitats, smaller claypans and some saline lakes. Aquatic plants, with the exception of fringing emergents, are also relatively rare in channel habitats of Australia's arid rivers (Beadle 1981; Casanova 1999). Elsewhere, Australia's desert riverine landscapes support a wide diversity of vegetation types including woodlands, shrublands, grasslands, sedgelands and a variety of forb associations.

Submerged vegetation, often growing in dense beds, occurs in both freshwater and saline waterbodies and is typically dominated by charophytes (e.g. *Lamprothamnium* and *Nitella* spp.) and aquatic forbs such as *Ruppia* spp., *Marsilea* spp., *Myriophyllum* spp., *Lepilaena* spp. and *Glossostigma diandrum* (Marchant and Halse 1990; Duguid *et al.* 2005; Porter *et al.* 2007). Most waterbodies and some channel reaches, often in more sheltered locations, also support fringing vegetation communities comprising emergent sedges (e.g. *Cyperus* spp. and *Eleocharis* spp.), rushes (e.g. *Juncus* spp. and *Typha* spp.), grasses (e.g. *Pseudoraphis spinescens* and *Eragrostis australasica*) or forbs (e.g. *Ludwigia peploides*).

Woodlands, usually dominated by one or two common species, frequently *Eucalyptus camaldulensis*, *E. coolabah* and/or *Acacia stenophylla*, fringe most major watercourses and lakes throughout arid Australia. Locally common species, however, dominate the woody riparian vegetation in some places, e.g. *E. ochrophloia* in the Paroo Lakes, *Acacia tephrina* at Lake Gregory or *Melaleuca* spp. in parts of the Northern Territory. In the ephemeral rivers of central Australia (e.g. the Finke River), woodland communities comprising *E. camaldulensis*, *E. coolabah* and *Acacia estrophiolata* (ironwood) also occur on generally dry channel beds (Duguid *et al.* 2005). In other places, woodlands, although typically much sparser than those lining channels, can extend considerable distances onto floodplains, often reflecting the inundation extent of past flood events, for example, *E. coolabah* and *Acacia cambagei* (gidgee) on the Cooper Creek floodplain (Capon 2005). Riparian woodlands can also have shrub layers of *Duma florulenta*, various chenopods or *Acacia* spp., and exhibit a wide range of understorey assemblages variously dominated by grasses, forbs, sedges and rushes or a mixture of these. The fringing vegetation of some waterbodies lacks trees and can instead be dominated by shrubs (e.g. *D. florulenta*) or emergents. Many saline lakes are fringed by salt-tolerant samphire vegetation, (e.g. *Halosarcia* spp. and *Tecticornia* spp.), with other fleshy-leaved taxa including *Sarcozona praecox* and *Gunniopsis quadrifida* (Sturt's pigface).

Australia's desert floodplains support a range of vegetation types including shrublands, grasslands, forblands and mixed associations of grasses, sedges and forbs. Trees are often absent from these habitats or only sparsely distributed in isolated patches or as scattered individuals (Brock *et al.* 2006). Swampy, low-lying floodplain habitats often support extensive canegrass (*Eragrostis australasica*) or dense shrubby vegetation. Lignum swamps, dominated by *D. florulenta*, are particularly prevalent in the eastern parts of the region (Capon 2005), but smaller patches also occur in central Australia (Duguid *et al.* 2005). In contrast, bluebush (*Chenopodium auricomum*) swamps are widespread in the Northern Territory (Duguid *et al.* 2005), but occur in more elevated and drier habitats than lignum swamps in the larger eastern floodplains (Capon 2005). Other chenopod shrubs (e.g. *Atriplex nummularia*) can also be locally abundant in some swamps. Rarely inundated marginal floodplain areas are often dominated by low sparse shrubland mainly comprising chenopods such as *Sclerolaena* spp., *Atriplex* spp. and *Maireana* spp. Sparse, low shrublands dominated by *Frankenia* species are prevalent in drier parts of the Cooper Creek's southern floodplains (Capon 2005).

## Vegetation dynamics

Surface water hydrology has an overriding influence on temporal and spatial patterns in the vegetation of desert river habitats, with geomorphology playing an important, but generally secondary role (Brock *et al.* 2006). All stages of plant life histories, as well as plant interactions, are affected both directly and indirectly by flooding, which, depending on the plants and habitats involved, can act as a disturbance (e.g. mechanically damaging plants or scouring them from channel habitats), a stressor (e.g. limiting oxygen and light availability to submerged plants) or a subsidy (e.g. providing soil moisture). At a landscape scale, flooding tends to homogenise vegetation (Capon 2003), probably promoted by the widespread dispersal of plant propagules by wind, floodwaters and waterbirds (Capon and Brock 2006; Green *et al.* 2008). Divergence between vegetation patches occurs during dry phases as the influence of local factors (e.g. soil characteristics, weather) grows (Capon 2003; Brock *et al.* 2006).

In habitats with standing water, vegetation composition and structure tend to vary in relation to water permanence, salinity and turbidity, all of which interact spatially and temporally (Porter *et al.* 2007). The typically low abundance and diversity of macrophytes in Australia's desert river channels, for instance, are widely attributed to high levels of turbidity and scouring (Bunn *et al.* 2003). In general, temporary wetlands display higher plant diversity than wetlands, which retain water for much longer periods. In wetlands of the Bulloo and Paroo catchments, the composition of aquatic plant communities, as well as soil seed banks, differ between freshwater and saline, and permanent and temporary wetlands, with temporary freshwater wetlands displaying higher levels of diversity (Porter *et al.* 2007). In the Coongie Lakes in northern South Australia, the density of fringing vegetation also declines in relation to decreasing water permanence of lakes from upstream to the end of the systems (Timms 2001a).

The composition, structure and condition of vegetation in desert riparian and floodplain habitats can change dramatically in response to flooding and drying, with massive increases in species richness and plant abundance typically occurring following the recession of floodwaters (Capon 2003; Capon and Brock 2006). Flooding can cause mortality of flood-intolerant species (e.g. some chenopod shrubs) as well as triggering the rapid growth of flood-tolerant perennial species (e.g. *Duma florulenta* or *Chenopodium auricomum*) and germination from the soil seed bank or other sources of propagules (e.g. buried rhizomes or plant fragments dispersed by floodwaters or waterbirds) of a wide diversity of ephemeral, annual and perennial species (e.g. Capon 2003). Vegetation responses to flooding vary widely in relation to the attributes of a particular flood pulse. Time since the last flood event, for instance, will influence the initial floristic composition available to respond to flooding. For example, longer durations of drying may result in a greater abundance of terrestrial invaders that may be drowned (Capon 2003). The magnitude of flooding will determine the area of habitat within which vegetation will respond, while the duration of inundation and subsequent waterlogging can affect the abundance of plants emerging from soil seed banks (Capon 2007). The seasonal timing of flooding may also be important. On the Cooper Creek floodplain, for example, summer flooding tends to promote the establishment of monocots while germination of forbs is favoured by winter flooding (Capon 2007).

Variation in vegetation composition and structure in desert riverine habitats is strongly affected by spatial differences in flood history (e.g. flood frequency), which in turn are strongly shaped by topography and position in the landscape (Brock *et al.* 2006). Riparian trees such as *E. coolabah*, for example, grow preferentially in habitats that experience frequent flushing by streamflow, i.e. tops of channel banks (Costelloe *et al.* 2008). The population structure and composition of riparian woodlands and shrublands also typically reflect flood history (Capon 2005; Brock *et al.* 2006;

Thoms *et al.* 2007). On floodplains, frequently flooded areas often support flood-tolerant plants and annual species that are able to quickly reproduce between flood events, while rarely flooded areas are dominated by perennial species (Capon 2003). Species richness and vegetation cover also tend to vary in relation to flood frequency. On the Cooper Creek floodplain, for example, both species richness and understorey plant cover decline with decreasing flood frequency (Capon 2003, 2005). Patterns in the composition and abundance of soil seed banks are also shaped by flood history, with higher abundance and species richness often occurring in intermediately flooded habitats, such as regularly inundated floodplains (Capon and Brock 2006; James *et al.* 2007).

Many complex feedbacks occur between plants and their environments in desert river landscapes, some of which have important implications for vegetation resilience. On the very low gradient floodplains of the Lake Eyre Basin, for instance, plants can have a significant influence on the distribution of shallow floodwater and, therefore, the extent of habitats inundated. Plants also affect fluvial processes and, therefore, the physical shape and structure of their habitats. In the ephemeral Marshall River of central Australia, for example, colonisation of flat, sandy channel beds by *Melaleuca glomerata* (inland tea tree) promotes the formation and expansion of ridges which eventually develop into channel anabranches that are considered to be critical for the maintenance of efficient water and sediment dynamics (Tooth and Nanson 2000).

## MANAGEMENT ISSUES

### Ecosystem goods and services

People derive a wide range of direct and less tangible benefits from the plants and vegetation communities of Australia's desert river landscapes and have done so for many thousands of years. Indigenous Australians have a long history of utilising a diversity of plant species from these environments for food, wood, fibre and other materials (e.g. adhesives) as well as for drinking water (e.g. from fleshy roots or leaves) during drier periods (Latz 1999). More recent settlers adopted few edible native plants into their diet but have continued to harvest timbers from riverine habitats, e.g. *E. camaldulensis*, particularly for firewood (Mac Nally *et al.* 2011). Honey is harvested from floodplain eucalypt species including Yapunyah (*Eucalyptus ochrophloia*) on the Paroo River and river red gum (*E. camaldulensis*) throughout the Murray–Darling Basin. The dominant land use across most of arid and semi-arid Australia is now livestock grazing, mainly of cattle and sheep, and native floodplain grasses and forbs are highly valued as pastures. The channel country floodplains of the Lake Eyre Basin, for example, are touted as providing some of the best cattle pastures in the world after flooding (Capon 2003). Vegetation is also critical to the many supporting and regulating ecosystem services provided by riverine habitats, for example, regulation of instream temperature and light regimes (Capon *et al.* 2013). Furthermore, the capacity of some riverine habitats to act as evolutionary and ecological refuges for fauna in arid Australia is likely to be influenced by ecological functions provided by vegetation (e.g. nesting hollows and shade) (Davis *et al.* 2013). Finally, vegetation contributes to the many cultural services provided by riverine ecosystems including the high wilderness, aesthetic and scientific values placed on Australia's arid zone. Riverine environments in central Australia are also increasingly popular locations for recreational and tourism activities (e.g. camping, fishing).

### Human pressures

Central Australia supports a very small human population and its riverine landscapes are therefore often considered to be substantially less degraded and in a relatively good, if not 'reference', condition in comparison with many coastal catchments, particularly in Australia's south-east and south-west (e.g. Jenkins *et al.* 2005). Water resources development is widely acknowledged as the greatest threat to riverine habitats in dryland Australia, although current levels of development are minimal in most places, with the exception of the semi-arid catchments in the west of the Murray–Darling Basin, where considerable regulation of rivers and wetland systems (e.g. Menindee Lakes) has occurred since the mid 19th century (Kingsford 2000; Capon 2003; Brock *et al.* 2006; Sheldon *et al.* 2010). Significant floodplain developments to capture overland flows for irrigated agriculture have also occurred widely in the northern Murray–Darling Basin. There is increasing pressure to develop both surface and water resources for irrigation and other activities throughout the region (Duguid *et al.* 2005; Jenkins *et al.* 2005). Additionally, other infrastructure developments (e.g. road construction) have the potential to significantly alter surface water pattern because of the low topographic gradient of many desert riverine landscapes (Brock *et al.* 2006). The effects of flow alteration on riverine vegetation are discussed in Chapter 14 of this volume. In general, however, hydrological changes have the potential to result in significant changes in the composition, structure and condition of riverine vegetation including biodiversity loss and the streamwards migration of vegetation zones (Brock *et al.* 2006). Widespread declines in the health of mature riparian trees in response to drought and river regulation, especially among *E. camaldulensis* and *E. largiflorens* populations, as well as low levels of recruitment, are of particular public and political concern in semi-arid catchments of the Murray–Darling Basin (Mac Nally *et al.* 2011). Degradation of soil and water quality resulting from hydrological alterations (e.g. dryland salinity and acidification) can also have serious implications for riverine vegetation including tree mortality (Baldwin and Capon 2011).

Other human activities in Australia's deserts have further affected riverine vegetation, with impacts generally concentrated in the eastern parts where agriculture, forestry, mining and human settlements are more prevalent. Selective timber harvesting in riparian woodlands, for example, has resulted in even-aged stands that lack large spreading trees (Mac Nally *et al.* 2011). Lake bed cropping is also relatively widespread. Although this has clear implications for native lake bed vegetation during dry phases, the ecological effects on lake ecosystems during periods of inundation are mostly unknown (Jenkins *et al.* 2005). Mining can impact on aquatic and riparian vegetation both directly, via clearing and excavation, and indirectly as a result of pollution and changes to water quality and hydrology, such as through the release of waste waters (e.g. Duguid *et al.* 2005). Recreation and tourism in riverine habitats also have the potential to affect vegetation both locally (e.g. through trampling) and at larger scales (e.g. via promoting the spread of invasive plants).

The most widespread existing human impacts on riverine vegetation in Australia's deserts are likely to relate to livestock grazing, although the severity of impacts is often difficult to quantify. Grazing can have a range of direct (e.g. herbivory) and indirect (e.g. soil compaction and bank erosion) effects on vegetation (see Chapter 17), which may result in changes to vegetation cover and composition. Recruitment of riparian trees may also be limited by grazing (Robertson and Rowling 2000). A 10-fold reduction in the area of fringing trees around Lake Gregory between 1953 and 1988, for example, has been attributed to the limiting effects on regeneration of grazing (Marchant and Halse 1990). Similarly, reductions in bands of littoral trees at Coongie Lakes have been associated with grazing (Marchant and Halse 1990). In the channel country of the Lake Eyre Basin, grazing is also acknowledged to affect the composition and structure of floodplain vegetation during dry phases, although such effects tend to be mitigated by the enormity of the germination responses and biomass production that ensue after flooding (Edmonston 2001). Furthermore, livestock are typically incapable of accessing vegetation on inundated or waterlogged ground, which allows plants to complete their life cycles and replenish soil seed banks before grazing occurs. The provision of artificial watering points for livestock may also impact riverine vegetation both through altering local hydrology as well as by attracting feral animals such as goats and pigs (James *et al.* 1999). Feral pigs are particularly notorious for causing severe local impacts in riparian habitats as a result of their habit of ploughing furrows (i.e. rooting) in soft soils. Introduced fish such as carp have become widespread and, in addition to their well-documented impacts on aquatic fauna, affect light penetration and aquatic plants by increasing turbidity and nutrients (Lougheed *et al.* 1998; Miller and Crowl 2006).

Plant invasions have occurred throughout Australia's desert river landscapes, often as a result of deliberate introductions for pasture or erosion control (see Chapter 18). Weed infestations tend to be particularly prevalent in riparian areas, but have also encroached on swamps and other wetland habitats in places. Common and widespread noxious weeds in riverine habitats include *Xanthium occidentale* (Noogoora burr), *Argemone ocrholecua* (Mexican poppy), *Phyla canescens* (lippia) and *Parkinsonia aculeata* (parkinsonia). *Cynodon dactylon* (couch grass), widely planted as a lawn species as well as to control erosion, is perceived to be one of the most significant threats to native flora in desert wetlands of central Australia, where it is attributed with displacing native riparian species, such as *Cyperus bulbosus* (nut grass), an important food source for the endangered bilby (Duguid *et al.* 2005). The introduced pasture grass *Cenchrus ciliaris* (buffel grass) is also a major problem on many desert floodplains, particularly due to its effects on the severity of fires as a result of greater fuel loads, which consequently increases mortality of woodland canopy species from fire (Miller *et al.* 2010). Tamarisk (*Tamarix aphylla*), a common shade tree planted at homesteads throughout central Australia, is also a significant weed species in riverine habitats and is attributed with displacing native riparian eucalypts where it has established in some locations such as on the Finke River (Griffin *et al.* 1989).

Finally, climate change poses a significant threat to many riverine and wetland ecosystems of central Australia, both directly (e.g. as a result of warming and rainfall changes) and indirectly as a result of effects on a wide range of biogeochemical processes as well as via human response to climate change (see Chapter 21; Hughes 2003; Capon *et al.* 2013). Climate change is expected to increase the frequency of intense fires as well as salinity in dryland Australia with significant implications for vegetation. Negative effects of warming and reduced rainfall in grasslands, however, may be mitigated by $CO_2$ enrichment. Encroachment of grass and forb communities by woody vegetation, especially shrubs, is also widely anticipated (Hughes 2003).

## Management approaches

Significant areas of Australia's deserts, including their riverine ecosystems, are protected under the national reserve system. Major parks include the Lake Eyre National Park, Simpson Desert Regional Reserve, which encompasses a large area of playa lakes in central Australia, Strzelecki Regional Reserve and Diamantina National Park in south-west Queensland, which contains numerous wetlands and floodplains of the Diamantina River. Several Ramsar sites also exist in Australia's arid and semi-arid zones including: Coongie Lakes in north-east South Australia; Lake Pinaroo, Paroo River Wetlands, Narran Lake Nature Reserve and the Macquarie Marshes in north-west New South Wales; and Currawinya Lakes (in Currawinya National Park) in south-west Queensland (Fig. 13.1). Under the Ramsar convention, Australia's governments have a commitment to protecting the ecological character of these significant wetland sites.

Off-reserve land and water management practices are increasingly recognised for their significance to biodiversity conservation, particularly with respect to growing threats posed by climate change (Dunlop and Brown 2008). Appropriate grazing management, for example, is particularly important for the diversity of much riverine vegetation across central Australia where, in many places, the sheer scale of riparian habitats prohibits the restriction of livestock via fencing as can be done successfully in other environments. Alternative strategies often involve the management of stocking rates including sporadic resting of paddocks. Provision of alternative watering points located away from channels and wetland areas can also alleviate grazing pressure (Jansen and Robertson 2001).

Water resources management, however, is arguably the most important area of human intervention with respect to the conservation of riverine vegetation diversity in inland Australia. Water management plans that regulate water diversions and extraction and protect particular riverine habitats are in place for many inland catchments, especially those in the Murray–Darling and Lake Eyre basins. Environmental watering strategies to supplement natural flows as well as to sustain important wetland refuges during drought have also been developed, particularly in the Murray–Darling Basin (Capon 2014). In some places, such as Toorale station at the junction of the Warrego and Darling rivers in north-west New South Wales, decommissioning of existing water resource infrastructure (e.g. weirs) is also being considered as a means of removing instream barriers to biodiversity movement and reinstating more natural flows. Where vegetation has established in response to long-term changes to flooding patterns, however, such interventions risk the loss of valuable extant vegetation communities with no guarantee that this will be offset by vegetation establishing downstream in response to reinstated flows.

## CONCLUSIONS

Australia's desert riverine habitats support a wide diversity of plants exhibiting a range of forms and life histories that enable their persistence under conditions of extreme hydrological variability and unpredictability. The vegetation is characterised by high levels of spatial and temporal heterogeneity driven primarily by patterns of flooding and drying. Often perceived as near pristine, Australia's desert river landscapes face an increasing number of pressures including mineral extraction and climate change, many of which threaten the diversity and function of riparian vegetation communities as well as the ecosystem goods and services they provide. Ongoing research and monitoring of these landscapes is therefore critical to support management approaches that extend beyond those provided by protected areas and promote sustainable use.

## REFERENCES


Alpert P, Simms EL (2002) The relative advantages of plasticity and fixity in different environments: when is it good for a plant to adjust? *Evolutionary Ecology* **16**, 285–297. doi:10.1023/A:1019684612767

Baldwin DS, Capon SJ (2011) 'Sulfidic sediments in inland waterways.' Waterlines report. National Water Commission, Canberra.

Barlow BA 1981. The Australian flora: its origin and evolution. In *Flora of Australia, Vol. 1.* (Ed. A George) pp. 25–76. Commonwealth of Australia, Canberra.

Baskin CC, Baskin JM (1998) *Seeds: Ecology, Biogeography and Evolution of Dormancy and Germination.* Academic Press, San Diego, CA.

Beadle NCW (1981) *The Vegetation of Australia.* Cambridge University Press, Cambridge, UK.

Box JB, Duguid A, Read RE, Kimber RG, Knapton A, Davis J, Bowland AE (2008) Central Australian waterbodies: the importance of permanence in a desert landscape. *Journal of Arid Environments* **72**(8), 1395–1413. doi:10.1016/j.jaridenv.2008.02.022

Brock M, Capon S, Porter J (2006) Disturbance of plant communities dependent on desert rivers. In *Ecology of Desert Rivers.* (Ed. RT Kingsford) pp. 100–132. Cambridge University Press, Cambridge, UK.

Bunn SE, Davies PM, Winning M (2003) Sources of organic carbon supporting the food web of an arid zone floodplain river. *Freshwater Biology* **48**(4), 619–635. doi:10.1046/j.1365-2427.2003.01031.x

Bunn SE, Thoms MC, Hamilton SK, Capon SJ (2006) Flow variability in dryland rivers: boom, bust and the bits in between. *River Research and Applications* **22**(2), 179–186. doi:10.1002/rra.904

Bureau of Meteorology (2012) 'Australian water resources assessment 2012.' The Bureau of Meteorology, Melbourne.

Byrne M, Yeates DK, Joseph L, Kearney M, Bowler J, Williams MAJ, Cooper S, Donnellan SC, Keogh JS, Leys R, Melville J, Murphy DJ, Porch N, Wyrwoll KH (2008) Birth of a biome: insights into the assembly and maintenance of the Australian arid zone biota. *Molecular Ecology* **17**, 4398–4417. doi:10.1111/j.1365-294X.2008.03899.x

Capon SJ (2003) Plant community responses to wetting and drying in a large arid floodplain. *River Research and Applications* **19**(5, 6), 509–520. doi:10.1002/rra.730

Capon SJ (2005) Flood variability and spatial variation in plant community composition and structure on a large arid floodplain. *Journal of Arid Environments* **60**(2), 283–302. doi:10.1016/j.jaridenv.2004.04.004

Capon SJ (2007) Effects of flooding on seedling emergence from the soil seed bank of a large desert floodplain. *Wetlands* **27**(4), 904–914. doi:10.1672/0277-5212(2007)27[904:EOFOSE]2.0.CO;2

Capon S (2014) Adapting water resources management to drought and water scarcity in Australia's Murray–Darling Basin. In *Adaptation to Climate Change Through Water Resources Management: Capacity, Equity and Sustainability*. (Eds D Stucker and E Lopez-Gunn) pp. 110–133. Routledge, Abingdon, UK.

Capon SJ, Brock MA (2006) Flooding, soil seed bank dynamics and vegetation resilience of a hydrologically variable desert floodplain. *Freshwater Biology* **51**(2), 206–223. doi:10.1111/j.1365-2427.2005.01484.x

Capon SJ, James CS, Williams L, Quinn GP (2009) Responses to flooding and drying in seedlings of a common Australian desert floodplain shrub: *Muehlenbeckia florulenta* Meisn. (tangled lignum). *Environmental and Experimental Botany* **66**(2), 178–185. doi:10.1016/j.envexpbot.2009.02.012

Capon SJ, Chambers LE, Mac Nally R, Naiman RJ, Davies P, Marshall N, Pittock J, Reid M, Capon T, Douglas M, Catford J, Baldwin DS, Stewardson M, Roberts J, Parsons M, Williams SE (2013) Riparian ecosystems in the 21st century: hotspots for climate change adaptation? *Ecosystems* **16**(3), 359–381. doi:10.1007/s10021-013-9656-1

Carolin RC (1982) A review and critique of studies on the phytogeography of arid Australia. In *Evolution of the Flora and Fauna of Arid Australia*. (Eds WR Barker and PJM Greenslade) pp. 119–123. Peacock Publications, Frewville, South Australia.

Casanova MT (1999) Plant establishment in Paroo wetlands: the importance of water regime. In *A Free-Flowing River: the Ecology of the Paroo River*. (Ed. RT Kingsford) pp. 138–48. New South Wales National Parks and Wildlife Service, Hurstville, NSW.

Chaves MM, Maroco JP, Pereira JS (2003) Understanding plant responses to drought – from genes to the whole plant. *Functional Plant Biology* **30**(3), 239–264. doi:10.1071/FP02076

Costelloe JF, Hudson J, Pritchard J, Puckridge JT, Reid J (2004) 'ARIDFLO scientific report: environmental flow requirements of arid zone rivers with particular reference to the Lake Eyre drainage basin.' Technical report prepared for the South Australian Department of Water, Land and Biodiversity Conservation.

Costelloe JF, Powling J, Reid JR, Shiel RJ, Hudson P (2005) Algal diversity and assemblages in arid zone rivers of the Lake Eyre Basin, Australia. *River Research and Applications* **21**(2, 3), 337–349. doi:10.1002/rra.851

Costelloe JF, Shields A, Grayson RB, McMahon TA (2007) Determining loss characteristics of arid zone river waterbodies. *River Research and Applications* **23**, 715–731. doi:10.1002/rra.991

Costelloe JF, Payne E, Woodrow IE, Irvine EC, Western AW, Leaney FW (2008) Water sources accessed by arid zone riparian trees in highly saline environments, Australia. *Oecologia* **156**(1), 43–52. doi:10.1007/s00442-008-0975-4

Costelloe JF, Irvine EC, Western AW, Herczeg AL (2009) Groundwater recharge and discharge dynamics in an arid zone, ephemeral lake system, Australia. *Limnology and Oceanography* **54**, 86–100. doi:10.4319/lo.2009.54.1.0086

Davies BR, Thoms MC, Walker KF, O'Keefe JH, Gore JA (1994) Dryland rivers: their ecology, conservation and management. In *The Rivers Handbook, Vol. 2.* (Eds P Calow and GE Petts) pp. 484–511. Blackwell Scientific, Oxford, UK.

Davis J, Pavlova A, Thompson R, Sunnucks P (2013) Evolutionary refugia and ecological refuges: key concepts for conserving Australian arid zone freshwater biodiversity under climate change. *Global Change Biology* **19**(7), 1970–1984. doi:10.1111/gcb.12203

Duguid A, Barnetson J, Clifford B, Pavey C, Albrecht D, Risler J, McNellie M (2005) 'Wetlands in the arid Northern Territory.' A report to Environment Australia on the inventory and significance of wetlands in the arid NT. Parks and Wildlife Commission of the Northern Territory, Alice Springs, NT.

Dunlop M, Brown PR (2008) *Implications of Climate Change for Australia's National Reserve System: a Preliminary Assessment.* Department of Climate Change, Canberra, Australia.

Edmonston V (2001) *Managing the Channel Country Sustainably: Producers' Experiences.* Queensland Department of Primary Industries, Brisbane.

Fensham RJ, Fairfax RJ, Pocknee D, Kelley J (2004) Vegetation patterns in permanent spring wetlands in arid Australia. *Australian Journal of Botany* **52**, 719–728. doi:10.1071/BT04043

Green AJ, Jenkins KM, Bell D, Morris PJ, Kingsford RT (2008) The potential role of waterbirds in dispersing invertebrates and plants in arid Australia. *Freshwater Biology* **53**(2), 380–392.

Griffin GF, Smith DS, Morton SR, Allan GE, Masters KA, Preece N (1989) Status and implications of the invasion of tamarisk (*Tamarix aphylla*) on the Finke River, Northern Territory, Australia. *Journal of Environmental Management* **29**(4), 297–315.

Halford D, Fensham RJ (2001) A new species of *Myriophyllum* L. (Haloragaceae) from artesian springs in Queensland. *Austrobaileya* **6**, 133–137.

Halse SA, Pearson GB, Kay WR (1998) Arid zone networks in time and space: waterbird use of Lake Gregory in North-Western Australia. *International Journal of Ecology and Environmental Sciences* **24**, 207–222.

Harris N (2003) *Atlas of the World's Deserts.* Fitzroy Dearborn, New York, NY.

Hughes L (2003) Climate change and Australia: trends, projections and impacts. *Austral Ecology* **28**(4), 423–443. doi:10.1046/j.1442-9993.2003.01300.x

James CD, Landsberg J, Morton SR (1999) Provision of watering points in the Australian arid zone: a review of effects on biota. *Journal of Arid Environments* **41**(1), 87–121. doi:10.1006/jare.1998.0467

James CS, Capon SJ, White MG, Rayburg SC, Thoms MC (2007) Spatial variability of the soil seed bank in a heterogeneous ephemeral wetland system in semi-arid Australia. *Plant Ecology* **190**(2), 205–217. doi:10.1007/s11258-006-9201-9

Jansen A, Robertson AI (2001) Relationships between livestock management and the ecological condition of riparian habitats along an Australian floodplain river. *Journal of Applied Ecology* **38**(1), 63–75. doi:10.1046/j.1365-2664.2001.00557.x

Jenkins KM, Boulton AJ, Ryder DS (2005) A common parched future? Research and management of Australian arid-zone floodplain wetlands. *Hydrobiologia* **552**(1), 57–73. doi:10.1007/s10750-005-1505-6

Jobson RW (2013) Five new species of *Utricularia* (Lentibulariaceae) from Australia. *Telopea* **15**, 127–142. doi:10.7751/telopea2013017

Jolly ID, Walker GR (1996) Is the field water use of *Eucalyptus largiflorens* F. Muell. affected by short-term flooding? *Australian Journal of Ecology* **21**(2), 173–183. doi:10.1111/j.1442-9993.1996.tb00598.x

Keith DA (2004) *Ocean Shores to Desert Dunes – the Native Vegetation of New South Wales and the ACT.* Department of Environment and Conservation, Hurstville, Sydney.

Kingsford RT (2000) Ecological impacts of dams, water diversions and river management on floodplain wetlands in Australia. *Austral Ecology* **25**(2), 109–127. doi:10.1046/j.1442-9993.2000.01036.x

Kingsford RT, Porter J (1999) Wetlands and waterbirds of the Paroo and Warrego Rivers. In *A*

*Free-Flowing River: the Ecology of the Paroo River*. (Ed. RT Kingsford) pp. 23–50. New South Wales National Parks and Wildlife Service, Hurstville, NSW.

Kingsford RT, Jenkins KM, Porter JL (2004) Imposed hydrological stability on lakes in arid Australia and effects on waterbirds. *Ecology* **85**, 2478–2492.

Kingsford RT, Jenkins KM, Porter JL (2004) Imposed hydrological stability on lakes in arid Australia and effects on waterbirds. *Ecology* **85**, 2478–2492.

Knighton AD, Nanson GC (1994) Waterholes and their significance in the anastomosing channel system of Cooper Creek, Australia. *Geomorphology* **9**(4), 311–324. doi:10.1016/0169-555X(94)90052-3

Kondo T, Crisp MD, Linde C, Bowman DM, Kawamura K, Kaneko S, Isagi Y (2012) Not an ancient relic: the endemic *Livistona* palms of arid central Australia could have been introduced by humans. *Proceedings of the Royal Society B: Biological Sciences* **279**(1738), 2652–2661.

Kotwicki V, Allan R (1998) La Niña de Australia: contemporary and palaeo-hydrology of Lake Eyre. *Palaeogeography, Palaeoclimatology, Palaeoecology* **144**, 265–280. doi:10.1016/S0031-0182(98)00122-9

Lake PS (2000) Disturbance, patchiness, and diversity in streams. *Journal of the North American Benthological Society* **19**, 573–592.

Larcher W (2003) *Physiological plant ecology: ecophysiology and stress physiology of functional groups*. Springer Science & Business Media.

Latz P (1999) *Pocket Bushtucker: a Field Guide to the Plants of Central Australia and their Traditional Uses*. IAD Press, Alice Springs, NT.

Leigh C, Sheldon F, Kingsford RT, Arthington AH (2010) Sequential floods drive 'booms' and wetland persistence in dryland rivers: a synthesis. *Marine and Freshwater Research* **61**(8), 896–908. doi:10.1071/MF10106

Lougheed VL, Crosbie B, Chow-Fraser P (1998) Predictions on the effect of common carp (*Cyprinus carpio*) exclusion on water quality, zooplankton, and submergent macrophytes in a Great Lakes wetland. *Canadian Journal of Fisheries and Aquatic Sciences* **55**, 1189–1197. doi:10.1139/f97-315

Mac Nally R, Cunningham SC, Baker PJ, Horner GJ, Thomson JR (2011) Dynamics of Murray–Darling floodplain forests under multiple stressors: the past, present, and future of an Australian icon. *Water Resources Research* **47**(12). doi:10.1029/2011WR010383

Marchant NG, Halse SA (1990) The flora of Lake Gregory. In *The Natural Features of Lake Gregory: a Preliminary Review*. (Ed. SA Halse) pp. 12–16. Department of Conservation and Land Management, Perth.

Martin HA (2006) Cenozoic climatic change and the development of the arid vegetation in Australia. *Journal of Arid Environments* **66**(3), 533–563. doi:10.1016/j.jaridenv.2006.01.009

Miller SA, Crowl TA (2006) Effects of common carp (*Cyprinus carpio*) on macrophytes and invertebrate communities in a shallow lake. *Freshwater Biology* **51**, 85–94. doi:10.1111/j.1365-2427.2005.01477.x

Miller G, Friedel M, Adam P, Chewings V (2010) Ecological impacts of buffel grass (*Cenchrus ciliaris* L.) invasion in central Australia – does field evidence support a fire-invasion feedback? *The Rangeland Journal* **32**, 353–365. doi:10.1071/RJ09076

Morton SR, Stafford Smith DM, Dickman CR, Dunkerley DL, Friedel MH, McAllister RRJ, Reid JRW, Roshier DA, Smith MA, Walsh FJ, Wardle GM, Watson IW, Westoby M (2011) A fresh framework for the ecology of arid Australia. *Journal of Arid Environments* **75**(4), 313–329. doi:10.1016/j.jaridenv.2010.11.001

Pellow BJ, Porter JL (2005) A new species of *Goodenia* (Goodeniaceae) from Nocoleche Nature Reserve, Far Western Plains, New South Wales. *Telopea* **11**, 35–41. doi:10.7751/telopea20055702

Poff NL, Zimmerman JKH (2010) Ecological responses to altered flow regimes: a literature review to inform the science and management of environmental flows. *Freshwater Biology* **55**, 194–205. doi:10.1111/j.1365-2427.2009.02272.x

Porter JL, Kingsford RT, Brock MA (2007) Seed banks in arid wetlands with contrasting flooding, salinity and turbidity regimes. *Plant Ecology* **188**(2), 215–234. doi:10.1007/s11258-006-9158-8

Puckridge JT, Sheldon F, Walker KF, Boulton AJ (1998) Flow variability and the ecology of large rivers. *Marine and Freshwater Research* **49**(1), 55–72. doi:10.1071/MF94161

Puckridge JT, Walker KF, Costelloe JF (2000) Hydrological persistence and the ecology of dryland rivers. *Regulated Rivers: Research and Management* **16**(5), 385–402. doi:10.1002/1099-1646(200009/10)16:5<385::AID-RRR592>3.0.CO;2-W

Read IG (1994) *The Bush: a Guide to the Vegetated Landscapes of Australia.* UNSW Press, Kensington, NSW.

Robertson AI, Rowling RW (2000) Effects of livestock on riparian zone vegetation in an Australian dryland river. *Regulated Rivers: Research and Management* **16**(5), 527–541. doi:10.1002/1099-1646(200009/10)16:5<527::AID-RRR602>3.0.CO;2-W

Roshier DA, Whetton PH, Allan RJ, Robertson AI (2001) Distribution and persistence of temporary wetland habitats in arid Australia in relation to climate. *Austral Ecology* **26**(4), 371–384. doi:10.1046/j.1442-9993.2001.01122.x

Sainty GR, Jacobs SW (2003) *Waterplants in Australia.* 4th edn. Sainty and Associates, Griffith, NSW.

Sheldon F, Bunn SE, Hughes JM, Arthington AH, Balcombe SR, Fellows CS (2010) Ecological roles and threats to aquatic refugia in arid landscapes: dryland river waterholes. *Marine and Freshwater Research* **61**(8), 885–895. doi:10.1071/MF09239

Sim LL, Chambers JM, Davis JA (2006) Ecological regime shifts in salinised wetland systems. I. Salinity thresholds for the loss of submerged macrophytes. *Hydrobiologia* **573**, 89–107. doi:10.1007/s10750-006-0267-0

Stafford-Smith MS, Cribb J (2009) *Book Review Dry Times: Blueprint for a Red Land.* CSIRO Publishing, Melbourne.

Stern H, de Hoedt G (2000) Objective classification of Australian climates. *Australian Meteorological Magazine* **49**, 87–96.

Steward AL, von Schiller D, Tockner K, Marshall JC, Bunn SE (2012) When the river runs dry: human and ecological values of dry riverbeds. *Frontiers in Ecology and the Environment* **10**, 202–209. doi:10.1890/110136

Stromberg JC (2001) Restoration of riparian vegetation in the south-western United States: importance of flow regimes and fluvial dynamism. *Journal of Arid Environments* **49**, 17–34.

Thoms MC, Southwell M, McGinness HM (2005) Floodplain–river ecosystems: fragmentation and water resources development. *Geomorphology* **71**(1, 2), 126–138. doi:10.1016/j.geomorph.2004.10.011

Thoms M, Capon S, James C, Padgham M, Rayburg S (2007) 'The Narran ecosystem project: the response of a terminal wetland system to variable wetting and drying.' Final report to the Murray–Darling Basin Commission, MDBC publication, (40/08).

Timms BV (2001a) Large freshwater lakes in arid Australia: a review of their limnology and threats to their future. *Lakes and Reservoirs: Research and Management* **6**, 183–196. doi:10.1046/j.1440-1770.2001.00132.x

Timms BV (2001b) Limnology of the intermittent pools of Bells Creek, Paroo, arid Australia, with special reference to biodiversity of invertebrates and succession. *Proceedings of the Linnean Society of New South Wales* **123**, 193–213.

Timms BV, Boulton AJ (2001) Typology of arid-zone floodplain wetlands of the Paroo River, inland Australia and the influence of water regime turbidity and salinity on their aquatic invertebrate assemblages. *Archiv für Hydrobiologie* **153**, 1–27.

Tooth S, Nanson GC (2000) The role of vegetation in the formation of anabranching channels in an ephemeral river, northern plains, arid central Australia. *Hydrological Processes* **14**(16, 17), 3099–3117.doi:10.1002/1099-1085(200011/12)14:16/17<3099::AID-HYP136>3.0.CO;2-4

Valladares F, Gianoli E, Gomez JM (2007) Ecological limits to plant phenotypic plasticity. *New Phytologist* **176**, 749–763. doi:10.1111/j.1469-8137.2007.02275.x

van Etten EJB (2009) Inter-annual rainfall variability of arid Australia: greater than elsewhere? *Australian Geographer* **40**, 109–120.

Walker KF, Puckridge JT, Blanch SJ (1997) Irrigation development on Cooper Creek, central Australia – prospects for a regulated economy in a boom-and-bust ecology. *Aquatic Conservation: Marine and Freshwater Ecosystems* **7**, 63–73. doi:10.1002/(SICI)1099-0755(199703)7:1<63::AID-AQC218>3.0.CO;2-5

Walker KF, Sheldon F, Puckridge JT (1995) A perspective on dryland river ecosystems. *Regulated Rivers: Research and Management* **11**(1), 85–104.

Williams WD (1998) Dryland wetlands. In *Wetlands for the Future: INTECOL's Fifth International Conference on Wetlands.* (Eds AJ McComb and JA Davis) pp. 33–47. Gleneagles Publishing, Adelaide.

# Section 4

Management concerns

# 14

# Flow alteration and its effects on Australia's riverine vegetation

*Richard Kingsford*

## INTRODUCTION

Rivers are complex systems whose flow regimes are the conductors that determine spatial and temporal interactions between their abiotic and biotic components (Walker *et al.* 1995). As a result, the composition and dynamics of flow-dependent biota tend to reflect the spatial and temporal distribution of water in riverine landscapes, but with a range of complex interdependencies. Hydrologic disturbance regimes, governed primarily by the interaction of geomorphological environments and flow, are thus critically important and provide the template for pattern and process in riverine landscapes and their dependent vegetation.

Broadly, most river systems comprise a few large-scale habitat types including the river channel (along with billabongs, waterholes), floodplains, swamps and lakes (saline and freshwater), and estuaries (or inland deltas and terminal lakes for internally draining systems). These different habitats clearly have considerable spatial and temporal variability across Australia's riverine landscapes, with river channels, waterholes and billabongs comprising the more permanently wet habitats. Lakes, both freshwater and saline, vary in size from small claypans, filled by local rainfall, to large inland lakes that can hold water for several years or longer (Kingsford *et al.* 2001). Floodplains are the most transitory habitats in terms of inundation duration (Roshier *et al.* 2001), but are also the most extensive habitats (Kingsford *et al.* 2004a) for riverine vegetation. Significantly, floodplains in Australia have also been the most neglected with respect to the management of flows. It has often been much easier and politically expedient to leave floodplains out of definitions and legislation of the riverine environment in water management. For example, water legislation in Queensland in the 1990s did not define floodplains as part of rivers and recent wetland mapping does not include floodplains as wetlands, even though a floodplain digital layer exists (DERM 2009).

Vegetation communities form clear patterns across major riverine habitat types in relation to their hydrological disturbance regime (henceforth, the 'flow regime'; Brock *et al.* 2006). Vegetation composition and structure, as well as plant growth, survival, reproduction and recruitment within riverine habitats all tend to reflect the frequency and duration of river flows (Roberts and Marston 2011). It is against this broad background that I examine the effects of river management on Australia's riverine landscapes and how it has influenced vegetation habitats, communities and dynamics, creating increasing challenges for managers and policymakers. River management is defined in this chapter as all the ways in which humans have controlled and altered flow regimes, the most extreme form of which is regulation of rivers by dam building. The environmental impacts of river regulation are widely recognised as having resulted in altered and degraded states within channels, terminal wetlands (Kingsford and Thomas 1995, 2004) and floodplain vegetation communities (Horner *et al.* 2009; Mac Nally *et al.* 2011). Widespread public concern about the condition of riverine landscapes and their vegetation has been instrumental in shifting the focus of river management towards greater consideration of environmental objectives rather than the historically narrower focus on increasing the availability of reliable water supplies to agriculture and human settlements. In this chapter, I briefly summarise different forms of river regulation, their effects on flow and riverine vegetation, and effective management of regulated rivers for environmental objectives.

## FLOW ALTERATION IN AUSTRALIA

Before European colonisation, Aboriginal people managed rivers primarily to harvest biota, such as altering flow paths to form fish traps. In the 20th century, Australia's river management shifted to controlling and capturing river flows for irrigation, generation of hydroelectric power, cooling coal-fired power stations, drinking water or industrial supply (WCIC 1971; Kingsford 1995), and it was predominantly focused in the south-east of the continent (Fig. 14.1). During this period of expansion and exploitation, river management has been dominated by three approaches not necessarily mutually exclusive in purpose: building of dams and weirs, diversion of water and floodplain development.

### Dams and weirs

Most of Australia's water storage capacity was established with the building of dams between 1950 and 1970 (Fig. 14.2). Most dams were built in the south-east of the continent with relatively little water stored in dams in the northern or central Australian river basins (Fig. 14.1). Dams built for hydroelectricity generation usually did not divert water and were predominantly built in Tasmania (Kingsford 2000) and the Snowy Mountains (Erskine

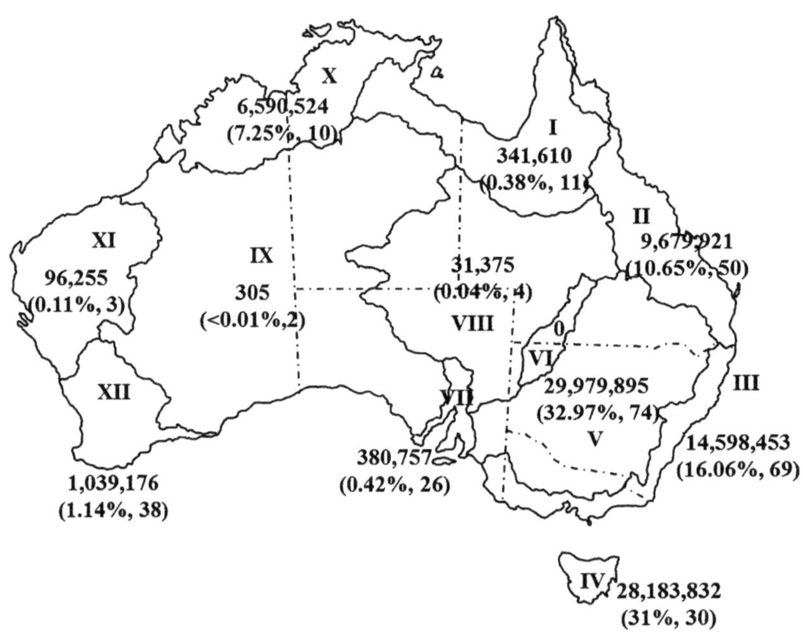


**Figure 14.1:** Australia's 12 major river basins, giving the cumulative capacity (ML) with the percentage of continental storage capacity and number of storages in parentheses. River basins are (I) Gulf of Carpentaria, (II) north-east, (III) south-east, (IV) Tasmania, (V) Murray–Darling Basin, (VI) Bulloo-Bancannia, (VII) South Australian gulf, (VIII) Lake Eyre, (IX) Western Plateau, (X) Timor Sea, (XI) Indian Ocean and (XII) south-west. Data sourced from Geoscience waterbody layer, 1:250 000 (1857–1993).

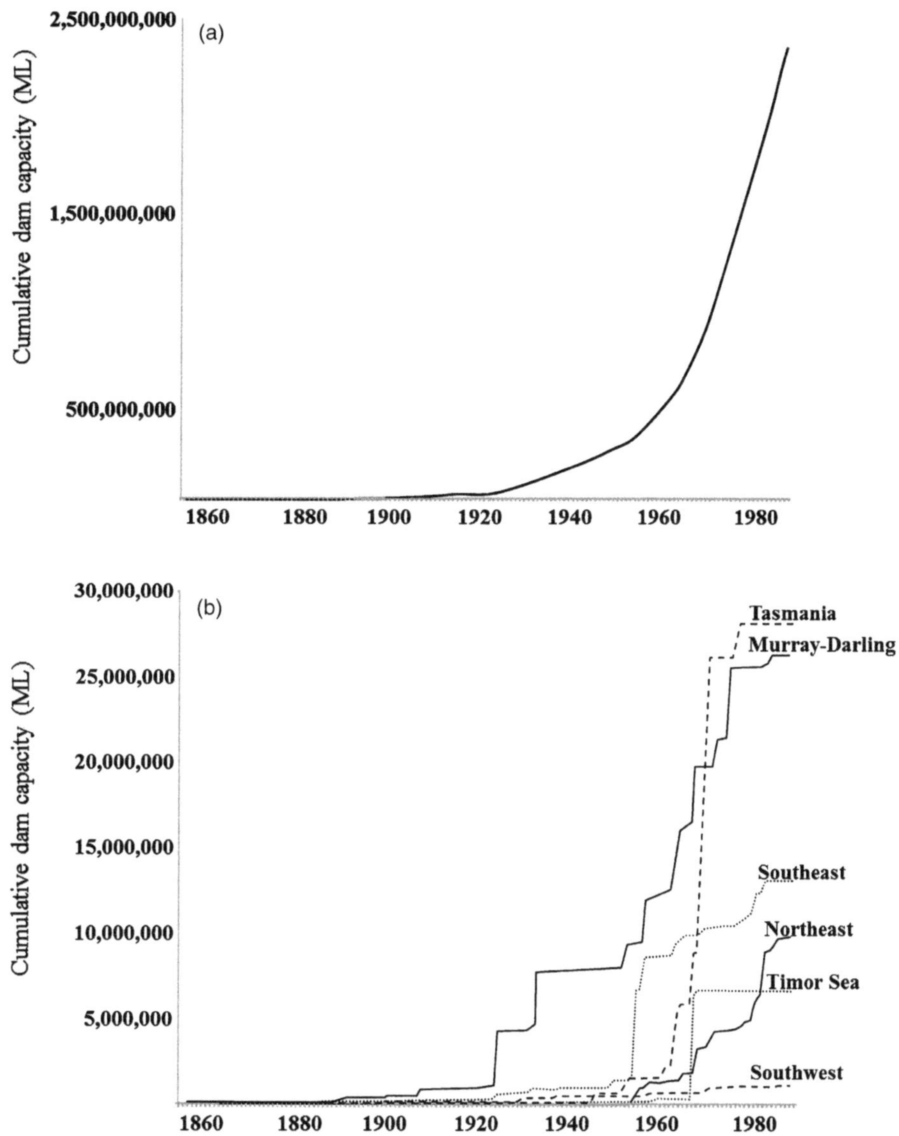


**Figure 14.2:** Top: total cumulative capacity (ML) of dams in Australia, 1857–1993. Bottom: the six major river basins with the highest cumulative capacity of dams, 1857–1993. Data sourced from Geoscience waterbody layer, 1:250 000 (1857–1993). Total estimates differ slightly from those in Fig. 14.1, because date of building dams was not always available.

*et al.* 1999). Poor flow frequency and volume in Australia's arid interior, as well as a lack of appropriate sites for large dams, mostly prevented the building of large public dams on dryland rivers. About 90% of Australia's water storage capacity in large dams is in the north-east, south-east, Murray–Darling Basin and Tasmanian river basins (Fig. 14.2).

Dam construction in Australia during the last century varied between river basins. River management in the Murray–Darling Basin followed a reasonably steady trajectory of increasing storage until the building of the Snowy Mountains hydro-electric scheme in the 1960s and 1970s (Fig. 14.2). Dam building was most rapid in Tasmania where high storage capacity was established after 1960 (Figs 14.1, 14.2). The Timor Sea Basin is dominated by Lake Argyle on the Ord River, which is the single largest storage in Australia (10 763 000 ML), built in 1971 (Fig. 14.2). The river basins that mostly supply drinking water (i.e. in the most populous centres of Brisbane, Melbourne, Perth and Sydney) showed a reasonably steady trajectory of increasing storage capacity after 1950 with their capacity to store flow reflecting population size (Fig. 14.2). Plans for further large dam construction and other river management structures to provide drinking water supplies and support irrigation continue to be proposed, particularly in the south-east and north-east basins. In an effort to increase regulation of rivers, Australian governments also converted many lakes near rivers into storages by building dams or weirs (known as re-regulating storages) which allowed water to be diverted and stored (e.g. Menindee Lakes, Lake Victoria) (Kingsford 2000; Kingsford *et al.* 2004b). As well as these large dams, there are many thousands of small weirs and structures across rivers that have been installed for a variety of purposes and contribute to anthropogenic flow alteration. In the Macquarie Marshes, for instance, only 36% of barriers identified during an aerial survey were listed in the government database (T. Rayner, unpubl. data).

### Water diversion

In most of Australia's river basins, large volumes of water can also be pumped directly from rivers into water storages that have been built on floodplains. Mostly privately operated, these off-river storages allow for the diversion of river flows not controlled by large dams (Steinfeld and Kingsford 2013) and facilitate greater levels of water diversion from more variable inland rivers, many of which are not regulated by large dams. In the northern rivers of the Murray–Darling Basin, for example, there is significant private capacity to store water (i.e. 2 435 000 ML) (Kingsford 2004).

### Floodplain development

The most poorly understood type of riverine flow management occurs on floodplains via the building of earthworks (e.g. levees, off-river storages, channels) to manipulate overland flows. Such structures can effectively capture overbank flows and redirect these into storages, often with significant physical and hydrological impacts. More than 2000 km of floodplain earthworks, for instance, occur on the Macquarie floodplain (Steinfeld and Kingsford 2013), including earthworks built within floodways. The government developed guidelines in 1978–82 (WRC 1978, 1982) that these floodways should be free of structures, allowing water flow downstream to the Macquarie Marshes, but 470 km of earthwork structures were subsequently built within these floodways.

## EFFECTS OF FLOW ALTERATION ON RIVERINE VEGETATION

The major impacts of flow alteration on freshwater aquatic ecosystems and their dependent biota broadly fall into three main categories. The most extensive of these involves the reduction in river flows delivered to floodplain ecosystems and to those at the lower end of river systems as a result of large storage dams and upstream diversions (Kingsford 2000; Arthington and Pusey 2003). Second, river regulation typically increases levels of inundation in wetlands that are used to store water or in those that occur in close proximity to river channels (Kingsford *et al.* 2004b). Third, the presence of regulating structures within channels and on floodplains alters connectivity between riverine habitats, both longitudinally and latitudinally between rivers and their floodplains, potentially alienating different flow-dependent ecosystems from their water supply. Local impacts on vegetation of infrastructure construction are also one of the most immediate and obvious manifestations of water resources developments. Such impacts can be exacerbated by other threatening processes including invasive species (Greet *et al.* 2013), grazing and timber harvesting (Horner *et al.* 2012), and climate change (Cunningham *et al.* 2011). River regulation affects riverine vegetation by fundamentally altering the flow regime, changing flood frequency, timing, duration and extent across a range of different spatial and temporal scales.

Initial consequences of dam construction for vegetation include the sudden and dramatic loss of local vegetation, especially submerged terrestrial communities where the dam is built. With the altered flow regime, downstream effects of river regulation on flow-dependent vegetation may not be obvious initially, but are similarly profound. Such impacts can extend from the river channels immediately downstream of a dam to floodplain systems and estuaries up to thousands of kilometres downstream, occurring laterally as well as longitudinally. Immediately downstream of dams, riverine vegetation communities can either increase or decrease because of altered erosion and sedimentation processes resulting from river regulation (Cluett 2005). For example, in the Cudgegong River, channel pools and banks were invaded by river oaks (*Casuarina cunninghamiana*), *Vallisneria* spp., *Phragmites* spp. and *Juncus* spp. after Windamere Dam was built (Benn and Erskine 1994). Also, small regulated floods allowed colonisation by dense and small-girthed stands of *Melaleuca leucadendra* on the lower Ord River in north-western Australia, which large natural floods would have removed (Doupé and Petitt 2002). The diversion of 99% of the Snowy River's flow in south-eastern Australia changed vegetation in the river channel, allowing invasion of the riparian zone by exotic trees (e.g. *Salix* spp., willows) and terrestrial trees (e.g. *Callitris glaucophylla*) (Erskine *et al.* 1999).

The effects of river management and diversions on the floodplains of regulated rivers are considerable, primarily because of changes to the timing, frequency, extent and duration of flooding (Kingsford 2000; Kingsford and Thomas 2004; Kingsfords. 2006; Cunningham *et al.* 2009; Horner *et al.* 2009; Mac Nally *et al.* 2011; Thomas *et al.* 2011).

Timing of flows alters vegetation extent, germination and processes of water-borne seed dispersal (Greet *et al.* 2013). Altered seasonality affects riverine vegetation, as flow regimes shift from a spring to summer regime to provide water for irrigation (Maheshwari *et al.* 1995). Aquatic macrophyte diversity and production can decrease with changes in flow timing (Nielsen and Chick 1997; Robertson *et al.* 2001). Germination is often highest in river red gum (*Eucalyptus camaldulensis*) forests in south-eastern Australia as spring and early summer floods recede (Dexter 1978), but viability of germination reduces with dry conditions in the summer (Jensen *et al.* 2008). Ultimately, seasonal responses change community composition and structure of river red gum forests (Bren 1992; Robertson *et al.* 2001). Reduced flow may also reduce colonisation by annual plants because of the poor establishment prospects (Roberts and Ludwig 1991).

When frequency of flooding is reduced, increasing drying, communities transition to more terrestrial communities (Roberts and Marston 2000; Brock *et al.* 2006). For example, coolibah (*E. coolabah*) requires flooding for about two or three weeks every five years while tangled lignum (*Duma florulenta*) appears to need flooding about once every three years (Roberts and Marston 2000). Further, black box (*E. largiflorens*) experiences poor condition where rainfall and groundwater levels are low and flooding occurs less than every two years (McGinness *et al.* 2013). Consequently, when flooding or drying requirements for a vegetation community are exceeded, there is a high potential for significant mortality and a shift in vegetation composition and structure.

Reductions in frequency are also usually related to similar reduction in extent and duration. The most common response of riverine vegetation to this reduced flooding is a shift in composition towards species with increased tolerance of desiccation. Increased flooding is generally associated with increased diversity of native plant species (Horner *et al.* 2012). Such transitions may occur with the building of dams and upstream water diversion or when floodplain vegetation is alienated by structures that inhibit hydrologic connectivity (Steinfeld and Kingsford 2013). Over time, recruits of different floodplain vegetation communities colonise contours of a changed flow regime according to their water requirements. As connectivity decreases, species most tolerant of drying occur on the boundaries of floodplains and replace species requiring more frequent flooding (Brock *et al.* 2006; Thoms *et al.* 2006). Where floods no longer occur, terrestrial vegetation colonises the floodplain (Catford *et al.* 2011), such as chenopods in the Macquarie Marshes. Sometimes such changes can take hundreds of years. For example, black box communities dependent on river flows from the Fiddlers Creek system on the Lowbidgee floodplain are now largely dead, except for a few individuals with some leaves. This distributary system was probably most affected by the building of Burrinjuck Dam and the establishment of the Murrumbidgee irrigation area in the 1920–1930s (Kingsford 2003).

Most of our knowledge of the effects of river regulation on vegetation communities in Australia concerns the iconic river red gum and the communities that they dominate. River red gums supplied with sufficient flows are usually larger, have better canopies and leaves with higher moisture content and area than river red gums on drier parts of the floodplain (Stone and Bacon 1994, 1995). As flooding regimes have decreased across floodplains, river red gums have invaded plant communities that historically occurred in habitats with more frequent flooding, such as the Moira grass (*Pseudoraphis spinescens*) community of the Barmah–Millewa Forest of the Murray River (Bren 1992). Correspondingly, where flooding is no longer frequent enough to support river red gums, sometimes indicated by groundwater levels, there is considerable decline in condition as well as mortality (Cunningham *et al.* 2009; Horner *et al.* 2009; Catelotti *et al.* 2015). Up to 95% of the eucalypts on the floodplain of the lower Murray River are dying from reduced flows and salinisation (Jolly *et al.* 1993; MDBC 2003), and 45–70% of river red gum forest have died back on the floodplain of the Murray River in Victoria (Cunningham *et al.* 2010, 2011). These changes are probably exacerbated by insect herbivory, which can be higher on trees in poor condition (Stone and Bacon 1995). Declines in flow frequency also reduce opportunities for germination, growth and recruitment of river red gums (Stone and Bacon 1994). Seed release from water-stressed floodplain trees, such as river red gums and black box, can be nine times less where natural flood regimes occur (Jensen *et al.* 2008).

In some places in Australia, increased flooding has dramatically affected riverine vegetation through the permanent inundation of wetlands and river channels, by re-regulatory storages or because low flows constantly connect them to the river. Riparian trees, while dependent on periodic flooding, die under conditions of permanent inundation as demonstrated by the many dead trees present in lakes of regulated rivers (e.g. Menindee Lakes, Lake Victoria, Lake Mokoan, Kow swamp). River red gums die when their base is submerged for 2–4 years (Briggs and Maher 1983; Chesterfield 1986; Bren 1987) while black box flooded for 12–18 months usually die (Shepheard 1992; Briggs and Townsend 1993; Roberts and Marston 2000). Other common perennial riparian plants of south-eastern Australia (e.g. tangled lignum, coolibah) also die when flooded for too long (Roberts and Marston 2000). Additionally, permanent flooding decreases diversity and composition of aquatic macrophytes (Nielsen and Chick 1997). Vegetation communities of weir pools along the Murray River, for example, have become more lentic than lotic, reflecting the reduction in variability of water levels imposed in such habitats by river regulation (Blanch *et al.* 1999, 2000). The stable water levels also reduce germination and growth of many aquatic plants that favour more by variable water levels (Blanch *et al.* 2000).

Effects of alteration of flow, and consequently riverine vegetation, significantly affects many other organisms dependent on vegetation for food and habitat. Aquatic and riparian vegetation provides considerable organic matter for rivers which, when inundated, results in primary and secondary production which ultimately benefits high order organisms such as fish and waterbirds. Riparian woodlands are also important for invertebrate communities (Thomson *et al.* 2012). Waterbirds, in particular, depend on riverine vegetation with certain structural characteristics (e.g. size and density of tangled lignum shrubs), as well as high levels of aquatic productivity, for successful breeding and recruitment (Kingsford and Johnson 1998). There is evidence that woodland bird communities also fundamentally change as river red gums die, with species more tolerant of terrestrial landscapes replacing those that depend on healthy river red gums (Blackwood 2009; McGinness *et al.* 2010).

## ENVIRONMENTAL FLOW MANAGEMENT

Over the past few decades, there has been an increasing focus on the importance of managing rivers for environmental flows in Australia as governments try to rehabilitate rivers in response to growing public pressure. For example, one of the world's more ambitious river management programs has involved the buy-back of water from the irrigation industry (at a cost of $3.1 billion) by the Australian Government for the rivers of the Murray–Darling Basin. Consequently, some of the storage capacity in many dams, particularly in the Murray–Darling Basin, is now dedicated to providing environmental flows. In some river systems, the environmental flow component may equate to a substantial quantum of the storage of the dam. For example, there is an environmental flow of >300 000 ML on the Macquarie River (~30% of Burrendong Dam when full) that can be actively managed for environmental purposes. Accordingly, there is increasing pressure for good environmental management of these rivers and accountability to demonstrate the effects of environmental water management.

Removal or rehabilitation of structures that intercept flows, such as floodplain earthworks, is also being conducted or considered (e.g. Toorale station, terminus of the Warrego River; Macquarie Marshes) (Steinfeld and Kingsford 2013). The construction of 'environmental works and measures', including the construction of new levees, channels and other small infrastructure, however, has also occurred in recent years as a means of increasing the efficiency of environmental delivery to key wetlands. This may dramatically affect connectivity to vegetation communities, either increasing or decreasing their resilience. This potential problem is rarely acknowledged or managed. The provision of environmental flows in the Murray–Darling Basin rivers is aimed at restoring the health of dependent ecosystems, often targeted floodplain communities. There are considerable challenges in adequately meeting many of the environmental objectives on floodplains given insufficient volumes of water, originally estimated to be 3000–4000 GL for the basin but with a revised target of 2750 GL. This may be even further reduced with works and measures designed to improve the targeting of environmental flows, but inevitably reducing the resilience of other flow-dependent vegetation communities.

In regulated rivers, environmental flows can include the base flow used to deliver water to users, as well as licenced water specifically for the environment that may be released under prescribed rules for a particular dam. Some of this flow is released from the dam to mimic flows of tributaries flowing into storage with little analysis of ecological responses. Opportunities to release this type of flow are also confined to times when the river is not being run as a delivery channel for irrigation, at which times the channel may be full and so not able to carry the environmental component. Knowledge of the ecological requirements of vegetation is essential for the management of environmental flows, potentially becoming an important measure of success or otherwise. For example, environmental watering can trigger germination of floodplain trees but with limited survival, which could be increased if delivery complemented rainfall or short flood events (Jensen *et al.* 2008).

Ultimately, it is important to have a measure of the likely effects of different release patterns on target ecosystems. This may involve defining the composition of different flood-dependent vegetation communities that exist in a wetland under different conditions. Such decisions also need to be informed by an understanding of the extent of vegetation communities prior to river regulation, as well as the conditions the communities require to be self-sustaining. Explicit objectives for managing the entire ecosystem, including vegetation (Biggs and Rogers 2003), are also required to guide the management and operation of the river in order to learn about and improve this management. A particular challenge relates to the management of floodplain vegetation on rivers where a significant amount of upstream flow is currently diverted (CSIRO 2008). In such cases, active management is necessary to ensure sufficient water exists to meet the requirements of a particular vegetation community. Sometimes environmental flows may need to be 'piggy-backed' on top of a delivery flow for irrigation to ensure that sufficient flows reach different parts of the floodplain.

As well as using such management options, there is a need to clearly define the desired state of a wetland and objectives for different vegetation communities that reflect this condition (Biggs and Rogers 2003). Such objectives need to be built around establishment and maintenance of vegetation communities as self-sustaining populations with sufficient water for reproduction and recruitment. Knowledge of how much can be achieved is equally important; some objectives for floodplain vegetation communities may not be achievable. More sophisticated scenario modelling is required to ensure the appropriate timing and quantity of flow delivery to achieve such objectives. Environmental flow management therefore places significant demands on scientific understanding of the basic biology of different flood-dependent vegetation communities and how much is achievable is highly dependent on the amount of environmental flow available.

## CONCLUSIONS

River management for much of Australia's history, particularly after European settlement, has primarily focused on the development of water resources for irrigated agriculture and other human uses. Governments built dams and other structures on the river to control flow so that it could be made available for diversion. In addition, structures on floodplains enabled protection of crops but also the capture of overland flows that could be diverted into off-river storages for use in irrigated agriculture, simultaneously alienating parts of the floodplain and their vegetation communities. Such river management has considerably altered vegetation communities reliant on river flows and flooding. The greatest impact has been the contraction of floodplain vegetation so that its distribution and extent now primarily reflect reduced flooding patterns and decreased connectivity. Impacts have included widespread death and declines in condition of iconic floodplain eucalypts, many of which may never be re-established. Development of water resources for agriculture is increasingly underway on the rivers in northern Australia. The likely changes to dependent floodplain vegetation are predictable and now well known from current understanding of impacts within the Murray–Darling Basin. It is important that such costs are assessed well before major water resource development of these rivers occurs.

The prospect of increasing environmental flows in rivers through the buy-back of water and increasing water use efficiency provides considerable opportunity for rehabilitation of some vegetation communities, but has considerable challenges. There is relatively poor knowledge of the flow requirements for most vegetation communities in terms of producing self-sustaining populations. Another major challenge is ensuring that there are clear objectives in place with adaptive management processes that measure whether environmental flows actually result in self-sustaining ecosystems. Different and imaginative ways of managing environmental water are likely to be needed.

## ACKNOWLEDGEMENTS

I thank Kate Brandis for assistance with figures and Daniela Binder for help with references. I also thank Sam Capon for the opportunity to contribute. This work was supported by the University of New South Wales and an ARC Linkage Research Grant LP0453694.

## REFERENCES

Arthington AH, Pusey BJ (2003) Flow restoration and protection in Australian rivers. *River Research and Applications* **19**, 377–395. doi:10.1002/rra.745

Benn PC, Erskine WD (1994) Complex channel response to flow regulation: Cudgegong River below Windamere Dam, Australia. *Applied Geography (Sevenoaks, England)* **14**, 153–168. doi:10.1016/0143-6228(94)90058-2

Biggs HC, Rogers KH (2003) An adaptive system to link science, monitoring, and management in practice. In *The Kruger Experience: Ecology and Management of Savanna Heterogeneity*. (Eds JT du Toit, KH Rogers and HC Biggs) pp. 59–80. Island Press, Washington, DC.

Blackwood AJ (2009) The effect of river red gum decline on woodland birds in the Macquarie Marshes. Honours thesis. University of NSW, Sydney.

Blanch SJ, Ganf GG, Walker KF (1999) Tolerance of riverine plants to flooding and exposure indicated by water regime. *Regulated Rivers: Research and Management* **15**, 43–62. doi:10.1002/(SICI)1099-1646(199901/06)15:1/3<43::AID-RRR535>3.0.CO;2-Q

Blanch SJ, Walker KF, Ganf GG (2000) Water regimes and littoral plants in four weir pools of the River Murray, Australia. *Regulated Rivers: Research and Management* **16**, 445–456. doi:10.1002/1099-1646(200009/10)16:5<445::AID-RRR596>3.0.CO;2-L

Bren LJ (1987) The duration of inundation in a flooding river red gum forest. *Australian Forest Research* **17**, 191–202.

Bren LJ (1992) Tree invasion of an intermittent wetland in relation to changes in the flooding frequency of the River Murray, Australia. *Australian Journal of Ecology* **17**, 395–408. doi:10.1111/j.1442-9993.1992.tb00822.x

Briggs SV, Maher MT (1983) Litter fall and leaf decomposition in a river red gum (*Eucalyptus camaldulensis*) swamp. *Australian Journal of Botany* **31**, 307–316. doi:10.1071/BT9830307

Briggs S, Townsend G (1993) 'Restoration and management of Nearie Lake.' Final report, Project N209, Natural Resources Management Strategy.

Brock MA, Capon SJ, Porter JL (2006) Disturbance of plant communities dependent on desert rivers. In *Ecology of Desert Rivers*. (Ed. RT Kingsford) pp. 203–247. Cambridge University Press, Cambridge, UK.

Catelotti K, Kingsford RT, Bino G, Bacon P (2015) Inundation requirements for persistence and recovery of river red gums (*Eucalyptus camaldulensis*) in semi-arid Australia. *Biological Conservation* **184**, 346–356.

Catford JA, Downes BJ, Gippel CJ, Vesk PA (2011) Flow regulation reduces native plant cover and facilitates exotic invasion in riparian wetlands. *Journal of Applied Ecology* **48**, 432–442. doi:10.1111/j.1365-2664.2010.01945.x

Chesterfield EA (1986) Changes in the vegetation of the river red gum forest at Barmah, Victoria. *Australian Forestry* **49**, 4–15. doi:10.1080/00049158.1986.10674458

Cluett L (2005) The role of flooding in morphological changes in the regulated lower Ord River in tropical north-western Australia. *River Research and Applications* **21**, 215–227. doi:10.1002/rra.842

CSIRO (2008) 'Water availability in the Murray–Darling Basin.' A report from CSIRO to the Australian Government from the CSIRO Murray–Darling Basin Sustainable Yields Project. CSIRO, Canberra.

Cunningham SC, Nally RM, Read J, Baker PJ, White M, Thomson JR, Griffioen P (2009) A robust technique for mapping vegetation condition across a major river system. *Ecosystems* **12**, 207–219. doi:10.1007/s10021-008-9218-0

Cunningham SC, Thomson JR, Read J, Baker PJ, Mac Nally R (2010) Does stand structure influence susceptibility of eucalypt floodplain forests to dieback? *Austral Ecology* **35**, 348–356. doi:10.1111/j.1442-9993.2009.02043.x

Cunningham SC, Thomson JR, Mac Nally R, Read J, Baker PJ (2011) Groundwater change forecasts widespread forest dieback across an extensive floodplain system. *Freshwater Biology* **56**, 1494–1508. doi:10.1111/j.1365-2427.2011.02585.x

DERM (2009) *Wetland Info*. Department of Natural Resources and Mines, Brisbane, <http://www.epa.qld.gov.au/wetlandinfo/site/WetlandDefinitionstart.html>.

Dexter BD (1978) Silviculture of the river red gum forests of the central Murray floodplain. *Proceedings of the Royal Society of Victoria* **90**, 175–194.

Doupé RG, Petitt NE (2002) Ecological perspectives on regulation and water allocation for the Ord

River, Western Australia. *River Research and Applications* **18**, 307–320. doi:10.1002/rra.676

Erskine WD, Terrazzolo N, Warner RF (1999) River rehabilitation from the hydrogeomorphic impacts of a large hydro-electric power project: Snowy River, Australia. *Regulated Rivers: Research and Management* **15**, 3–24. doi:10.1002/(SICI)1099-1646(199901/06)15:1/3<3::AID-RRR532>3.0.CO;2-R

Greet J, Cousens RD, Webb JA (2013) Flow regulation is associated with riverine soil seed bank composition within an agricultural landscape: potential implications for restoration. *Journal of Vegetation Science* **24**, 157–167. doi:10.1111/j.1654-1103.2012.01445.x

Horner GJ, Baker PJ, Nally RM, Cunningham SC, Thomson JR, Hamilton F (2009) Mortality of developing floodplain forests subjected to drying climate and water extraction. *Global Change Biology* **15**, 2176–2186. doi:10.1111/j.1365-2486.2009.01915.x

Horner GJ, Cunningham SC, Thomson JR, Baker PJ, Mac Nally R (2012) Forest structure, flooding and grazing predict understorey composition of floodplain forests in southeastern Australia. *Forest Ecology and Management* **286**, 148–158. doi:10.1016/j.foreco.2012.08.023

Jensen AE, Walker KF, Paton DC (2008) The role of seedbanks in restoration of floodplain woodlands. *River Research and Applications* **24**, 632–649. doi:10.1002/rra.1161

Jolly ID, Walker GR, Thorburn PJ (1993) Salt accumulation in semi-arid floodplain soils with implications for forest health. *Journal of Hydrology* **150**, 589–614. doi:10.1016/0022-1694(93)90127-U

Kingsford RT (1995) Ecological effects of river management in New South Wales. In *Conserving Biodiversity: Threats and Solutions*. (Eds RA Bradstock, TD Auld, DA Keith, RT Kingsford, D Lunney and DP Sivertsen) pp. 144–161. Surrey Beatty & Sons, Sydney.

Kingsford RT (2000) Ecological impacts of dams, water diversions and river management on floodplain wetlands in Australia. *Austral Ecology* **25**, 109–127. doi:10.1046/j.1442-9993.2000.01036.x

Kingsford RT (2003) Social, institutional and economic drivers for water resource development – a case study of the Murrumbidgee River, Australia. *Aquatic Ecosystem Health & Management* **6**, 69–79. doi:10.1080/14634980301480

Kingsford RT (2004) Wetlands and waterbirds of the Darling River. In *The Darling*. (Eds R Breckwoldt, R Boden and J Andrew) pp. 234–259. Murray–Darling Basin Commission, Canberra.

Kingsford RT, Johnson W (1998) Impact of water diversions on colonially nesting waterbirds in the Macquarie Marshes in arid Australia. *Colonial Waterbirds* **21**, 159–170. doi:10.2307/1521903

Kingsford RT, Thomas RF (1995) The Macquarie Marshes and its waterbirds in arid Australia: a 50-year history of decline. *Environmental Management* **19**, 867–878. doi:10.1007/BF02471938

Kingsford RT, Thomas RF (2004) Destruction of wetlands and waterbird populations by dams and irrigation on the Murrumbidgee River in arid Australia. *Environmental Management* **34**, 383–396. doi:10.1007/s00267-004-0250-3

Kingsford RT, Thomas RF, Curtin AL (2001) Challenges for the conservation of wetlands on the Paroo and Warrego Rivers. *Pacific Conservation Biology* **7**, 21–33.

Kingsford RT, Brandis K, Thomas RF, Knowles E, Crighton P, Gale E (2004a) Classifying landform at broad landscape scales: the distribution and conservation of wetlands in New South Wales, Australia. *Marine and Freshwater Research* **55**, 17–31. doi:10.1071/MF03075

Kingsford RT, Jenkins KM, Porter JL (2004b) Imposed hydrological stability on lakes in arid Australia and effect on waterbirds. *Ecology* **85**, 2478–2492. doi:10.1890/03-0470

Kingsford RT, Lemly AD, Thompson JR (2006) Impacts of dams, river management and diversions on desert rivers. In *Ecology of Desert Rivers*. (Ed. RT Kingsford) pp. 203–247. Cambridge University Press, Cambridge, UK.

Mac Nally R, Cunningham SC, Baker PJ, Horner GJ, Thomson JR (2011) Dynamics of Murray–Darling floodplain forests under multiple stressors: the past, present, and future of an

Australian icon. *Water Resources Research* **47**(12), doi:10.1029/2011WR010383

Maheshwari BL, Walker KF, McMahon TA (1995) Effect of regulation on the flow regime of the River Murray, Australia. *Regulated Rivers: Research and Management* **10**, 15–38. doi:10.1002/rrr.3450100103

McGinness HM, Arthur AD, Reid JRW (2010) Woodland bird declines in the Murray–Darling Basin: are there links with floodplain change? *The Rangeland Journal* **32**, 315–327. doi:10.1071/RJ10016

McGinness HM, Arthur AD, Davies M, Mcintyre S (2013) Floodplain woodland structure and condition: the relative influence of flood history and surrounding irrigation land use intensity in contrasting regions of a dryland river. *Ecohydrology* **6**, 201–213. doi:10.1002/eco.1259

MDBC (2003) 'Preliminary investigations into observed river red gum decline along the River Murray below Euston.' Technical report 03/03. Murray–Darling Basin Commission, Canberra.

Nielsen DL, Chick AJ (1997) Flood-mediated changes in aquatic macrophyte community structure. *Marine and Freshwater Research* **48**, 153–158. doi:10.1071/MF96090

Roberts J, Ludwig JA (1991) Riparian vegetation along current exposure gradients in floodplain wetlands of the River Murray *Australian Journal of Ecology* **79**, 117–127. doi:10.2307/2260787

Roberts J, Marston F (2000) 'Water regime of wetland and floodplain plants in the Murray–Darling Basin: a source book of ecological knowledge.' Technical report. CSIRO Land & Water, Canberra.

Roberts J, Marston F (2011) 'Water regime for wetland and floodplain plants: a source book for the Murray–Darling Basin.' National Water Commission, Canberra.

Robertson AI, Bacon P, Heagney G (2001) The responses of floodplain primary production to flood frequency and timing. *Journal of Applied Ecology* **38**, 126–136. doi:10.1046/j.1365-2664.2001.00568.x

Roshier DA, Whetton PH, Allan RJ, Robertson AI (2001) Distribution and persistence of temporary wetland habitats in arid Australia in relation to climate. *Austral Ecology* **26**, 371–384. doi:10.1046/j.1442-9993.2001.01122.x

Shepheard M (1992) 'Flooding frequency and *Eucalyptus largiflorens* (black box) wetland health in the Wah Wah district.' Technical report 92/14. NSW Department of Water Resources, Parramatta, NSW.

Steinfeld CMM, Kingsford RT (2013) Disconnecting the floodplain: earthworks and their ecological effect on a dryland floodplain in the Murray–Darling Basin, Australia. *River Research and Applications* **29**, 206–218. doi:10.1002/rra.1583

Stone C, Bacon P (1994) Relationships among moisture stress, insect herbivory, foliar cineole content and the growth of river red gum *Eucalyptus camaldulensis*. *Journal of Applied Ecology* **31**, 604–612. doi:10.2307/2404151

Stone C, Bacon PE (1995) Leaf dynamics and insect herbivory in a *Eucalyptus camaldulensis* forest under moisture stress. *Australian Journal of Ecology* **20**, 473–481. doi:10.1111/j.1442-9993.1995.tb00566.x

Thomas RF, Kingsford RT, Lu Y, Hunter SJ (2011) Landsat mapping of annual inundation (1979–2006) of the Macquarie Marshes in semi-arid Australia. *International Journal of Remote Sensing* **32**, 4545–4569. doi:10.1080/01431161.2010.489064

Thoms MC, Beyer PH, Rogers KH (2006) Variability, complexity and diversity – the geomorphology of river ecosystems in dryland regions. In *Ecology of Desert Rivers*. (Ed. RT Kingsford) pp. 47–75. Cambridge University Press, Cambridge, UK.

Thomson JR, Bond NR, Cunningham SC, Metzeling L, Reich P, Thompson RM, Mac Nally R (2012) The influences of climatic variation and vegetation on stream biota: lessons from the Big Dry in southeastern Australia. *Global Change Biology* **18**, 1582–1596. doi:10.1111/j.1365-2486.2011.02609.x

Walker KF, Sheldon F, Puckridge JT (1995) A perspective on dryland river ecosystems. *Regulated Rivers: Research and Management* **11**, 85–104. doi:10.1002/rrr.3450110108

WCIC (1971) *Water Resources of NSW*. Water Conservation and Irrigation Commission, Sydney.

WRC (1978) *Guidelines for Floodplain Development, Macquarie River, Narromine to Warren*. Water Resources Commission, Sydney.

WRC (1982) *Guidelines for Floodplain Development, Macquarie River, Warren to Oxley*. Water Resources Commission, Sydney.

15

# Salinisation and its impacts on Australia's riverine vegetation

*George Ganf and Kaylene Morris*

## INTRODUCTION

Salinity is an ancient, global environmental problem that has caused the demise of civilisations and reduced vibrant cites to hamlets. The collapse of the Sumerian civilisation almost 4500 years ago, for example, is largely attributed to rising salinity levels which forced these early agriculturalists to abandon wheat as a major crop and replace it with barley. As Jacobsen and Adams (1958) point out, 'probably there is no historical event of this magnitude for which a single explanation is adequate, but that growing salinity played an important part in the breakup of the Sumerian civilisation seems beyond question'. In the 21st century, salinity continues to pose a major global problem (Gelburd 1985) with 1000 million hectares affected (Szabolcs 1994), an area 1.3 times the land mass of Australia. In Australia, salinity problems are particularly insidious. In his book *Collapse: How Societies Choose To Fail Or Survive*, Diamond (2005) devotes an entire chapter to the mining of Australia and in particular the problem of salinity.

Australia is surrounded by oceans. As sea levels have risen and fallen over time, these oceans have advanced and retreated across Australia's coastal regions, leaving behind salt deposits that are subject to natural erosion. In addition, salt-laden ocean winds deposit salt across the continent. Although the ocean is the ultimate source of sodium chloride (NaCl), substantial changes occur as salt passes through eons of hydrological cycles, from rainfall to soil water to groundwater. Using stable isotopes, Herczeg *et al.* (2001) demonstrated that all of the original seawater in groundwaters of the Murray–Darling Basin has been flushed out of its aquifers. Their results indicate that ~1.5 million tons of new salt, however, are deposited in the Murray–Darling Basin each year by rainfall. Interestingly, the chemical composition of saline groundwater in the Murray–Darling Basin is remarkably comparable to that of seawater (Herczeg *et al.* 2001).

In Western Australia, wheat is grown on land that has accumulated salt over millennia due to the delivery and deposition of around 20 to 200 kg of salt per hectare per year by salt-laden winds from the Indian Ocean. As a result, there is around 5000 to 10 000 tonnes of salt per hectare that, if mobilised, could cause major salinity problems. In this region, irrigation and the replacement of deep-rooted perennial native plant species with shallow-rooted annual agricultural crops have caused the water table to rise, consequently bringing salt to the surface. This 'secondary salinisation' has impacted on the nature of many wetland plant communities in the region, with a loss of more salt-sensitive species such as *Lemna* and *Marsilea* and their replacement by salt-tolerant species such as *Ruppia* and *Lepilaena* (Sim *et al.* 2006). Similar problems are widespread throughout Australia.

Wetland surveys by Smith *et al.* (2009) confirmed the earlier work of Brock and colleagues (Brock 1981a; Brock and Lane 1983; Brock and Shiel 1983) and suggest that at salinities as low as 1.5 dS/m (1000 mg/L) 40–50% fewer plant species may occur in wetlands compared with the freshest wetlands surveyed. At 6.0 dS/m (4000 mg/L) species number may decline by as much as 60–70%. This chapter focuses on plant responses to salinity. However, because life stages of both flora and fauna differ in their response to salinity (Ye *et al.* 2010; Beatty *et al.* 2011; Kefford *et al.* 2011), management should recognise the diversity of salinity tolerances of all riverine biota (Hart *et al.* 2003).

In this chapter, we discuss salt in Australia's riverine landscapes including processes of primary and secondary salinisation. We examine the effects of salt on plants, particularly aquatic macrophytes, and discuss how other environmental variables interact with salinity to affect plant performance. Methods of measuring salinity and experimental considerations in assessing the effects of salinity on plants are outlined in Box 15.1 and Box 15.2, respectively. A case study of the Coorong at the mouth of the Murray River in South Australia is also provided to illustrate the importance of understanding the salinity tolerances of different plant life stages (Box 15.3). Finally we discuss trends in salinisation in Australia's rivers and catchments, strategies to manage salinity and the implications of climate change.

## PRIMARY AND SECONDARY SALINISATION IN AUSTRALIA

Various terms are used to describe land salinisation, the most common refer to whether salinisation has occurred through natural phenomena (primary salinisation) or via human interference with the hydrological balance (secondary salinisation). The saline lakes that occur in the low-lying areas of northern South Australia represent an example of primary salinisation. These lakes act as the terminal drainage points for rivers including the Cooper, Diamantina and Georgina that drain into Lake Eyre, 15.2 m below sea level. Discharges of these rivers are extremely variable, but with exceptional monsoonal rainfall in northern Australia freshwater enters Lake Eyre triggering a surge of biological activity (Bayly 1976). Water levels in Lake Eyre quickly recede, however, as evaporation becomes the dominant process and salinity of the drying lake rapidly exceeds that of seawater. The lake eventually dries out leaving a flat expanse of salt.

Secondary salinisation occurs as a consequence of changes in land use and hydrology that cause the water table to rise. The main factor producing secondary salinisation is leakage (Eberbach 2003), i.e. water passing through the root zone and accumulating in the water table causing it to rise. Prior to European settlement, the Australian landscape was relatively balanced hydrologically. Evapotranspiration by native vegetation, evaporation and runoff were in balance with rainfall so that little, if any, water penetrated (i.e. leaked) below the root zone. European agriculture, however, has replaced native vegetation with relatively shallow-rooted annual crops which permit water to pass below the root zone. Leakage is difficult to measure because its magnitude depends on climate, rainfall, vegetation cover and soil characteristics (Walker *et al.* 1999). Walker *et al.* (1999) suggest that annual leakage rates under native vegetation may be between 0.5 and 5 mm whereas leakage under annual crops and pastures is likely to be 50 mm up to as high as 150 mm. Furthermore, water is probably accumulating at a rate an order of magnitude higher than it did before European agricultural developments. As a general rule, leakage under native perennial deep-rooted vegetation (i.e. trees) ranges from 0–1% of annual rainfall compared with 0–10% under perennial deep-rooted pasture, such as lucerne, and between 30–40% under shallow-rooted annual crops and pastures.

## Box 15.1: Measuring salinity

### Water salinity

Salinity refers to the sum total of all ion concentrations, not just NaCl, and is referred to as total soluble salt (TSS). It is impractical to measure the concentration of all the salts present in a solution, so a simpler approach is typically used whereby residual weight is measured after the evaporation of a known volume of water that has been filtered to remove particulate matter (Peck *et al.* 1983). The measurement produced by this method is referred to as the total dissolved salt (TDS) and closely approximates TSS. Most often, however, water salinity is determined indirectly by measuring conductivity at 25°C. Modern conductivity meters are simple to use in the field and allow the direct determination of surface water conductivity. There are inherent problems, however, in inferring salinity from conductivity measures. For example, the relationship between conductance and solute concentration changes at high salt concentrations and can vary depending on which ions are dominant (Williams and Sherwood 1994; Talling 2009). As such, conductivity measurements may be inaccurate when salt concentration exceeds seawater or when NaCl is not the dominant ion (Williams and Sherwood 1994).

Levels of salinity are reported in a confusing range of measurements and unit, from TDS as parts per million (ppm) or mg/L or electrical conductivity (EC) as µS/cm, dS/m, mS/cm or mm NaCl. Salinity derived from measurements of electrical conductivity should use units of conductivity proposed by The International System of Units (i.e. Siemens per unit length, e.g. dS/m = mS/cm; 1000 µS/cm). Hart *et al.* (1991) recommend a conversion factor of 0.68 to covert mg/L to µS/cm. Seawater, for example, has a typical conductivity of 54 dS/m or 34 560 ppm or mg/L. Caution should be applied when converting between these measures of salinity.

### Soil salinity

The measurement of soil salinity is more complicated than that for water and requires mixing soil with water at a ratio of one part soil to five parts water by weight (Slavich and Petterson 1993). Soil samples are air-dried, ground and sieved to <2 mm before analysis. Distilled water is added and the solution shaken for 1 h before electrical conductivity is measured. Soil salinity measured in this way is expressed as conductivity of a 1:5 extract ($EC_{1:5}$) and gives a measure of salt concentration per unit weight of soil (Slavich and Petterson 1993).

To assess the influence of salinity on plant growth, it is best to estimate the concentration of salts in terms of kg of salts/kg of soil water based on a standard water content, typically the saturation water content. Salinity measured in this way is expressed as conductivity of the saturated paste ($EC_e$) (see Slavich and Petterson 1993). A variety of conversion factors, depending on soil type, are available to convert $EC_{1:5}$ to $EC_e$ (see Slavich and Petterson 1993; Zhang *et al.* 2005; Sonmez *et al.* 2008). Commercial soil salinity sensors, consisting of a porous ceramic conductivity cell and a thermistor are available and measure the conductivity of the soil directly. The underlying assumption is that the soil solution that imbibes the cell is in equilibrium with the ionic constituents of the adjacent soil (Aragüés 1982). The sensor is analogous to a standard conductivity cell and conversion to salinity suffers from the inherent problems described above.

### Landscape salinity

In many parts of Australia, salinity is a large-scale problem. Mapping salinity over large spatial scales is best achieved via airborne electromagnetic (AEM) surveys, which rely on electromagnetic pulses from a transmitter towed by an aircraft flying 75–150 m above the ground (Cresswell *et al.* 2007). The transmitted pulse induces an electromagnetic response from the ground that gives a three-dimensional image of the conductivity of the ground. The pulse received by the aircraft receiver is sensitive to changes in the conductivity of the ground. These data are converted, using specialised software, into underground three-dimensional maps that illustrate where salt is and the likely trajectory it will take as water moves through the landscape. Airborne electromagnetics are currently quite expensive (around $10 per hectare) and are best suited to landscapes where the ground is relatively flat and salinity gradients relatively simple and easily defined.

## Box 15.2: Experimental approaches

Experimental protocols for assessing plant responses to salinity vary widely among researchers and may influence the observed tolerance of plants. In designing salinity experiments on plants, thought should be given to: (i) the ionic composition of the salt solution; (ii) the rate at which salinities are increased; (iii) the duration of salt exposure; and (iv) measuring soil salinities to account for the potential for salts to accumulate in the root zone through evaporation and transpiration and exceed the salinity of the applied salt solution. These points are discussed in more detail below.

The relative proportion of the major cations and anions that comprise saline inland waters vary geographically (Radke *et al.* 2002, 2003). As the sensitivity of plants to NaCl can be modified by the presence of other cations (particularly Ca), the ionic composition of experimental salt solution needs some consideration. Using seawater as a standard may be appropriate in many studies and also permits comparison among studies. The ionic composition of seawater generally reflects that of the majority of saline inland waters in south-eastern Australia (Williams 1967), as well as coastal systems subject to saline intrusions. Partridge and Furey (2002) point out that while the exact composition of saline groundwater in the Western Australian wheat belt varies over small spatial scales it generally reflects that of seawater. In regions where the ionic composition of saline water differs substantially from that of seawater, and where the objective is to understand the salinity tolerance of the local flora, then the ionic composition of the experiment salt solution should reflect that of the study region.

The capacity of plants to osmotically adjust to salt is influenced by the rate at which salinities are increased to target levels and this differs widely across experiments. Gradual increases in salinity have been demonstrated to improve salinity tolerance (Heth and MacRae 1993), but the rate of salinisation had little effect on the persistence of macrophyte communities developing from soil seed banks (Nielsen *et al.* 2008). The rate at which salinities are increased experimentally needs to be assessed based on: (i) the time it will take to reach the final salt concentration; (ii) the sensitivity of the species examined and; (iii) the question being addressed by the study. Testing plant responses to salt fluxes that occur naturally will provide the greatest insights into the likely outcomes of elevated salinities on vegetation communities.

The duration of exposure to salinity will also influence tolerance. Salinity tolerance generally decreases over time as ions accumulate within the plant to toxic levels (Yeo *et al.* 1991; Munns 1993). The time it takes to observe the effects of salinity on plant growth will vary with the level of salt applied and the salt sensitivity of the species.

Increased soil salinities resulting from evaporation and transpiration should be considered when designing, conducting and interpreting experiments and field studies. Evaporation and transpiration can cause soil salinities to increase above levels measured in the surface water or groundwater. In exposed sediments, the evaporation of saline water causes salts to accumulate at the soil surface. Brock *et al.* (2005) found that sediments exposed to salinity under damp conditions developed higher salt loads than sediment under flooded conditions, causing lower macrophyte emergence for the soil seed bank. As plants exclude salt from the transpiration stream, salts can accumulate around the root zone and reach higher concentrations than the bulk soil (Smith *et al.* 1989; Passioura *et al.* 1992). The extent that salts accumulate in the soil profile through evaporation and transpiration depend not only on temperature and vapour pressure deficits, which govern rates of evaporation and transpiration but on soil properties, inundation, groundwater depth and the frequency of flushing events such as rainfall.

Secondary salinisation is also classified as either dryland salinity or irrigated land salinity. The former involves the movement of water and salt to the land surface in areas that are not irrigated whereas the latter occurs in areas that are, and is further increased if irrigation water itself is slightly saline. Areas at high risk of secondary salinisation occur in Western Australia, South Australia, Victoria, southern NSW and Tasmania. The southern boundaries of areas at high risk of dryland salinity are climate driven and depend on the intensity and timing of rainfall (Cresswell 2008). In northern Australia, high and variable seasonal rainfall coincides with maximum potential evaporation whereas in southern Australia, rainfall occurs mainly in winter, coinciding with

## Box 15.3: The Coorong: a case study

The Coorong is a permanent coastal lagoon located in the south-east of South Australia and is one of Australia's largest wetland systems, covering an area of 467 km$^2$ (Fig. 15.1). It receives inflows from the Murray River via the Lower Lakes (Lake Alexandrina and Lake Albert). The Coorong and Lower Lakes are wetlands of international importance under the Ramsar convention (Ramsar convention on wetlands, http://www.ramsar.org/), largely due to the abundance and diversity of the Coorong's waterbird community. The climate of the region is Mediterranean with hot and dry summers (December–February) and cool and wet winters (June–August).

The Coorong comprises two main lagoons (north and south) separated by a narrow channel at Parnka Point. Flow to the Coorong is controlled by five barrages that separate the north lagoon of the Coorong from Lake Alexandrina, a large freshwater lake fed by the Murray River. Historically, the salinity at the northern end of the north lagoon in the vicinity of the Murray mouth was fresh to brackish dependent on freshwater input. Due to development within the Murray–Darling Basin, inflows to the Coorong are now 40% of natural inflows (Brookes *et al.* 2009). The lack of freshwater inflows and evapo-concentration causes the salinity to increase from the Murray mouth, where salinities vary from fresh to saline, to saline to hyper-saline in the south lagoon.

*Ruppia* spp. are the key plant species of the Coorong and Lower Lakes and represent a significant food source for resident and migratory birds, which form the basis of the Ramsar listing of these sites. The lack of freshwater inflows as a consequence of recent drought (2006–10) caused salinity to increase in the Coorong and adversely alter the distribution of *R. tuberosa* and *R. megacarpa* (Nicol 2005; Rogers and Paton 2009; Frahn *et al.* 2012). *Ruppia tuberosa* disappeared from the hypersaline southern end of the south lagoon (164 g/L), but began to colonise the less saline waters further north (36 g/L), while *R. megacarpa* disappeared from the north lagoon. Previous Coorong studies implicated water depth, sediment type, light and salinity as causes of the decline of *Ruppia* (Nicol 2005; Paton *et al.* 2009; Frahn *et al.* 2012; Kim *et al.* 2013). However, in the nearby ephemeral saline lakes, Pipe Clay Lake (26–124 g/L) and Lake Cantara north and south (19–70 g/L), *R. tuberosa* continued to flourish. Kim (2013) compares and contrasts habitats where *R. tuberosa* flourished (Pipe Clay and Lake Cantara) with habitats where *Ruppia* declined (the Coorong).

### *Ruppia* and salinity

*Ruppia* is a cosmopolitan species inhabiting brackish to hypersaline habitats. Both *R. megacarpa* and *R. tuberosa* have occurred in the Coorong (Dick *et al.* 2011), but have different reported salinity tolerances: *R. tuberosa* 13–230 g/L and *R. megacarpa* 2–50 g/L. *Ruppia megacarpa* is a robust rhizomatous perennial with stems normally 20–30 cm long but may reach 2.5 m. *Ruppia tuberosa* reproduces both sexually and asexually (via turions) but is non-rhizomatous. It has thin thread-like leaves 5–10 cm long. Both are important food sources and refuges for birds, fish and invertebrates.

Under optimal conditions the life cycle of *R. tuberosa* takes ~3 months to complete via germination, growth, flowering, seed set and/or turion development. The optimal conditions for each of these phases (Fig. 15.2) encompass the salinity range 30–55 g/L. Salinities <20–30 g/L offer the opportunity for severe competition from filamentous algae, which results in poor plant performance (Frahn *et al.* 2012). At salinities >54 g/L, shoot growth, flowering and seed set are compromised but turion formation is maintained up to 77 g/L (115 mS/cm). Growth, flowering and seed set are very poor >74 g/L and non-existent once salinities rise beyond ~101 g/L. The poor performance of *R. tuberosa* at elevated salinities is correlated with a fall in quantum yield, an increase in tissue Na concentration, a fall in K and the resultant increase in the Na:K ratio (Kim 2013).

*Ruppia tuberosa* has evolved strategies to counteract rising salinities as temperatures rise, evaporation increases and the input of freshwater lessens. Seeds generally germinate at salinities <90 g/L but not above 90 g/L. However, seeds that don't germinate at >90 g/L are able to germinate when transferred to lower salinities (Kim 2013). This suggests that *R. tuberosa* has a capacity to use its seed bank to bet-hedge against rising salinities (Ooi *et al.* 2009). Successful turion formation and germination extends well beyond the salinity range optimal for growth, flowering, seed set and germination. This may explain the sparse and patchy distribution of *R. tuberosa* reported by Frahn *et al.* (2012). In contrast, *R. megacarpa* does not appear to have evolved specific reproductive strategies to overcome

(Continued)

## Box 15.3: (Continued)

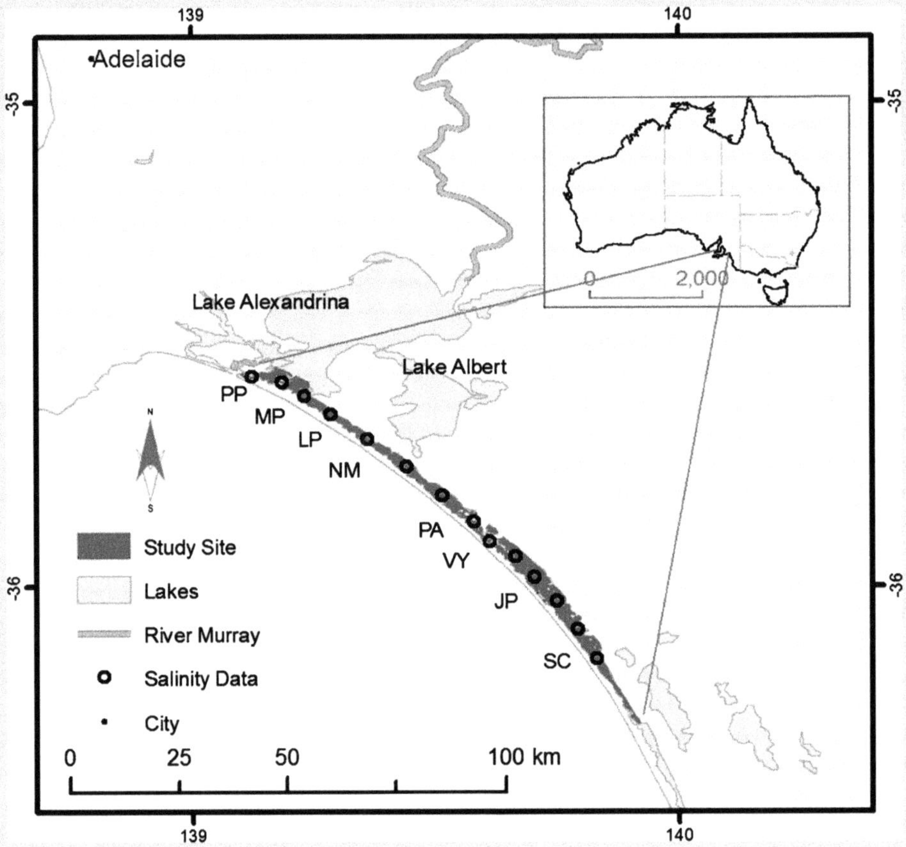


**Figure 15.1:** The Coorong and neighbouring Lake Pipe Clay and Lake Cantara, South Australia. PP, Pelican Point; MP, Mark Point; LP, Long Point; NM, Noonameena; PA, Parnka Point; VY, Villa Dei Yumpa; JP, Jack Point; SC, Salt Creek. Modified from Kim 2013.

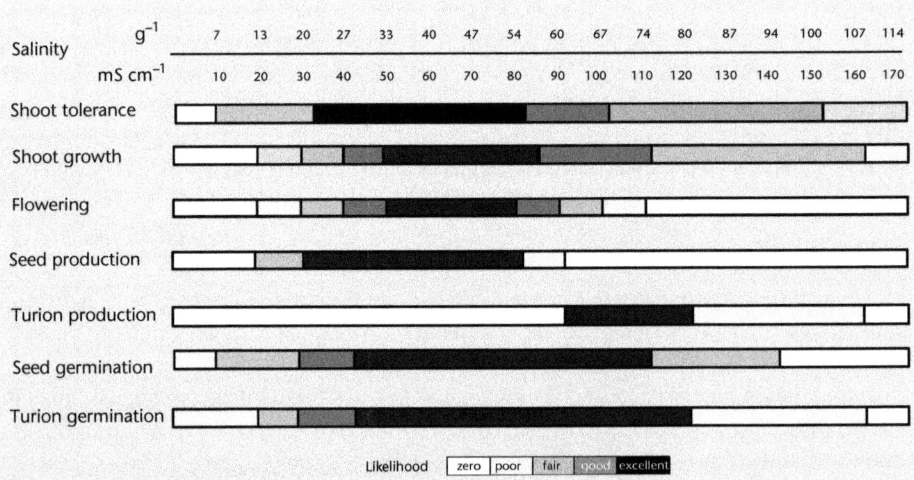


**Figure 15.2:** The likelihood of finding *Ruppia tuberosa* shoots (shoot tolerance), successful shoot growth, flowering, seed and turion production and germination across a range of salinities. Likelihood based on Kim 2013.

## Box 15.3: (Continued)

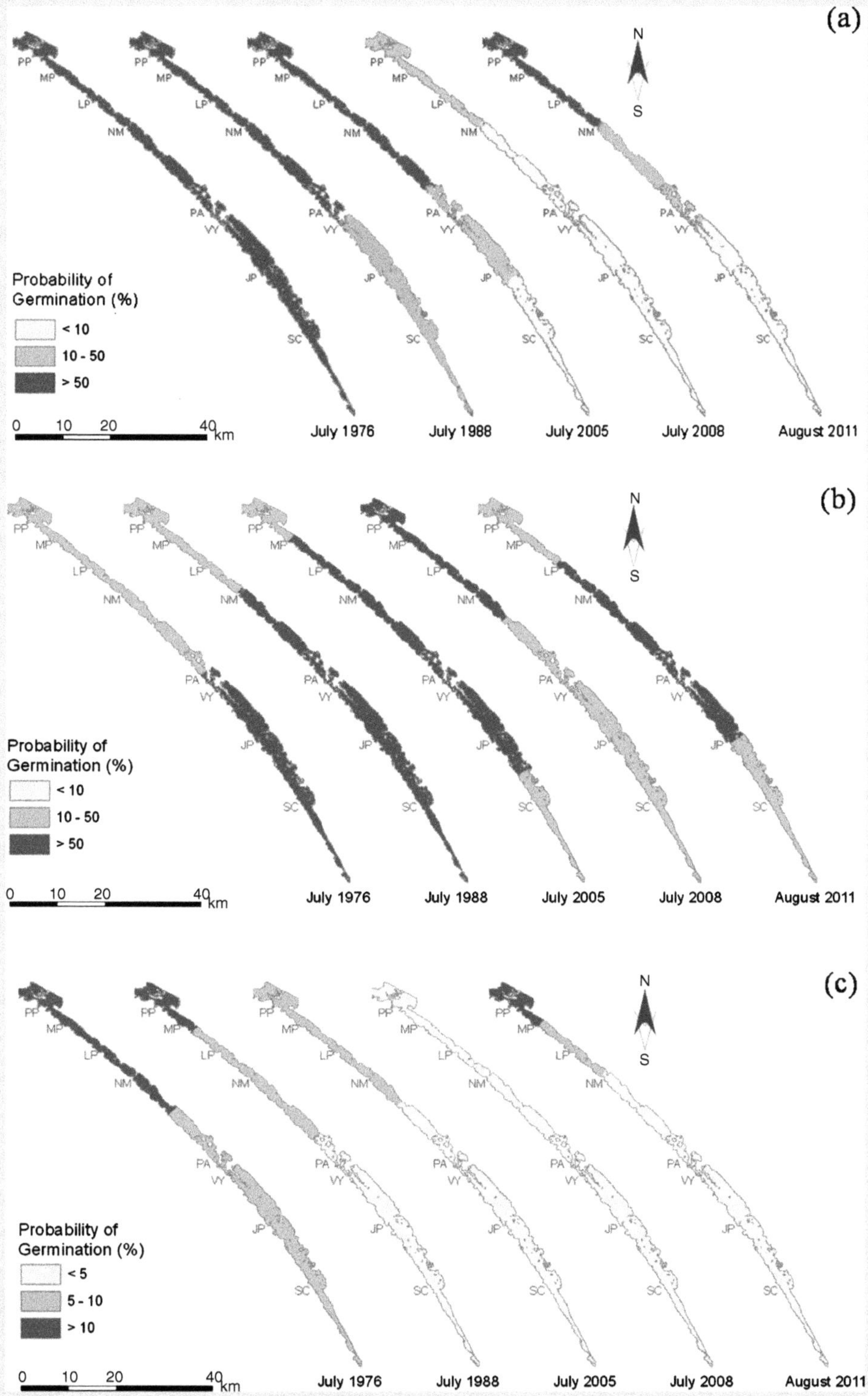


**Figure 15.3:** The probability of (a) *Ruppia tuberosa* seed, (b) turion and (c) *R. megacarpa* seed germination in the Coorong in July 1976, July 1988, July 2005, August 2008 and August 2011. PP, Pelican Point; MP, Mark Point; LP, Long Point; NM, Noonameena; PA, Parnka Point; VY, Villa Dei Yumpa; JP, Jack Point; SC, Salt Creek. Copied from Kim 2013.

(Continued)

**Box 15.3: (Continued)**

increasing salinity. *Ruppia megacarpa* reproduces via rhizomes and seeds. Seeds germinate at salinities of 0–30 g/L but do not germinate at salinities >30 g/L.

Successful colonisation of new areas by both *Ruppia* species is likely to occur via the dispersal of seeds and/or turions and their subsequent germination. Germination is governed by salinity, which in the Coorong shows a substantial north–south gradient that varies annually depending on freshwater inputs. Kim *et al.* (2013) combined seed and turion germination data with Coorong salinities for a low salinity year (1976), a high salinity year (2008) and when freshwater inputs occurred after the 2011 drought (Kim *et al.* 2013). The probability of germination (Fig. 15.3) of both *R. tuberosa* and *R. megacarpa* was high during low salinity (1976), but severely restricted during high salinity (2008). Frahn *et al.* (2012) report that after a freshwater input in 2011 there was patchy re-colonisation of the south lagoon and in the southern end of the north lagoon by *R. tuberosa*. This observation is in general agreement with the modelled germination response for *R. tuberosa* and emphasises the crucial role salinity would play in the re-colonisation of the Coorong.

**Management**

There is much debate on how to manage the Coorong (Paton 2013; MDBA 2014) for the re-establishment of *R. tuberosa* and *R. megacarpa*. This case study shows that for the optimal re-establishment of *R. tuberosa* salinities should be maintained between 30 and 60 g/L for a period of not less than three months in the south lagoon. Salinities between 2 and 50 g/L in the north lagoon would encourage the re-establishment *R. megacarpa*.

The latest plans (MDBA 2014) recommend salinities of <35 g/L in 30% of the estuary area and north lagoon with salinities of <19 g/L to maintain or improve *R. megacarpa* colonisation and reproduction. In the south lagoon, salinities of 60–100 g/L are recommended to enable *R. tuberosa* to colonise and reproduce. This would be achieved by freshwater flows of either 6000 GL/yr at a frequency of 1 in 3 years or flows of 10 000 GL/yr at a frequency of 1 in 7 years. In addition, the South Australian Government initiated a project to transplant *R. tuberosa* seeds from Lake Cantara to the south lagoon when the salinity is within the optimal range for germination, growth, flowering and seed set (Paton 2013). These science-driven recommendations are consistent with observations of the distribution and abundance of both species of *Ruppia* as well as other biota including fish and benthic invertebrates.

minimal evaporation. As a result of the balance between rainfall and evaporation, dryland salinity in southern Australia tends to occur where annual rainfall is between 400 and 800 mm while in northern Australia areas receiving between 700 and 1100 mm per year are most at risk. Cresswell (2008) suggests that the long-term variability of rainfall determines whether or not a region is likely to exhibit an increased or decreased risk of dryland salinity. In northern Australia, an increased risk is associated with higher than average rainfall whereas in southern Australia dryland salinity is promoted by positive rainfall balances. Intermittent wetlands, such as those located on floodplains, are also at particular risk of secondary salinisation because salts are concentrated during the drawdown of floodwaters (James *et al.* 2009).

## Processes of salinisation in riverine systems

Salts mobilised through primary and secondary land salinisation are ultimately expressed in riverine systems through the seepage of groundwater into watercourses or through surface runoff from salinised land. Changes in river or wetland hydrology through human modification or natural events can greatly influence the extent of salinisation in riverine systems.

### River regulation

The construction of weirs and locks along the Murray River has created a unique hydrological condition. The weight and depth of impounded water behind the weirs pushes saline groundwater downwards, but in downstream reaches groundwater is forced upwards and enters the river (Walker 2006). As the floodplain has accumulated salts over time, lowering water levels within the weir pools is expected to cause short-term increases in river salinity and would need to be carefully managed to control the rate that stored salts are flushed (Walker 2006).

### Floodplain attenuation

Floodplains can accumulate salt when saline groundwater is accessible to the roots of floodplain vegetation. Although groundwater may be lost through evapotranspiration, the salt it contains is left behind and accumulates over time unless flushed by floods (Elmahdi *et al.* 2008). In this way, floodplains can attenuate the seepage of salts into river systems. The effect is highly variable from site to site, however, depending on hydrology, floodplain width and rates of evapotranspiration (Elmahdi *et al.* 2008). While the attenuation of salt loads by floodplains can help to protect rivers from rising salinity, floodplain health is ultimately compromised.

### Freshwater flows

Freshwater flows contribute to the dilution and flushing of salts that have entered riverine systems. In regulated systems, the frequency of high river flows are often reduced with potential implications for salt balances in affected systems.

### Salt water intrusions

In coastal rivers and wetlands, increasing salinities can be caused by hydrological changes brought about as freshwater flows are significantly reduced due to drought, impoundments and/or the destruction of coastal dune systems or severe modification of low-lying coastal wetlands and mud flats (Knighton *et al.* 1991; Lee *et al.* 2006; Cobb *et al.* 2007). The Murray River mouth in South Australia, Lake Wellington in Victoria, Hawkesbury River in New South Wales, the Alligator Rivers in the Northern Territory, and the Swan River in Western Australia are all susceptible to salt water intrusions. Excessive groundwater pumping in low-lying coastal areas can also cause seawater intrusions, as has occurred on the lower Burdekin aquifer near Townsville in Queensland (Narayan *et al.* 2007).

## IMPACTS OF SALINITY ON PLANT GROWTH

Plants vary widely in their tolerance to salinity, from halophytic species which can tolerate salinities in excess of 300 mm NaCl (~25 dS/m) with only small reductions in growth (Tester and Davenport 2003), to glycophytic species which can show profound growth reductions or mortality at much lower salinities. The effects of salinity on plants are characterised by reductions in height, shoot number, leaf number, leaf size, flowering effort, crop yield and, ultimately, biomass. Even for taxa that are considered to be salt-tolerant, salinity can alter growth without necessarily reducing it (Tester and Davenport 2003). In many species, for instance, salinity causes the root to shoot ratio to decrease (Greenway and Munns 1980; Munns and Termaat 1986). Three key mechanisms contribute to the effects of salinity on plant performance: (i) low external water potential at the roots, (ii) ion toxicity and (iii) nutrient imbalance.

### Low external water potential at roots

The first effects of salinity on plant growth occur due to the presence of salt outside the plant (Munns and Tester 2008). In the soil, the presence of salt lowers the external water potential, causing roots to experience a water deficit which can reduce cell turgor. Although turgor is rapidly restored in many plants, this osmotic stress triggers the roots to send a signal to the shoot that causes a persistent reduction in growth (Munns 1993; Munns and Cramer 1996). The signal sent to the shoot instigates rapid reductions in cell elongation and division as well as stomatal aperture. These changes slow the rate of leaf production and yield small and often thick leaves (Thiel *et al.* 1988; Cramer and Bowman 1991; Yeo *et al.* 1991; Munns 1993; Munns and Tester 2008). Levels of salt above 40 mm NaCl (~3.4 dS/m) are sufficient to trigger this response in most plants, although lower salt levels may trigger the response in salt-sensitive species (Munns and Tester 2008). These effects are not considered to be salt specific, as they can be produced with other osmoticum and occur before high foliar ion concentrations are reached (Munns 1993). The most likely root messengers belong to a group of negative growth regulators called DELLA proteins, which play a central role in coordinating plant growth responses to the environment (Munns and Tester 2008).

## Ion toxicity

In the longer term, further growth reductions will ensue due to the presence of salt within the plant (Munns and Tester 2008). Mechanisms to control sodium (Na) and chlorine (Cl) within the plant impose energy costs and limit growth. Where these mechanisms fail, toxic levels are reached in leaves causing premature senescence. Most plants have the capacity to exclude Na and Cl from the transpiration stream, but this capacity declines as salt concentrations increase. Epidermal and cortical root cells can limit the entry of salts to the xylem by sequestering salts in the vacuole and exporting salts back out to the soil solution (Läuchli *et al.* 2008; Munns and Tester 2008). Despite these control mechanisms, a portion of the Na and Cl will move across the endodermis to stellar cells, which then move into the stellar apoplast and subsequently into the xylem. Once in the xylem, salts may be retrieved back into stellar cells before reaching leaf cells. The capacity to control the uptake of salt at the root, as well as the ability to retrieve salts from the xylem contributes significantly to salt tolerance (Munns and Tester 2008).

Once salts have reached leaf cells, sensitive cytoplasmic enzymes are protected by the storage of Na and Cl in the vacuole (Greenway and Munns 1980). For a variety of agricultural plants, Na appears to be sequestered in the vacuole via a Na/H antiporter in the tonoplast (Blumwald *et al.* 1987; Martinez and Läuchli 1993; Reuveni 1993; Ballesteros *et al.* 1997). Greenway and Munns (1980) have speculated that species highly sensitive to NaCl may lack the capacity to sequester ions into the vacuole. This has been demonstrated in *Plantago* where Na/H antiport activity at the tonoplast was found in *P. maritimus*, a salt-tolerant species, but not in *P. media*, a salt-sensitive species (Staal *et al.* 1991).

Compartmentalisation of Na and Cl within the vacuole will increase its osmotic potential, necessitating osmotic adjustment of the cytoplasm. This is achieved with non-toxic organic compounds, typically sucrose, proline and glycinebetaine (Munns and Tester 2008). In halophytic aquatic plants such as *Ruppia tuberosa* a degree of osmoregulation is achieved via the accumulation of proline (Brock 1981b). In *Phragmites australis* (Zhen and Ma 2009) proline also acts in osmoregulation but only at salinities of <200 mM NaCl. Salt-sensitive aquatics such as *Salvinia natans* accumulate low concentrations of proline, which plays a very minor role in osmoregulation (Jampeetong and Brix 2009). The contribution of glycinebetaine in osmotic adjustment is more convincing than proline, which tends to accumulate only under severe stress (Greenway and Munns 1980). The highest concentrations of glycinebetaine are found in halophytes. In glycophytes, the level of glycinebetaine present is to some extent reflective of a species' sensitivity to NaCl (Storey and Wyn Jones 1977). Although Na and Cl are compartmentalised within vacuoles, their sink capacity will eventually be exhausted. Ions will then become concentrated in the cytoplasm or the cell wall resulting in toxicity and desiccation, respectively. Cell death ensues and the longevity of leaves is shortened (Oertli 1968; Greenway and Munns 1980; Yeo and Flowers 1986).

## Nutrient imbalance

Salinity can interfere with nutrient acquisition in several ways: by interfering with nutrient uptake processes (Lynch *et al.* 1987; Grattan and Grieve 1992), by inhibiting the growth of roots and mycorrhiza, or by altering soil structure (Tester and Davenport 2003). Here we discuss the significance of calcium (Ca), potassium (K) and nitrogen (N) in the response of vegetation to salinity.

Calcium is required in plants for cell division, cell wall development, and as a secondary messenger (Marschner 1995). Calcium is also bound to cell membranes where it plays an essential role in maintaining the integrity of cell membrane transport systems. Under saline conditions, Na ions can displace Ca that is bound in cell membranes, impairing ion selectivity. As a result, more Na enters the cell, hastening the onset of ion toxicity (LaHaye and Epstein 1969; Cramer *et al.* 1985; Lynch and Läuchli 1988; Zidan *et al.* 1991). In some plants, the displacement of membrane-bound Ca can also cause leakage of cytosolic K which may lead to K deficiencies (Epstein 1961; LaHaye and Epstein 1969). The effects of Ca imbalance produced by salinity are most pronounced in experiments where little or no Ca is added, and are minimised if the Na:Ca ratio is kept below 17, a level not uncommon is saline soils (LaHaye and Epstein 1969; Greenway and Munns 1980).

Plants require K to maintain cell turgor and hence cell expansion, and as a co-factor for many enzymes. Exposure to NaCl has been found to reduce foliar K concentrations in some plants (Storey and Wyn Jones 1978; Ball *et al.* 1987; Bohra and Doerffling 1993). Low tissue K concentrations occur because Na competes with K for uptake sites on the plasma membrane (Epstein 1961) and because ion selectivity is impaired, causing leakage of cytosolic K. The synthesis of the D1 protein, an integral component of photosystem II (PS II), appears to be affected by low foliar K in the mangrove *Avicennia marina*, giving rise to reductions in quantum yield (Ball *et al.* 1987). The ameliorative effect of enhanced K nutrition in plants subjected to saline conditions has been demonstrated in several agricultural species (Helal *et al.* 1975; Bohra and Doerffling 1993; Garg *et al.* 1993).

Nitrogen assimilation is not only essential for plant growth and development, but also in adaptation to salinity stress (Aslam *et al.* 1984; Mansour 2000). For plants that use glycinebetaine or proline in adaptation to saline conditions, the demand for N may increase significantly as these compounds contain 12% N (Cavalieri 1983). Salinity has also been found to inhibit the uptake of $NO_3$ and/or $NH_4$ (Helal *et al.* 1975; Aslam *et al.* 1984; Bradley and Morris 1991; Hawkins and Lewis 1993a, 1993b; Lea-Cox and Syvertsen 1993). Where the requirement for N is increased by salinity, or where the uptake of N is competitively inhibited, plant performance may benefit from N fertilisation. Field studies have found growth of *Spartina alterniflora* to be enhanced by N application (Gallagher 1975; Buresh *et al.* 1980), but this can be dependent on phosphorus (P) supply (Broome *et al.* 1975).

## Impacts on whole plant performance

Plants grow when the amount of carbon assimilated by the shoots, a function of the total leaf area and the rate of photosynthesis, exceeds the respiratory demands of the whole plant. Salinity can lower leaf area by reducing leaf expansion, resulting in smaller leaves, and by increasing the rate of leaf senescence, reducing the total number of leaves.

Reductions in photosynthesis under saline conditions have been reported for both glycophytes and halophytes (Downton 1977; Ball and Farquhar 1984; Seemann and Critchley 1985; Yeo *et al.* 1985; Pezeshki and DeLaune 1993; Walker *et al.* 1993). Such reductions may be mediated by effects of salinity on the biochemical apparatus and via stomatal limitation of $CO_2$ uptake. Stomatal conductance is frequently reduced by salinity, mediated initially by the rapid synthesis of abscisic acid in leaf tissue upon exposure to salinity. Abscisic acid concentration, however, declines within hours of initial exposure to salinity and it is thought that root signals are responsible for the persistent reduction in stomatal conductance (Munns and Tester 2008). Reductions in stomatal conductance can limit the flux of $CO_2$ into the leaf, leading to a decline in photosynthesis. Munns (1993) also proposed that lower rates of photosynthesis may arise from feedback inhibition in response to the accumulation of carbohydrates in leaves, which probably occurs as a result of salinity blocking their utilisation (Munns 1993).

Maintenance respiration is the respiration that occurs when growth rate is zero, i.e. the cost of re-synthesising degraded biomass (Schwarz and Gale 1981). With greater maintenance respiration, less carbon is available for growth. Salinity is thought to increase maintenance respiration by increasing the demand for vacuolar compartmentalisation, secretion and repair processes (Greenway and Munns 1980; Schwarz and Gale 1981). Schwarz and Gale (1981) found that maintenance respiration increased with rising salinity at non-toxic levels. Once salinities produced symptoms of toxicity, however, maintenance respiration declined.

## Impacts on freshwater macrophytes

Hart *et al.* (1991) reviewed the effects of increasing salinities on Australian freshwater systems and concluded that freshwater biota would be negatively impacted if salinities reached 1.5 dS/m (1000 mg/L). Field surveys carried out by Brock and colleagues (Brock 1981a; Brock and Lane 1983; Brock and Shiel 1983) in south-east South Australia and south-west Western Australia showed that aquatic plant species more commonly associated with freshwater habitats tend to be absent from sites with salinities in excess of 5.9 dS/m (4000 mg/L), although some species were absent when salinities exceeded 1.5 dS/m (1000 mg/L). As salinities increased, species diversity declined and once salinities exceeded 14.7 dS/m (10 000 mg/L) only halophytic species (*Ruppia* spp. and *Lepilaena* spp.) were recorded.

More recently, these salinity thresholds have been supported by vegetation surveys of 58 wetlands varying in salinity from 0.06 to 173.9 dS/m (40–118 000 mg/L) in the Wimmera region of western Victoria (Smith *et al.* 2009). In Smith *et al.*'s (2009) study, a Bayesian modelling approach was used to assess the salt sensitivity of the aquatic plant taxa surveyed. The results suggested that 25 of the 76 freshwater species recorded would be unlikely to occur above 1 dS/m (680 mg/L), signifying a very low threshold for species loss. A series of seed bank studies examining the emergence and survival of freshwater plant communities under a range of salinities demonstrate the same trend (Nielsen *et al.* 2003; Brock *et al.* 2005; Nielsen *et al.* 2008). At 1.5 dS/m (1000 mg/L), the number of plants emerging from the soil seed bank declined and once salinity reached 7.3 dS/m (5000 mg/L) very few aquatic plants persisted.

Although many aquatic plants may be restricted to sites where salinities are less than 4000 mg/L (~5.9 dS/m), there are a range of non-halophytic aquatic species that are tolerant of higher salinities. For example, Brock and Lane (1983) found that *Lemna minor* and *Stuckenia pectinata* were present at ephemeral sites where salinities ranged from 7.3 to 10.3 dS/m (5000–7000 mg/L). In salinity trials, Warwick and Bailey (1997) found that *Amphibromus fluitans* was unaffected by salinities of 8.8 dS/m (6000 mg/L), and Salter *et al.* (2008) report that *Vallisneria australis* could persist at salinities of 18 dS/m (11 500 mg/L) for at least 26 weeks. Other species, such as *Phragmites australis* and *Melaleuca ericifolia* have a high salt tolerance, respectively 22–39.7 dS/m (Lissner and Schierup 1997; Adams and Bate 1999; Talley and Levin 2001) and 37–44 dS/m (Bird 1962; Salter *et al.* 2007).

The long-term persistence of plant populations under saline conditions is determined by the salt tolerance of each phase of the life cycle. Although many aquatic plants persist by clonal growth, new populations establish from vegetative fragments and seed, and these life stages may differ in salinity tolerance to that of mature plants. Only a few studies have compared the salinity tolerance of a species at a range of life stages. Choudhuri (1968) found that five steppe species, including *Typha latifolia*, were more sensitive to salinity at germination than at the seedling stage. Similarly, Zedler *et al.* (1990) found salinity tolerance increased with plant age in both *Juncus kraussii* and *Typha orientalis*. This trend is also apparent among more salt-tolerant taxa. *Melaleuca ericifolia*, for example, persists at sites with salinities ranging from 37 to 44 dS/m (25–30 g/L; Bird 1962) while less than 15% of seeds germinate at 23.5 dS/m (16 g/L) and no germination occurs at 47 dS/m (32 g/L; Robinson *et al.* 2006). Sensitivity of earlier life stage to salinity suggests that establishment in saline regions will depend on periods of low salinity which permit germination and seedling establishment. Differences in salinity tolerance exhibited at each life history stage can also shift the competitive balance between species, giving rise to compositional changes in aquatic communities.

## Salinity tolerance and avoidance

Plants may persist in saline environments because they have mechanisms that impart tolerance or because they have strategies to avoid high salinity. Salinity tolerance is determined by a plant's capacity to tolerate osmotic stress, limit Na transport to its leaves, and to compartmentalise Na (and sometimes Cl) within the plant, limiting its toxic effects on cellular function (Munns and Tester 2008). Variation in salinity tolerance is observed not only between species but also within a species. Considerable variability in the salt tolerance of seedlings has been reported between populations of *Melaleuca thyoides* (van der Moezel and Bell 1987) and *Eucalyptus camaldulensis* (Heth and MacRae 1993; Akilan *et al.* 1997). Seeds sourced from a *M. thyoides* population growing in saline soil were more salt tolerant than seeds sourced from populations growing in non-saline soil (van der Moezel and Bell 1987). *Phragmites communis* plants grown from seed sourced from a salt mine demonstrated high salinity tolerance, with growth unaffected at 300 mm NaCl (25 dS/m). Salt tolerance was attributed to the re-translocation of Na from the shoot to the root maintaining low foliar Na concentrations (Matoh *et al.* 1988). Exploiting genetic variability in salt tolerance within species may prove a useful tool in restoring vegetation in salt-affected areas.

Networks of interconnected ramets that span across salinity gradients may permit access to less saline water, allowing persistence beyond the normal tolerance range. Evans and Whitney (1992) demonstrated that ramets of *Hydrocotyle bonarieni* (a sand dune perennial) that were connected to other ramets grown under non-saline conditions were able to proliferate despite localised salt exposure and did not demonstrate any visible effects of salinity. In contrast, where the connection with the non-salinised ramet was severed, high mortality was demonstrated under saline conditions. A similar response has also been observed in *Spartina patens* by Hester *et al.* (1994). These findings suggest that where salinity is spatially heterogeneous, connections between ramets can prove beneficial to plant growth (Evans and Whitney 1992; Hester *et al.* 1994).

For flood-tolerant trees associated with riparian zones and aquatic habitats, salinity may be avoided by accessing alternative water sources of lower salinity. Kennedy *et al.* (2003) report that comparisons between Na:K ratios of river and soil water with foliar ratios of the two willow species common along the lower Murray River indicate that river water is used in preference to soil water. Mensforth *et al.* (1994) identified water sources of *Eucalyptus camaldulensis* growing within 15 km of a river by comparing isotopic signatures of water extracted for the xylem tissue and soil, as well as shoot and soil water potentials. The study found that *E. camaldulensis* accessed river water in summer, avoiding the highly saline groundwater.

Many aquatic macrophytes have underground organs that store carbohydrates and enable them to remain dormant during unfavourable conditions, particularly during drying events. These structures may also impart some resilience to salinity. Salinity was found to alter biomass allocation patterns in *Bolboschoenus medianus* (Morris and Ganf 2001) and *Baumea arthrophylla* (Morris 1998), with rising salinity resulting in shifts in biomass allocation from culms to tubers and from culms to rhizomes, respectively. Such responses may ensure sufficient carbohydrate stores are acquired to permit persistence during winter dormancy and/or when conditions become more favourable. A similar response was found in *Vallisneria australis*, which produced smaller leaves but more ramets when grown at 18 dS/m (Salter *et al.* 2008). Favouring ramet production over increases in leaf length may enable plants to locate areas of lower salinity.

## Interactions between salinity and other environmental drivers

Plant responses to salinity are modified by a wide range of environmental factors including temperature, water regime, soil pH and nutrient availability. In many regions, salinisation and water logging co-occur due to the presence of shallow water tables, impermeable soils or seawater intrusion (Barrett-Lennard 2003; Halse *et al.* 2003). Under saline conditions, flooding increases the flux of Na and Cl to leaves increasing salt sensitivity. Salt uptake is thought to increase under flooded conditions because energy-dependant membrane transport mechanisms in the root are impaired by oxygen deficits (Barrett-Lennard 2003). Negative interactions between flooding and salinity have been demonstrated for several emergent plants and woody wetland trees (Froend *et al.* 1987; Akilan *et al.* 1997; Kozlowski 1997; Barrett-Lennard 2003; Halse *et al.* 2003; Salter *et al.* 2007, 2008). Carter *et al.* (2006) found that under saline conditions foliar Na and Cl concentrations were higher in *Casuarina obesa* in flooded than in drained conditions. In contrast, under saline conditions foliar Na and Cl concentration did not differ between flooded and well-drained *Melaleuca cuticularis* plants. The ability of *M. cutucularis* plants to produce aerenchyma and adventitious roots may improve root aeration and hence ion selectivity (Carter *et al.* 2006).

The capacity of freshwater macrophytes to tolerate fluctuations in the frequency and duration of wetting and drying cycles that characterise many Australian wetlands are also likely to be modified by salinity. Salter *et al.* (2008) tested the effect of increasing salinity on the capacity of the wetland tree *Melaleuca ericifolia* and the submerged monocot *Vallisneria australis* to tolerate flooding and drying, respectively (i.e. water regimes considered unsuited to their growth form). For both species, the capacity to tolerate the unfavourable water regime was reduced by salinity. For example, under fresh conditions some *V. australis* plants regenerated from underground stolons after soil was dried, but under saline conditions, drying the soil completely inhibited regeneration.

In some riverine habitats, salinisation is coupled with increased nutrient levels. Morris (1998) examined the interaction between salinity and nutrients and found that nutrient enrichment enhanced growth under fresh and saline conditions in the fast-growing species *Bolboschoenus medianus* and *Typha domingensis*, but not in the slow-growing species *Baumea arthrophylla*. At high levels of salinity the growth responses of *T. domingensis* and *B. medianus* to fertilisation diminished, but no negative effects of fertilisation were observed, even at the highest loading. Similarly, Macek and Rejmánková (2007) found that *Typha domingensis* and *Eleocharis cellulosa* showed a strong growth increase in response to nutrient addition, while *Cladium jamaicense*, a stress-tolerant species, did not. The results indicate that the growth of *Typha* would be constrained under low levels of P and high salinity, thus preventing the competitive exclusion of co-occurring species such as *C. jamaicense* (Macek and Rejmánková 2007).

## MANAGEMENT IMPLICATIONS

### Trends in salinisation in Australia

It is only over the past two decades that salinity has become recognised as a national problem in Australia, although it was documented in Western Australia 85 years ago by Wood (1924). This was partly because the salinity data required to provide a comprehensive audit were lacking.

The approach taken to this nationwide problem involves assessment of trends in river salinity via the application of catchment salt balances in relation to a range of physical processes such as rainfall, river flow and the extent of vegetation clearance. The influence of vegetation clearance on stream salinity was amply demonstrated by Peck and Hatton (2003). The Blackwood River catchment was cleared extensively whereas the Donnelly catchment was largely uncleared – in the 1940s the Blackwood River's salinity was <500 μS/cm but increased to >2000 μS/cm by the 1980s whereas the Donnelly's salinity remained below 300 μS/cm.

Whether a catchment is accumulating or losing salt is estimated by calculating the salt export ratio (E/I). Values <1 indicate salt is accumulating whereas values >1 suggest salt is exported from the system.

Superficially this seems a simple calculation, but Biggs *et al.* (2013) point out that the absolute values of E/I vary depending on the availability of data, the magnitude of base flow and the method of calculation used to estimate E/I ratios. Using this approach, Biggs *et al.* (2013) conclude that in contrast with the lower Murray–Darling Basin, the Queensland Murray–Darling Basin generally accumulates salts, although the eastern upland catchments are generally net salt exporters because base flow occurs and stream salinity is higher as a result of long-term land use.

Jolly *et al.* (2001) undertook an extensive survey of historical stream salinity trends and catchment salt balances across the Murray–Darling Basin, in four broad geographical regions. The northern and western dryland showed no significant salt imbalances; the southern and eastern dryland had significantly rising stream salinity trends in 50% of the tributaries, but the change in the salt balance was minimal in areas with rainfall <800 mm per year compared with areas with rainfall of 500–800 mm per year where trends were significant and rising. Irrigation regions occur on the lower reaches of rivers with an annual rainfall of <500 mm. Over 40% of these streams had increasing salinity while none exhibited a decreasing trend. The lower Murray region showed either slightly decreasing or no significant stream salinity trends. The mean salinities at 90 stations throughout the four regions (Jolly *et al.* 2001) indicate that the river water salinity is unlikely to represent a current threat to aquatic and/or riparian vegetation.

Elmahdi *et al.* (2008) integrated models of groundwater, floodplain salinity and surface water to provide a predictive analysis of the salinity levels at Morgan, South Australia, 315 km upstream of the Murray mouth. The results suggested that salinity was likely to rise by 200 to 400 µS/cm by 2057. An increase at Morgan of 400 µS/cm above 800 µS/cm, the current daily target for 95% of the time (MDBA 2012), would be detrimental to aquatic flora and fauna (Smith *et al.* 2009). However, a comparison of the 1975 modelled 'no further remedial action' daily salinities at Morgan with observed salinities (1986–2011) showed that the cumulative benefits of salt interception works and changed irrigation practices, particularly during periods of low flows, amounted to a reduction in salinity of between 0 and 672 µS/cm for 2011 (MDBA 2012). Indeed, during the drought years 2006 to 2010 the modelled no intervention scenario predicted salinities at Morgan would reach 1600 µS/cm compared with the maximum observed of ~780 µS/cm. Landscape-wide salt mitigation works and practices are a weapon against the likelihood of an increase in secondary salinisation as the climate becomes more variable.

## Management approaches

Strategies to manage salinity fall into two broad categories: (1) those aimed at reducing salinity levels in the environment and (2) those that focus on ways to live with elevated salinity.

### Strategies to reduce salinity

Management strategies to reduce salinity in the environment primarily aim to reduce saline groundwater levels through: (1) revegetating groundwater recharge and transmission zones with salt-tolerant trees or other perennial, high-water-use vegetation; (2) preventing further clearing of vegetation; (3) adopting efficient water irrigation practices; and (4) preventing water seepage from water supply channels (MDBMC 2001; Pannell and Ewing 2006). The responsiveness of the water table to management interventions is variable, and achieving catchment-scale changes is difficult. For example, in many cases more than 50% of land needs to be under perennial vegetation to lower water tables at the catchment scale (George *et al.* 1999; Pannell and Ewing 2006).

While widespread tree planting has the potential to lower groundwater tables and reduce saline groundwater discharge to streams, in some circumstance it may also reduce fresh surface inflows and result in an initial increase in salinity (Heron *et al.* 2003; Campbell 2008). As such, large-scale plantings need to be carefully planned and sited to avoid substantial reductions in streamflow in drier salt-affected catchments (Campbell 2008).

Where saline groundwater tables cannot be lowered by managing land use, or where there is a need to achieve a more immediate reduction in salinity, salt interception and disposal schemes are implemented. Salt interception works pump or drain saline groundwater and dispose of the salt water, typically in evaporation basins. Several salt interception works have been implemented along the Murray River since 1988. Each year it is estimated that this scheme prevents more than half a million tonnes of salt entering the Murray River each year (MDBA 2010). For riverine systems, salinity management activities also aim to dilute salts through the delivery of environmental flows and to protect water inputs originating from high rainfall areas (MDBA 2010).

In coastal areas, when river flows are substantially reduced by river regulation, seawater intrusions in lower river reaches can migrate significant distances upstream, particularly during periods of prolonged low rainfall. The release of environmental flows from upstream water storages can help reduce the upstream extent of saline intrusions and freshen the top layers of the water column. In saline river reaches, where water is artificially delivered to floodplain wetlands through water off-takes from the river, management strategies should be aimed at ensuring off-takes are placed above stratified saline layers or by delivering fresh flows to dilute salts in the river before delivering river water to the wetland. Where freshwater aquifers are hydraulically connected to seawater, the over-extraction of groundwater for irrigation can lead to seawater intrusion. In these cases salinity management aims to artificially replenish groundwater to control seawater intrusion (Narayan *et al.* 2007).

For wetlands where the main source of salinity is due to shallow saline groundwater, unnaturally long periods of drying should be avoided to prevent increases in soil salinity. During periods of high water availability, draining saline water from the wetland and replacing it with freshwater can reduce the salt load in the wetland.

### Living with salt

In some instances it may not be possible to reduce salinity levels and management may instead focus on managing the transition to a more saline state to optimise production and/or biodiversity values (O'Connell *et al.* 2006; Pannell and Roberts 2010). These approaches can include: (1) planting with salt-tolerant species, varieties or genotypes; (2) amelioration of sodic and saline-sodic soils with gypsum; and (3) revegetating on mounds at sites prone to waterlogging (Ritson and Pettit 1992).

## CONCLUSIONS

Land salinisation and its impact on riverine systems is a pervasive and complex environmental problem in Australia. Salinity is difficult to manage due to the scale of intervention required to lower groundwater and the variability in the responsiveness of different groundwater flow systems to interventions, with some systems responding over decades to centuries. Climate change adds to the uncertainty and complexity of managing salinity. In areas where climate change produces a warmer and drier climate, groundwater levels will fall, reducing the transport of salts and lessening salinity. However, water inflow to streams will also decline resulting in increased stream salinity. In the Murrumbidgee, for example, models predict that under a hotter and drier climate water yields will drop as much as 48% by 2070, reducing salt accumulation on the land surface by ~30%. However, reductions in water inflow under a drier scenario are expected to increase stream salinity concentrations by 11% (Jenet *et al.* 2006; Campbell 2008).

Salinisation of rivers is a threat to aquatic biodiversity worldwide (Cañedo-Argüelles *et al.* 2013). Because salinity tolerance differs among the life stages of a species and among species (Kefford *et al.* 2007), further research is required to identify the salinity tolerance of life stages of riverine biota and how this interacts with other environmental variables such as temperature, flow, nutrients and turbidity. Although high salinities are demonised, it is worth remembering that there are a group of highly specialised aquatic organisms that have evolved mechanisms to overcome even extreme salinities and their role in the ecosystem and the habitats that support them should not be forgotten.

Where rising salinity cannot be contained and salinity levels exceed the tolerance of freshwater biota it is these very salt-tolerant organisms that we will look to, to colonise and enhance the biodiversity and production values of these degraded systems (Nielsen and Brock 2009).

## REFERENCES


Adams JB, Bate GC (1999) Growth and photosynthetic performance of *Phragmites australis* in estuarine waters: a field and experimental evaluation. *Aquatic Botany* **64**, 359–367. doi:10.1016/S0304-3770(99)00063-7

Akilan K, Farrell RCC, Bell DT, Marshall JK (1997) Responses of clonal river red gum (*Eucalyptus camaldulensis*) to waterlogging by fresh and salt water. *Australian Journal of Experimental Agriculture* **37**, 243–248. doi:10.1071/EA96072

Aragüés R (1982) Factors affecting the behaviour of soil salinity sensors. *Irrigation Science* **3**, 133–147. doi:10.1007/BF00264856

Aslam M, Huffaker RC, Rains DW (1984) Early effects of salinity on nitrate assimilation in barley seedlings. *Plant Physiology* **76**, 321–325. doi:10.1104/pp.76.2.321

Ball MC, Farquhar GD (1984) Photosynthetic and stomatal responses of two mangrove species, *Aegiceras corniculatum* and *Avicennia marina*, to long term salinity and humidity conditions. *Plant Physiology* **74**, 1–6. doi:10.1104/pp.74.1.1

Ball MC, Chow WS, Anderson JM (1987) Salinity-induced potassium deficiency causes loss of functional photosystem II in leaves of the grey mangrove, *Avicennia marina*, through depletion of the atrazine-binding polypeptide. *Australian Journal of Plant Physiology* **14**, 351–361. doi:10.1071/PP9870351

Ballesteros E, Blumwald E, Donaire JP, Belver A (1997) $Na^+/H^+$ antiport activity in tonoplast vesicles isolated from sunflower roots induced by NaCl stress. *Physiologia Plantarum* **99**, 328–334. doi:10.1111/j.1399-3054.1997.tb05420.x

Barrett-Lennard EG (2003) The interaction between waterlogging and salinity in higher plants: causes, consequences and implications. *Plant and Soil* **253**, 35–54.

Bayly IAE (1976) The plankton of Lake Eyre. *Australian Journal of Marine and Freshwater Research* **27**, 661–665. doi:10.1071/MF9760661

Beatty SJ, Morgan DL, Rashnavadi M, Lymbery AJ (2011) Salinity tolerances of endemic freshwater fishes of south-western Australia: implications for conservation in a biodiversity hotspot. *Marine and Freshwater Research* **62**, 91–100.

Biggs AJW, Silburn DM, Power RE (2013) Catchment salt balances in the Queensland Murray–Darling Basin Australia. *Journal of Hydrology* **500**, 104–113. doi:10.1016/j.jhydrol.2013.07.027

Bird ECF (1962) The swamp paper-bark. *The Victorian Naturalist* **79**, 72–81.

Blumwald E, Cragoe EJ, Jr, Poole RJ (1987) Inhibition of $Na^+/H^+$ antiport activity in sugar beet tonoplast by analogs of amiloride. *Plant Physiology* **85**, 30–33. doi:10.1104/pp.85.1.30

Bohra JS, Doerffling K (1993) Potassium nutrition of rice *Oryza sativa* L. varieties under sodium chloride salinity. *Plant and Soil* **152**(2), 299–303. doi:10.1007/BF00029100

Bradley PM, Morris JT (1991) Relative importance of ion exclusion, secretion and accumulation in *Spartina alterniflora* Loisel. *Journal of Experimental Botany* **42**(12), 1525–1532. doi:10.1093/jxb/42.12.1525

Brock MA (1981a) The ecology of halophytes of the south-east of South Australia. *Hydrobiologia* **81, 82**, 23–32. doi:10.1007/BF00048703

Brock MA (1981b) Accumulation of proline in a submerged aquatic halophyte, *Ruppia* L. *Oecologia* **51**, 217–219. doi:10.1007/BF00540604

Brock MA, Lane JAK (1983) The aquatic macrophyte flora of saline wetlands in Western Australia in relation to salinity and permanence. *Hydrobiologia* **105**, 63–76. doi:10.1007/BF00025177

Brock MA, Nielsen DL, Crosslé K (2005) Changes in biotic communities developing from freshwater wetland sediments under experimental salinity and water regimes. *Freshwater Biology* **50**, 1376–1390. doi:10.1111/j.1365-2427.2005.01408.x

Brock MA, Shiel RJ (1983) The composition of aquatic communities on saline wetlands in

Western Australia. *Hydrobiologia* **105**, 77–84. doi:10.1007/BF00025178

Brookes JD, Lamontagne S, Aldridge KT, Benger S, Bissett A, Bucater L, Cheshire AC, Cook PLM, Deegan BM, Dittmann S, Fairweather PG, Fernandes MB, Ford PW, Geddes MC, Gillanders BM, Grigg NJ, Haese RR, Krull E, Langley RA, Lester RE, Loo M, Munro AR, Noell CJ, Nayar S, Paton DC, Revill AT, Rogers DJ, Rolston A, Sharma SK, Short DA, Tanner JE, Webster IT, Wellman NR, Ye Q (2009) 'An ecosystem assessment framework to guide management of the Coorong.' Final report of the The Coorong, Lower Lakes and Murray Mouth ecology research cluster. CSIRO, Water for a Healthy Country National Research Flagship, Canberra.

Broome SW, Woodhouse WW, Seneca ED (1975) The relationship of mineral nutrients to growth of *Spartina alterniflora* in North Carolina: II. The effects of N, P, and Fe fertilizers. *Soil Science Society of American Proceedings* **39**, 301–307. doi:10.2136/sssaj1975.03615995003900020022x

Buresh RJ, DeLaune RD, Patrick WH, Jr (1980) Nitrogen and phosphorus distribution and utilization by *Sparina alterniflora* in a Louisiana gulf coast marsh. *Estuaries* **3**(2), 111–121. doi:10.2307/1351555

Campbell A (2008) *Managing Australian Landscapes in a Changing Climate: a Climate Change primer for Regional Natural Resource Management Bodies.* Department of the Environment, Water, Heritage and the Arts, Canberra, Australia.

Cañedo-Argüelles M, Kefford BJ, Piscart C, Prat N, Schäfer RB, Schulz C-J (2013) Salinisation of rivers: an urgent ecological issue. *Environmental Pollution* **173**, 157–167. doi:10.1016/j.envpol.2012.10.011

Carter JL, Colmer TD, Veneklaas EJ (2006) Variable tolerance of wetland tree species to combined salinity and waterlogging is related to regulation of ion uptake and production of organic solutes. *New Phytologist* **169**, 123–134. doi:10.1111/j.1469-8137.2005.01552.x

Cavalieri AJ (1983) Proline and glycinebetaine accumulation by *Spartina alterniflora* Loisel. in response to NaCl and nitrogen in a controlled environment. *Oecologia* **57**, 20–24. doi:10.1007/BF00379556

Choudhuri GN (1968) Effect of soil salinity on germination and survival of some steppe plants in Washington. *Ecology* **49**, 465–471. doi:10.2307/1934113

Cobb SM, Saynor MJ, Eliot M, Eliot I, Hall R (2007) 'Saltwater intrusion and mangrove encroachment of coastal wetlands in the Alligator Rivers Region, Northern Territory, Australia.' Supervising scientist report 191, Darwin.

Cramer GR, Bowman DC (1991) Kinetics of maize leaf elongation. *Journal of Experimental Botany* **42**(11), 1417–1426. doi:10.1093/jxb/42.11.1417

Cramer GR, Läuchli A, Polito VS (1985) Displacement of $Ca^{2+}$ by $Na^{+}$ from the plasmalemma of root cells. *Plant Physiology* **79**, 207–211. doi:10.1104/pp.79.1.207

Cresswell RG (2008) Cresswell drives salinity! A personal view of dryland salinity in Australia. In *2nd International Salinity Forum: Salinity, Water and Society – Global Issues, Local Action*. 31 March to 3 April, Adelaide.

Cresswell RG, Mullen IC, Kingham R, Kellett J, Dent DL, Jones GL (2007) Airborne electromagnetics supporting salinity and natural resource management decisions at the field scale in Australia. *International Journal of Applied Earth Observation and Geoinformation* **9**, 91–102. doi:10.1016/j.jag.2006.05.005

Diamond J (2005) *Collapse: How Societies Choose to Fail or Succeed.* Revised edn. Penguin, New York, NY.

Dick J, Haynes D, Tibby J, Garcia A, Gell P (2011) A history of aquatic plants in the Coorong, a Ramsar-listed coastal wetland, South Australia. *Journal of Paleolimnology* **46**, 623–635. doi:10.1007/s10933-011-9510-4

Downton WJS (1977) Photosynthesis in salt stressed grapevines. *Australian Journal of Plant Physiology* **4**, 183–192. doi:10.1071/PP9770183

Eberbach PL (2003) The eco-hydrology of partly cleared, native ecosystems in southern Australia: a review. *Plant and Soil* **257**, 357–369. doi:10.1023/A:1027392703312

Elmahdi A, Connor J, Doble R, Stenson M, Kirby M, Walker G (2008) Forecasting lower Murray River salinity under climatic uncertainty and reduced dilution flow. In *2nd International Salinity Forum: Salinity, Water and Society – Global Issues, Local Action.* 31 March to 3 April, Adelaide. pp. 1–5.

Epstein E (1961) The essential role of calcium in selective cation transport by plant cells. *Plant Physiology* **36**, 437–444. doi:10.1104/pp.36.4.437

Evans JP, Whitney S (1992) Clonal integration across a salt gradient by a nonhalophyte *Hydrocotyle bonariensis* (Apicaceae). *American Journal of Botany* **79**(12), 1344–1347. doi:10.2307/2445132

Frahn K, Nicol J, Strawbridge A (2012) 'Current distribution and abundance of *Ruppia tuberosa* in the Coorong, South Australia.' SARDI publication no. F2012/000074–1. SARDI research reports series no. 625. South Australian Research and Development Institute (Aquatic Sciences), Adelaide.

Froend RH, Heddle EM, Bell DT, McComb AJ (1987) Effects of salinity and waterlogging on the vegetation of Lake Toolibin, Western Australia. *Australian Journal of Ecology* **12**, 281–298. doi:10.1111/j.1442-9993.1987.tb00949.x

Gallagher JL (1975) Effects of an ammonium nitrate pulse on the growth and elemental composition of natural stands of *Spartina alterniflora* and *Juncus roemerianus*. *American Journal of Botany* **62**(6), 644–648. doi:10.2307/2441945

Garg BK, Vyas SP, Kathju S, Lahiri AN, Mali PC, Sharma PC (1993) Salinity-fertility interaction on growth, mineral composition and nitrogen metabolism of Indian mustard. *Journal of Plant Nutrition* **16**(9), 1637–1650. doi:10.1080/01904169309364639

Gelburd DE (1985) Managing salinity lessons from the past. *Journal of Soil and Water Conservation* **40**, 329–331.

George RJ, Nulsen RA, Ferdowsian R, Raper GP (1999) Interactions between trees and groundwaters in recharge and discharge areas – a survey of Western Australian sites. *Agricultural Water Management* **39**, 91–113.

Grattan SR, Grieve CM (1992) Mineral element acquisition and growth response of plants grown in saline environments. *Agriculture, Ecosystems & Environment* **38**(4), 275–300. doi:10.1016/0167-8809(92)90151-Z

Greenway H, Munns R (1980) Mechanisms of salt tolerance in nonhalophytes. *Annual Review of Plant Physiology* **31**, 149–190. doi:10.1146/annurev.pp.31.060180.001053

Halse SA, Ruprecht JK, Pinder AM (2003) Salinisation and prospects for biodiversity in rivers and wetlands of south-west Western Australia. *Australian Journal of Botany* **51**(6), 673–688. doi:10.1071/BT02113

Hart BT, Bailey P, Edwards R, Hortle K, James K, McMahon A, Meredith C, Swadling K (1991) A review of the salt sensitivity of the Australian freshwater biota. *Hydrobiologia* **210**, 105–144. doi:10.1007/BF00014327

Hart BT, Lake PS, Webb JA, Grace MR (2003) Ecological risk to aquatic ecosystems from salinity increases. *Australian Journal of Botany* **51**, 689–702. doi:10.1071/BT02111

Hawkins HJ, Lewis OAM (1993a) Combination effect of sodium chloride salinity, nitrogen form and calcium concentration on the growth, ionic content and gaseous exchange properties of *Triticum aestivum* L. cv. Gamtoos. *New Phytologist* **124**(1), 161–170. doi:10.1111/j.1469-8137.1993.tb03806.x

Hawkins HJ, Lewis OAM (1993b) Effects of NaCl salinity, nitrogen form, calcium and potassium concentration on nitrogen uptake and kinetics in *Triticum aestivum* L. cv. Gamtoos. *New Phytologist* **124**(1), 171–177. doi:10.1111/j.1469-8137.1993.tb03807.x

Helal M, Koch K, Mengal K (1975) Effect of salinity and potassium on the uptake of nitrogen and on nitrogen metabolism in young barley plants. *Physiologia Plantarum* **35**, 310–313. doi:10.1111/j.1399-3054.1975.tb03911.x

Herczeg AL, Dogramacia SS, Leaney FWJ (2001) Origin of dissolved salts in a large, semi-arid groundwater system: Murray Basin, Australia. *Marine and Freshwater Research* **52**, 41–52. doi:10.1071/MF00040

Herron N, Davis R, Dawes W, Evans R (2003) Modelling the impacts of strategic tree plantings on salt loads and flows in the Macquarie River Catchment, NSW, Australia. *Journal of Environmental Management* **68**(1), 37–50.

Hester MW, McKee KL, Burdock DM, Koch MS, Flynn KM, Patterson S, Mendelssohn IA (1994) Clonal integration in *Spartina patens* across a nitrogen and salinity gradient. *Canadian Journal of Botany* **72**, 767–770. doi:10.1139/b94-096

Heth D, MacRae S (1993) Selection of salt tolerant clones of *Eucalyptus camaldulensis* Dehn. *South African Forestry Journal* **164**, 21–26. doi:10.1080/00382167.1993.9629373

Jacobsen T, Adams RM (1958) Salt and silt in ancient Mesopotamian agriculture. *Science* **128**, 1251–1258. doi:10.1126/science.128.3334.1251

James KR, Hart BT, Bailey PCE, Blinn DW (2009) Impact of secondary salinisation on freshwater ecosystems: effect of experimentally increased salinity on an intermittent floodplain wetland. *Marine and Freshwater Research* **60**, 246–258. doi:10.1071/MF08099

Jampeetong A, Brix H (2009) Effects of NaCl salinity on growth, morphology, photosynthesis and proline accumulation of *Salvinia natans. Aquatic Botany* **91**(3), 181–186. doi:10.1016/j.aquabot.2009.05.003

Jenet A, Zhang L, Jones R, Durack P, Dawes W, Hairsine P (2006) 'Implications of climate change for natural resource management: an assessment of the impact of climate change on catchment salt and water balances in the Murray–Darling Basin.' Report to the Australian Greenhouse Office, Canberra.

Jolly ID, Williamson DR, Gilfedder M, Walker GR, Morton R, Robinson G, Jones H, Zhang L, Dowling TI, Dyce P, Nathan RJ, Nandakumar N, Clarke R, McNeill V (2001) Historical stream salinity trends and catchment salt balances in the Murray–Darling Basin, Australia. *Marine and Freshwater Research* **52**, 53–63. doi:10.1071/MF00018

Kefford MJ, Fields EJ, Clay C, Nugegoda D (2007) The salinity tolerance of riverine microinvertebrates from the southern Murray–Darling Basin *Marine and Freshwater Research* **58**, 1019–1031. doi:10.1071/MF06046

Kefford BJ, Marchant R, Schafer RB, Metzeling L, Dunlop JE, Choy SC, Goonan P (2011) The definition of species richness used by species sensitivity distributions approximates observed effects of salinity on stream macroinvertebrates. *Environmental Pollution* **159**, 302–310. doi:10.1016/j.envpol.2010.08.025

Kennedy SA, Ganf GG, Walker KF (2003) Does salinity influence the distribution of exotic willows (*Salix* spp.) along the lower River Murray? *Marine and Freshwater Research* **54**(7), 825–831. doi:10.1071/MF03035

Kim DH (2013) The influence of salinity on various life stages of *Ruppia tuberosa* and implications for its distribution in the Coorong, South Australia. PhD thesis. The University of Adelaide, South Australia.

Kim DH, Aldridge KT, Brookes JD, Ganf GG (2013) The effect of salinity on the germination of *Ruppia tuberosa* and *Ruppia megacarpa* and implications for the Coorong: a coastal lagoon of southern Australia. *Aquatic Botany* **111**, 81–88. doi:10.1016/j.aquabot.2013.06.008

Knighton AD, Mills K, Woodroffe CD (1991) Tidal-creek extension and saltwater intrusion in northern Australia. *Geology* **19**, 831–834. doi:10.1130/0091-7613(1991)019<0831:TCEASI>2.3.CO;2

Kozlowski TT (1997) Responses of woody plants to flooding and salinity. *Tree Physiology Monograph* **1**, 1–17.

LaHaye PA, Epstein E (1969) Salt toleration by plants: enhancement with calcium. *Science* **166**, 395–396. doi:10.1126/science.166.3903.395

Läuchli A, James RA, Huang CX, McCully M, Munns R (2008) Cell-specific localization of $Na^+$ in roots of durum wheat and possible control points for salt exclusion. *Plant, Cell & Environment* **31**, 1565–1574. doi:10.1111/j.1365-3040.2008.01864.x

Lea-Cox JD, Syvertsen JP (1993) Salinity reduces water use and nitrate nitrogen use efficiency of citrus. *Annals of Botany* **72**(1), 47–54. doi:10.1006/anbo.1993.1079

Lee SY, Dunn RJK, Young RA, Connolly RM, Dale PER, Dehayr R, Lemckert CJ, McKinnon S, Powell B, Teasdale PR, Welsh DT (2006) Impact of urbanization on coastal wetland structure and function. *Austral Ecology* **31**(2), 149–163. doi:10.1111/j.1442-9993.2006.01581.x

Lissner J, Schierup H-H (1997) Effects of salinity on the growth of *Phragmites australis*. *Aquatic Botany* **55**, 247–260. doi:10.1016/S0304-3770(96)01085-6

Lynch J, Läuchli A (1988) Salinity affects intracellular calcium in corn root protoplasts. *Plant Physiology* **87**, 351–356. doi:10.1104/pp.87.2.351

Lynch J, Cramer GR, Läuchli A (1987) Salinity reduces membrane-associated calcium in corn root protoplasts. *Plant Physiology* **83**, 390–394. doi:10.1104/pp.83.2.390

Macek P, Rejmánková E (2007) Response of emergent macrophytes to experimental nutrient and salinity additions. *Functional Ecology* **21**, 478–488. doi:10.1111/j.1365-2435.2007.01266.x

Mansour MMF (2000) Nitrogen containing compounds and adaptation of plants to salinity stress. *Biologia Plantarum* **43**, 491–500. doi:10.1023/A:1002873531707

Marschner H (1995) *Mineral Nutrition of Higher Plants.* 2nd edn. Academic Press, London.

Martinez V, Läuchli A (1993) Effects of $Ca^{2+}$ on the salt-stress response of barley roots as observed by in-vivo $^{31}$P-nuclear magnetic resonance and in-vitro analysis. *Planta* **190**, 519–524. doi:10.1007/BF00224791

Matoh T, Matsushita N, Takahashi E (1988) Salt tolerance of the reed plant *Phragmites cummunis*. *Physiologia Plantarum* **72**, 8–14. doi:10.1111/j.1399-3054.1988.tb06615.x

MDBA (2010) 'Murray–Darling Basin Authority annual report 2009–2010.' Murray–Darling Basin Authority, Canberra.

MDBA (2012) 'Basin salinity management strategy 2010–11 annual implementation report.' Murray–Darling Basin Authority publication no. 21/12. Murray–Darling Basin Authority, Canberra.

MDBA (2014) 'Lower Lakes, Coorong and Murray Mouth environmental water management plan.' Under a Creative Commons Attribution 3.0 Australia Licence from the Murray–Darling Basin Authority, Canberra.

MDBMC (2001) 'Basin salinity management strategy: 2001–2015.' Murray–Darling Basin Ministerial Council, Canberra.

Mensforth LJ, Thorburn PJ, Tyerman SD, Walker GR (1994) Sources of water used by riparian *Eucalyptus camaldulensis* overlying highly saline groundwater. *Oecologia* **100**, 21–28. doi:10.1007/BF00317126

Morris K (1998) Salinity and nutrients; growth and water use of aquatic macrophytes under controlled and natural conditions. PhD thesis. University of Adelaide.

Morris K, Ganf GG (2001) The response of an emergent sedge *Bolboschoenus medianus* to salinity and nutrients. *Aquatic Botany* **70**, 311–328. doi:10.1016/S0304-3770(01)00152-8

Munns R (1993) Physiological processes limiting plant growth in saline soils: some dogmas and hypotheses. *Plant, Cell & Environment* **16**, 15–24. doi:10.1111/j.1365-3040.1993.tb00840.x

Munns R, Cramer GR (1996) Is coordination of leaf and root growth mediated by abscisic acid? Opinion. *Plant and Soil* **185**, 33–49. doi:10.1007/BF02257563

Munns R, Termaat A (1986) Whole plant responses to salinity. *Australian Journal of Plant Physiology* **13**, 143–160. doi:10.1071/PP9860143

Munns R, Tester M (2008) Mechanisms of salinity tolerance. *Annual Review of Plant Biology* **59**, 651–681. doi:10.1146/annurev.arplant.59.032607.092911

Narayan KA, Schleeberger C, Bristow KL (2007) Modelling seawater intrusion in the Burdekin Delta irrigation area, North Queensland, Australia.

*Agricultural Water Management* **89**(3), 217–228. doi:10.1016/j.agwat.2007.01.008

Nicol J (2005) 'The ecology of *Ruppia* spp. with reference to the Coorong.' SARDI Aquatic Science publication number RD 04/0247–2. South Australian Research and Development Institute (Aquatic Sciences), Adelaide.

Nielsen DL, Brock MA (2009) Modified water regime and salinity as a consequence of climate change: prospects for wetlands of southern Australia *Climatic Change* **95**, 523–533. doi:10.1007/s10584-009-9564-8

Nielsen DL, Brock MA, Crosslé K, Harris K, Healey M, Jarosinski I (2003) The effects of salinity on aquatic plant germination and zooplankton hatching from two wetland sediments. *Freshwater Biology* **48**, 2214–2223. doi:10.1046/j.1365-2427.2003.01146.x

Nielsen DL, Brock MA, Vogel M (2008) From fresh to saline: a comparison of zooplankton and plant communities developing under a gradient of salinity with communities developing under constant salinity levels. *Marine and Freshwater Research* **59**, 549–559. doi:10.1071/MF07166

O'Connell M, Young J, Kingwell R (2006) The economic value of saltland pastures in a mixed farming system in Western Australia. *Agricultural Systems* **89**, 371–389. doi:10.1016/j.agsy.2005.10.003

Oertli JJ (1968) Extracellular salt accumulation; a possible mechanism for salt injury in plants. *Agrochemica* **12**, 46–69.

Ooi MKJ, Auld TD, Denham AJ (2009) Climate change and bet-hedging: interactions between increased soil temperatures and seed bank persistence. *Global Change Biology* **15**, 2375–2386. doi:10.1111/j.1365-2486.2009.01887.x

Pannell DJ, Ewing MA (2006) Managing secondary dryland salinity: options and challenges. *Agricultural Water Management* **80**(1–3), 41–56. doi:10.1016/j.agwat.2005.07.003

Pannell DJ, Roberts AM (2010) Australia's national action plan for salinity and water quality: a retrospective assessment. *The Australian Journal of Agricultural and Resource Economics* **54**, 437–456. doi:10.1111/j.1467-8489.2010.00504.x

Partridge GJ, Furey A (2002) Culturing snapper in Dumbleyung – a case study for determining the potential for inland saline groundwater to grow marine fish in Western Australia. In *8th National Conference and Workshop on the Productive Use and Rehabilitation of Saline Lands (PURSL).* pp. 16–20. Promaco, Perth, Australia, September. Fremantle, Western Australia, Australia.

Passioura JB, Ball MC, Knight JH (1992) Mangroves may salinize the soil and in so doing limit their transpiration rate. *Functional Ecology* **6**, 476–481. doi:10.2307/2389286

Paton D (2013) Open letter from Associate Professor David Paton to the head of South Australian Environment, Water and Natural Resources Department, http://www.abc.net.au/news/2013-11-18/open-letter-from-assoc-prof-david-paton/5098868.

Paton D, Rogers DJ, Aldridge K, Deegan B, Brookes J (2009) A future for the Coorong and Lower Lakes. *Pacific Conservation Biology* **15**, 7–10.

Peck AJ, Hatton T (2003) Salinity and the discharge of salts from catchments in Australia. *Journal of Hydrology* **272**, 191–202. doi:10.1016/S0022-1694(02)00264-0

Peck AJ, Thomas JF, Williamson DR (1983) *Salinity Issues: Effects of Man on Salinity in Australia.* Australian Government Publishing Service, Canberra.

Pezeshki SR, DeLaune RD (1993) Effects of soil hypoxia and salinity on gas exchange and growth of *Spartina patens*. *Marine Ecology Progress Series* **96**(1), 75–81. doi:10.3354/meps096075

Radke LC, Howard KWF, Gell PA (2002) Chemical diversity in south-eastern Australian saline lakes. I. Geochemical causes. *Marine and Freshwater Research* **53**, 941–959. doi:10.1071/MF01231

Radke LC, Juggins S, Halse SA, De Deckker P, Finston T (2003) Chemical diversity in south-eastern Australian saline lakes. II. Biotic implications. *Marine and Freshwater Research* **54**, 895–912. doi:10.1071/MF03021

Reuveni M (1993) Changes in vacuolar pH of carrot cells in suspension culture grown under saline conditions. *Biologia Plantarum (Prague)* **35**, 1–9. doi:10.1007/BF02921109

Ritson P, Pettit NE (1992) Double ridge mounds improve tree establishment in saline seeps. *Forest Ecology and Management* **48**, 89–98. doi:10.1016/0378-1127(92)90123-Q

Robinson RW, Boon PI, Bailey P (2006) Germination characteristics of *Melaleuca ericifolia* Sm (swamp paperbark) and their implication for the rehabilitation of coastal wetlands. *Marine and Freshwater Research* **57**, 703–711. doi:10.1071/MF06006

Rogers DJ, Paton DC (2009) *Changes in the Distribution and Abundance of* Ruppia tuberosa *in the Coorong*. Water for a Healthy Country National Research Flagship, CSIRO, Canberra.

Salter J, Morris K, Bailey PCE, Boon PI (2007) Interactive effects of salinity and water depth on the growth of *Melaleuca ericifolia* Sm. (swamp paperbark) seedlings. *Aquatic Botany* **86**, 213–222. doi:10.1016/j.aquabot.2006.10.002

Salter J, Morris K, Boon PI (2008) Does salinity reduce the tolerance of two contrasting wetland plants, the submerged monocot *Vallisneria australis* and the woody shrub *Melaleuca ericifolia*, to wetting and drying? *Marine and Freshwater Research* **59**, 291–303. doi:10.1071/MF07147

Schwarz M, Gale J (1981) Maintenance respiration and carbon balance of plants at low levels of sodium chloride salinity. *Journal of Experimental Botany* **32**(5), 933–941. doi:10.1093/jxb/32.5.933

Seemann JR, Critchley C (1985) Effects of salt stress on the growth, ion content, stomatal behaviour and photosynthetic capacity of a salt sensitive species, *Phaseolus vulgaruis* L. *Planta* **164**, 151–162. doi:10.1007/BF00396077

Sim LL, Chambers JM, Davis JA (2006) Ecological regime shifts in salinised wetland systems. I. Salinity thresholds for the loss of submerged macrophytes. *Hydrobiologia* **573**, 89–107. doi:10.1007/s10750-006-0267-0

Slavich PG, Petterson GH (1993) Estimating the electrical conductivity of saturated paste extracts from 1:5 soil-water suspensions and texture. *Australian Journal of Soil Research* **31**, 73–81. doi:10.1071/SR9930073

Smith JAC, Popp M, Lüttge U, Cram WJ, Diaz M, Griffiths H, Lee HSJ, Medina E, Schafer C, Stimmel K-H, Thonke B (1989) Ecophysiology of xerophytic and halophytic vegetation of a coastal alluvial plain in northern Venezuela. VI. Water relations and gas exchange of mangroves. *New Phytologist* **111**, 293–307. doi:10.1111/j.1469-8137.1989.tb00693.x

Smith MJ, Ough KM, Scroggie MP, Schreiber SG, Kohout M (2009) Assessing changes in macrophyte assemblages with salinity in non-riverine wetlands: a Bayesian approach. *Aquatic Botany* **90**, 137–142. doi:10.1016/j.aquabot.2008.08.004

Sonmez S, Buyuktas D, Okturen F, Citak S (2008) Assessment of different soil to water ratios (1:1, 1:2.5, 1:5) in soil salinity studies. *Geoderma* **144**(1, 2), 361–369. doi:10.1016/j.geoderma.2007.12.005

Staal M, Maathuis JM, Elzenga TM, Overbeek JHM, Prins HBA (1991) $Na^+/H^+$ antiport activity in tonoplast vesicles from roots of the salt-tolerant *Plantago maritima* and the salt-sensitive *Plantago media*. *Physiologia Plantarum* **82**, 179–184. doi:10.1111/j.1399-3054.1991.tb00078.x

Storey R, Wyn Jones RG (1977) Quaternary ammonium compounds in plants in relation to salt resistance. *Phytochemistry* **16**(4), 447–453. doi:10.1016/S0031-9422(00)94326-7

Storey R, Wyn Jones RG (1978) Salt stress and comparative physiology in the Gramineae. I. Ion relations of two salt-and water-stressed barley cultivars, California Mariout and Arimar. *Australian Journal of Plant Physiology* **5**, 801–816. doi:10.1071/PP9780801

Szabolcs I (1994) Soils and salinisation. In *Handbook of Plant and Crop Stress*. (Ed. M Pessarakli) pp. 3–11. Marcel Dekker, New York, NY.

Talley TS, Levin LA (2001) Modification of sediment and macrofauna by an invasive marsh plant. *Biological Invasions* **3**, 51–68.

Talling JF (2009) Electrical conductance – a versatile guide in freshwater. *Freshwater Reviews* **2**, 65–78. doi:10.1608/FRJ-2.1.4

Tester M, Davenport R (2003) $Na^+$ tolerance and $Na^+$ transport in higher plants. *Annals of Botany* **91**, 503–527. doi:10.1093/aob/mcg058

Thiel G, Lynch J, Läuchli A (1988) Short-term effects of salinity on the turgor and elongation of growing barley leaves. *Journal of Plant Physiology* **132**, 38–44. doi:10.1016/S0176-1617(88)80180-9

van der Moezel PG, Bell DT (1987) The effect of salinity on the germination of some Western Australian *Eucalyptus* and *Melaleuca* species. *Seed Science and Technology* **15**(1), 239–246.

Walker K (2006) Serial weirs, cumulative effects: the lower River Murray, Australia. In *Ecology of Desert Rivers*. (Ed. RT Kingsford) pp. 265. Cambridge University Press, Cambridge, UK.

Walker RR, Blackmore DH, Qing S (1993) Carbon dioxide assimilation and foliar ion concentrations in leaves of lemon (*Citrus limon* L.) trees irrigated with sodium chloride or sodium sulfate. *Australian Journal of Plant Physiology* **20**(2), 173–185. doi:10.1071/PP9930173

Walker G, Gilfedder M, Williams J (1999) *Effectiveness of Current Farming Systems in the Control of Dryland Salinity*. CSIRO Land & Water, Canberra.

Warwick NWM, Bailey PCE (1997) The effect of increasing salinity on the growth and ion content of three non-halophytic wetland macrophytes. *Aquatic Botany* **58**, 73–88. doi:10.1016/S0304-3770(96)01104-7

Williams WD (1967) The chemical composition of lentic surface waters in Australia. In *Australian Inland Waters and Their Fauna*. (Ed. AH Weatherley) pp. 18–77. ANU Press, Canberra.

Williams WD, Sherwood JE (1994) Definition and measurement of salinity in salt lakes. *Journal of Salt Lake Research* **3**, 53–63. doi:10.1007/BF01990642

Wood WE (1924) Increase of salt in soil and streams following the destruction of native vegetation. *Journal of the Royal Society of Western Australia* **10**, 35–47.

Ye Q, Cheshire KJ, McNeil DG (Eds) (2010) 'Influences of salinity, water quality and hydrology on early life stages of the fishes of the lower River Murray, South Australia.' SARDI publication no. F2009/000470-1-6 SARDI research reports series no. 446. South Australian Research and Development Institute (Aquatic Sciences), Adelaide.

Yeo AR, Flowers TJ (1986) Salinity resistance in rice (*Oryza sativa* L.) and a pyramiding approach to breeding varieties for saline soils. *Australian Journal of Plant Physiology* **13**, 161–173. doi:10.1071/PP9860161

Yeo AR, Caporn SJM, Flowers TJ (1985) The effect of salinity upon photosynthesis in rice (*Oryza sativa* L.): gas exchange by individual leaves in relation to their salt content. *Journal of Experimental Botany* **36**(8), 1240–1248. doi:10.1093/jxb/36.8.1240

Yeo AR, Lee KS, Izard PJ, Boursie PJ, Flowers TJ (1991) Short- and long-term effects of salinity on leaf growth in rice (*Oryza sativa* L). *Journal of Experimental Botany* **42**(7), 881–889. doi:10.1093/jxb/42.7.881

Zedler JB, Paling E, McComb A (1990) Differential responses to salinity help explain the replacement of native *Juncas kraussii* by *Typha orientalis* in Western Australian salt marshes. *Australian Journal of Ecology* **15**, 57–72. doi:10.1111/j.1442-9993.1990.tb01021.x

Zhang H, Schroder JL, Pittman JJ, Wang JJ, Payton ME (2005) Soil salinity using saturated paste and 1:1 soil to water extracts. *Soil Science Society of America Journal* **69**(4), 1146–1151. doi:10.2136/sssaj2004.0267

Zhen WB, Ma QH (2009) Proline metabolism in response to salt stress in common reed *Phragmites australis* (Cav.) Trin. ex Steud. *Botanica Marina* **52**(4), 341–347. doi:10.1515/BOT.2009.007

Zidan I, Jacoby B, Ravina I, Neuman PM (1991) Sodium does not compete with calcium in saturating plasma membrane sites regulating $Na^{22}$ influx in salinised maize roots. *Plant Physiology* **96**, 331–334. doi:10.1104/pp.96.1.331

# 16

# Fire in Australia's riparian landscapes

*Michael Douglas, Neil Pettit and Samantha Setterfield*

## INTRODUCTION

Fire is a widespread natural disturbance in Australian ecosystems and its impacts and management in terrestrial ecosystems have been widely examined. In comparison, the impact of fires on riparian zones, and its consequent impacts on stream ecosystems, has received little attention in Australia or globally (Douglas *et al.* 2003; Pettit and Naiman 2005; Stromberg and Rychener 2010). This chapter provides an overview of fire regimes and their impacts in riparian landscapes by focussing on northern Australia where fire is a very frequent and widespread disturbance. We also present results from a study that represents one of the few where quantitative assessments of fire impacts on riparian zones have been undertaken.

We begin by summarising factors influencing fire behaviour and fire regimes with respect to riverine landscapes, including how fire interacts with river flow. Impacts of fire on riparian vegetation composition and structure are then discussed with a focus on riparian zones and floodplains of northern Australia. Fire effects on plant regeneration are summarised, as are the interactive effects of fire and invasive species. Finally, we consider approaches to fire management in northern Australian riparian zones.

## FIRE REGIMES AND FIRE BEHAVIOUR IN RIPARIAN ZONES

How fires affect the ecology of riparian zones depends to a large extent on the interactions between the particular characteristics of fires and the riparian biophysical environment in which they occur (Fig. 16.1; Pettit and Naiman 2007a). The ecological impacts of individual fires should not be confused with the impacts of fire regimes, which are described in terms of fire frequency (i.e. the average time before fire re-burns an area), fire intensity, and spatial scale and pattern. The short- and long-term impacts of fire regimes depend on both fire behaviour, that is, the physical properties of particular fire events (e.g. fire intensity, rate of spread, flame height), and fire frequency. These factors are influenced by climatic conditions (e.g. air temperature, relative humidity, wind speed), characteristics of the physical riparian environment (e.g. slope, the quantity and moisture content of fuel) and patterns in other disturbance events (e.g. flooding, grazing).

There are currently very few published quantitative studies characterising fires in riparian zones either in Australia or around the world (Douglas *et al.* 2003; Dwire and Kauffman 2003; Pettit and Naiman 2007b). Fire regimes in riparian areas are generally considered to be of lower intensity and frequency than in surrounding uplands (Agee 1988; Benda *et al.* 1998; Patten 1998; Dwire and Kauffman 2003). Fire typically enters the riparian zone from uplands and its penetration into riparian zones along higher order streams is often patchy or completely absent in all but the most intense fires, due to certain characteristics of the riparian zone

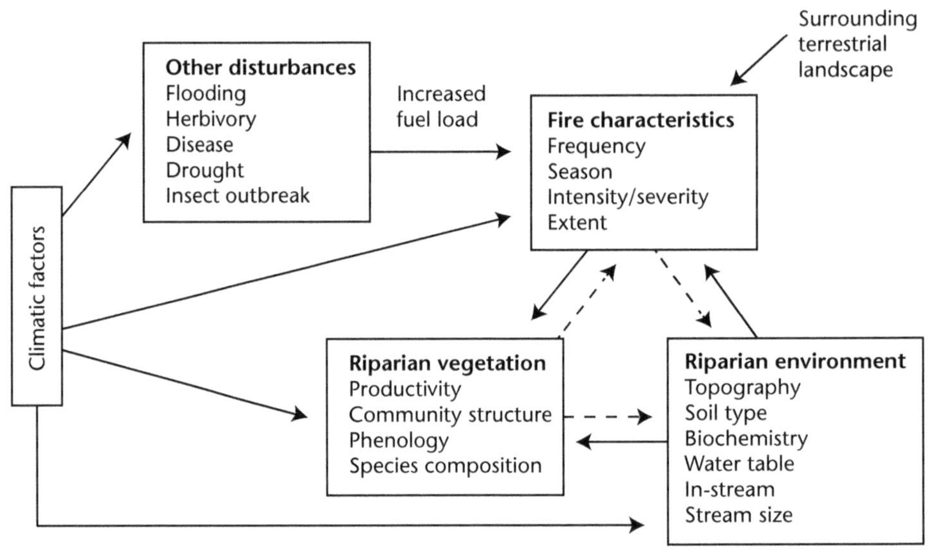


**Figure 16.1:** Fire-related processes in the riparian zone. Climate has a strong influence on the characteristics of a fire while the physical riparian environment largely determines the effects of fire on riparian vegetation, as well as the condition and type of the surrounding landscape. Riparian vegetation, in turn, affects the characteristics of a fire and the physical riparian environment. Other disturbances, such as flooding, can create synergistic fire characters via increased fuel load, which increases fire intensity and severity (adapted from Pettit and Naiman 2007a).

(Kellman and Tackaberry 1993). Leaf litter moisture levels of >20%, for instance, will prevent ignition in eucalypt forests in Australia, and these conditions often occur in riparian zones (Cheney and Sullivan 2008). In contrast, fires of moderate intensity are able to move through entire riparian zones on lower order streams, although in some instances headwater springs, seeps and small wetlands can mediate local fire effects (Pettit and Naiman 2007a). Fire effects on streams tend to be greatest in headwaters and dissipate progressively downstream (Minshall *et al.* 1989). Consequently, fire frequency and severity in riparian zones usually decreases as channel size increases.

Riparian vegetation exerts considerable influence on local microclimates and the fuel characteristics that influence fire behaviour (Dwire and Kauffman 2003). Where riparian vegetation is characterised by dense, closed canopies, evaporation tends to be reduced and high relative humidity is often maintained compared with adjacent uplands (Brosofske *et al.* 1997). This is likely to keep fuel moisture levels higher. Also contributing to higher humidity levels is the close proximity of water, which often sustains mesic vegetation in riparian zones by keeping surface soils moist. In addition, shading lowers air temperatures, reducing drying rates of fuel, and dense riparian vegetation reduces wind speeds thus helping to maintain high humidity and subsequently lowering fire severity.

The topography and geomorphology of riparian zones within catchments can also influence fire characteristics. Wind speeds will typically be lower in valleys, particularly in more steeply sided riparian areas, which will slow the rate of spread of a fire front. For these reasons, the riparian zone often acts as a natural firebreak for low intensity fires (Douglas *et al.* 2003; Dwire and Kauffman 2003). However, in more severe fire weather conditions and when high fuel loads are present, the riparian zone can act as a corridor for fire, supporting high intensity fires. In the Pacific northwest of the US, for example, where modern forestry practices have resulted in periods of fire exclusion, increased fuel loads often lead to severe episodic fire events (Agee 1998).

## Interaction between river flows and fire

Flooding interacts with fire, providing an important element of the riparian fire disturbance regime and influencing ecosystem-scale outcomes. Flooding can lead, for example, to the destruction of riparian forest and the accumulation of wrack and woody fuel (Pettit *et al.* 2005). This opens up the area to greater radiant heat (Cochrane 2003), thereby increasing the drying-out of fuels including surviving vegetation, ground litter and woody debris. This combination of increased fuel load from flood debris and the drying out of fuels increases the likelihood of subsequent fires. After this initial period of higher fire risk after flooding, however, terrestrial fires can be prohibited from entering the riparian zone by greener vegetation and increased soil and fuel moisture levels resulting from increased productivity and the replenishment of water tables and soil sediments and nutrients. Frequent flooding can also inhibit vegetation growth and the build-up of fuel, thus reducing the risk and intensity of fire. For a low rainfall savanna in Botswana, however, fire frequency was higher on floodplains than in adjacent uplands due to higher biomass production and associated fuel loads (Heinl *et al.* 2006). Such observations demonstrate that the relationship between fire and flood frequency is non-linear (Pettit and Naiman 2007a).

A large impact of fire on rivers in general is the removal by fire of riparian vegetation, which can lead to increased bank erosion and channel migration due to a loss of root strength and bank cohesion (Owens *et al.* 2013). Loss of vegetation cover will also result in reductions in evapotranspiration and soil infiltration capacity leading to greatly increased runoff during high intensity rainfall events after fire. After large severe fires, catchment runoff and stream peak discharge is likely to increase, resulting in increased channel erosion, mass movement and deposition of sediment and organic material into the channel as well as general catchment erosion (Coombs and Melack 2013). These fire effects are greatest in the initial period after fire but may last for several years (Owens *et al.* 2013); however, effects may be greatly reduced if riparian areas are patchily burnt. Unburnt patches are highly effective at reducing hydrologic connectivity from upslope burnt areas even in high severity fires (Cawson *et al.* 2013).

## IMPACTS OF FIRE ON RIPARIAN VEGETATION

The effects of fire on riparian vegetation depend on the characteristics of the fire and of the vegetation, both of which vary considerably between regions. A replicated catchment-scale study of the effect of fires regimes on low order, seasonal streams in a tropical savanna area of northern Australia was undertaken as part of the Kapalga fire experiment (Andersen *et al.* 2003) and it remains one of the few experimental studies of the effects of fire on riparian vegetation (Douglas 1999; Douglas *et al.* 2003; Andersen *et al.* 2005). The study compared three unburnt catchments with three catchments subjected to annual fires lit late in the dry season which resulted in relatively high intensity fires. These fires burnt most of the catchment including the riparian zone and the (dry) streambed. Some comparisons were also made with three catchments burnt early in the dry season which experienced lower intensity fires that only burnt patches of the riparian zone. The study demonstrated that late dry season fires had dramatic effects on the structure, floristic composition and regenerative potential of the riparian vegetation, which, in turn, affected the stream ecosystem. The results are discussed here.

After five years the density of woody species in unburnt riparian zones was about three times higher than in late-burnt catchments. Not all trees were affected equally: burning reduced the abundance of small and medium-sized trees (<5 m which accounted for the majority of the trees in the riparian zones), while large trees were mostly unaffected. Burning reduced total basal area, although not in all reaches of these stream catchments, a result that reflected the size-specific nature of reductions in abundance. A few very large trees (that were unaffected by burning) in the middle reaches of some burnt streams contributed greatly to total basal area. Consequently, basal area only tended to be reduced in the upper and lower reaches of the streams. Canopy cover in the burnt riparian zones was about half that in the unburnt riparian zones due to the reduction in density of trees and shrubs and the reduced canopy width of the surviving riparian trees. Reduced tree canopy cover led to an increase in cover of fire-tolerant life forms, such as grasses, which may further increase fuel load and hence fire frequency and intensity (Douglas *et al.* 2003). This pattern is similar to the increase in grass cover that occurs following frequent fire in upland savanna in northern Australia (Scott *et al.* 2012).

In addition to changes in riparian structure, community composition was also affected by fire. Average species richness of woody plants in burnt riparian zones was about half that in unburnt riparian zones and about half of the species recorded in the study occurred only in the unburnt riparian zones. Almost all of the woody plant species that occurred in both treatments were more abundant in the unburnt riparian zones. Similarly, vine species richness was more than three times greater in unburnt riparian zones. Many of the plant species in unburnt riparian zones were characteristic of monsoon vine forests rather than savanna habitats.

This important study demonstrated several impacts of fire on riparian vegetation that have been reported elsewhere. Smaller riparian trees are more susceptible to fire while larger trees generally escape the deleterious effects of burning (Hopkins 1965; Afolayan 1978; Isichei *et al.* 1986). Some riparian species are particularly sensitive to fire, such as trees with thin bark (Cochrane 2003) as well as many species of vine (Laurance 2003). Consequently, higher frequency fires on lower order streams can lead to the loss of fire-sensitive plant species. Studies of riparian fire in a South African savanna indicate that fire has a greater effect on particular life forms, with several woody perennials and vine species recorded in unburnt sites absent from burnt sites (Pettit and Naiman 2007b). This is not unexpected, as in wet tropical gallery forests fire-sensitive trees and lianas are destroyed by fire and susceptible to local extinction (Kellman *et al.* 1998). Community-level floristic changes can also be substantial after fire, with woody trees, shrubs and lianas predominant in unburnt sites while annual herbs and grasses tend to be more abundant at burnt sites. Such patterns were apparent three years after fire on a savanna river in South Africa (Pettit and Naiman 2007b).

## Impacts on vegetation responses to flooding

Fires can also alter pathways of riparian vegetation succession after floods by destroying surviving plants as well as triggering the regeneration of vegetation, creating a mosaic of alternate successional states (Pettit and Naiman 2007b). Indeed, a riparian fire shortly after flooding can kill regenerating seedlings and greatly reduce the biomass of vegetation recovering from the loss of mature individuals in a recent flood. In contrast, flooding during the immediate period after fire can result in increased bank erosion and damage to regenerating vegetation. After fire, the trajectory of stream ecosystem succession may depend to a large extent on levels of stream flooding in subsequent years (Minshall *et al.* 1989). Life history adaptations of riparian vegetation to flooding (such as the ability to re-sprout), may also confer some resilience in riparian plants to the effects of fire.

## Impacts on plant regeneration

There has been little investigation of life history traits possessed by riparian plants that provide resilience to fire, but examples from upland vegetation give important insights about potential riparian plant responses. In fire-adapted savanna communities, for example, there is a relationship between fire history and plant life history, so the phenology of the vegetation becomes an important determinant of subsequent fire impacts (Williams *et al.* 2002). This is seen in key phenological phases such as leaf flush, flowering and seed set, which for most savanna woody plants occur during the dry season, coinciding with the main fire season (Setterfield and Williams 1996; Williams *et al.* 2003). In these systems, fire has a significant impact on reproductive success. This was demonstrated in the Kapalga fire experiment where burning late in the dry season greatly reduced flowering and fruiting of the dominant riparian tree, *Eucalyptus alba* (Douglas *et al.* 2003). Only 10% the

*E. alba* in the late-burnt catchments bore fruit, compared with 87% in the unburnt catchments. Therefore, burning not only reduced the abundance of riparian vegetation by increasing mortality of established trees, but it also reduced the capacity for vegetation to regenerate from seed after fire.

Elsewhere, the combination of fire and flood may promote seed germination and seedling development by riparian plants. Fires in south-eastern Australia enhance the germination of soil-stored seed of the riparian shrub *Grevillea rivularis* by helping break seed dormancy. Seedlings of this species emerge after fires that are followed by local flooding (Pickup *et al.* 2003). Stimulation of riparian plant regeneration has been reported for the upper Murray River and rivers in central Queensland where fire is used intentionally to promote re-growth of riparian grasses (Parson 1991).

The most widespread riparian tree in Australia, river red gum (*Eucalyptus camaldulensis*), is sensitive to even low-intensity fires, particularly when young (Dexter 1978). Consequently, fires lit by Indigenous Australians before European settlement are likely to have suppressed regeneration of river red gum seedlings, favoured grassy woodland development and maintained forest–grassland boundaries (Chesterfield 1986). Some riparian species may be killed outright by fire, but the death of adult trees usually results in prolific seedfall and mass regeneration from seed, as is the case in Tasmania in the riparian tree *Callitris oblonga* (Harris and Kirkpatrick 1991).

### Impacts on floodplain vegetation

In northern Australia, fire is an important disturbance in the extensive floodplain ecosystems that flank many of the large river systems. From 1996 to 2000, for example, 25% of floodplains in Kakadu National Park in the Northern Territory were burnt (Edwards *et al.* 2003). Bayliss *et al.* (2006) estimated that fire frequency on the Magela floodplain in Kakadu was one in five years between 1980 and 2006, although this frequency is lower than adjacent savanna woodland, where frequency is one in two years. The difference in fire frequency reflects differences in substrate moisture conditions and fuel accumulation. Most fires, particularly on floodplains, occur in the late dry season and are typically of relatively high intensity as fuel is cured and more severe fire weather conditions prevail at this time of the year (Andersen *et al.* 2003; Price *et al.* 2005). The area of burning reportedly increased on Kakadu floodplains in the late 1980s to early 1990s after the eradication of water buffalo (Gill *et al.* 2000) and the subsequent increase in above-ground grass biomass (Petty *et al.* 2007). It is unlikely, however, that these factors restored the frequency of burns carried out by Aboriginal people before the establishment of the national park (Petty *et al.* 2007). Further significant change occurred from the 1990s due to the spread of invasive plant species in some areas (Douglas and O'Connor 2004; Ferdinands *et al.* 2005).

## INTERACTIONS BETWEEN FIRE AND INVASIVE SPECIES

Invasive plants are driving significant changes in fire regimes in ecosystems globally (Brooks *et al.* 2004) and many examples are evident in Australian riparian systems. Arguably, one of the most ecologically destructive invasive plants in northern Australia is gamba grass (*Andropogon gayanus*), which increases fuel loads and fire severity dramatically compared with native grasses, with gamba-fuelled fires resulting in rapid reductions in overstorey vegetation (Brooks *et al.* 2010; Setterfield *et al.* 2010). Investigation of patterns of gamba grass invasion indicate that riparian zones are a key pathway for its spread, with the grass invading along the highly favourable riparian communities and out into the upland savanna vegetation (Petty *et al.* 2012). Gamba grass has the potential to invade 70% of north Australia's tropical savanna and therefore represents a major threat to riparian communities in this region. Stream corridors in northern Australia are also being heavily invaded by a range of other high-impact grass species, including *Pennisetum* spp. (mission grasses), *Themeda quadrivalvis* (grader grass), *Melinus minutifolia* (molasses grass) and *Megathyrsus maximus* (Guinea grass). Vast areas of tropical floodplain have been invaded by several high-biomass, aquatic invasive grasses, particularly *Urochloa mutica* (para grass), *Hymenachne amplexicaulis* (olive hymenachne) and *Echinochloa polystachion* (aleman grass).

In general, plant invasions lead to greater fuel loads and, in turn, higher fire intensities (Douglas and O'Connor 2004). In some areas, the rainforest margin vegetation and areas of *Melaleuca* spp. (paperbarks) have been severely damaged as a result of plant invasions and their effects on fire behaviour. The subsequent ecological and cultural impacts have not been documented, but anecdotal information suggests that increased fire intensity has impacted on freshwater turtles that aestivate in the dry sediments on the floodplain during the dry season, with more severe fires increasing surface and sub-surface temperatures which may be lethal for these animals. Increased fire risk on the floodplain can also endanger the populations of other animals that use the floodplain in the dry season such as dusky rats and water pythons. The invasion of para grass and the altered fire regime are also likely to affect the carbon budget for the floodplain (Pettit *et al.* 2011) and may be exacerbated by future climate change (Catford *et al.* 2013).

Woody weeds are also a significant threat to savanna riparian zones by displacing native biodiversity and altering fire regimes. *Cryptostegia grandiflora* (rubber vine), is a multi-stemmed woody vine from Madagascar that threatens biodiversity in large areas in north-east Queensland, particularly riparian zones (Valentine *et al.* 2007). Rubber vine can grow to 5 m unsupported or can climb trees to 30 m, forming a fuel ladder, promoting crown fires in areas that do not typically support crown fires (Radford *et al.* 2008), while suppressing grasses and reducing the frequency of low-intensity ground fires (Grice 1997). Repeated burning is recommended as a tool to manage populations of this invasive weed, although experimental assessment by Valentine *et al.* (2012) showed that species richness and overall abundance of birds were lower in the sites burnt twice compared with sites that were either not burnt or were burnt once. The abundance of frugivorous and insectivorous birds was lower in the twice-burnt sites, therefore fire frequency needs to be considered in the management of this invasive threat. In contrast, the shrub *Mimosa pigra* (mimosa) has invaded vast areas of tropical Australia's floodplains and is associated with reduced cover of native grasses and sedges. As a result, mimosa is likely to suppress fire, although the change in fire frequency and subsequent impacts on biodiversity have not been quantified (Braithwaite *et al.* 1989; Lonsdale and Miller 1993).

## FIRE MANAGEMENT IN THE RIPARIAN ZONE

As riparian fires usually originate in adjacent uplands, the management of fire regimes that develop in uplands clearly has an impact on riparian areas. The distinct vegetation community that exists in the riparian zone and the relationship that exists between flooding and fire suggests, however, that there is some case for the management of fire in riparian zones as a separate and distinct ecosystem, at least for higher order streams (Pettit and Naiman 2007a). The riparian zone is frequently protected from fire to 'preserve' aquatic ecosystems; however, under conditions of severe drying weather and inherently high fuel loads (Gregory *et al.* 2003), riparian areas can act as a corridor for fire (Segura and Snook 1992; Agee 1998) and can then carry a more severe fire than upland areas, particularly when upland fuels have been depleted by recent burning.

Preferred approaches to fire management may include replicating natural fire regimes as well as producing a mosaic of fire patches to create and maintain biodiversity and, thereby, ecosystem resilience. There is generally poor knowledge of fire regime effects on Australia's riparian biota and ecosystems, and little understanding of natural riparian fire regimes. An adaptive management approach to fire management is therefore advocated (Andersen *et al.* 2003; Bisson *et al.* 2003) whereby management decisions are based on developing a thorough ecological understanding, with outcomes from management prescriptions continuously refined through feedback from monitoring.

## CONCLUSIONS

Riparian areas and their associated streams are pervasive features of landscapes. Consequently, riparian fires can have spatially broad consequences extending beyond the actual location of burning, as well as for riparian areas themselves. The nature of riparian fires is highly episodic, depending on the influence of particular conditions of fuel load and climate. There is also vast variation in the nature of riparian zones themselves and the effects of fire on these are therefore highly unpredictable. The heterogeneous fire conditions found in riparian zones, created largely by topography, canopy gaps, moisture-gathering areas, distribution of wood fuel and viability in fuel moisture content create a mix of crown fires, surface fires and unburnt patches – thus contributing to an even greater degree of heterogeneity within the riparian zone. Increased heterogeneity and biodiversity, often brought about by disturbances such as fire, may, in turn, offer riparian ecosystems greater resilience to disturbance by providing flexibility and more options for responses to subsequent disturbances (Pettit and Naiman 2005). The trajectory of ecosystem re-establishment after fires is, however, somewhat unpredictable and most vulnerable when seedlings are developing and ecosystem components and processes are reorganising if there is a subsequent disturbance such as flooding. Further burning at such a stage of reorganisation may, for instance, change the successional trajectory of the riparian community to a dramatically different state.

Most knowledge of riparian fires relates to short-term responses to burning, but there is also a need to consider long-term fire regimes. Future research should focus on variability within and between habitats in fire responses and the resultant effects on habitat quality and diversity, as well as on acknowledging elements and processes important for ecosystem resilience (Andersen *et al.* 2003). For higher order streams, at least, where riparian areas typically represent distinct vegetation community types, there is a strong case for focused fire management in the riparian zone. Implicit in this recommendation is a requirement for greater understanding of the natural recovery processes of riparian plants and vegetation after fire.

## REFERENCES


Afolayan TA (1978) Savanna burning in Kainji Lake National Park, Nigeria. *East African Wildlife Journal* **16**, 245–255. doi:10.1111/j.1365-2028.1978.tb00445.x

Agee JK (1988) Successional dynamics in forest riparian zones. In *Streamside Management: Riparian Wildlife and Forestry Interactions*. (Ed. KJ Raedeke) pp. 31–43. Contribution no. 59, University of Washington Institute of Forest Resources, Seattle, WA.

Agee JK (1998) The landscape ecology of western forest fire regimes. *Northwest Science* **72**, 24–34.

Andersen AN, Cook GD, Williams RJ (Eds) (2003) *Fire in Tropical Savanna: the Kapalga Experiment*. Springer-Verlag, New York, NY.

Andersen AN, Cook GD, Corbett LK, Douglas MM, Eager RW, Russell-Smith J, Setterfield SA, Williams RJ, Woinarski JCZ (2005) Fire frequency and biodiversity conservation in Australian tropical savannas: implications from the Kapalga fire experiment. *Austral Ecology* **30**, 155–167. doi:10.1111/j.1442-9993.2005.01441.x

Bayliss P, van Dam R, Boyden J, Walden D (2006) Ecological risk assessment of Magela floodplain to differentiate mining and non-mining impacts. In *ERISS Research Summary 2004–2005*. (Eds KG Evans, J Rovis-Hermann, A Webb and DR Jones) Supervising scientist report 189, Darwin.

Benda LE, Miller DJ, Dunne T, Reeves GH, Agee JK (1998) Dynamic landscape systems. In *River Ecology and Management: Lessons from the Pacific Coastal Ecoregion*. (Eds RJ Naiman and RE Bilby) pp. 261–288. Springer, New York, NY.

Bisson PA, Rieman BE, Luce C, Hessberg PF, Lee DC, Kershner JL, Reeves GH, Gresswell RE (2003) Fire and aquatic ecosystems of the western USA: current knowledge and key questions.

*Forest Ecology and Management* **178**, 213–229. doi:10.1016/S0378-1127(03)00063-X

Braithwaite RW, Lonsdale WM, Estbergs JA (1989) Alien vegetation and native biota in tropical Australia: the impact of *Mimosa pigra*. *Biological Conservation* **48**, 189–210. doi:10.1016/0006-3207(89)90118-3

Brooks ML, D'Antonio CM, Richardson DM, Grace JB, Keeley JE, DiTomaso JM, Hobbs RJ, Pellant M, Pyke D (2004) Effects of invasive alien plants on fire regimes. *Bioscience* **54**, 677–688. doi:10.1641/0006-3568(2004)054[0677:EOIAPO]2.0.CO;2

Brooks KJ, Setterfield SA, Douglas MM (2010) Exotic grass invasions: applying a conceptual framework to the dynamics of degradation and restoration in Australia's tropical savannas. *Restoration Ecology* **18**, 188–197. doi:10.1111/j.1526-100X.2008.00470.x

Brosofske KD, Chen J, Naiman RJ, Franklin JF (1997) Harvesting effects on microclimatic gradients from small streams to uplands in western Washington *Ecological Applications* **7**, 1188–1200. doi:10.1890/1051-0761(1997)007[1188:HEOMGF]2.0.CO;2

Catford JA, Naiman RJ, Chambers LE, Roberts J, Douglas MM, Davies PM (2013) Predicting novel riparian ecosystems in a changing climate. *Ecosystems* **16**, 382–400. doi:10.1007/s10021–012–9566–7

Cawson JG, Sheridan GJ, Smith HG, Lane PNJ (2013) Effects of fire severity and burn patchiness on hillslope-scale surface run-off, erosion and hydrologic connectivity in a prescribed burn. *Forest Ecology and Management* **310**, 219–233. doi:10.1016/j.foreco.2013.08.016

Cheney P, Sullivan A (2008) *Grassfires: Fuel, Weather and Fire Behaviour.* CSIRO, Canberra.

Chesterfield EA (1986) Changes in the vegetation of the river red gum forest at Barmah, Victoria. *Australian Forestry* **49**, 4–15. doi:10.1080/00049158.1986.10674458

Cochrane MA (2003) Fire science for rainforests. *Nature* **421**, 913–919. doi:10.1038/nature01437

Coombs JS, Melack JM (2013) Initial impacts of a wildfire on hydrology and suspended sediments and nutrient export in California chaparral watershed. *Hydrological Processes* **27**, 3842–3851. doi:10.1002/hyp.9508

Dexter BD (1978) Silviculture of the river red gum forests of the central Murray flood plain. *Royal Society of Victoria Proceedings* **90**, 175–191.

Douglas MM (1999) The effect of fire regime on riparian vegetation. PhD thesis. Charles Darwin University, Darwin.

Douglas MM, Townsend SA, Lake PS (2003) Streams. In *Fire in Tropical Savanna: the Kapalga Experiment*. (Eds AN Andersen, GD Cook and RJ Williams) Springer, New York, NY.

Douglas MM, O'Connor RA (2004) Weed invasion changes fuel characteristics: para grass (*Urochloa mutica* (Forssk.) T.Q. Nguyen) on a tropical floodplain. *Ecological Management & Restoration* **5**, 143–145. doi:10.1111/j.1442-8903.2004.201-7.x

Dwire KA, Kauffman JB (2003) Fire and riparian ecosystems in landscapes of the western USA. *Forest Ecology and Management* **178**, 61–74. doi:10.1016/S0378-1127(03)00053-7

Edwards A, Kennett R, Price O, Russell-Smith J, Spiers G, Woinarski J (2003) Monitoring the impacts of fire regimes on vegetation in northern Australia: an example from Kakadu National Park. *International Journal of Wildland Fire* **12**, 427–440. doi:10.1071/WF03031

Ferdinands K, Beggs K, Whitehead P (2005) Biodiversity and invasive grass species: multiple-use or monoculture? *Wildlife Research* **32**, 447–457.

Gill AM, Ryan PG, Moore PHR, Gibson M (2000) Fire regimes of world heritage Kakadu National Park. *Austral Ecology* **25**, 616–625.

Gregory SV, Boyer K, Gurnell A (2003) *The Ecology and Management of Wood in World Rivers*. American Fisheries Society, Bethesda, MD.

Grice A (1997) Post-fire regrowth and survival of the invasive tropical shrubs *Cryptostegia grandiflora* and *Ziziphus mauritiana*. *Austral Ecology* **22**, 49–55. doi:10.1111/j.1442-9993.1997.tb00640.x

Radford IJ, Grice AC, Abbott BN, Nicholas DM, Whiteman L (2008) Impacts of changed fire regimes on tropical riparian vegetation invaded by an exotic vine. *Austral Ecology* **33**, 151–167. doi:10.1111/j.1442-9993.2007.01803.x

Harris S, Kirkpatrick JB (1991) The distribution, dynamics and ecological differentiation of *Callitris* species in Tasmania. *Australian Journal of Botany* **39**, 187–202. doi:10.1071/BT9910187

Heinl M, Neuenschwander A, Sliva J, Vanderpost C (2006) Interactions between fire and flooding in a southern African floodplain system (Okavango Delta, Botswana). *Landscape Ecology* **21**, 699–709. doi:10.1007/s10980-005-5243-y

Hopkins B (1965) Observations on savanna burning in the Olokomeji Forest reserve, Nigeria. *Journal of Applied Ecology* **2**, 367–381. doi:10.2307/2401486

Isichei AO, Ekeleme F, Jimoh BA (1986) Changes in a secondary forest in southwestern Nigeria following a ground fire. *Journal of Tropical Ecology* **2**, 249–256. doi:10.1017/S0266467400000870

Kellman M, Tackaberry R (1993) Disturbance and tree species coexistence in tropical riparian forest fragments. *Global Ecology and Biogeography Letters* **3**, 1–9. doi:10.2307/2997557

Kellman M, Tackaberry R, Rigg L (1998) Structure and function in two tropical gallery forest communities: implications for forest conservation in fragmented systems. *Journal of Applied Ecology* **35**,195–206.doi:10.1046/j.1365-2664.1998.00300.x

Laurance WF (2003) Slow burn: the insidious effects of surface fires on tropical forests. *Trends in Ecology & Evolution* **18**, 209–212. doi:10.1016/S0169-5347(03)00064-8

Lonsdale WM, Miller IL (1993) Fire as a management tool for a tropical woody weed: *Mimosa pigra* in northern Australia. *Journal of Environmental Management* **39**, 77–87. doi:10.1006/jema.1993.1055

Minshall GW, Brock JT, Varley JD (1989) Wildfires and Yellowstone's stream ecosystems. *Bioscience* **39**, 707–715. doi:10.2307/1311002

Owens PN, Giles TR, Pettigrew EL, Leggat MS, Moore RD, Eaton BC (2013) Muted responses to streamflow and suspended sediment flux in a wildfire-affected watershed. *Geomorphology* **202**, 128–139. doi:10.1016/j.geomorph.2013.01.001

Parson A (1991) *Conservation and Ecology of Riparian Tree Communities in the Murray–Darling Basin, New South Wales: Literature Review.* NSW National Parks and Wildlife Service, Sydney.

Patten DT (1998) Riparian ecosystems of semi-arid North America: diversity and impacts. *Wetlands* **18**, 498–512. doi:10.1007/BF03161668

Pettit NE, Naiman RJ (2005) Flood-deposited wood debris and its contribution to heterogeneity and regeneration in a semi-arid riparian landscape. *Oecologia* **145**, 434–444. doi:10.1007/s00442-005-0143-z

Pettit NE, Naiman RJ, Rogers KH, Little JE (2005) Post-flooding distribution and characteristics of large woody debris piles along the semi-arid Sabie River, South Africa. *River Research and Applications* **21**, 27–38. doi:10.1002/rra.812

Pettit NE, Naiman RJ (2007a) Fire in the riparian zone: characteristics and ecological consequences. *Ecosystems* **10**, 673–687. doi:10.1007/s10021-007-9048-5

Pettit NE, Naiman RJ (2007b) Post-fire response of riparian vegetation and soils and the effect of flood-deposited wood in a semi-arid landscape. *Ecology* **88**, 2094–2104. doi:10.1890/06-1270.1

Pettit NE, Bayliss P, Davies PM, Hamilton SH, Warfe DM, Bunn SE, Douglas MM (2011) Seasonal contrasts in carbon resources and ecological processes on a tropical floodplain. *Freshwater Biology* **56**, 1047–1064. doi:10.1111/j.1365-2427.2010.02544.x

Petty AM, Werner PA, Lehmann CER, Riley JE, Banfai DS, Elliot LP (2007) Savanna responses to feral buffalo in Kakadu National Park, Australia. *Ecological Monographs* **77**, 441–463. doi:10.1890/06-1599.1

Petty AM, Setterfield SA, Ferdinands KB, Barrow P (2012) Inferring habitat suitability and spread patterns from large scale distributions of an exotic invasive pasture grass in north Australia. *Journal of Applied Ecology* **49**, 742–752.

Pickup M, McDougall KL, Whelan RJ (2003) Fire and flood: soil-stored seed bank and germination ecology in the endangered Carrington Falls *Grevillea* (*Grevillea rivularis*, Proteaceae). *Austral Ecology* **28**, 128–136. doi:10.1046/j.1442-9993.2003.01255.x

Price O, Edwards A, Connors G, Woinarski J, Ryan G, Turner A, Russel-Smith J (2005) Fire heterogeneity in Kakadu National Park, 1980–2000. *Wildlife Research* **32**, 425–433. doi:10.1071/WR03043

Scott K, Setterfield SA, Douglas MM, Parr C, Schatz J, Andersen AN (2012) Does long-term fire exclusion in an Australian tropical savanna result in a biome shift? A test using the re-introduction of fire. *Austral Ecology* **37**, 693–711. doi:10.1111/j.1442-9993.2012.02379.x

Segura G, Snook LC (1992) Stand dynamics and regeneration patterns of a pinyon pine forest in east-central Mexico. *Forest Ecology and Management* **47**, 175–194. doi:10.1016/0378-1127(92)90273-C

Setterfield SA, Williams RJ (1996) Patterns of flowering and seed production in *Eucalyptus miniata* and *Eucalyptus tetrodonta* in a tropical savanna woodland, northern Australia. *Australian Journal of Botany* **44**, 107–122. doi:10.1071/BT9960107

Setterfield SA, Rossiter-Rachor NA, Hutley LB, Douglas MM, Williams RJ (2010) Turning up the heat: the impacts of *Andropogon gayanus* (gamba grass) invasion on fire behaviour in northern Australian savannas. *Diversity & Distributions* **16**, 854–861. doi:10.1111/j.1472-4642.2010.00688.x

Stromberg JC, Rychener TJ (2010) Effects of fire on riparian forests along a free-flowing dryland river. *Wetlands* **30**, 75–86. doi:10.1007/s13157-009-0008-z

Valentine LE, Roberts B, Schwarzkopf L (2007) Mechanisms driving avoidance of non-native plants by lizards. *Journal of Applied Ecology* **44**, 228–237. doi:10.1111/j.1365-2664.2006.01244.x

Valentine LE, Schwarzkopf L, Johnson CN (2012) Effects of a short fire-return interval on resources and assemblage structure of birds in a tropical savanna. *Austral Ecology* **37**, 23–34. doi:10.1111/j.1442-9993.2011.02244.x

Williams RJ, Griffiths AD, Allan GE (2002) Fire regimes and biodiversity in the savannas of northern Australia. In *Flammable Australia: the Fire Regimes and Biodiversity of a Continent*. (Eds RA Bradstock, JE Williams and AM Gill) pp. 281–304. Cambridge University Press, Cambridge, UK.

Williams RJ, Müller WJ, Wahren C, Setterfield SA, Cusack J (2003) Vegetation. In *Fire in Tropical Savannas: the Kapalga Experiment*. (Eds AN Andersen, GD Cook and RJ Williams) pp. 79–106. Springer, New York, NY.

# 17

# Grazing

*Christopher Jones and Peter Vesk*

## INTRODUCTION

Grazing by livestock represents a major threat to riverine landscapes throughout the world and their vegetation in particular. Australia has no evolutionary history of ungulate grazers and a relatively short exposure to livestock grazing in comparison with many other countries. Over the past 200 years or so, livestock grazing has become the most extensive land use in Australia and now covers around 50% of the continent (NLWRA 2002). While grazing provides a major source of food, materials and income for Australia's human population, these benefits have come at considerable environmental cost. National land and water resources audit (NLWRA 2002) figures indicate that almost half of Australia's grazing land is affected by salinity, acidity or sodicity. Such soil degradation is attributable to the loss of vegetation, erosion and nutrient changes across the landscape. Direct removal of vegetation by grazers and the practices that facilitate grazing (e.g. vegetation clearing, sowing exotic species for pasture and damming of watercourses for stock watering and irrigation) are among the greatest causes of post-European erosion increases, contributing significantly to changes in nutrient cycles across the continent (Young 2000; NLWRA 2001a).

Australia has many native herbivores. Native macropods, such as kangaroos, wallabies and wallaroos, can influence the structure and floristic composition of vegetation through grazing (Fensham and Skull 1999). However, it is the broad-scale grazing by millions of introduced mammalian grazers, both farmed and feral, that has caused the most serious impacts on the landscape. Introduced grazers include buffalo, camels, horses, rabbits, pigs, goats, deer, and many breeds of sheep and cattle. The impacts these animals have on vegetation depend on body size and shape, foot type (i.e. hoofed or padded), population size, preferred habitat and foraging technique. Rabbits in particular have played a large part in recent vegetation changes in Australia with their population reaching an estimated 600 million by the mid 20th century, causing major erosion problems through burrowing. After the release of the myxomatosis virus in 1950 the population size dropped to around 100 million before resistance stabilised the population (Williams *et al.* 1995). The number of rabbits in Australia declined again after the calicivirus was released in 1996. However, the vast numbers and large body size of hoofed domestic livestock (in particular sheep and cattle) have had, and continue to have, a large effect on the Australian environment (Wilson 1990).

In this chapter, we review the occurrence and impacts of grazing on vegetation in Australia's riverine landscapes. Specifically, we consider the ways in which grazing alters vegetation and review the literature reporting direct and indirect effects of grazing on riparian habitats. Potential management options are also discussed.

## THE OCCURRENCE OF GRAZING IN AUSTRALIA'S RIVERINE LANDSCAPES

Livestock grazing in Australia has had a relatively short history compared with many other countries (Hobbs and Hopkins 1990). Following European settlement, however, expansion in livestock numbers, particularly sheep, has been rapid due to land availability and the export potential of meat and wool. From only 6124 sheep in 1800, Australia's total sheep population rose to 20 million in 1860 and to over 100 million by 1890 (Fig. 17.1). Through agricultural developments such as wire fencing in the 1850s (Williams 1988), horse-powered machinery in the early 1900s, and the tractor in the 1920s, Australia quickly developed into a major exporter of livestock, meat and wool. Depression and drought through the mid 20th century affected agricultural productivity, but with the introduction of affordable feed supplements and improvements in animal husbandry and genetics, numbers rose to 180 million sheep and 18 million cattle by 1970. Livestock ratios in recent times have changed with demand and farming conditions, resulting in sharp decreases in sheep numbers but a marginal growth in cattle production (ABS 1907, 2000, 2006, 2011).

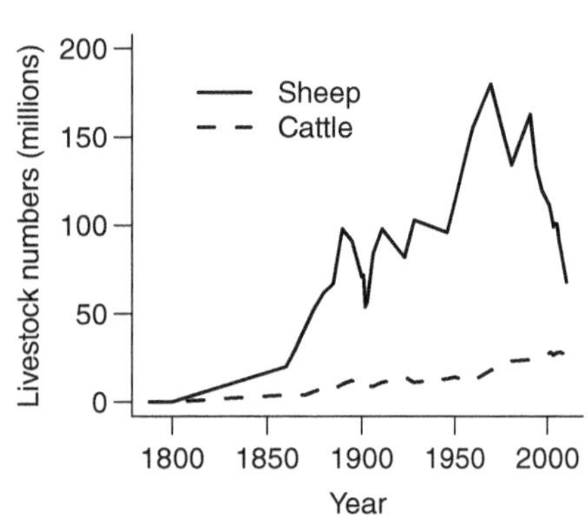


**Figure 17.1:** Livestock numbers in Australia during European occupation. Data sourced from ABS: 1907, 2000, 2006 and 2011.

Australian landscapes are broadly characterised by low and irregular rainfall, high evaporation rates and generally ancient, weathered, shallow and infertile soils. The most productive landscapes in Australia are often reserved for highly profitable crops, whereas grazing occupies much of the subtropical and semi-arid grassland areas of Australia where resources can support pasture growth (Wilson 1990). Since less than 1% of Australia's grazing land is irrigated (ABS 2009), forage quantity and quality are highly reliant on rainfall. Consequently, vegetation near natural water sources such as creeks, rivers, lakes and wetlands can be a forage refuge in low rainfall periods. In some landscapes, vegetation growth supported by these water sources is essential to productive agriculture. This is particularly true for areas with low rainfall but semi-permanent or permanent waterbodies, such as the channel country in south-west Queensland and the Ord River district in Western Australia. Riverine landscapes are therefore highly valued in Australia for grazing and are often preferred by livestock because of their vegetation, shade and water (Bryant *et al.* 1983; James *et al.* 1999; Lovett and Price 1999). As a consequence, there is often a disproportionate use of riverine landscapes for grazing. A study conducted in montane USA, for instance, found that over 80% of forage intake from livestock grazing was gleaned from riparian strips, which made up less than 2% of the total area available (Roath and Krueger 1982). Similar results could be expected in equivalent Australian riparian landscapes.

Over-exploitation of vegetation resources in riverine landscapes has occurred throughout Australia to support livestock, resulting in many and widespread vegetation changes (Fig. 17.2). Riparian vegetation, although relatively poorly adapted to livestock grazing disturbance, is adapted to disturbances from floods, erosion and upland influences (Naiman *et al.* 1993). Combined effects of grazing and natural disturbances, however, have resulted in large-scale degradation of Australian waterways and riverine ecosystems (McIntyre *et al.* 2002).

**Figure 17.2:** (a) Selective clearing of riparian vegetation for grazing along the Macalister River, Victoria, and (b) bank erosion from clearing and grazing along Glenmaggie Creek, Victoria.

## THE ROLE OF GRAZING IN SHAPING VEGETATION

Throughout the world, grazing is associated with vegetation change, and this grazer–vegetation relationship has been extensively researched (see Box 17.1). In a global meta-analysis, Milchunas and Lauenroth (1993) found that the impacts of grazing on vegetation are greatest at sites where the evolutionary history of grazing by ungulates is relatively short, productivity is low, and current grazing intensity is high – all typical features of many grazed Australian landscapes. These broad descriptors are influenced by site-specific attributes and management. In a study along the Murrumbidgee River in south-eastern Australia that investigated current site attributes only (i.e. did not specifically include the history of grazing or average rainfall), the primary drivers of compositional change in riparian vegetation were stocking rate, distance upstream from a major river confluence, relative periods of paddock rest and grazing, proportion of bank accessible to stock, and the presence of off-river water in the paddock (Jansen and Robertson 2001). It is notable that all but one of these attributes (i.e. distance upstream) concern livestock grazing rather than natural environmental characteristics. The impact of livestock on vegetation near water depends on animal physiology, population numbers and the time spent near the water source. Native grazers, such as kangaroos, are efficient

### Box 17.1: Research themes in the ecology of grazing effects on vegetation

Investigation into the response of plant species to grazing has a long history. Charles Darwin, and others, developed ideas about the effects of grazing on plant diversity, which persist to this day. One such example for temperate grasslands was that high productivity could lead to a small number of competitive species excluding numerous subordinate species; however, removal of biomass and creation of gaps through grazing then allows these subordinate species a refuge from competition, leading to increased diversity under (intermediate) grazing. The intermediate disturbance hypothesis was built on this example (Grime 1973).

**Range condition model**

In the context of monitoring and managing pasture condition for sustained productivity (or range and rangeland as it is sometimes called in extensive grazing landscapes), a different approach was formed by the North American school of plant ecologists led by Clements and Weaver (Clements 1916; Weaver and Clements 1938). The range condition model posited that grazing was an opposing force preventing the natural successional tendency of

(Continued)

## Box 17.1: (Continued)

grasslands to develop into forest (Dyksterhuis 1949). Species were classified as increasers or decreasers according to their frequency of occurrence in ungrazed or heavily grazed vegetation. The condition of a pasture could then be measured by the relative abundances of increaser and decreaser species. The focus of this work was on using the response of individual plant species as unambiguous indicators of vegetation condition in a specified region, with the assumption that increasing or decreasing stocking rate would cause a linear decline or improvement in condition, respectively. The range condition concept has since been modified and is still used in contemporary vegetation condition studies (Joyce 1993).

**State and transition models**

It was soon apparent, however, that vegetation did not always respond as simply as early range condition models predicted. Vegetation condition was observed to respond inconsistently to grazing pressure as environmental and other factors varied. These observations, coupled with a developing understanding of non-equilibrium ecology, led Westoby *et al.* (1989) to develop state and transition models. These models were developed to assist grazing managers understand the possible ways in which vegetation might occur (states) and the possible transitions between those states in a specified region. Managers can then identify hazards and opportunities for change in different states and manage appropriately. This framework is useful for comparing sites and setting management objectives, and has been widely adopted for grazing management. For further information on more current use of state and transition models see Bestelmeyer *et al.* (2003), Briske *et al.* (2005), McIntyre and Lavorel (2007) and Rumpff *et al.* (2011).

**A global contextual model**

A general conceptual model for the response of grassland communities to grazing, developed by North and South American ecologists (Sala *et al.* 1986; Milchunas *et al.* 1988), provides a framework for understanding responses globally in reference to the productivity of the site (high above-ground net primary production) and the evolutionary history of grazing at that site. A meta-analysis confirmed most of the model's predictions (Milchunas and Lauenroth 1993). Importantly, this research theme emphasises the need for an understanding of vegetation responses and context to model responses to grazing at a global scale.

**Plant trait-based models**

Extrapolating plant responses to grazing across broad areas involves the comparison of responses from species that may only occur at a subset of sites. The need for a shared species component, coupled with the practical limitations of evaluating multiple individual species' responses to grazing, led workers to group species into plant functional types (McIntyre *et al.* 1999). Functional types enable generalisation based on the expectation that specific plant traits and combinations of traits will influence the response to grazing consistently across species and contexts (McIntyre *et al.* 1999; Díaz *et al.* 2007). Commonly used traits when considering grazing impacts include plant height, life span and leaf morphology, but may vary between studies depending on the data available and expected responses. Plant traits and functional types can be effective for examining and predicting the responses of plants to grazing pressure (McIntyre *et al.* 1999; Vesk and Westoby 2001; Dorrough and Scroggie 2008). However, plant trait analyses are not a panacea. Species can respond inconsistently across situations (e.g. due to differences in environment, grazing pressure) in ways that traits alone cannot predict, and some situations may be inherently less predictable than others (Vesk and Westoby 2001; Vesk *et al.* 2004).

water users so are able to spend more time further away from water sources, resulting in less impact on the vegetation around water. The foraging distance of an animal from water and proportion of time spent there will also vary depending on factors such as climate, vegetation availability and terrain.

Local environmental characteristics, including soil, slope, nutrient availability and local climatic conditions, can have a greater influence on local vegetation characteristics than the intensity of grazing (Milchunas and Lauenroth 1993; Robertson and Rowling 2000). An understanding of the history

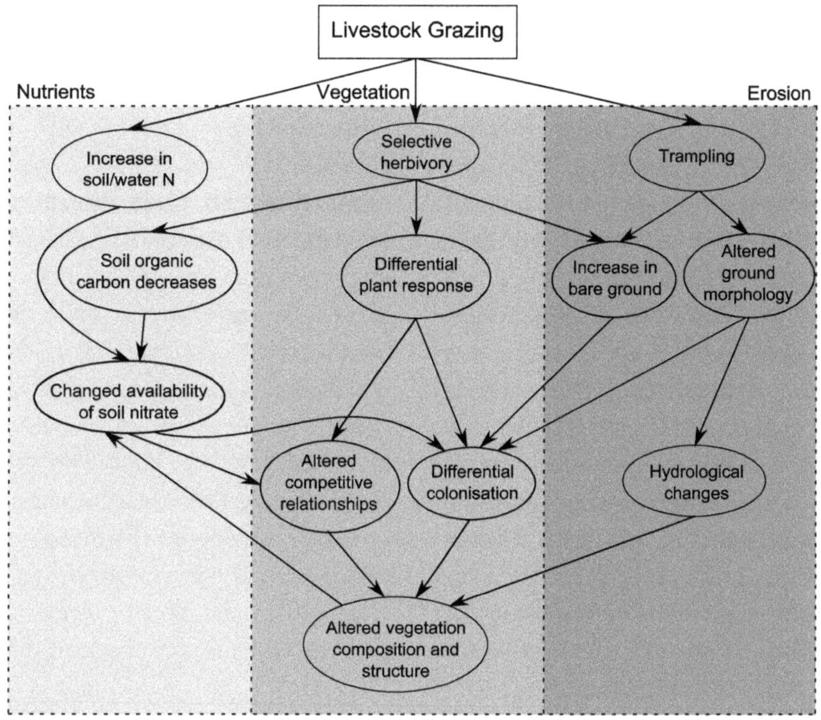


**Figure 17.3:** A conceptual model of the effects of livestock grazing on vegetation of riverine landscapes. Coincident drivers acting outside the direct influence of grazing, such as pasture improvement through addition of fertilisers are not included here. 'Differential plant responses' includes tolerance, growth, fecundity and mortality; 'altered ground morphology' includes in-channel, bank and soil compaction effects. Feedbacks from altered vegetation on grazing are not shown here, but may occur.

of grazing and other environmental characteristics, including relative rainfall, is therefore required to interpret influences of livestock grazing on vegetation.

Grazing effects on vegetation can occur either directly, through plant consumption and trampling, or indirectly through effects on nutrient cycling and soil structure. Figure 17.3 presents a conceptual model of grazing effects on environmental attributes in riverine landscapes, which we refer to in the following sections.

## DIRECT EFFECTS OF GRAZING ON PLANTS: HERBIVORY AND TRAMPLING

Grazers choose what they eat and selectivity varies between species and with total grazing pressure. Grazing preference also changes in relation to the relative abundance of plant species (Wilson 1990). A given species is preferred when more palatable species are less abundant, but preference is reduced by lack of feed, to the point that grazers will eat almost anything when there is little available. When grazing pressure is light, grazers can be choosier.

### Plant responses to grazing

The capacity of individual plants to achieve regrowth after being grazed or trampled varies. The mode of response that a species displays will influence its population growth under grazing pressure. For an individual plant, responses include reduced root growth, increased tillering, altered physiology and changed reproductive allocation (Briske 1991). Water depth and availability influence this response in aquatic plants (Blanch and Brock 1994; Crosslé and Brock 2002; Hayball and Pearce 2004). Broadly speaking, plants that do well relative to neighbouring plants under grazing pressure are either tolerant (e.g. they have fast regrowth or rapid seed maturation) or are able to avoid grazing (e.g. they are unpalatable as a result of thorns, toxins, etc. or are inaccessible) (Briske 1991). Plant species that perform poorly under grazing disturbance relative to co-occurring species, because they have slower regrowth or a longer period between flowering and seed maturation, may experience population decline and possible local extinction in the grazed landscape.

Changes in relative performance of plant species under grazing lead to changes in species competitiveness, relative abundance and hence vegetation composition. Some species respond in a way that results in plants becoming relatively more competitive than they were before grazing (increasers), others respond insufficiently and become less competitive (decreasers), and some remain relatively neutral. Whether a species behaves as an increaser or decreaser is influenced by the presence of specific plant traits that are common among species with similar life forms. Life forms known to show positive responses to grazing include many annual grasses and herbs (Biddiscombe 1953; Pettit *et al.* 1995; Prober and Thiele 1995; Fensham *et al.* 1999; McIntyre and Lavorel 2001; Landsberg *et al.* 2003; Dorrough and Scroggie 2008), species that can both re-sprout after damage and reproduce via seed (Pettit *et al.* 1995), and unpalatable woody shrubs (Harrington *et al.* 1979; Wilson 1990). These life forms have one or more traits favoured by grazing, such as short life spans (annuals), low-growing habits (perennials), small seeds, high regrowth potential, plasticity in response to grazing and high fecundity (Landsberg *et al.* 1999; McIntyre and Lavorel 2001).

Aside from particularly unpalatable woody species, perennials commonly decrease under grazing due to their relatively slow growth and reproduction. Life forms and other groups of plants that commonly decrease in abundance under grazing are perennial shrubs and herbs (Pettit *et al.* 1995), perennial grasses (Biddiscombe 1953; Milchunas and Lauenroth 1993; Fensham *et al.* 1999; McIntyre and Lavorel 2001), perennial geophytes, ferns and shrubs (Dorrough and Scroggie 2008), trees (as a result of reduced regeneration (Robertson and Rowling 2000; Dorrough and Moxham 2005) and instream vegetation (Jansen and Healey 2003).

Over time, intense grazing results in vegetation dominated by species with traits that convey a competitive advantage under grazing. In some cases, this also alters the ratio of native and exotic species, as many exotics are increasers and natives are often decreasers (Robertson and Rowling 2000; McIntyre and Martin 2001; McIntyre *et al.* 2003). Species richness may increase with low levels of grazing at a local scale as grazing creates opportunities for invasion (Landsberg *et al.* 2002). High levels of grazing, however, generally reduces species richness (Milchunas and Lauenroth 1993).

Although general trends do exist for these plant groupings, the variation in responses, between species within life forms, and between studies for a given a species, is substantial (Noy-Meir *et al.* 1989; Díaz *et al.* 2001; Vesk and Westoby 2001; Crosslé and Brock 2002; Díaz *et al.* 2007). Therefore, to understand and reliably predict responses of species to grazing over large spatial scales, there is a need to determine what drives these inconsistencies. Species' responses may be contingent on their relative palatability, which depends on co-occurring plant species and grazing pressure. Inconsistencies may also result from differences in local abiotic and biotic context in space and time (Vesk and Westoby 2001). Drought, for example, compromises the ability of species to cope with grazing (Hodgkinson and Müller 2005); similarly, higher rainfall sites offering greater potential for regrowth than drier sites (Vesk and Westoby 2001). Furthermore, differences in optimal moisture conditions for regrowth between species, even within life forms, can affect regrowth potential and competitive outcomes between seemingly similar species.

## Plant invasions in grazing systems

Riverine landscapes experience frequent disturbance due to the movement of water. In Australia, highly variable rainfall and surface water flows create particularly dynamic environments for riverine vegetation (Friedel *et al.* 1990; Thoms and Sheldon 2000). Although riverine species are adapted to such frequently changing environmental conditions (Blanch and Brock 1994; Naiman and Decamps 1997; Crosslé and Brock 2002), disturbances by extremes of rainfall variation (i.e. flooding or drought) in concert with grazing and erosion create considerable opportunities for plant invaders (Humphries *et al.* 1991).

Disturbance creates opportunities for plant invasion through changes in resource availability, such as space or nutrients, which can provide a competitive advantage for opportunistic species (Davis *et al.* 2000). Some disturbances occur regularly and with relative predictability such that the existing vegetation is adapted to those disturbances. Indeed, changes to a particular disturbance regime may be more influential on vegetation than the disturbances themselves (Moles *et al.* 2012). Exotic annual grasses and herbs are particularly quick to take advantage of changes in resource availability, and are therefore often observed to increase under grazing (Biddiscombe 1953; Pettit *et al.* 1995; Prober and Thiele 1995; Fensham *et al.* 1999; McIntyre and Lavorel 2001; Landsberg *et al.* 2003; Dorrough and Scroggie 2008). The success of these plant groups is largely due to their capacity for colonising newly formed gaps along with rapid growth in any relatively nitrogen-rich soils (Prober *et al.* 2002b). In addition to being a relatively disturbance-prone environment, riparian zones are particularly vulnerable to invasions due to their typically high resource availability, which is favoured by many exotic species (Swincer 1986; Humphries *et al.* 1991; Stohlgren *et al.* 1998).

Exotic propagules are sourced from gardens or agriculture, as well as the aquarium trade for instream plants, and dispersal vectors include wind, water and animals. Exotic pasture species to support livestock, predominantly grasses and legumes, have been deliberately introduced into Australia since the 1880s (Mott 1986; Lonsdale 1994). The ability of these plants to withstand grazing disturbance has made them ideal invaders of native grazing lands. Some of Australia's worst environmental weeds were introduced as pasture grasses (Low 1997). Large areas of pasture and crop land in catchments also contribute high volumes and diversity of seed from exotic species to adjacent riverine habitats (Williams *et al.* 2008).

## INDIRECT EFFECTS OF GRAZING ON VEGETATION

Riparian vegetation acts as a buffer strip for waterways through its ability to intercept nutrients and sediments across and through soils. Grazing and its associated influences, such as trampling, alter the effectiveness of vegetation to act as a filter and to provide structural support against erosion through stream bank stabilisation. Such changes can create complex feedbacks, affecting riparian habitat quality for plants and exacerbating impacts of grazing on riverine vegetation.

### Nutrients

Changes in nutrient concentrations in riparian areas can negatively impact riparian vegetation, which, in turn, affects instream water quality and habitat (Butler *et al.* 2008; Vidon *et al.* 2008), which can potentially cause further impacts on the vegetation. Nutrients can accumulate or be removed from soils through grazing. The most widely studied elements in agriculture include carbon, phosphorus and nitrogen. Carbon is a product of photosynthesis whereas phosphorus and nitrogen are accumulated from natural sources and are added artificially to soils to increase plant growth. These and other nutrients are removed from grazing landscapes by farming of meat, milk or wool, in runoff and via chemical breakdown.

The primary source of nitrogen in soil is the atmosphere, which is mediated through fixation by microbes in soil or in roots of legumes, although nitrogen-based fertiliser usage is rapidly increasing (NLWRA 2001a). Digestion and excretion of plant matter can be concentrated within faeces such that local soil nitrate concentrations, unlike phosphorus, can be directly attributable to stock presence (Prober *et al.* 2002a). The addition of nitrogen to grazing systems on a local scale, through faeces, fixation or fertilisers, may have a positive feedback on soil nitrate. Increased nitrogen relative to carbon causes inefficient nitrogen use by plants, which results in greater nitrogen accumulation in plants and therefore high nitrogen concentrations in litter, and further additions to soils (Wedin 1999). Increasing soil and water nitrogen concentrations alters competition between plants, benefitting short-lived exotic species and instream algae.

Phosphorus occurs naturally in soil, but phosphate fertiliser application to pastures can be detrimental to native plant populations (Dorrough *et al.* 2006; Dorrough and Scroggie 2008). Annual grasses and forbs are generally more tolerant of both grazing and high phosphorous so they invade sites where fertiliser has been added, for instance, exotic vegetation increases and native perennial grasses decrease under increasing phosphorous concentrations (Allcock 2002; Garden *et al.* 2003).

Soil microbes influence both soil structure and its ability to hold and transport water and minerals (Northup *et al.* 1999). Microbes feed on organic matter produced by the plants such that the highest concentrations of microbes are found around live plants. Plants have moderately stable root to shoot ratios that reflect supply and demand of water and minerals by roots, and carbon by shoots. When grazing reduces shoot biomass, root biomass often decreases correspondingly. Grazing can therefore significantly reduce the activity of microbes in the soil and hence the availability of other soil nutrients to plants. Correspondingly, decomposition rates from microbe activity have been shown to increase following grazing exclusion (Lindsay and Cunningham 2009).

### Erosion

The level of erosion in any riverine system is dependent on rainfall, vegetation cover, wind, slope (i.e. length and gradient), soil type and disturbance (NLWRA 2001b; Bramley and Roth 2002). Grazing facilitates erosion through vegetation changes (Robertson and Rowling 2000), soil compaction (Bezkorowajnyj *et al.* 1993; Clary 1995), track formation, soil moisture changes (Milchunas and Lauenroth 1993), pugging (i.e. animal tracks in soft soil) and changes in bank morphology (Kauffman *et al.* 1983) (Fig. 17.2). Vegetation and litter inhibit erosion by decreasing the velocity of streamflow and surface flow (McKergow *et al.* 2006), and trees and shrubs with larger roots provide structural support to banks (Abernethy and Rutherford 2001).

Since European occupation, grazing-induced soil erosion in Australia has more than doubled (Young 2000; NLWRA 2001a). Such changes have implications for the plant species that are able to establish and persist in affected riparian habitats. Australia's soil conservation legislation for the mainland states, introduced in the 1940s, illustrates the long-term awareness of the negative effects of livestock grazing. Yet, little was done to negate these effects before 1970 (Conacher and Conacher 1988).

## GRAZING MANAGEMENT

Large-scale action to counter degradation of Australian riverine landscapes did not occur until the 1970s. Management options such as fencing of frontages and off-stream watering points for livestock were not widely used before this time, to the detriment of both the natural values of Australia's riverine landscapes and their productive capacity for agriculture. Currently, many riverine landscapes are covered by Australian federal and state biodiversity conservation policies and strategies, but policy implementation has often been lacking (Brock 2003).

### Active grazing management

The most commonly used methods to manage grazer access to riverine landscapes are exclusion (i.e. fencing), or control of grazing levels. The latter may be either forced (i.e. licensing) or owner-driven rotational, timed, seasonal or crash grazing, such as restricting access in the wetter or dryer times when damage is intensified (see Box 17.2) (Conacher and Conacher 1988; McIntyre *et al.* 2002; Nelson *et al.* 2011).

Many factors influence the ability of vegetation to recover after grazing has ceased or been reduced. A system will not necessarily return to its pre-grazed

## Box 17.2: Case study: Barmah Forest

Barmah Forest is located ~225 km north of Melbourne in the southern Murray–Darling Basin. Covering 29 500 ha of Murray River floodplain on the Victorian side of the Barmah–Millewa parks system, the Barmah Forest represents the largest river red gum (*Eucalyptus camaldulensis*) forest in the world (66 000 ha). Regular flooding of the river has resulted in a variety of permanent and temporary wetlands throughout the forest. In 1971, Barmah Forest was declared a wetland of international significance under the Ramsar convention (UNESCO 1971). Grazing and timber harvesting in combination with regulation of river flows have dramatically altered the flooding regime of the river, severely impacting much of the river red gum forest (Mac Nally *et al.* 2011).

While still under management by Aboriginal people, the forest mostly consisted of woodland and open forest with occasional open grassland patches. Livestock grazing was introduced to the forest in the 1840s, initially mostly sheep but cattle dominated after 1885 (DCE 1992). During this time many other exotic grazers became resident in the forest, including rabbits, hares, deer and horses. The impact of rabbits was severe before the release of myxomatosis, which dramatically reduced their numbers (Chesterfield 1986). Kangaroos are also in the forest and contribute to overall grazing impacts. The introduction of livestock grazing coincided with extensive timber harvesting of the river red gums. This early grazing reduced tree and shrub regeneration, reduced palatable understorey species, and increased abundance of exotic plant species (Chesterfield 1986). Construction of the Hume Weir (1934) and Dartmouth Dam (1980) upstream of the forest altered flooding patterns on the Barmah Forest floodplain. This river regulation has subsequently altered vegetation composition throughout the forest (Chesterfield 1986).

In 1987, a desire to protect the forest and control the impact of grazing resulted in the Victorian Government proclaiming an area of 7900 ha of the forest as Barmah State Park, as well as two 'reference areas' totalling 280 ha. Grazing continued throughout the forest under licences in the state park and primarily under agistments in the remaining state forest. Stock numbers were strictly controlled in this period under the Department of Sustainability and Environment, Parks Victoria and the Barmah Forest Advisory Committee. Average stock numbers during this time were around 2000 head in summer and 800 in winter depending on weather conditions such as floods (VEAC 2006). Unfortunately, since the review by Chesterfield (1986), there have been few published studies of vegetation in the forest to quantify any subsequent changes or attribute them to altered grazing patterns. Most of the work undertaken in Barmah Forest since 1986 has been focused of flooding regime effects (e.g. Bren 1992).

A comparison between grazed and ungrazed sites in nearby Gulpa Island Forest (part of the Barmah–Millewa parks system) over a 12-year period showed only minor differences in understorey plant cover and species composition (Lunt *et al.* 2007). Rainfall changes had a greater influence on cover relative to altered grazing practices. It was concluded for Gulpa Island that removal of grazing was unlikely to cause a dramatic vegetation response due to low resource levels and previous degradation.

Public opinion on the presence of grazing in the forest remains divided. Grazing is considered an important part of the heritage of the forest as well as a tool to reduce weeds and fuel for bushfires. Conversely, grazing is believed to spread weeds, cause erosion and pugging, and negatively affect the native flora and fauna (VEAC 2006). However, based on several studies detailing decreases in riparian vegetation condition under grazing along the floodplains of the Murray and Murrumbidgee rivers and other areas, as well as several other considerations, most of the Barmah State Park and Barmah State Forest became a national park in 2009 (Robinson and Mann 1998; Robertson and Rowling 2000; Jansen and Robertson 2001; Jansen and Healey 2003; Dorrough *et al.* 2004; Martin and Possingham 2005). The Barmah National Park covers 28 505 ha in which livestock grazing is prohibited.

state after grazing exclusion, hence the development of models that accommodate such observations (Box 17.1). Species that decrease in abundance under grazing will only return to the system if propagules are available from the soil seed bank or nearby dispersal sources. Exclosure experiments are used to study the recovery (or change) of vegetation with grazing removal. In some situations, exclosure results in decreasing bare ground (Carr and Turner 1959; Popolizio *et al.* 1994), increasing tree regeneration (Robertson and Rowling 2000), increasing litter and vegetation cover mainly due to recovery of perennial shrubs and herbs (Popolizio *et al.* 1994; Pettit *et al.* 1995; Robertson and Rowling 2000), and richer floristic composition with the return of more grazing-sensitive species (Lunt and Morgan 1999; Price *et al.* 2010). Predicting the vegetation response to grazing removal requires an understanding of the impacts of the current level of grazing as well as mechanisms of recovery.

## Making appropriate management decisions

Management of grazing land should account for the variation in site conditions, vegetation composition, grazing levels and land ownership. Broad-scale management plans, although cost and time efficient, are unlikely to be equally effective across the entire area (Hancock *et al.* 1996; Lunt *et al.* 2007). Appropriate management strategies depend on the aims or focus of management (e.g. vegetation, erosion and eutrophication) as well as cost. Resources to undertake management are typically limited, so resource allocation decisions and trade-offs must be made and are often difficult. The following are two examples of distinct aims that require different management approaches:

(1) If the aim is to maximise species diversity, a study of wetland species in a lagoon in northern New South Wales by Blanch and Brock (1994) suggests that low intensities of grazing may help support a high diversity of species. Alternatively, a range of grazing levels may be applied, providing conditions that benefit both grazing-sensitive species (low grazing intensity) and grazing-tolerant species (high grazing intensity) (McIntyre *et al.* 2003). In either case, maintaining a high diversity of species may be aided by creating a mosaic of plant and habitat types (Brock 2003).

(2) If bank erosion is the management focus, then removal of grazer foot traffic and increased plant cover should be the priority. In this case, addition of any vegetation may be positive, whether native or exotic, diverse or homogeneous (Jansen and Robertson 2001). A single management tool, such as exclusion of grazing, while necessary is rarely sufficient to achieve success. Successful outcomes may require a combination of disturbances or actions to stimulate recruitment (Clarke 2002), for instance, planting and eradicating exotic mammalian herbivores such as goats and rabbits (Chesterfield and Parsons 1985).

Most management decisions can be assisted by the use of models (see Box 17.1). Regardless of the model used (e.g. range condition or state and transition), knowledge of species composition and species responses to grazing under specific conditions is essential for managing grazing in these landscapes (Vesk and Westoby 2001). Appropriate management of grazing can result in positive outcomes such as reduced stream bank erosion (McKergow *et al.* 2003) and enhanced native tree regeneration (Chesterfield and Parsons 1985). The time required to reach successful recovery to pre-grazing condition will vary depending on the level of degradation, climate, distance to appropriate propagule source, and current management actions. Achieving a full recovery to pre-grazing or target conditions is slow and may take many decades (Robertson and Rowling 2000; Price *et al.* 2010), or in some cases may never be fully achieved.

Reducing the level of grazing on a site is not necessarily detrimental for the landholder. There is a compromise between maximising stock numbers and maintaining fodder quality. Moderate stocking rates can generate more profitable returns (MacLeod and McIntyre 1997). Observations in tropical systems suggest that clearing of all paddock trees increases carrying capacity of that land but can reduce pasture quality (Ash and McIvor 1998). Despite the known benefits of sustainable farming practices in riverine landscapes, a postal survey of waterway frontage owners in the Goulburn-Broken catchment in Victoria showed a limited adoption of current recommended practices such as off-stream stock watering, fencing and revegetation (Curtis and Robertson 2003). Reasons provided for the low adoption of recommended approaches included the commonly held beliefs that grazing exclusion promotes weeds and fires, is expensive and removes a valuable asset due to the high productivity of riverine vegetation (McIntyre *et al.* 2002). A better understanding by landholders of riparian vegetation and its functions is required for increased adoption of recommended practices (Curtis and Robertson 2003). There also needs to be recognition by the relevant authorities that riparian buffer zones will require ongoing active management and commitment particularly in heavily modified landscapes. In areas where grazing has ceased due to changes in land use from grazing to cropping, such as the Burdekin irrigation area of northern Queensland, a lack of active management of riparian zones has resulted in significant riparian zone degradation (Tait and Perna 2001). As landholders have direct and frequent access to their own land and adjacent public land, they may be highly valuable as riparian land managers if given appropriate understanding of methods and theory (Wilson *et al.* 2006), and resources.

## CONCLUSIONS

Despite the short history of livestock grazing in Australia, its effects on riverine landscapes have been extensive and sometimes severe. Typical responses of vegetation to livestock grazing are a reduction in grazing-sensitive species (i.e. often perennials), increases in exotic annual species (i.e. commonly invaders from agriculture), decreases in native tree regeneration and an overall loss in native species richness. Environmental effects associated with vegetation change are increased bare ground, decreased soil microbe activity, soil nutrient changes, erosion and soil compaction. The difficulty of managing grazing in riverine landscapes arises from the great variation in vegetation responses to grazing and its removal. At the site level, understanding the context, including environmental and site-specific conditions, species present (flora and fauna) and the history of grazing, is crucial to achieving successful outcomes. Grazing effects have been formally studied since Darwin's investigations in 1859 and much is now known about vegetation responses to grazing. However, the details of grazing and vegetation relationships across the full range of Australian landscapes are still unclear. Further information is needed to maximise management successes and create a sustainable future for our billion-dollar agricultural industries. Appropriate management of our grazing systems is crucial for preserving our native biodiversity and the health of our waterways and their vegetation.

## REFERENCES


Abernethy B, Rutherford ID (2001) The distribution and strength of riparian tree roots in relation to riverbank reinforcement. *Hydrological Processes* **15**, 63–79. doi:10.1002/hyp.152

ABS (1907) *Official Year Book of the Commonwealth of Australia 1901–1907.* Australian Bureau of Statistics, Canberra.

ABS (2000) *Year Book Australia: a Hundred Years of Agriculture*. Australian Bureau of Statistics, Canberra.

ABS (2006) *Year Book Australia*. Australian Bureau of Statistics, Canberra.

ABS (2009) *Water use on Australian farms, 2007–08.* Australian Bureau of Statistics, Canberra.

ABS (2011) *Australian Farming in Brief.* Australian Bureau of Statistics, Canberra.

Allcock KG (2002) Effects of phosphorus on growth and competitive interactions of native and introduced species found in white box woodlands. *Austral Ecology* **27**, 638–646. doi:10.1046/j.1442-9993.2002.01225.x

Ash AJ, McIvor JG (1998) How season of grazing and herbivore selectivity influence monsoon tall-grass communities of northern Australia. *Journal of Vegetation Science* **9**, 123–132. doi:10.2307/3237230

Bestelmeyer BT, Brown JR, Havstad KM, Alexander R, Chavez G, Herrick JE (2003) Development and use of state-and-transition models for rangelands. *Journal of Range Management* **56**, 114–126. doi:10.2307/4003894

Bezkorowajnyj PG, Gordon AM, McBride RA (1993) The effect of cattle foot traffic on soil compaction in a silvo-pastoral system. *Agroforestry Systems* **21**, 1–10. doi:10.1007/BF00704922

Biddiscombe EF (1953) A survey of the natural pastures of the Trangie district, New South Wales, with particular reference to the grazing factor. *Australian Journal of Agricultural Research* **4**, 1–30. doi:10.1071/AR9530001

Blanch SJ, Brock MA (1994) Effects of grazing and depth on two wetland species. *Australian Journal of Marine and Freshwater Research* **45**, 1387–1394. doi:10.1071/MF9941387

Bramley RGV, Roth C (2002) Land-use effects on water quality in an intensively managed catchment in the Australian humid tropics. *Marine and Freshwater Research* **53**, 931–940. doi:10.1071/MF01242

Bren LJ (1992) Tree invasion of an intermittent wetland in relation to changes in the flooding frequency of the River Murray, Australia. *Australian Journal of Ecology* **17**, 395–408. doi:10.1111/j.1442-9993.1992.tb00822.x

Briske DD (1991) Developmental morphology and physiology of grasses. In *Grazing Management: an Ecological Perspective*. (Eds RK Heitschmidt and JW Stuth) pp. 85–108. Timber Press, Portland, OR.

Briske DD, Fuhlendorf SD, Smeins FE (2005) State-and-transition models, thresholds, and rangeland health: a synthesis of ecological concepts and perspectives. *Rangeland Ecology and Management* **58**, 1–10. doi:10.2111/1551-5028(2005)58<1:SMTARH>2.0.CO;2

Brock MA (2003) Australian wetland plants and wetlands in the landscape: conservation of diversity and future management. *Aquatic Ecosystem Health & Management* **6**, 29–40. doi:10.1080/14634980301479

Bryant JP, Chapin FS, III, Klein DR (1983) Carbon/nutrient balance of boreal plants in relation to vertebrate herbivory. *Oikos* **40**, 357–368. doi:10.2307/3544308

Butler DM, Ranells NN, Franklin DH, Poore MH, Green JT, Jr (2008) Runoff water quality from manured riparian grasslands with contrasting drainage and simulated grazing pressure. *Agriculture, Ecosystems & Environment* **126**, 250–260.

Carr SGM, Turner JS (1959) The ecology of the Bogong High Plains. *Australian Journal of Botany* **7**, 34–63. doi:10.1071/BT9590034

Chesterfield EA (1986) Changes in the vegetation of the river red gum forest at Barmah, Victoria. *Australian Forestry* **49**, 4–15. doi:10.1080/00049158.1986.10674458

Chesterfield CJ, Parsons RF (1985) Regeneration of three tree species in arid southeastern Australia. *Australian Journal of Botany* **33**, 715–732. doi:10.1071/BT9850715

Clarke PJ (2002) Experiments on tree and shrub establishment in temperate grassy woodlands: seedling survival. *Austral Ecology* **27**, 606–615. doi:10.1046/j.1442-9993.2002.01221.x

Clary WP (1995) Vegetation and soil response to grazing simulation on riparian meadows. *Journal of Range Management* **48**, 18–25. doi:10.2307/4002499

Clements FE (1916) *Plant Succession: an Analysis of the Development of Vegetation*. Carnegie Institute of Washington, Washington, DC.

Conacher A, Conacher J (1988) The exploitation of soils. In *The Australian Experience: Essays in Australian Land Settlement and Resource Management*. (Ed. RL Heathcote) Chap. 10. Longman Cheshire, Melbourne.

Crosslé K, Brock MA (2002) How do water regime and clipping influence wetland plant establishment from seed banks and subsequent reproduction? *Aquatic Botany* **74**, 43–56. doi:10.1016/S0304-3770(02)00034-7

Curtis A, Robertson A (2003) Understanding landholder management of river frontages: the Goulburn Broken. *Ecological Management & Restoration* **4**, 45–54. doi:10.1046/j.1442-8903.2003.t01-1-00137.x

Davis MA, Grime JP, Thompson K (2000) Fluctuating resources in plant communities: a general theory of invasibility. *Journal of Ecology* **88**, 528–534. doi:10.1046/j.1365-2745.2000.00473.x

DCE (1992) *Management Plan for Barmah State Park and Barmah State Forest*. Department of Conservation and Environment, Melbourne.

Díaz S, Noy-Meir I, Cabido M (2001) Can grazing response of herbaceous plants be predicted from simple vegetative traits? *Journal of Applied Ecology* **38**, 497–508. doi:10.1046/j.1365-2664.2001.00635.x

Díaz S, Lavorel S, McIntyre SUE, Falczuk V, Casanoves F, Milchunas DG, Skarpe C, Rusch G, Sternberg M, Noy-Meir I, Landsberg J, Zhang WEI, Clark H, Campbell BD (2007) Plant trait responses to grazing – a global synthesis. *Global Change Biology* **13**, 313–341. doi:10.1111/j.1365-2486.2006.01288.x

Dorrough J, Moxham C (2005) Eucalypt establishment in agricultural landscapes and implications for landscape-scale restoration. *Biological Conservation* **123**, 55–66. doi:10.1016/j.biocon.2004.10.008

Dorrough J, Scroggie P (2008) Plant responses to agricultural intensification. *Journal of Applied Ecology* **45**, 1274–1283. doi:10.1111/j.1365-2664.2008.01501.x

Dorrough J, Yen A, Turner V, Clark SG, Crosthwaite J, Hirth JR (2004) Livestock grazing management and biodiversity conservation in Australian temperate grassy landscapes. *Australian Journal of Agricultural Research* **55**, 279–295. doi:10.1071/AR03024

Dorrough J, Moxham C, Turner V, Sutter G (2006) Soil phosphorus and tree cover modify the effects of livestock grazing on plant species richness in Australian grassy woodland. *Biological Conservation* **130**, 394–405. doi:10.1016/j.biocon.2005.12.032

Dyksterhuis EJ (1949) Condition and management of range land based on quantitative ecology. *Journal of Range Management* **2**, 104–115. doi:10.2307/3893680

Fensham RJ, Skull SD (1999) Before cattle: a comparative floristic study of *Eucalyptus* savanna grazed by macropods and cattle in north Queensland, Australia. *Biotropica* **31**, 37–47.

Fensham RJ, Holman JE, Cox MJ (1999) Plant species responses along a grazing disturbance gradient in Australian grassland. *Journal of Vegetation Science* **10**, 77–86. doi:10.2307/3237163

Friedel MH, Foran BD, Stafford Smith DM (1990) Where the creeks run dry or ten feet high: pastoral management in arid Australia. *Proceedings of the Ecological Society of Australia* **16**, 185–194.

Garden DL, Ellis NJS, Rab MA, Langford CM, Johnston WH, Shields C, Murphy T, Holmberg M, Dassanayake KB, Harden S (2003) Fertiliser and grazing effects on production and botanical composition of native grasslands in south-east Australia. *Australian Journal of Experimental Agriculture* **43**, 843–859. doi:10.1071/EA02216

Grime JP (1973) Competitive exclusion in herbaceous vegetation. *Nature* **242**, 344–347. doi:10.1038/242344a0

Hancock CN, Ladd PG, Froend RH (1996) Biodiversity and management of riparian vegetation in Western Australia. *Forest Ecology and Management* **85**, 239–250. doi:10.1016/S0378-1127(96)03761-9

Harrington GN, Oxley RE, Tongway DJ (1979) The effects of European settlement and domestic livestock on the biological system in poplar box (*Eucalyptus populnea*) lands. *The Australian Rangeland Journal* **1**, 271–279. doi:10.1071/RJ9790271

Hayball N, Pearce M (2004) Influences of simulated grazing and water-depth on the growth of juvenile *Bolboschoenus caldwellii*, *Phragmites australis* and *Schoenoplectus validus* plants. *Aquatic Botany* **78**, 233–242. doi:10.1016/j.aquabot.2003.10.004

Hobbs RJ, Hopkins AJM (1990) From frontier to fragments: European impact on Australia's vegetation. *Proceedings of the Ecological Society of Australia* **16**, 93–114.

Hodgkinson KC, Müller WJ (2005) Death model for tussock perennial grasses: a rainfall threshold for survival and evidence for landscape control of death in drought. *The Rangeland Journal* **27**, 105–115. doi:10.1071/RJ05009

Humphries SE, Groves RH, Mitchell DS (1991) *Plant Invasions: the Incidence of Environmental Weeds in Australia*. Australian National Parks and Wildlife Service, Canberra.

James CD, Landsberg J, Morton SR (1999) Provision of watering points in the Australian arid zone: a review of effects on biota. *Journal of Arid Environments* **41**, 87–121. doi:10.1006/jare.1998.0467

Jansen A, Healey M (2003) Frog communities and wetland condition: relationships with grazing by domestic livestock along an Australian floodplain river. *Biological Conservation* **109**, 207–219. doi:10.1016/S0006-3207(02)00148-9

Jansen A, Robertson AI (2001) Relationships between livestock management and the ecological condition of riparian habitats along an Australian floodplain river. *Journal of Applied Ecology* **38**, 63–75. doi:10.1046/j.1365-2664.2001.00557.x

Joyce LA (1993) The life cycle of the range condition concept. *Journal of Range Management* **46**, 132–138. doi:10.2307/4002269

Kauffman JB, Krueger WC, Vavra M (1983) Impacts of cattle on streambanks in northeastern Oregon. *Journal of Range Management* **36**, 683–685. doi:10.2307/3898184

Landsberg J, Lavorel S, Stol J (1999) Grazing response groups among understorey plants in arid rangelands. *Journal of Vegetation Science* **10**, 683–696. doi:10.2307/3237083

Landsberg J, James CD, Maconochie J, Nicholls AO, Stol J, Tynan R (2002) Scale-related effects of grazing on native plant communities in an arid rangeland region of South Australia. *Journal of Applied Ecology* **39**, 427–444. doi:10.1046/j.1365-2664.2002.00719.x

Landsberg J, James CD, Morton SR, Muller WJ, Stol J (2003) Abundance and composition of plant species along grazing gradients in Australian rangelands. *Journal of Applied Ecology* **40**, 1008–1024. doi:10.1111/j.1365-2664.2003.00862.x

Lindsay EA, Cunningham SA (2009) Livestock grazing exclusion and microhabitat variation affect invertebrates and litter decomposition rates in woodland remnants. *Forest Ecology and Management* **258**, 178–187. doi:10.1016/j.foreco.2009.04.005

Lonsdale WM (1994) Inviting trouble: introduced pasture species in northern Australia. *Australian Journal of Ecology* **19**, 345–354. doi:10.1111/j.1442-9993.1994.tb00498.x

Lovett S, Price P (1999) *Riparian Land Management Technical Guidelines*. Land and Water Resources Research and Development Corporation, Canberra.

Low T (1997) Tropical pasture plants as weeds. *Tropical Grasslands* **31**, 337–343.

Lunt ID, Morgan JW (1999) Vegetation changes after 10 years of grazing exclusion and intermittent burning in a *Themeda triandra* (Poaceae) grassland reserve in south-eastern Australia. *Australian Journal of Botany* **47**, 537–552. doi:10.1071/BT98011

Lunt ID, Jansen A, Binns DL, Kenny SA (2007) Long-term effects of exclusion of grazing stock on degraded herbaceous plant communities in a riparian *Eucalyptus camaldulensis* forest in south-eastern Australia. *Austral Ecology* **32**, 937–949. doi:10.1111/j.1442-9993.2007.01782.x

Mac Nally R, Cunningham SC, Baker PJ, Horner GJ, Thomson JR (2011) Dynamics of Murray–Darling floodplain forests under multiple stressors: the past, present, and future of an Australian icon. *Water Resources Research* **47**, W00G05. doi:10.1029/2011WR010383

MacLeod ND, McIntyre S (1997) Stocking rate impacts on the production and economic performance of steers grazing black speargrass pastures. *The Rangeland Journal* **19**, 174–189. doi:10.1071/RJ9970174

Martin T, Possingham HP (2005) Predicting the impact of livestock grazing on birds using forage height data. *Journal of Applied Ecology* **42**, 400–408. doi:10.1111/j.1365-2664.2005.01012.x

McIntyre S, Lavorel S (2001) Livestock grazing in subtropical pastures: steps in the analysis of attribute response and plant functional types. *Journal of Ecology* **89**, 209–226. doi:10.1046/j.1365-2745.2001.00535.x

McIntyre S, Lavorel S (2007) A conceptual model of land use effects on the structure and function of herbaceous vegetation. *Agriculture, Ecosystems & Environment* **119**, 11–21. doi:10.1016/j.agee.2006.06.013

McIntyre S, Lavorel S, Landsberg J, Forbes TD (1999) Disturbance response in vegetation: towards a global perspective on functional traits. *Journal of Vegetation Science* **10**, 621–630.

McIntyre S, Martin TG (2001) Biophysical and human influences on plant species richness in grasslands: comparing variegated landscapes in subtropical and temperate regions. *Austral Ecology* **26**, 233–245. doi:10.1046/j.1442-9993.2001.01108.x

McIntyre S, McIvor JG, Heard KM (2002) *Managing and Conserving Grassy Woodlands*. CSIRO Publishing, Melbourne.

McIntyre S, Heard KM, Martin TG (2003) The relative importance of cattle grazing in subtropical grasslands: does it reduce or enhance plant biodiversity? *Journal of Applied Ecology* **40**, 445–457. doi:10.1046/j.1365-2664.2003.00823.x

McKergow LA, Weaver DM, Prosser IP, Grayson RB, Reed AEG (2003) Before and after riparian management: sediment and nutrient exports from a small agricultural catchment, Western Australia. *Journal of Hydrology* **270**, 253–272. doi:10.1016/S0022-1694(02)00286-X

McKergow LA, Prosser IP, Weaver DM, Grayson RB, Reed AEG (2006) Performance of grass and *Eucalyptus* riparian buffers in a pasture catchment, Western Australia, part 2: water quality. *Hydrological Processes* **20**, 2327–2346. doi:10.1002/hyp.6054

Milchunas DG, Lauenroth WK (1993) Quantitative effects of grazing on vegetation and soils over a global range of environments. *Ecological Monographs* **63**, 327–366. doi:10.2307/2937150

Milchunas DG, Sala OE, Lauenroth WK (1988) A generalized model of the effects of grazing by large herbivores on grassland community structure. *American Naturalist* **132**, 87–106. doi:10.1086/284839

Moles AT, Flores-Moreno H, Bonser SP, Warton DI, Helm A, Warman L, Eldridge DJ, Jurado E, Hemmings FA, Reich PB, Cavender-Bares J, Seabloom EW, Mayfield MM, Sheil D, Djietror JC, Peri PL, Enrico L, Cabido MR, Setterfield SA, Lehmann CER, Thomson FJ (2012) Invasions: the trail behind, the path ahead, and a test of a disturbing idea. *Journal of Ecology* **100**, 116–127. doi:10.1111/j.1365-2745.2011.01915.x

Mott RM (1986) Planned invasions of Australian tropical savannas. In *Ecology of Biological Invasions*. (Eds RH Groves and JJ Burdon) pp. 89–96. Cambridge University Press, Cambridge, UK.

Naiman RJ, Decamps H (1997) The ecology of interfaces: riparian zones. *Annual Review of Ecology and Systematics* **28**, 621–658. doi:10.1146/annurev.ecolsys.28.1.621

Naiman RJ, Decamps H, Pollock M (1993) The role of riparian corridors in maintaining regional biodiversity. *Ecological Applications* **3**, 209–212. doi:10.2307/1941822

Nelson KS, Gray EM, Evans JR (2011) Finding solutions for bird restoration and livestock management: comparing grazing exclusion levels. *Ecological Applications* **21**, 547–554. doi:10.1890/10-0032.1

NLWRA (2001a) *National Land and Water Resources Audit: Australian Agriculture Assessment*. National Heritage Trust, Canberra.

NLWRA (2001b) *National Land and Water Resources Audit: Australian Native Vegetation Assessment*. National Heritage Trust.

NLWRA (2002) *National Land and Water Resources Audit: Australians and National Resource Management*. National Heritage Trust.

Northup BK, Brown JR, Holt JA (1999) Grazing impacts on the spatial distribution of soil microbial biomass around tussock grasses in a tropical grassland. *Applied Soil Ecology* **13**, 259–270. doi:10.1016/S0929-1393(99)00039-6

Noy-Meir I, Gutman M, Kaplan Y (1989) Responses of Mediterranean grassland plants to grazing and protection. *Journal of Ecology* **77**, 290–310. doi:10.2307/2260930

Pettit NE, Froend RH, Ladd PG (1995) Grazing in remnant woodland vegetation: changes in species composition and life form groups. *Journal of Vegetation Science* **6**, 121–130. doi:10.2307/3236263

Popolizio CA, Goetz H, Chapman PL (1994) Short-term response of riparian vegetation to four grazing treatments. *Journal of Range Management* **47**, 48–53. doi:10.2307/4002840

Price JN, Wong NK, Morgan JW (2010) Recovery of understorey vegetation after release from a long history of sheep grazing in a herb-rich woodland. *Austral Ecology* **35**, 505–514. doi:10.1111/j.1442-9993.2009.02057.x

Prober SM, Thiele KR (1995) Conservation of the grassy white box woodlands: relative contributions of size and disturbance to floristic composition and diversity of remnants. *Australian Journal of Botany* **43**, 349–366. doi:10.1071/BT9950349

Prober SM, Lunt ID, Thiele KR (2002a) Determining references conditions for management and restoration of temperate grassy woodlands: relationships among trees, topsoils and understorey flora in little-grazed remnants. *Australian Journal of Botany* **50**, 687–697. doi:10.1071/BT02043

Prober SM, Thiele KR, Lunt ID (2002b) Identifying ecological barriers to restoration in temperate grassy woodlands: soil changes associated with different degradation states. *Australian Journal of Botany* **50**, 699–712. doi:10.1071/BT02052

Roath LR, Krueger WC (1982) Cattle grazing influence on a mountain riparian zone. *Journal of Range Management* **35**, 100–104. doi:10.2307/3898531

Robertson AI, Rowling RW (2000) Effects of livestock on riparian zone vegetation in an Australian dryland river. *Regulated Rivers: Research and Management* **16**, 527–541. doi:10.1002/1099-1646(200009/10)16:5<527::AID-RRR602>3.0.CO;2-W

Robinson D, Mann S (1998) *Effects of Grazing, Fencing and Licencing on the Natural Values of Crown Land Frontages in the Goulburn-Broken Catchment.* Goulburn Valley Environmental Group, Shepparton, Victoria.

Rumpff L, Duncan DH, Vesk PA, Keith DA, Wintle BA (2011) State-and-transition modelling for adaptive management of native woodlands. *Biological Conservation* **144**, 1224–1236. doi:10.1016/j.biocon.2010.10.026

Sala OE, Osterheld M, Léon RC, Soriano A (1986) Grazing effects upon plant community structure in subhumid grasslands of Argentina. *Vegetatio* **67**, 27–32.

Stohlgren TJ, Bull KA, Otsuki Y, Villa CA, Lee M (1998) Riparian zones as havens for exotic plant species in the central grasslands. *Plant Ecology* **138**, 113–125. doi:10.1023/A:1009764909413

Swincer DE (1986) Physical characteristics of sites in relation to invasions. In *Ecology of Biological Invasions: an Australian Perspective.* (Eds RH Groves and JJ Burdon) pp. 67–76. Australian Academy of Science, Canberra.

Tait J, Perna C (2001) Fish habitat management challenges on an intensively developed tropical floodplain. *RipRap* **19**, 14–21.

Thoms MC, Sheldon F (2000) Lowland rivers: an Australian introduction. *Regulated Rivers: Research and Management* **16**, 375–383. doi:10.1002/1099-1646(200009/10)16:5<375::AID-RRR591>3.0.CO;2-#

UNESCO (1971) *Convention on Wetlands of International Importance Especially as Waterfowl Habitat.* Ramsar Convention, Iran.

VEAC (2006) 'River red gum forests investigation discussion paper.' Victorian Environmental Assessment Council, Melbourne

Vesk PA, Leishman MR, Westoby M (2004) Simple traits do not predict grazing response in Australian dry shrublands and woodlands. *Journal of Applied Ecology* **41**, 22–31.

Vesk PA, Westoby M (2001) Predicting plant species' responses to grazing. *Journal of Applied Ecology* **38**, 897–909. doi:10.1046/j.1365-2664.2001.00646.x

Vidon P, Campbell MA, Gray M (2008) Unrestricted cattle access to streams and water quality in till landscape of the Midwest. *Agricultural Water Management* **95**, 322–330. doi:10.1016/j.agwat.2007.10.017

Weaver DM, Clements FE (1938) *Plant Ecology.* McGraw-Hill, New York, NY.

Wedin DA (1999) Nitrogen availability, plant-soil feedbacks and grassland stability. In *Sixth International Rangeland Congress Proceedings.* 19–23 July, Townsville, Queensland. (Eds D Eldridge and D Freudenberger) pp. 193–197.

Westoby M, Walker B, Noy-Meir I (1989) Opportunistic management for rangelands not at equilibrium. *Journal of Range Management* **42**, 266–274. doi:10.2307/3899492

Williams M (1988) The clearing of the woods. In *The Australian Experience: Essays in Australian Land Settlement and Resource Management.* (Ed. RL Heathcote) pp. 115–126. Longman Cheshire, Melbourne.

Williams CK, Parer I, Coman BJ, Burley J, Braysher ML (1995) *Managing Vertebrate Pests: Rabbits.* Bureau of Resource Sciences/CSIRO Division of Wildlife and Ecology. Australian Government Publishing Service, Canberra.

Williams L, Reich P, Capon SJ, Raulings E (2008) Soil seed banks of degraded riparian zones in southeastern Australia and their potential contribution to the restoration of understorey vegetation. *River Research and Applications* **24**, 1002–1017. doi:10.1002/rra.1123

Wilson AD (1990) The effect of grazing on Australian ecosystems. *Proceedings of the Ecological Society of Australia* **16**, 235–244.

Wilson A, Jansen A, Curtis A, Robertson A (2006) Measuring riparian condition: a comparison of assessments by landholders and scientists. *Ecological Management & Restoration* **7**, 123–129. doi:10.1111/j.1442-8903.2006.00266.x

Young A (2000) *Environmental Change in Australia Since 1788.* 2nd edn. Oxford University Press, Melbourne.

# 18

# Alien plant invasions in Australia's riparian zones

*Jane Catford and Garreth Kyle*

## INTRODUCTION

Biological invasions have been identified as one of the main components of global environmental change (Millenium Ecosystem Assessment 2005) and are considered the second largest threat to biodiversity after habitat destruction (Sala *et al.* 2000). Invasive plant species can alter ecosystem structure and function (Richardson and Pysek 2006), compromising the services that ecosystems provide. Accounting for 13% of Australia's flora, there are currently over 3000 alien plant species growing wild in Australia, with a further 6000 species predicted to naturalise in the future (Randall 2007). Alien plants cost more than $4 billion each year in Australia (Sinden *et al.* 2005) and put native biodiversity directly at risk: of 1302 plant species listed under Australia's *Environment Protection and Biodiversity Conservation Act 1999*, more than two-thirds are threatened by alien plants (DSEWPC 2012). The extent and effects of alien plant invasion are particularly apparent in riparian zones.

Riparian zones often experience high levels of invasion compared with other ecosystems (Fig. 18.1; Stohlgren *et al.* 2002; Richardson *et al.* 2007; Catford *et al.* 2012b). For example, of 24 habitat types examined in Catalonia, Spain, riparian ecosystems were among the four most invaded habitat types with respect to alien species richness (Vilà *et al.* 2007). Based on the proportion of the alien species pool that invades a given vegetation type, Catford and Jansson (2014) found that riparian scrub was the third most invaded of 20 vegetation types in Victoria, Australia, mirroring global trends. As well as having significant cultural and socio-economic values, riparian zones are crucial for nutrient processing, water purification and for the provision of resources for native plants and animals. Although these ecosystem functions may be compromised by plant invasions, traditional weed control techniques may further erode them, presenting a dilemma for natural resource managers.

In this chapter, we focus on aspects of alien plant invasion most relevant to Australia's riparian zones. We discuss the primary mechanisms and pathways of invasion, the functional characteristics of invading species, impacts of alien species, management options and key knowledge gaps. Though some authors contend that native species should be included in discussions of biological invasions (Valéry *et al.* 2013), we focus on species that are alien (i.e. exotic, foreign, non-indigenous, non-native, introduced) to Australia. We note that there are both invasive and non-invasive alien species. Though definitions vary (Colautti and MacIsaac 2004), alien species are generally considered invasive when they spread beyond the area where they were initially introduced and become dominant in a community (Richardson *et al.* 2000).

**Figure 18.1:** Rivers provide great environments for recreation but are very prone to weed invasion. Much of the Murray River in Barmah National Park, Victoria, is fringed with invasive weeping willow (*Salix babylonica*) and delta arrowhead (*Sagittaria platyphylla*), pictured here. (Source: J.A. Catford.)

## STAGES AND MECHANISMS OF PLANT INVASION

Plant invasion is the result of three interdependent factors, all of which must be accommodating, if not favourable, for invasion to occur (Catford *et al.* 2009). The first is propagule availability or propagule pressure, defined as the number of individuals (including seeds and clonal fragments) introduced per event multiplied by the frequency of introduction events (Eppstein and Molofsky 2007). Propagule pressure is a prerequisite for invasion and the greater the number of propagules introduced, the greater the likelihood of successful invasion (Simberloff 2009). Second, the physical, or abiotic, characteristics of a habitat must be within the environmental tolerance range of an alien species for invasion to occur. Third, invasion outcomes are affected by the biological characteristics of, and interactions among, the invaders and the recipient community. The relative importance of these three factors varies in space and time.

### Propagule transport and introduction

Invasion of alien species requires the transport, either intentional or unintentional, of plant propagules (i.e. adult plants, seeds, spores or vegetative fragments) to a new location. Alien species may be imported as a commodity, arrive with a transport vector, or may spread from another invaded region (Hulme *et al.* 2008). Global patterns suggest that the majority of terrestrial and aquatic plants (both of which invade riparian ecosystems) escape into natural ecosystems after being intentionally introduced, most often for ornamental purposes or use in aquaria (Hulme *et al.* 2008). In Australia, most alien plant species have been introduced for domestic gardens, agriculture and pastoralism (Randall 2007). *Annona glabra* (pond apple), for example, was originally imported as a rootstock for custard apples, while *Ulex europaeus* (gorse) was introduced as an ornamental hedge, and *Rubus fruticosus* (blackberry) for its fruit. All of these species have since escaped and represent significant problems in riparian or wetland systems (Anon 2008). The weeds of national significance (WoNS) ranks problematic species according to invasiveness, potential for spread, impact, and ecological and socio-economic values of species (Anon 2008). Fifteen of the 32 WoNS are known to be successful in riparian or aquatic systems and were introduced into Australia intentionally (AWC 2012). However, only a few were deliberately introduced into riparian zones (e.g. *Hymenachne amplexicaulis* for ponded pasture, Smith 1995; and *Salix nigra* for soil stabilisation, Ladson 1997).

Alien species may escape into riparian zones because human activities often concentrate in and around rivers (Fig. 18.1). Riparian systems, being effectively linear, have a large edge to core ratio, which makes them particularly vulnerable to species invading from surrounding ecosystems (Cumming 2002). Towns are frequently located near rivers, as are areas of agriculture, which benefit from the fertile soil and proximate water supply. Organic waste from gardens and agriculture may be dumped in riparian zones, and alien species used in aquaria may be released into waterways. Centres of human activity thus provide a ready source of propagules.

### Colonisation and establishment

After arriving at a site, introduced plants must colonise and survive. Native and alien species undergo the same colonisation process (Meiners *et al.* 2004) and both need space to obtain resources. Riparian ecosystems typically have high availability of the resources (especially nutrients and water) that are necessary for the establishment and persistence of plants. Resource availability in riparian zones fluctuates because river systems are temporally and spatially dynamic (Richardson *et al.* 2007). Pulses in resource availability may provide an opportunity for alien species to colonise and establish (Davis *et al.* 2000). Riparian zones experience high levels of disturbance from floods and human activities, such as logging and grazing. By reducing the uptake of resources (e.g. via removing standing vegetation) and increasing resource supply (e.g. through the deposition of nutrient-rich alluvium), such disturbance events can increase overall resource availability, providing space and opportunities for colonisation (Bornette and Amoros 1996).

Invasion typically occurs more frequently during the early stages of succession, which are characterised by high resource availability and low competition (Catford *et al.* 2012a). Accordingly, many invasive species are early successional species adept at colonisation (Rejmanek and Richardson 1996), although this may also reflect a bias towards good colonisers in the introduced species pool (Colautti *et al.* 2006). By reaching reproductive maturity quickly, invading species can exploit fluctuations in resource availability and become naturalised. For species to become naturalised, pioneer populations must be self-sustaining and not rely on continued input of propagules.

As transition zones between aquatic and terrestrial habitats that are subject to frequent flood disturbance, riparian habitats have high levels of environmental heterogeneity. As a consequence, riparian zones can also host species with a wide range of life histories and environmental preferences (Catford and Jansson 2014). If alien species are functionally distinct from native species, they may be able to inhabit an 'empty niche' unoccupied by resident species (Hierro *et al.* 2005). The potential for unoccupied niches is greater in heterogeneous environments such as riparian zones (Melbourne *et al.* 2007).

Although disturbance is a leading explanation for invasion (Moles *et al.* 2012), native riparian species would presumably be well adapted to frequent disturbance from floods (Catford *et al.* 2014). As such, individual disturbance events may not be a key driver of invasion in riparian ecosystems. Instead, changes to disturbance regimes may be more important in facilitating invasions (Moles *et al.* 2012). Human activities can alter nutrient levels, geomorphic processes and light availability in riparian ecosystems, but the flow regime is arguably the most important force structuring these systems (Merritt *et al.* 2010).

Modification of the timing, duration, frequency, magnitude and predictability of flooding and drying may alter the environmental conditions and the types of niches available in riparian zones (Catford *et al.* 2011). If native species are adapted to historical flow regimes and are unable to tolerate modified conditions, invaders may benefit from reduced competition and an increase in resource availability (Catford *et al.* 2011). Invasion will be enhanced if alien species are pre-adapted (or adapt rapidly) to the modified conditions (Mack 2003). For example, the success of alien *Tamarix* species in riparian zones of regulated rivers in western USA can be at least partly attributed to their greater tolerance of modified flood regimes compared with native riparian tree species (Stromberg *et al.* 2007). Along the regulated Murray River in south-eastern Australia, modification of wetland flooding regimes appears to have inhibited native species adapted to historical conditions and facilitated alien plant invasion (Catford *et al.* 2011). Trait-based analyses suggest that alien species are not specifically adapted to the modified conditions; rather, alien species may indirectly benefit from a decline in competition from native species that are poorly adapted to the new conditions (Catford *et al.* 2014).

Invading species may become established at a site because of favourable biotic interactions. Regardless of the type of ecosystem, invaders can lose and gain biotic relationships on entry into a new range (Mitchell *et al.* 2006). While the loss of mutualists can impede alien species' establishment, the loss of pathogens and herbivores may be of benefit (Keane and Crawley 2002). 'Enemy release' can reduce mortality and enable alien species to re-allocate resources from defence to growth and reproduction (Joshi and Vrieling 2005). Factors that facilitate naturalisation may enable invaders to outcompete native species, shifting the relationship from coexistence to invader dominance.

## Population spread

Although localised dominance by an alien species can be ecologically important (Colautti and MacIsaac 2004), species are only considered invasive when their populations spread beyond the site where first introduced. There is commonly a lag phase between introduction and spread (Kowarik 1995). The lag phase is typically attributed to the time required for alien species to genetically adapt to the conditions of its 'new' region (Joshi and Vrieling 2005) or for propagule pressure to increase and environmental conditions to become more suitable for the invading species (Keane and Crawley 2002; Richardson and Pysek 2006). However, the lag phase may simply reflect exponential population growth where it is initially slow, later increasing at a rapid rate (D. Tilman, pers. comm.).

Riparian zones offer an additional method of dispersal for plants in the way of hydrochory, or water-borne dispersal. Although many alien species are not specifically adapted to hydrochory (Stokes *et al.* 2010), they may opportunistically use this dispersal method, increasing the chance of their success (Eschtruth and Battles 2011). Based on monthly surveys of seeds caught in drift nets along two regulated and two free-flowing rivers in Victoria over the course of a year, Greet *et al.* (2012) found that more exotic species used hydrochory in summer and autumn, whereas native seeds were mostly found in the drift in winter and spring, reflecting the time at which they produce and release seeds. In the two regulated rivers, the timing of peak flows has changed from winter–spring to summer–autumn. Given that most propagules are dispersed during peak flows, Greet *et al.* (2012) concluded that plant phenology (i.e. timing of seed release), hydrochory and flow regulation may jointly facilitate increased dominance and spread of alien plants in regulated rivers.

As well the initial transport and introduction of alien species, humans can facilitate spread of alien species by acting as additional dispersal vectors and by increasing their abundance in the landscape through planting in horticulture and agriculture (Hodkinson and Thompson 1997; Catford *et al.* 2012a). The frequent use of rivers and riparian zones by humans increases the chance of alien species spread. At least half of the riparian WoNS benefit from secondary anthropogenic dispersal (e.g. *Alternanthera philoxeroides* can be dispersed in mud attached to machinery or in turf; Parsons and Cuthbertson 1992). Aside from enabling geographic spread, higher dispersal success and propagule supply can increase the dominance of alien species by enabling alien species to dominate the seed bank (Catford *et al.* 2012a). With high numbers of propagules and frequent disturbance, invaders can become established and dominant based on numbers alone.

In summary, riparian ecosystems may be subject to disproportionately high levels of invasion compared with other ecosystems because they have high edge to core ratios, high rates of introduction because of their proximity to human activities, additional dispersal opportunities provided by water and human vectors, and they experience frequent flood disturbance, which increases resource availability and maintains high habitat heterogeneity (Richardson *et al.* 2007; Eschtruth and Battles 2011; Catford and Jansson 2014). Alteration of flow regimes and other environmental conditions can prompt a decline in native species and may directly and indirectly facilitate the establishment of alien species better adapted to the modified conditions (Catford *et al.* 2011).

# TRAITS OF INVASIVE PLANTS

Since Baker (1965) first suggested the profile of an 'ideal weed', the search for plant functional traits associated with invasive species has been the focus of much research (Pysek and Richardson 2007; van Kleunen *et al.* 2010). Isolating a syndrome of traits associated with successful invaders is seen as a practical alternative to compiling an exhaustive list of potentially invasive species. Although a single invasive syndrome has proved elusive (Pysek and Richardson 2007; Moles *et al.* 2008), invasive alien species tend to exhibit higher performance than non-invasive aliens and native species, as indicated by growth rate, plant size, leaf-area allocation, shoot allocation, fitness and physiology (van Kleunen *et al.* 2010).

Many of the characteristics associated with invasiveness are particularly advantageous in frequently disturbed ecosystems with high resource availability, like riparian zones. For example, vegetative reproduction, persistent seed banks and an annual life history help plants cope with flood disturbance (Catford and Jansson 2014), and many of these traits are common among invasive species (Pysek and Richardson 2007; van Kleunen *et al.* 2010). Rather than being riparian specialists, it seems more likely that most invasive species opportunistically invade riparian ecosystems. Indeed, of the 170 plant species in the global invasive species database (http://www.issg.org/database) that invade riparian ecosystems, only two invade riparian ecosystems exclusively (Catford and Jansson 2014). Patterns in Victoria, Australia are similar: of 518 alien plants that invade riparian scrubs, only 10 species invade this vegetation type exclusively; most species have a much broader range (Catford and Jansson 2014). We briefly describe some of the plant traits likely to facilitate invasion in riparian ecosystems. A more complete review of plant adaptations to riparian environments is provided in Catford and Jansson (2014).

## Propagules and dispersal

High fecundity, where a large number of propagules are produced per plant, increases the probability of colonisation and is therefore key to facilitating invasion. *Sagittaria platyphylla*, a macrophyte native to North America that invades wetlands, riverbanks and irrigation channels in south-eastern Australia, produces 6900 seeds per inflorescence, and seeds float for seven days (Fig. 18.2; Adair *et al.* 2012). The species can also reproduce from vegetative fragments, which are buoyant, and can spread laterally via ramets and tubers at the terminal end of tubers (Adair *et al.* 2012). Its high rate of seed production, capacity for clonal spread and re-sprouting ability makes its populations very difficult to control and contain.

The capacity for asexual reproduction is an important trait for invasive species in riparian systems (Boedeltje *et al.* 2003; Lake and Leishman 2004). This is because clonal spread and reproduction from vegetative fragments are independent of flowering and fruiting, can enhance dispersal downstream (especially as most fragments are buoyant; Catford and Jansson 2014) and enable quicker recovery after burial (Xiong *et al.* 2001). Many invasive willows (*Salix* spp.) cannot reproduce sexually in Australia because they are dioecious and only a single sex was introduced to the country (exceptions include *S. cineara*, grey sallow, and *S. nigra*, black willow; Holland-Clift and Davies 2007). Instead, most willows spread via vegetation fragments (e.g. twigs, branches), which can disperse great distances downstream (Holland-Clift and Davies 2007).

Long-distance dispersal typically facilitates rapid colonisation of the landscape and allows species to disperse far enough to avoid localised disturbance, ensuring the persistence of the genotype at the landscape level (Pausas and Lavorel 2003). The capacity for long-distance dispersal is therefore important for plant invasions. Compared with other dispersal vectors, a large proportion of hydrochores, especially when buoyant, can disperse long distances (Andersson *et al.* 2000; Catford and Jansson 2014). Humans, wind and animals also act as important dispersal vectors of invasive plants (Lake and Leishman 2004; Taylor *et al.* 2012).

## Rapid growth

Factors that influence capacity for rapid growth have been associated with increased invasive

**Figure 18.2:** Many invasive species can reproduce sexually and asexually and use water as a dispersal vector. (a) Delta arrowhead (*Sagittaria platyphylla*), invasive in south-eastern Australia, produces 6900 seeds per inflorescence, which help it to increase in population size and spread downstream (b) starting new, satellite populations; (c) it also spreads clonally. (Source: J.A. Catford.)

success among alien species. Rapid population growth enables a population to reach reproductive maturity quickly, providing an additional source of propagules that will accelerate rates of spread, especially when the population is at the frontier of the species' current range. Rapid growth also helps individuals to complete their life cycle while environmental conditions are suitable. As riparian zones have high resource availability and are temporally dynamic, they are particularly suitable for species with high growth rates. High growth rates are also important in riparian systems because they allow plants to grow to sufficient size to withstand the physical stresses associated with flooding. Characteristics often associated with rapid population growth include high relative growth rates, rapid maturation, short generation times and lifespans, high fecundity, small seeds, fast leaf economics (e.g. high specific leaf area, low leaf dry matter content) and low root biomass (Grime 1977).

In Australian riparian systems, alien species have been found to have a higher specific leaf area than native species (Lake and Leishman 2004; Kyle and Leishman 2009; Catford *et al.* 2014). Specific leaf area is a measure of investment in leaf production, with a higher specific leaf area indicating lower leaf production cost and faster return on investment, and is associated with faster growth. Similarly, species with an annual or biennial life history colonise, establish and reproduce within one or two growing seasons, and therefore tend to have much faster growth rates than perennial plants. Unlike native riparian species, the majority of alien species invading riparian zones are often annual or biennial, not perennial (Catford *et al.* 2011). This has led some authors to posit that the faster life histories of alien annuals may facilitate their dominance in riparian zones after floodwaters recede (Lunt *et al.* 2012). However, a trait-based test of this hypothesis in Murray River wetlands found no support for it (specific leaf area did not influence the distributions of exotic and native species either; Catford *et al.* 2014).

## Tolerance of stressful conditions

Longer-lived alien species that inhabit riparian zones would need traits that allow them to persist through the challenges presented by riparian environments. Through several physiological adaptations, successful riparian invaders such as willows are capable of tolerating prolonged periods of inundation (Kozlowski 2002), allowing them to colonise frequently inundated zones along the river edge. The ability to tolerate highly saline soils has allowed *Tamarix aphylla* (athel pine) to invade hypersaline environments where *Eucalyptus camaldulensis* (river red gum) cannot survive (Griffin *et al.* 1989).

Invasion success is context-dependent and the functional traits associated with alien species cannot be considered in isolation from the ecosystems they invade. In riparian environments, traits associated with invasiveness may include rapid growth and resource acquisition, vegetative reproduction, high propagule output, and tolerance of burial and inundation. Unsurprisingly, these characteristics are similar to those possessed by native riparian species. A point of differentiation may occur when environmental conditions are modified such that native and alien species' adaptations differ in their suitability.

## ECOLOGICAL IMPACTS OF INVASIVE PLANTS

Most alien species do not impact the ecosystems they invade, but just add to species richness. However, invasive alien species can alter ecosystem structure and function, reduce biodiversity and compromise ecosystem resilience. They can also threaten human health, reduce visual amenity and cost the economy billions of dollars (Sinden *et al.* 2005; Richardson and Pysek 2006). Impacts of alien species are conceived as being a function of their abundance, range and per capita (or per biomass) effects (Parker *et al.* 1999). While abundance and range can be quantified relatively easily, effects take many different forms and are often hard to attribute to just one cause. Indeed, invasion may be a result or a cause of environmental change (MacDougall and Turkington 2005), and invasion by some species can be both. Plant species that alter the environmental characteristics of an ecosystem, typically to their own advantage, are referred to as transformer species (ecosystem engineers in the case of animals) (Richardson *et al.* 2000). It is often difficult to determine whether invasive species are drivers, passengers or transformers of environmental change, yet the management responses to each of these differ.

We acknowledge that alien species can have positive, neutral and negative impacts on the ecosystems they invade (Shackelford *et al.* 2013), but here we focus on some of the negative ecological effects of invasive alien species in riparian zones because these are typically the focus of management actions.

### Abiotic effects

Invasive plants can affect riparian or wetland systems through alteration of the physical environment, including changes in light levels, soil and water chemistry, hydraulics and sedimentation rates. Willows have been found to cause a reduction in sedimentation, leading to scour in extreme cases (Ladson 1997), whereas other species increase sediment deposition rates (e.g. *Sagittaria platyphylla*, a WoNS) by slowing the velocity of streamflow and increasing the capacity to trap sediment (Malanson 1993). In the Australian context, both athel pine (Griffin *et al.* 1989) and para grass (*Urochloa mutica*; Bunn *et al.* 1998) have been implicated in increased sedimentation rates. In the Northern Territory, increased sedimentation following the invasion of para grass has been shown to decrease channel capacity by up to 85% (Bunn *et al.* 1998). This has resulted in a substantial increase in the frequency of overbank flows and may have important implications for the spread and colonisation of para grass propagules from the river during these events.

Invasive species can also alter the chemical characteristics of an environment through several pathways including the addition or consumption of resources, such as nitrogen or oxygen (Richardson *et al.* 2000). In wetland systems, the consumption of large amounts of oxygen by invasive macrophytes can produce highly anoxic conditions with flow-on consequences for aquatic biota. For example, the removal of water hyacinth (*Eichhornia crassipes*) from the Burdekin delta in northern Queensland resulted in significantly improved levels of dissolved oxygen, dramatically improving the capacity of the lagoons to support fish species (Perna and Burrows 2005). Invasive plants are also capable of altering the chemical composition of the soil. Salt exudates from the leaves of athel pine can be up to 50 times the concentration of the groundwater (Berry 1970). This is believed to have been a contributing factor to the increase in abundance of salt-tolerant chenopod and grass species in regions invaded by athel pine along the Finke River in central Australia (Griffin *et al.* 1989).

## Biotic effects

Invasive plant species can have numerous effects on the ecological components of riparian systems, including profound changes in community composition. In the Northern Territory, invasive species have had a significant impact on the composition of native wetland plant communities. Mono-specific infestations of mimosa (*Mimosa pigra*) have replaced native sedgelands and have prevented the re-establishment of *Melaleuca* seedlings along watercourses (Walden *et al.* 2004). Similarly, the control of water buffalo has allowed para grass to successfully outcompete *Phragmites karka*, forming mono-specific swards (Finlayson *et al.* 1997). The invasion of para grass also has implications for the fire regime of the region, with larger fuel loads promoting hotter fires (Finlayson *et al.* 1997). These fires represent a significant threat for neighbouring *Melaleuca* wetlands and fire-sensitive monsoonal rainforest communities. Over time, alien plant invasion may lead to the extinction of native species (Gilbert and Levine 2013).

Invasive plants may affect native fauna directly and indirectly by altering habitat and food availability, environmental conditions and community interactions (e.g. changes in abundance of predators and mutualists). Griffin *et al.* (1989) documented a reduction in bird and lizard abundance and diversity in athel pine infestations compared with neighbouring river red gum. These differences were attributed to structural differences such as reduced abundance of woody debris and tree hollows, inadequate food supply due to smaller flowers, and increased salinity. Structural differences and insufficient resource provision were also implicated in the avoidance of riparian infestations of rubber vine (*Cryptostegia grandiflora*) by native lizards (Valentine *et al.* 2007). The abundance and diversity of native birds (Holland-Clift *et al.* 2011) and terrestrial arthropods (Greenwood *et al.* 2004) were lower in Victorian riparian zones dominated by invasive white-crack willow (*Salix* x *rubens*) than in those dominated by native vegetation. These findings were thought to reflect changes in habitat characteristics and food availability, including seasonal availability of resources (Greenwood *et al.* 2004; Holland-Clift *et al.* 2011).

# MANAGEMENT OF INVASIVE PLANTS

Preventing the introduction of invasive alien species will be the most effective strategy for minimising their impacts (Pheloung *et al.* 1999). Once invasive alien species are established though, their negative impacts can be ameliorated by reducing their population size and limiting spread. Weed control methods include mechanical or hand removal, mulch or weed mats, shading, fire and flame guns, biological control agents, slashing, ringbarking, controlled grazing, native plant competition or manipulation of environmental conditions. Ideally, potentially harmful alien species will be detected early, and eradicated while their populations are still confined and in the lag phase (Hobbs and Humphries 1995). Different control measures will be suitable depending on the stage and extent of a weed incursion, the target species, characteristics of the site, and resources available. For example, densely vegetated riparian zones can be hard to access, making manual and mechanical control challenging. Because of the sensitivity of freshwater ecosystems to chemicals, herbicide use is restricted. In Table 18.1, we provide a brief overview of the major methods of weed control and list their advantages and disadvantages for use in riparian zones. By targeting alien species with a suite of control measures, integrated weed management will often be more effective than a single approach (Rea and Storrs 1999). Active weed management may need to be sustained to yield lasting ecological outcomes (Harms and Hiebert 2006).

With any form of weed management, the indirect impact of alien species removal, as well as control, should be considered (Flory and Clay 2009). Impacts of weed control include erosion of bare banks, disruption of ecosystem processes, increase in water temperature after removal of vegetation, fouling of water by decomposing vegetation (Ainsworth and Bowcher 2005), loss of faunal habitat (Sogge *et al.* 2008), and disturbance of areas resulting in subsequent invasion by other alien species (Buckley *et al.* 2007). In some cases, mechanical control can result in fragmentation of plant material, resulting in further spread in waterbodies (e.g. *Alternanthera philoxeroides*; Parsons and Cuthbertson 1992).

Non-target impacts can be reduced by removing alien species gradually, using herbicides that target certain growth forms but not others (e.g. woody weeds but not grass) and leaving roots and stems intact to minimise erosion, then revegetating with native species (Ainsworth and Bowcher 2005). In limited cases, it may be inappropriate to control alien species because they provide an ecological function, such as provision of bird-breeding habitat (Sogge *et al.* 2008) or site colonisation in the early stages of secondary succession (Lee *et al.* 1986). Kyle *et al.* (2008) found that survival and growth of river red gum seedling tubestock was significantly higher when grown among alien willow and *Foeniculum vulgare* (fennel) compared with sites that had been slashed before planting. As succession progresses, it is expected that later-successional native species will shade and outcompete alien colonisers, reducing the need for specific weed management, though this is not guaranteed (Catford *et al.* 2012a).

Reflecting the myriad ways that alien species may become invasive in riparian ecosystems, riparian weed management is fraught with difficulty. The high frequency of disturbance and ready availability and dispersal of alien propagules means that management needs to be long term, sustained and catchment wide. There are currently 2739 alien plant species classified as invasive in Australia, many of which invade riparian ecosystems (Randall 2007). A species-based approach is clearly neither sustainable nor practical, though alien species that have particularly strong and undesirable impacts can be directly targeted in control.

Instead of species-specific measures, effort should be directed into maximising ecological resistance and resilience to invasion by rejuvenating native plant communities or manipulating environmental conditions. For example, provision of environmental flows that resemble historical conditions in regulated rivers has the potential to favour native species over alien ones (Catford *et al.* 2011). For ecological restoration, planting native species with similar traits as potential invaders is likely to increase biotic resistance and reduce the number of colonisation opportunities of alien species (Funk *et al.* 2008). Manipulating abiotic conditions may be a cost-effective weed control technique that yields multiple environmental benefits (Rood *et al.* 2003).

## CURRENT KNOWLEDGE GAPS AND FUTURE RESEARCH DIRECTIONS

There is considerable knowledge and understanding of riparian plant invasions in Australia. However, knowledge gaps, and research opportunities, remain. We outline three of these below.

(1) *Identifying and quantifying impacts of invasion* – Invasive alien species are primarily of concern because of their ecological and socio-economic impacts. However, it is very difficult to identify, isolate and quantify the magnitude of their impacts, especially when they are indirect, like changes to ecological processes and ecosystem services. Given that management budgets are limited, it is essential that the costs and threats of individual species can be compared so that

**Table 18.1.** Key approaches used for managing alien species and weed populations with recommendations for use in riparian zones

| Approach | Description | Advantages | Disadvantages | Resources | Examples | Recommendations |
|---|---|---|---|---|---|---|
| Physical | Physical destruction: machinery, hand weeding, mulch, weed mats, shading, ringbarking, tilling, slash, burn. | Can be used immediately after incursion; research often not required; hand weeding is targeted; eradication possible if treated early enough; can target single or multiple species. | Disturbance with potential for future invasion; access issues, especially with machines; potential soil erosion and bank destabilisation if not revegetated. | Labour, machinery, equipment. | Patchy mimosa infestations successfully treated (Rea and Storrs 1999); large-scale weed removal project in South Africa (Working for Water; Hobbs 2004). | Act early; suitable for small or patchy infestations; revegetate sites after weed removal and/or leave innocuous roots in ground. |
| Chemical | Herbicide application: spot spray, boom spray, aerial spray, cut stump, stem injection, basal bark application, granules, wick wiping. | Access issues vary depending on application method; effective at times; roots remain so bank destabilisation minimal; eradication possible if treated early; can target single or multiple species. | Non-target effects on native plants and aquatic organisms (death, bioaccumulation); potentially polluting (Ainsworth and Bowcher 2005); success dependent on application time and method; variable effectiveness; usually strict regulations. | Financial cost of herbicide; human effort varies depending on application method; regulatory approval. | Herbicide largely unsuccessful on mimosa (Lane *et al.* 1997); potential hazard of granulated herbicide use in the Top End (Rea and Storrs 1999). | Act early; only use during drawdown or in terrestrial zones; treat when least herbicide required; see Ainsworth and Bowcher (2005) for further information. |
| Biological | Plant pathogens; biological control agents (usually insects, native or alien, often from home range of the invader). | Targeted; very effective at times; agents can access areas inaccessible to humans and machines; reduction in alien species' population size; no physical disturbance; species-specific (stringent regulations); rigorous research and testing. | Long delay between alien species incursion and biocontrol agent release due to time required for rigorous research and risk assessment; chance of failure or non-target effects; very small risk of agent becoming invasive; cannot eradicate, only limit alien species; only targets one species. | Research and regulatory approval (resources and time); labour-intensive; need for continued releases (sometimes). | *Maravalia cryptostegiae* rust defoliates and kills *Crystostegia grandifolia* (Panetta *et al.* 1998); *Cyrtobagous salviniae* weevil can control *Salix* spp. (Julien and Griffiths 1998). | Only suitable for most invasive species due to expense and time of research; can effectively keep populations in check, especially where other methods have failed. |
| Biotic | Increase competition from native vegetation; enhance resistance and resilience of native vegetation. | No non-target risks; sites remain vegetated; ensures maintenance of ecosystem functions and processes (e.g. nutrient processing, bank stability). | Does not eradicate alien species, only reduces colonisation, dominance and spread; partial success likely, leading to coexistence of native and alien species. | Financial cost of revegetation (depending on scale and current condition). | Approach advocated by Funk *et al.* (2008) for restoration. | Ecosystem-level approach with multiple benefits. |

| Environmental | Manage abiotic conditions to discourage alien species and encourage native plant species. | Widespread control; not harmful to ecosystems if conditions are consistent with 'natural' conditions. | Socio-economic and political hurdles; changing one regime (e.g. flooding) may not control alien species if others remain (e.g. grazing); effects subtle, long-term, hard to detect; cost-benefit and effectiveness difficult to assess; requires healthy seedbank or suitable recruits in region. | Political and logistic costs; potential compensation for stakeholders (e.g. flooding of land or livestock removal). | Restoration of pre-regulation flow on regulated Murray River may decrease alien species in floodplain wetlands (Catford *et al.* 2011; Moles *et al.* 2008). | May not cause an immediate reduction in alien species, but offers long-term, ecosystem-wide benefits; particularly suitable for adaptive management approach. |
|---|---|---|---|---|---|---|
| Integrated | A range of control methods, tailored to a local problem and conditions, used concurrently (Rea and Storrs 1999). | Comprehensive; can be effective as alien species less able to adapt to different control practices; site and incursion-specific. | Most effective with detailed knowledge of problem and environment. | Depends on methods selected, but costs likely high (i.e. potentially all of the resources listed in this table). | Integrated management advocated as best approach to control mimosa (Paynter and Flanagan 2004) | Highly recommended and widely supported because it tackles alien species from multiple angles. |

management priorities can be objectively determined. However, species rarely have the same effect, so measuring these costs in a comparable way is challenging. Although some efforts have been made (e.g. Barney *et al.* 2013), this area is ripe for further research. Knowing the efficacy of management approaches in different contexts is also required.

(2) *Restoring historical environmental conditions and disturbance regimes* – If invasive species are a passenger of environmental change, restoring the historical environmental conditions, or changing them to favour natives over invaders, may be an effective way to manage alien species and facilitate native species. However, this requires detailed understanding of the system and the species in question. In the case of flow regulation, it often is unrealistic to aim for the complete restoration of historical flood regimes because of logistics, including limited water volumes. Environmental flows can be delivered to mimic historical conditions, but to get the most out of these flows, it is essential that the relative influence of flood magnitude, frequency, seasonality and duration on riparian vegetation is better understood (Catford *et al.* 2014).

(3) *Managing novel ecosystems* – Given global environmental change, including increasing pressure on water resources (Capon *et al.* 2013), it may not be possible to restore or maintain historical environmental conditions of riparian zones; novel ecosystems characterised by new species assemblages that result from human-induced ecological changes may be more likely (Hobbs *et al.* 2009; Catford *et al.* 2013). In novel ecosystems, dominance by invasive alien species may be inhibited by introducing native (or non-invasive alien) species that are capable of competing with invasive alien species under the modified conditions. Determining the types of species, based on their functional traits, that will be favoured in modified ecosystems will be essential for guiding species selection (Funk *et al.* 2008). If the functional composition of riparian flora shifts, this will likely prompt shifts in the properties, characteristics and functions of riparian ecosystems as a whole (Kominoski *et al.* 2013). It remains to be seen whether maintaining historical ecosystem functions is realistic in novel ecosystems.

## CONCLUSIONS

Riparian plant invasion is a major issue worldwide. Riparian systems seem particularly vulnerable to invasion by alien species because of their proximity to, and association with, human settlements, their large edge to core ratio, high rates of disturbance and high levels of habitat heterogeneity. Although most alien species do not become invasive or cause harm, a small proportion of alien plants can alter the structure and function of riparian ecosystems and threaten their biodiversity. Riparian ecosystems will never be free of alien species, but a range of management options is available to limit current and future invasions. Given the importance of riparian ecosystems now and in the future, effective understanding and management of invasive alien plant species in these systems is essential.

## ACKNOWLEDGEMENTS

We thank Michelle Leishman for suggesting we write this chapter and for her comments on an early draft. We also thank David Duncan and Steve Sinclair for helpful comments. Funding was provided by the Australian Research Council (DE120102221 to JAC) and the Commonwealth Environmental Research Facility – Landscape Logic (to GK).

## REFERENCES

Adair R, Keener B, Kwong R, Sagliocco J, Flower G (2012) The biology of Australian weeds 60. '*Sagittaria platyphylla*' (Engelmann) JG Smith and '*Sagittaria calycina*' Engelmann. *Plant Protection Quarterly* **27**, 47–58.

Ainsworth N, Bowcher A (2005) *Herbicides: Guidelines for Use In and Around Water.* Cooperative Research Centre for Australian Weed Management Guidelines, Adelaide.

Andersson E, Nilsson C, Johansson ME (2000) Plant dispersal in boreal rivers and its relation to the diversity of riparian flora. *Journal of Biogeography* **27**, 1095–1106. doi:10.1046/j.1365-2699.2000.00481.x

Anon (2008) *Weeds of National Significance – Update 2008*. Department of Primary Industries and Fisheries, Brisbane.

AWC (2012) *Weeds Australia*. The National Weeds Strategy Executive Committee, <http//www.weeds.org.au/nwsec.htm>.

Baker HG (1965) Characteristics and modes of origin of weeds. In *The Genetics of Colonizing Species*. (Eds HG Baker and GL Stebbins) pp. 147–168. Academic Press, New York, NY.

Barney JN, Tekiela DR, Dollete ESJ, Tomasek BJ (2013) What is the 'real' impact of invasive plant species? *Frontiers in Ecology and the Environment* **11**, 322–329. doi:10.1890/120120

Berry WL (1970) Characteristics of salts secreted by *Tamarix aphylla*. *American Journal of Botany* **57**, 1226–1230. doi:10.2307/2441362

Boedeltje G, Bakker JP, Bekker RM, Van Groenendael JM, Soesbergen M (2003) Plant dispersal in a lowland stream in relation to occurrence and three specific life-history traits of the species in the species pool. *Journal of Ecology* **91**, 855–866. doi:10.1046/j.1365-2745.2003.00820.x

Bornette G, Amoros C (1996) Disturbance regimes and vegetation dynamics: role of floods in riverine wetlands. *Journal of Vegetation Science* **7**, 615–622. doi:10.2307/3236373

Buckley YM, Bolker BM, Rees M (2007) Disturbance, invasion and re-invasion: managing the weed-shaped hole in disturbed ecosystems. *Ecology Letters* **10**, 809–817. doi:10.1111/j.1461-0248.2007.01067.x

Bunn SE, Davies PM, Kellaway DM, Prosser IP (1998) Influence of invasive macrophytes on channel morphology and hydrology in an open tropical lowland stream, and potential control by riparian shading. *Freshwater Biology* **39**, 171–178. doi:10.1046/j.1365-2427.1998.00264.x

Capon S, Chambers L, Mac Nally R, Naiman R, Davies P, Marshall N, Pittock J, Reid M, Capon T, Douglas M, Catford J, Baldwin D, Stewardson M, Roberts J, Parsons M, Williams S (2013) Riparian ecosystems in the 21st century: hotspots for climate change adaptation? *Ecosystems* **16**, 359–381. doi:10.1007/s10021-013-9656-1

Catford JA, Jansson R (2014) Drowned, buried and carried away: effects of plant traits on the distribution of native and alien species in riparian ecosystems. *New Phytologist* **204**, 19–36. doi:10.1111/nph.12951

Catford JA, Jansson R, Nilsson C (2009) Reducing redundancy in invasion ecology by integrating hypotheses into a single theoretical framework. *Diversity & Distributions* **15**, 22–40. doi:10.1111/j.1472-4642.2008.00521.x

Catford JA, Downes BJ, Gippel CJ, Vesk PA (2011) Flow regulation reduces native plant cover and facilitates exotic invasion in riparian wetlands. *Journal of Applied Ecology* **48**, 432–442. doi:10.1111/j.1365-2664.2010.01945.x

Catford JA, Daehler CC, Murphy HT, Sheppard AW, Hardesty BD, Westcott DA, Rejmánek M, Bellingham PJ, Pergl J, Horvitz CC, Hulme PE (2012a) The intermediate disturbance hypothesis and plant invasions: implications for species richness and management. *Perspectives in Plant Ecology, Evolution and Systematics* **14**, 231–241. doi:10.1016/j.ppees.2011.12.002

Catford JA, Vesk PA, Richardson DM, Pyšek P (2012b) Quantifying levels of biological invasion: towards the objective classification of invaded and invasible ecosystems. *Global Change Biology* **18**, 44–62. doi:10.1111/j.1365-2486.2011.02549.x

Catford JA, Naiman RJ, Chambers LE, Roberts J, Douglas M, Davies P (2013) Predicting novel riparian ecosystems in a changing climate. *Ecosystems* **16**, 382–400. doi:10.1007/s10021-012-9566-7

Catford JA, Morris WK, Vesk PA, Gippel CJ, Downes BJ (2014) Species and environmental characteristics point to flow regulation and drought as drivers of riparian plant invasion. *Diversity & Distributions* **20**, 1084–1096. doi:10.1111/ddi.12225

Colautti RI, MacIsaac HJ (2004) A neutral terminology to define 'invasive' species. *Diversity &*

*Distributions* **10**, 135–141. doi:10.1111/j.1366-9516.2004.00061.x

Colautti R, Grigorovich I, MacIsaac H (2006) Propagule pressure: a null model for biological invasions. *Biological Invasions* **8**, 1023–1037. doi:10.1007/s10530-005-3735-y

Cumming G (2002) Habitat shape, species invasions, and reserve design: insights from simple models. *Conservation Ecology* **6**, 3 [online]

Davis MA, Grime JP, Thompson K (2000) Fluctuating resources in plant communities: a general theory of invasibility. *Journal of Ecology* **88**, 528–534. doi:10.1046/j.1365-2745.2000.00473.x

DSEWPC (2012) *SPRAT Database*. Department of Sustainability Environment Water Population and Communities, <http://www.environment.gov.au/cgi-bin/sprat/public/sprat.pl>.

Eppstein MJ, Molofsky J (2007) Invasiveness in plant communities with feedbacks. *Ecology Letters* **10**, 253–263. doi:10.1111/j.1461-0248.2007.01017.x

Eschtruth AK, Battles JJ (2011) The importance of quantifying propagule pressure to understand invasion: an examination of riparian forest invasibility. *Ecology* **92**, 1314–1322. doi:10.1890/10-0857.1

Finlayson CM, Storrs MJ, Lindner G (1997) Degradation and rehabilitation of wetlands in the Alligator Rivers region of northern Australia. *Wetlands Ecology and Management* **5**, 19–36. doi:10.1023/A:1008271219441

Flory SL, Clay K (2009) Invasive plant removal method determines native plant community responses. *Journal of Applied Ecology* **46**, 434–442. doi:10.1111/j.1365-2664.2009.01610.x

Funk JL, Cleland EE, Suding KN, Zavaleta ES (2008) Restoration through reassembly: plant traits and invasion resistance. *Trends in Ecology & Evolution* **23**, 695–703. doi:10.1016/j.tree.2008.07.013

Gilbert B, Levine JM (2013) Plant invasions and extinction debts. *Proceedings of the National Academy of Sciences of the United States of America* **110**, 1744–1749. doi:10.1073/pnas.1212375110

Greenwood H, O'Dowd DJ, Lake PS (2004) Willow (*Salix* x *rubens*) invasion of the riparian zone in south-eastern Australia: reduced abundance and altered composition of terrestrial arthropods. *Diversity & Distributions* **10**, 485–492. doi:10.1111/j.1366-9516.2004.00104.x

Greet JOE, Cousens RD, Webb JA (2012) Flow regulation affects temporal patterns of riverine plant seed dispersal: potential implications for plant recruitment. *Freshwater Biology* **57**, 2568–2579. doi:10.1111/fwb.12028

Griffin GF, Smith DMS, Morton SR, Allan GE, Masters KA, Preece N (1989) Status and implications of the invasion of tamarisk (*Tamarix aphylla*) on the Finke River, Northern Territory, Australia. *Journal of Environmental Management* **29**, 297–315.

Grime JP (1977) Evidence for the existence of three primary strategies in plants and its relevance to ecological and evolutionary theory. *American Naturalist* **111**, 1169–1194. doi:10.1086/283244

Harms RS, Hiebert RD (2006) Vegetation response following invasive Tamarisk (*Tamarix* spp.) removal and implications for riparian restoration. *Restoration Ecology* **14**, 461–472. doi:10.1111/j.1526-100X.2006.00154.x

Hierro JL, Maron JL, Callaway RM (2005) A biogeographical approach to plant invasions: the importance of studying exotics in their introduced and native range. *Journal of Ecology* **93**, 5–15. doi:10.1111/j.0022-0477.2004.00953.x

Hobbs RJ (2004) The working for water programme in South Africa: the science behind the success. *Diversity & Distributions* **10**, 501–503. doi:10.1111/j.1366-9516.2004.00115.x

Hobbs RJ, Humphries SE (1995) An integrated approach to the ecology and management of plant invasions. *Conservation Biology* **9**, 761–770. doi:10.1046/j.1523-1739.1995.09040761.x

Hobbs RJ, Higgs E, Harris JA (2009) Novel ecosystems: implications for conservation and restoration. *Trends in Ecology & Evolution* **24**, 599–605. doi:10.1016/j.tree.2009.05.012

Hodkinson DJ, Thompson K (1997) Plant dispersal: the role of man. *Journal of Applied Ecology* **34**, 1484–1496. doi:10.2307/2405264

Holland-Clift S, Davies J (2007) *Willows National Management Guide: Current Management and*

*Control Options for Willows (*Salix *spp.) in Australia*. Victorian Department of Primary Industries, Geelong, Victoria.

Holland-Clift S, O'Dowd DJ, Mac Nally R (2011) Impacts of an invasive willow (*Salix* x *rubens*) on riparian bird assemblages in south-eastern Australia. *Austral Ecology* **36**, 511–520. doi:10.1111/j.1442-9993.2010.02178.x

Hulme PE, Bacher S, Kenis M, Klotz S, Kuhn I, Minchin D, Nentwig W, Olenin S, Panov V, Pergl J, Pysek P, Roques A, Sol D, Solarz W, Vila M (2008) Grasping at the routes of biological invasions: a framework for integrating pathways into policy. *Journal of Applied Ecology* **45**, 403–414. doi:10.1111/j.1365-2664.2007.01442.x

Joshi J, Vrieling K (2005) The enemy release and EICA hypothesis revisited: incorporating the fundamental difference between specialist and generalist herbivores. *Ecology Letters* **8**, 704–714. doi:10.1111/j.1461-0248.2005.00769.x

Julien MH, Griffiths MW (Eds) (1998) *Biological Control of Weeds: a World Catalogue of Agents and Their Target Weeds*. CABI Publishing, Wallingford, UK.

Keane RM, Crawley MJ (2002) Exotic plant invasions and the enemy release hypothesis. *Trends in Ecology & Evolution* **17**, 164–170. doi:10.1016/S0169-5347(02)02499-0

Kominoski JS, Shah JJF, Canhoto C, Fischer DG, Giling DP, González E, Griffiths NA, Larrañaga A, LeRoy CJ, Mineau MM, McElarney YR, Shirley SM, Swan CM, Tiegs SD (2013) Forecasting functional implications of global changes in riparian plant communities. *Frontiers in Ecology and the Environment* **11**, 423–432. doi:10.1890/120056

Kowarik I (1995) Time lags in biological invasions with regard to the success and failure of alien species. In *Plant Invasions General Aspects and Special Problems*. (Eds P Pysek, K Prach, M Rejmanek and M Wade) pp. 15–38. SPB Academic Publishers, Amsterdam.

Kozlowski TT (2002) Physiological-ecological impacts of flooding on riparian forest ecosystems. *Wetlands* **22**, 550–561. doi:10.1672/0277-5212(2002)022[0550:PEIOFO]2.0.CO;2

Kyle G, Leishman MR (2009) Functional trait differences between extant exotic, native and extinct native plants in the hunter river, NSW: a potential tool in riparian rehabilitation. *River Research and Applications* **25**, 892–903. doi:10.1002/rra.1192

Kyle G, Schmidt T, Cooke J, Leishman MR (2008) Growth and survival of riparian plantings in relation to willow canopy in the upper Hunter River. *Ecological Management & Restoration* **9**, 154–156. doi:10.1111/j.1442-8903.2008.00409.x

Ladson AR (1997) *Willows Along Victorian Waterways: Towards a Willow Management*. Strategy Department of Natural Resources and Environment, Melbourne.

Lake JC, Leishman MR (2004) Invasion success of exotic plants in natural ecosystems: the role of disturbance, plant attributes and freedom from herbivores. *Biological Conservation* **117**, 215–226. doi:10.1016/S0006-3207(03)00294-5

Lane AM, Williams RJ, Muller WJ, Lonsdale WM (1997) The effects of the herbicide tebuthiuron on seedlings of *Mimosa pigra* and native floodplain vegetation in northern Australia. *Australian Journal of Ecology* **22**, 439–447. doi:10.1111/j.1442-9993.1997.tb00695.x

Lee WG, Allen RB, Johnson PN (1986) Succession and dynamics of gorse (*Ulex europaeus* L) communities in the Dunedin ecological district, South Island, New Zealand. *New Zealand Journal of Botany* **24**, 279–292. doi:10.1080/0028825X.1986.10412678

Lunt ID, Jansen A, Binns DL (2012) Effects of flood timing and livestock grazing on exotic annual plants in riverine floodplains. *Journal of Applied Ecology* **49**, 1131–1139. doi:10.1111/j.1365-2664.2012.02176.x

MacDougall AS, Turkington R (2005) Are invasive species the drivers or passengers of change in degraded ecosystems? *Ecology* **86**, 42–55. doi:10.1890/04-0669

Mack MC (2003) Phylogenetic constraint, absent life forms, and preadapted alien plants: a prescription for biological invasions. *International Journal of Plant Sciences* **164**, S185–S196. doi:10.1086/368399

Malanson GP (1993) *Riparian Landscapes.* Cambridge University Press, Cambridge, UK.

Meiners SJ, Cadenasso ML, Pickett STA (2004) Beyond biodiversity: individualistic controls of invasion in a self-assembled community. *Ecology Letters* **7**, 121–126. doi:10.1111/j.1461-0248.2003.00563.x

Melbourne BA, Cornell HV, Davies KF, Dugaw CJ, Elmendorf S, Freestone AL, Hall RJ, Harrison S, Hastings A, Holland M, Holyoak M, Lambrinos J, Moore K, Yokomizo H (2007) Invasion in a heterogeneous world: resistance, coexistence or hostile takeover? *Ecology Letters* **10**, 77–94. doi:10.1111/j.1461-0248.2006.00987.x

Merritt DM, Scott ML, Poff NL, Auble GT, Lytle DA (2010) Theory, methods and tools for determining environmental flows for riparian vegetation: riparian vegetation-flow response guilds. *Freshwater Biology* **55**, 206–225. doi:10.1111/j.1365-2427.2009.02206.x

Millennium Ecosystem Assessment (2005) *Ecosystems and Human Well-being: Synthesis.* Island Press, Washington, DC.

Mitchell CE, Agrawal AA, Bever JD, Gilbert GS, Hufbauer RA, Klironomos JN, Maron JL, Morris WF, Parker IM, Power AG, Seabloom EW, Torchin ME, Vázquez DP (2006) Biotic interactions and plant invasions. *Ecology Letters* **9**, 726–740. doi:10.1111/j.1461-0248.2006.00908.x

Moles AT, Gruber MAM, Bonser SP (2008) A new framework for predicting invasive plant species. *Journal of Ecology* **96**, 13–17.

Moles AT, Flores-Moreno H, Bonser SP, Warton DI, Helm A, Warman L, Eldridge DJ, Jurado E, Hemmings FA, Reich PB, Cavender-Bares J, Seabloom EW, Mayfield MM, Sheil D, Djietror JC, Peri PL, Enrico L, Cabido MR, Setterfield SA, Lehmann CER, Thomson FJ (2012) Invasions: the trail behind, the path ahead, and a test of a disturbing idea. *Journal of Ecology* **100**, 116–127. doi:10.1111/j.1365-2745.2011.01915.x

Panetta FD, Groves RH, Shepherd RCH (1998) *The Biology of Australian Weeds.* R.G. and F.J. Richardson, Melbourne.

Parker IM, Simberloff D, Lonsdale WM, Goodell K, Wonham M, Kareiva PM, Williamson MH, Von Holle B, Moyle PB, Byers JE, Goldwasser L (1999) Impact: toward a framework for understanding the ecological effects of invaders. *Biological Invasions* **1**, 3–19. doi:10.1023/A:1010034312781

Parsons WT, Cuthbertson E (1992) *Noxious Weeds of Australia.* Inkata Press, Melbourne.

Pausas JG, Lavorel S (2003) A hierarchical deductive approach for functional types in disturbed ecosystems. *Journal of Vegetation Science* **14**, 409–416. doi:10.1111/j.1654-1103.2003.tb02166.x

Paynter Q, Flanagan G (2004) Integrated weed management – could we be doing better? Lessons from controlling the invasive wetland shrub, *Mimosa pigra.* In *XI International Symposium on Biological Control of Weeds.* 27 April–3 May, Canberra. (Eds JM Cullen, DT Briese, DJ Kriticos, WM Lonsdale, L Morin and JK Scott) pp. 361–369. CSIRO, Canberra.

Perna C, Burrows D (2005) Improved dissolved oxygen status following removal of exotic weed mats in important fish habitat lagoons of the tropical Burdekin River floodplain, Australia. *Marine Pollution Bulletin* **51**, 138–148. doi:10.1016/j.marpolbul.2004.10.050

Pheloung PC, Williams PA, Halloy SR (1999) A weed risk assessment model for use as a biosecurity tool evaluating plant introductions. *Journal of Environmental Management* **57**, 239–251. doi:10.1006/jema.1999.0297

Pysek P, Richardson DM (2007) Traits associated with invasiveness in alien plants: where do we stand? In *Biological Invasions.* (Ed. W Nentwig) pp. 98–111. Springer-Verlag, Berlin.

Randall RP (2007) *The Introduced Flora of Australia and its Weed Status.* CRC for Australian Weed Management, Department of Agriculture and Food, Western Australia, Glen Osmond, SA.

Rea N, Storrs MJ (1999) Weed invasions in wetlands of Australia's Top End: reasons and solutions. *Wetlands Ecology and Management* **7**, 47–62. doi:10.1023/A:1008425109473

Rejmanek M, Richardson DM (1996) What attributes make some plant species more invasive? *Ecology* **77**, 1655–1661. doi:10.2307/2265768

Richardson DM, Pysek P (2006) Plant invasions: merging the concepts of species invasiveness and community invasibility. *Progress in Physical Geography* **30**, 409–431. doi:10.1191/0309133306pp490pr

Richardson DM, Pysek P, Rejmanek M, Barbour MG, Panetta FD, West CJ (2000) Naturalization and invasion of alien plants: concepts and definitions. *Diversity & Distributions* **6**, 93–107. doi:10.1046/j.1472-4642.2000.00083.x

Richardson DM, Holmes PM, Esler KJ, Galatowitsch SM, Stromberg JC, Kirkman SP, Pysek P, Hobbs RJ (2007) Riparian vegetation: degradation, alien plant invasion, and restoration prospects. *Diversity & Distributions* **13**, 126–139. doi:10.1111/j.1366-9516.2006.00314.x

Rood SB, Gourley CR, Ammon EM, Heki LG, Klotz JR, Morrison ML, Mosley D, Scoppettone GG, Swanson S, Wagner PL (2003) Flows for floodplain forests: a successful riparian restoration. *Bioscience* **53**, 647–656. doi:10.1641/0006-3568(2003)053[0647:FFFFAS]2.0.CO;2

Sala OE, Chapin FS, Armesto JJ, Berlow E, Bloomfield J, Dirzo R, Huber-Sanwald E, Huenneke LF, Jackson RB, Kinzig A, Leemans R, Lodge DM, Mooney HA, Oesterheld M, Poff NL, Sykes MT, Walker BH, Walker M, Wall DH (2000) Biodiversity – global biodiversity scenarios for the year 2100. *Science* **287**, 1770–1774. doi:10.1126/science.287.5459.1770

Shackelford N, Hobbs RJ, Heller NE, Hallett LM, Seastedt TR (2013) Finding a middle-ground: the native/non-native debate. *Biological Conservation* **158**, 55–62. doi:10.1016/j.biocon.2012.08.020

Simberloff D (2009) The role of propagule pressure in biological invasions. *Annual Review of Ecology Evolution and Systematics* **40**, 81–102. doi:10.1146/annurev.ecolsys.110308.120304

Sinden J, Jones R, Hester S, Odom D, Kalisch C, James R, Cacho O (2005) *The Economic Impact of Weeds in Australia*. CRC for Australian Weed Management, Adelaide.

Smith N (1995) *Weeds of Natural Ecosystems: a Field Guide to Environmental Weeds of the Northern Territory*. Northern Territory Environment Centre, Darwin.

Sogge MK, Sferra SJ, Paxton EH (2008) *Tamarix* as habitat for birds: implications for riparian restoration in the southwestern United States. *Restoration Ecology* **16**, 146–154. doi:10.1111/j.1526-100X.2008.00357.x

Stohlgren TJ, Chong GW, Schell LD, Rimar KA, Otsuki Y, Lee M, Kalkhan MA, Villa CA (2002) Assessing vulnerability to invasion by non-native plant species at multiple spatial scales. *Environmental Management* **29**, 566–577. doi:10.1007/s00267-001-0006-2

Stokes K, Ward K, Colloff M (2010) Alterations in flood frequency increase exotic and native species richness of understorey vegetation in a temperate floodplain eucalypt forest. *Plant Ecology* **211**, 219–233. doi:10.1007/s11258-010-9833-7

Stromberg JC, Lite SJ, Marler R, Paradzick C, Shafroth PB, Shorrock D, White JM, White MS (2007) Altered stream-flow regimes and invasive plant species: the *Tamarix* case. *Global Ecology and Biogeography* **16**, 381–393. doi:10.1111/j.1466-8238.2007.00297.x

Taylor K, Brummer T, Taper ML, Wing A, Rew LJ (2012) Human-mediated long-distance dispersal: an empirical evaluation of seed dispersal by vehicles. *Diversity & Distributions* **18**, 942–951. doi:10.1111/j.1472-4642.2012.00926.x

Valentine L, Roberts B, Schwarzkopf L (2007) Mechanisms driving avoidance of non-native plants by lizards. *Journal of Applied Ecology* **44**, 228–237. doi:10.1111/j.1365-2664.2006.01244.x

Valéry L, Fritz H, Lefeuvre J-C (2013) Another call for the end of invasion biology. *Oikos* **122**, 1143–1146. doi:10.1111/j.1600-0706.2013.00445.x

van Kleunen M, Weber E, Fischer M (2010) A meta-analysis of trait differences between invasive and non-invasive plant species. *Ecology Letters* **13**, 235–245. doi:10.1111/j.1461-0248.2009.01418.x

Vilà M, Pino J, Font X (2007) Regional assessment of plant invasions across different habitat types. *Journal of Vegetation Science* **18**, 35–42. doi:10.1111/j.1654-1103.2007.tb02513.x

Walden D, van Dam R, Finlayson M, Storrs M, Lowry J, Kriticos D (2004) A risk assessment of the tropical wetland weed *Mimosa pigra* in

Northern Australia. In *Third International Symposium on the Management of* Mimosa pigra. Darwin, 22–28 September 2002 (Eds M Julien, G Flanagan, T Heard, B Hennecke, Q Paynter, C Wilson) pp. 11–22. Supervising Scientist, Darwin.

Xiong S, Nilsson C, Johansson ME, Jansson R (2001) Responses of riparian plants to accumulation of silt and plant litter: the importance of plant traits. *Journal of Vegetation Science* **12**, 481–490. doi:10.2307/3237000

# 19

# The ongoing challenge of restoring Australia's riparian zones

*Paul Reich, Laura Williams, Tim Cavagnaro and P. Sam Lake*

## INTRODUCTION

The need to protect and restore riparian systems has long been recognised in Australia. As far back as 1977, the Natural Resources Conservation League of Victoria published a special issue in their quarterly magazine *Victorian Resources* entitled 'Problems of streams and stream frontages' (Natural Resources Conservation League of Victoria 1977). In this issue several papers explored the principles that underpin some of the ecological benefits of intact riparian zones. Despite the infancy of the science relating to riparian systems at the time, many of these ideas have since been supported by a growing body of empirical evidence. Importantly, many of the authors highlighted the critical issues they saw impeding the progress towards successful protection and restoration of riparian zones. The impediments they identified overlap considerably with those issues that hinder the progress of river restoration more broadly (e.g. see Bernhardt *et al.* 2007). These include: balancing the competing needs of the environment with those of production landscapes; targeting protection and restoration activities at meaningful spatial scales; integrating restoration activities in ways that maximise the chances of achieving multiple environmental goals; and the development of research and monitoring programs that support improved riparian and stream management. Unfortunately, over 30 years later, many of these issues remain largely unresolved and continue to pose major challenges to natural resource managers and scientists alike, even though riparian restoration is now the dominant form of stream restoration (Bernhardt *et al.* 2007; Brooks and Lake 2007). Here we present an overview of problems and reiterate some of the major challenges ahead for restoring Australia's riparian zones. Although the general principles apply across the vegetation of all riverine landscapes, our focus is on the restoration of riparian zones along creeks and rivers, and we draw most strongly on examples from south-eastern Australia.

## THE DAMAGE DONE

Riparian zones in Australia have been subjected to widespread and severe degradation since European settlement. Approximately half (53%) of the river length assessed by the National Assessment of River Condition has substantially or severely modified riparian vegetation, to the extent that very little native riparian vegetation is left (Norris *et al.* 2001). A national land and water resource audit (Sattler and Creighton 2002) found that across more than two-thirds of Australia's subregions, riparian zones were degraded to such an extent that their recovery will at best require significant effort (38% of subregions) and may not even be possible in the medium term (31% of subregions). Moreover, the condition of 73% of riparian zones in Australia was in decline (Sattler and Creighton 2002).

Degradation is not uniform across Australia, or indeed the world. In Australia, some arid areas, parts of south-western Tasmania, Cape York Peninsula and the Northern Territory's Top End retain riparian zones in good condition, whereas the lowland riparian zones in the Murray–Darling Basin and cropping areas of Western Australia are among the most degraded (Sattler and Creighton 2002). Many of the most degraded riparian zones are in the lowland areas and coincide with locations of intensive agriculture (Sattler and Creighton 2002). For example, along the lowland streams of the Murray–Darling Basin, scattered canopy trees are often the only relic of native vegetation, with the understorey vastly altered such that often no shrub layer remains and exotics dominate the ground layer. Long-lived tree species (e.g. *Eucalyptus camaldulensis*) show little recruitment and, without appropriate action, may slowly become locally extinct (George *et al.* 2005; Cunningham *et al.* 2010).

The predominant agents of riparian degradation include land clearance, hydrological alteration, livestock access, elevated salinity and the invasion of riparian zones by exotic plant and animal species (Lovett and Price 1999; Norris *et al.* 2001; Sattler and Creighton 2002). Though many of these drivers of degradation are widespread, their distributions vary across the landscape and are linked to land use type, climate and topography. Flow regimes are a major determinant of riparian vegetation dynamics and artificial reductions in the frequency, extent and duration of floodplain inundation have all been shown to influence vegetation structure and composition (e.g. Bren 1988, 1992; Blanch *et al.* 1999; Kingsford 2000; Capon 2005; Horner *et al.* 2012). Livestock access to riparian zones represents one of the most pervasive threats to riparian vegetation, with those areas subjected to grazing having less regeneration of trees, fewer shrubs and less biomass of groundcover species (Robinson and Mann 1998; Robertson and Rowling 2000). These threats can often act in unison to favour the spread of exotic species. For instance, the spread of *Phyla canescens* (lippia), an exotic, perennial forb poses a major threat to floodplains in the northern Murray–Darling Basin (Earl 2003) and is thought to be promoted by a combination of heavy grazing and altered patterns of inundation due to river regulation. These conditions give lippia a competitive advantage over native perennial grasses and other groundcover, allowing it to form an impenetrable, dense mat (Taylor and Ganf 2005; Macdonald *et al.* 2012).

The condition of riparian zones also has well-recognised implications for the condition of adjacent aquatic and terrestrial ecosystems. In particular, degraded riparian zones are major contributors to the poor condition of many Australian rivers (Norris *et al.* 2001). The pathways by which riparian zones exert direct influence on stream condition are widely documented (e.g. Kauffman *et al.* 1997; Naiman and Décamps 1997). Degraded riparian vegetation has been linked to soil erosion, poor water quality, altered ecological function and the loss of terrestrial and aquatic biodiversity (see Naiman *et al.* 2005). The impacts of poor riparian condition on water quality are probably the most reliably documented: numerous Australian studies demonstrate increases in bank erosion, water temperature and nutrient levels with riparian degradation (e.g. Prosser *et al.* 2001; McKergow *et al.* 2003; Rutherford *et al.* 2004; McKergow *et al.* 2006). The removal of riparian tree species has also led to a reduction in the supply of coarse woody structure to rivers and streams – a critical resource for aquatic vertebrate and invertebrate fauna (e.g. Crook and Robertson 1999; Mac Nally and Horrocks 2002). Similarly, the loss of canopy species along lowland streams in south-eastern Australia has diminished the inputs of allocthonous leaf material – an important contributor of food and habitat to aquatic fauna (Reid *et al.* 2008a, 2008b). The invasion by exotic riparian canopy species can also negatively impact stream condition (e.g. Read and Barmuta 1999) and alter invertebrate and vertebrate terrestrial species assemblages (e.g. Greenwood *et al.* 2004; Holland-Clift and Davies 2007; Kramer *et al.* 2007).

## FIXING THE DAMAGE

Riparian restoration or 'streamside management' represents the major form of waterway management activity in many parts of Australia. Restoration efforts typically focus on re-establishing riparian vegetation through livestock exclusion and planting. Millions of seedlings are planted and thousands of kilometres of fencing erected each year across the country. For instance, in Victoria between 2002 and 2009, 7066 km of river and stream length was fenced to exclude livestock (Department of Sustainability and Environment Victoria 2010). In a review of stream management activities across four Victorian catchment management regions between 1999 and 2001, Brooks and Lake (2007) reported that nearly half of all projects were riparian related. The same pattern is mirrored in other parts of the world where riparian management is also the major focus of stream restoration (Bernhardt *et al.* 2005; Gilvear *et al.* 2012).

The focus on riparian systems is not only motivated by the degraded state of riparian systems, but also by a perception that restoring riparian zones will provide a suite of associated ecological benefits. Riparian restoration is thought to promise broad-ranging ecological benefits in degraded landscapes, including improvements to water quality, aquatic biodiversity, terrestrial biodiversity, resistance and resilience of plant and animal populations to environmental change, and the conservation of threatened species. Indeed, these motivations are understandable and conceptually sound given the evidence available to support the premise that intact riparian zones do have intrinsic values for terrestrial biodiversity (e.g. Sabo *et al.* 2005; Seavy *et al.* 2009) and also serve to mitigate some land use impacts on waterways (e.g. Lowrance *et al.* 1984; Gregory *et al.* 1991). However, our understanding of the ecological consequences of riparian degradation far outweighs the evidence required to demonstrate the effectiveness of our efforts to repair the damage.

The pathways to successful restoration of riparian zones are likely to be very different to those that led to degradation and may take decades to establish through carefully designed monitoring, on-ground works and ongoing revision (Gregory *et al.* 2007; Lake *et al.* 2007; Lindenmayer and Likens 2010). Unfortunately, monitoring and evaluation of riparian restoration projects is typically inadequate to provide the information necessary to assess performance (Bash and Ryan 2002; Bernhardt *et al.* 2007; Brooks and Lake 2007; but see Peel 2010). As a consequence, progress towards improved understanding has been slow and fundamental questions still remain unanswered. In many cases, we are still uncertain whether the damage to riparian plant communities is reversible and whether improving riparian condition will achieve the desired outcomes for waterways and terrestrial ecosystems.

Efforts to restore riparian zones typically focus on restoring vegetation structure. This may be a substantial challenge in itself, but represents only a small subset of the achievements necessary to restore a riparian zone and its ecological benefits within a landscape. Riparian restoration may be viewed as a series of challenges: if we can re-establish vegetation structure, is this sufficient to restore the composition and/or dynamics (e.g. self-sustaining populations and resilience) of the riparian plant community? If we can restore riparian plant communities, is this sufficient to restore entire riparian ecosystems (e.g. are soil, invertebrate and bird communities resilient and responsive to a restored plant community)? If we can restore a riparian ecosystem, is this sufficient to restore aquatic systems (e.g. are water quality and aquatic communities resilient and responsive to a restored riparian ecosystem)? If so, will restoration provide benefits to terrestrial systems (e.g. are opportunistic riparian species and aquatic–terrestrial food-web associations resilient and responsive to a restored riparian ecosystem)?

Despite strong established links between riparian condition and riverine ecosystems, the pathways to ecosystem recovery are likely to be substantially more complicated than simply restoring riparian zones. The damage wrought on aquatic systems reflects stresses (both past and contemporary) generated at several scales throughout a catchment (Allan 2004). In fact, several studies have attributed greater influence to catchment than to local land use on measurements of stream condition (e.g. Roth *et al.* 1996; Morley and Karr 2002 discussed below). Nonetheless, the restoration of riparian zones is often perceived to promise benefits disproportionately higher than alternative restorative activities (such as the re-instatement of natural flows, adjusting agricultural practices and/or widespread planting across entire catchments), which are practically more difficult to achieve and typically politically unpalatable (Bohn and Kershner 2002; Lake *et al.* 2007).

## THE BIG CHALLENGES

Restoration is about re-establishing an ecosystem that structurally, compositionally and/or functionally resembles its condition before degradation. Common goals of riparian restoration are to create a plant community that is native, diverse, autogenic and effective in performing a range of ecosystem functions (e.g. moderating stream temperature, mediating the input of sediment and nutrients into streams, providing habitat and food sources for aquatic organisms, contributing to terrestrial food webs, providing habitat for obligate and opportunistic riparian biota, and/or acting as dispersal corridors and refuges for terrestrial plants and animals) (Palmer *et al.* 2005; Gregory *et al.* 2007). Often targets of riparian restoration projects are diffuse, vaguely articulated and assume static endpoints. Successful restoration of riparian vegetation is underpinned by knowledge of vegetation dynamics and the factors that control these dynamics. This requires some understanding of how vegetation dynamics operate in both intact and degraded systems in order to formulate restoration targets and to identify and (hopefully) overcome constraints that can act to maintain a system in a degraded state and/or shift it towards an alternative undesired state despite restoration efforts (Suding *et al.* 2004; Richardson *et al.* 2007).

Restoration approaches may attempt to restore abiotic conditions or major structural attributes (e.g. canopy trees) with the assumption that this is sufficient to steer the degraded ecosystem along a trajectory towards the desired condition ('the field of dreams' hypothesis *sensu* Palmer *et al.* 1997). However, some degraded riparian systems may show little or no response to these efforts (e.g. Lunt *et al.* 2007) and sometimes restoration activities will lead to undesirable changes in a plant community by generating novel assemblages of species or monocultures of weeds (Suding *et al.* 2004; Richardson *et al.* 2007). Knowledge of vegetation dynamics and their controls in degraded systems is necessary to recognise and address mechanisms (stressors and their history) that lead to such undesired responses to restoration efforts (Patten 2006; Horner *et al.* 2012). Similarly, understanding the links between riparian zones and their adjacent environs will assist in planning and implementing riparian restoration in ways that maximise benefits to aquatic and terrestrial biota. Here we explore some common challenges faced when attempting to achieve the many and varied objectives of riparian restoration.

### Sources of colonists

One of the initial challenges faced in the restoration of a native, let alone diverse, riparian plant community is a source of propagules of desirable species. Even where suitable environmental conditions (e.g. water, nutrients and disturbance regimes for germination and establishment) exist, desired vegetation cannot develop without a propagule source. Propagules may occur at a site from three sources: the extant vegetation, in situ seed banks and/or dispersal (via wind, water, animals or aided by humans). Each may contribute to riparian restoration, but their importance differs among life forms and locations. In the case of shrubs and understorey species, human-aided dispersal of propagules (e.g. direct seeding or planting tubestock) may often be necessary to restore a structurally and compositionally diverse native community.

Extant vegetation can only contribute propagules from the species that are present. In degraded sites that are subject to restoration efforts, the extant vegetation often contains only a handful of native species – those that are the most tolerant or resilient to the impacts that have degraded the site over time – and may omit entire life forms, such as shrubs. Thus, the contribution of extant vegetation may be limited.

Canopy-stored seed may provide an important means for the regeneration of some remnant woody vegetation. The seeds of the most widespread riparian tree in Australia, *Eucalyptus camaldulensis* (river red gum), persist poorly in the soil and recruitment relies mostly on continual replenishment from canopy-stored seed (Jensen *et al.* 2008). Remnant trees are relatively common along streams even in otherwise degraded areas, such as in the Murray–Darling Basin, and may provide a seed source for natural regeneration. However, seedling establishment is often suppressed by grazing (Robertson and Rowling 2000; Jansen and Robertson 2001) or the interaction between grazing and seed-predating ants (Meeson *et al.* 2002). If the removal or reduction of grazing pressure coincides with sufficient water for germination and survival, successful recruitment of river red gums may occur which can supplement or even circumvent the need to plant tubestock (Meeson *et al.* 2002).

In situ soil seed banks usually do not contain the full complement of species required to restore a diverse native community. Woody riparian species can have poor soil seed banks (Pettit and Froend 2001; Jensen *et al.* 2008; Williams *et al.* 2008), although some shrub species (e.g. certain *Acacia* spp.) may possess persistent soil seed banks (e.g. Auld 1987). Furthermore, soil seed banks often contain a large proportion of exotics that may counteract restoration efforts (Williams *et al.* 2008). Nevertheless, in situ soil seed banks have the potential to assist the regeneration of some components of a native understorey. Understorey vegetation often receives little attention in riparian restoration efforts, but in many riparian systems, including river red gum-dominated lowland systems, the understorey represents most of the diversity (e.g. Williams *et al.* 2008) and contributes significantly to riparian functions such as the interception of nutrients and sediment (e.g. McKergow *et al.* 2006). Soil seed banks from several Australian riparian zones have been found to contain a diverse assemblage of understorey species, usually dominated by annual forbs and graminoids (Pettit and Froend 2001; Capon and Brock 2006; Williams *et al.* 2008).

If remnant vegetation that contains populations of desirable species exists within close proximity of the restoration site, then dispersal may provide a valuable source of propagules. However, in landscapes that have been extensively modified, the distance to remnant vegetation may exceed the dispersal distance of propagules from these desirable species. Therefore, the capacity for dispersal to contribute desirable propagules will largely depend on the landscape context.

## Invasive species

Invasive exotics have the potential to subvert desired trajectories of restoration, leading to communities that are neither diverse nor autogenic (Lake *et al.* 2007; Cramer *et al.* 2008). Riparian zones, as regions of high resource availability, are often highly susceptible to invasion by exotics (Planty-Tabacchi *et al.* 1996; Stohlgren *et al.* 1998, 2003; Richardson *et al.* 2007; Macdonald *et al.* 2012). Furthermore, some exotic species considered undesirable at a restoration site might have been an important crop or pasture plant and even actively promoted before restoration (e.g. grasses such as *Phalaris aquatica*; Williams *et al.* 2009). Exotics may be a symptom rather than a cause of poor riparian condition. Previous land use and current landscape context often mean that degraded riparian zones have disturbed soils with elevated nutrient content and high propagule pressure of exotic species (see 'Landscape context' section). These conditions may promote invasive species. Many invasive species are nitrophilous, with fast potential growth rates compared with many slower-growing native perennial species (McLendon and Redente 1991; Prober *et al.* 2005; Bidwell *et al.* 2006). Consequently, even where appropriate native species are present as seeds or planted species many will be rapidly outcompeted.

Exotic invaders of Australian riparian zones encompass the complete spectrum of life forms from trees (e.g. *Salix* spp., willow) to shrubs (e.g. *Rubus* spp., blackberry), vines (e.g. *Cryptostegia grandiflora*, rubber vine,), forbs (e.g. *Phyla canescens*) and graminoids (e.g. *Phalaris aquatica*, Toowoomba canary grass). Ecological effects of exotics may be equally wide-ranging. In some circumstances exotics or non-indigenous native plants may be largely benign or even assist restoration by acting as nurse plants (Kyle *et al.* 2008). For example, in the restoration of warm temperate riparian rainforest in south-eastern Australia, a native *Solanum* sp. has been used with success to create dense shade which controls the exotic grass *Pennisetum clandestinum* (kikuyu) and allows native species to recruit (Peel 2010). Other exotics, including those listed above, may impede restoration by exerting substantial influence on vegetation dynamics. For example, *Glyceria maxima*, an exotic perennial grass, may change nutrient and sediment dynamics to convert fast-flowing streams into swamp environments (Clarke *et al.* 2004). *Phalaris aquatica* can grow rapidly after livestock is excluded from lowland riparian zones and form dense swards, which may exclude and suppress the recruitment of other plant species, thus preventing the establishment of a native autogenic community (Williams *et al.* 2009).

Exotic riparian species can have implications for adjacent aquatic and terrestrial environments. For example, invasion by *Salix* spp. (willows) – deliberately introduced to the riparian zones of rivers in southern Australia to halt bank erosion – alters stream channel morphology (Ross 1994) and modifies the rate and timing of litter inputs, with subsequent changes to the composition and abundance of aquatic invertebrates (Read and Barmuta 1999). The impacts of invasive riparian plants on fauna are not limited to the stream channel and can also lead to substantial changes in both invertebrate and vertebrate terrestrial communities (e.g. Greenwood *et al.* 2004). Holland-Clift and Davies (2007) reported that willow-dominated riparian zones had lower habitat complexity and fewer native and woodland- dependant birds than comparable sites with native vegetation.

Exotic invaders may subvert restoration trajectories by promoting different disturbance regimes which can create conditions more favourable to exotics, thus instigating a positive feedback cycle that may be difficult to shift (D'Antonio and Vitousek 1992). For example, *Urochloa mutica* (para grass) is a perennial grass species that has invaded floodplains and wetlands in the Northern Territory (Douglas *et al.* 2001). Para grass can dominate the plant community, leading to significantly lower species richness and, because it has higher dry season fuel loads than the native plants it displaces, more intense fires (Douglas *et al.* 2001; Douglas and O'Connor 2004).

It is important to note that exotics may not always be bad. Depending on the specific objective of riparian management, exotic species may serve some beneficial roles over the alternative degraded state by controlling soil erosion and capturing sediment and/or excess nutrients before they enter waterways. That is, there may be cases where some vegetation is better than nothing and even cases where exotic species may be better than remnant native species at performing a given function. In North America, a three-zone buffer approach (the riparian forest buffer system) has been designed to capture, filter and process sediment and nutrients using a series of specific vegetation types and structures at different lateral distances from the channel (Welsch 1991). This approach emphasises the functional role of vegetation in achieving water quality objectives, which may not necessarily align favourably with the objective of restoring a plant assemblage comprising entirely of native species. It is also conceivable that exotic species could be exploited to aid restoration objectives in some circumstances. For example, where rapid establishment of fast-growing, nutrient-demanding exotics follows livestock exclusion, a carefully controlled grazing regime might help to reduce soil nutrients (e.g. Sanjari *et al.* 2008). Moreover, the addition of carbon (e.g. sugar, saw dust, molasses) can stimulate microbial drawdown of nitrogen (e.g. Morghan and Seastedt 1999), lowering soil nutrients to a level that favours less nutrient-demanding native species, at which point livestock could be permanently excluded and attempts to restore native ground cover can be initiated.

## Disturbance regimes

Vegetation dynamics are closely tied to disturbance regimes, that is, the intensity, frequency, size, predictability and type of disturbances that occur at a site (Pickett and White 1985). Disturbances play a major role in structuring plant communities by affecting germination, establishment and the relative competitive ability of species. Effects of human-driven disturbances on riparian vegetation, such as overgrazing by livestock, logging and dryland salinity, are well documented (Horner *et al.* 2009, 2012; Robertson and Rowling 2000; Jansen and Robertson 2001; Lymbery *et al.* 2003), as are human-induced changes to disturbance regimes, such as altered fire and flood frequency (Bren 1988, 1992; Pettit and Naiman 2007). Restoration efforts may be constrained if the effects of human-driven disturbances are insufficiently ameliorated, or if the disturbance regimes required to facilitate the desired native community are not reinstated. Furthermore, while current disturbances and their effects may be recognised and incorporated into the planning of riparian restoration, it is also crucial to gain an understanding of past disturbances – unrecognised legacies that may subvert restoration efforts. Legacies include high nutrient levels from past fertiliser use, contamination by long-lived pesticides and heavy metals, and the local extinction of key species (e.g. pollinators).

Self-sustaining populations are necessary to restore the spatial patterning of natural plant communities, such as the location, density, age structure and composition of plant species relative to the environmental conditions of a site. Even if seeds or propagules of appropriate native species are present, they require suitable conditions for their germination and survival. Disturbance regimes may be vital to provide cues and/or conditions for recruitment, and affect the relative competitive ability of species. At best, initial plantings strive to emulate the spatial patterning of natural plant communities. More often, plants are located haphazardly along parallel rip-lines. Several generations may be required to restore the spatial characteristics of plant communities, whereby dispersal and environmental conditions filter the germination, establishment and survival of individuals. Establishing an autogenic community depends on reinstating the conditions required for regeneration. Some species, particularly longer-lived woody species (e.g. *Acacia* spp.), may require a disturbance event such as fire or flood to trigger seed release, cue germination or provide environmental conditions required for establishment (e.g. Auld 1987; George *et al.* 2005).

Hydrological regimes encompass the timing, duration, size and predictability of flooding and drying events. These regimes drive the composition and spatial structure of riparian vegetation by creating spatial mosaics whereby recruitment opportunities, survival and competitive hierarchies differ among species (Tabacchi *et al.* 1998; van Coller *et al.* 2000; Horner *et al.* 2012). Hydrological regimes may also control functional attributes of plant communities such as production rates (e.g. growth rates of *Eucalyptus camaldulensis*) and nutrient dynamics. Alterations to the timing, frequency and size of floods may lead to the functional disconnection of floodplains from streams and are often implicated in the degradation of riparian vegetation. Restoration attempts that do not consider and reinstate hydrological regimes may be unsuccessful in creating a functional and autogenic riparian community. Moreover, in many cases heavily modified flow regimes will exert hydrological stresses on the aquatic biota that will far outweigh many of the assumed benefits of riparian restoration.

Disturbances are ubiquitous, but may change to form very different regimes. Capacity to reinstate past disturbance regimes may be challenged by the size, isolation and landscape surrounding restoration sites (i.e. Blanton and Marcus 2013), in addition to competing demands on resources (e.g. consumptive water) and changed conditions with a changing climate (e.g. drought). Instead, disturbances required for vegetation dynamics may need to be deliberately imposed (e.g. control burns, flood releases) or mimicked (e.g. mowing or grazing).

Successful restoration will need to consider (1) whether it is possible to reinstate past disturbance regimes, (2) how to establish plant communities and riparian systems that will cope with the current disturbance regime and (3) how to establish riparian zones that will cope with future disturbance regimes. Given the ubiquity of disturbance, restoration should aim to create resilient communities rather than static reconstructions of past plant assemblages.

## Climate and future environmental change

While many of the impacts that have degraded riparian zones are tightly linked to human activities, the capacity for restoration may also be constrained by climatic conditions and associated disturbances. Changing climates and environments provide a shifting background against which restoration efforts are taking place. This poses significant challenges for restoration. In the face of environmental change, including changes to climate and disturbance regimes, the challenge for restoration is to create resilient communities (Johnson *et al.* 2010). Species selection and restoration efforts might also consider a changing climate. For example, which long-lived tree species will grow under future projected conditions? Links should be tightened between efforts to reassemble plant communities, the functional traits of species within the community, and predicted future conditions. For example, species should be chosen according to their functional traits and matched to predicted future climate conditions (Hobbs and Cramer 2008). In particular, functional plant traits should form a crucial guide for the selection of species to plant or control. Future environmental change highlights the need for setting achievable dynamic targets that may be related more to functional composition than static targets based on taxonomic composition.

The timing of restoration efforts relative to prevailing climatic conditions and associated disturbances may be crucial to restoration success (Patten 2006). For example, drought, a 'ramp disturbance' (a disturbance that steadily increases in strength through time; *sensu* Lake 2000) that is driven by climatic phenomena, plays a large role in determining the capacity for restoration in some parts of Australia including the Murray–Darling Basin and Western Australian wheat belt (Hobbs and Norton 2004; Bond and Lake 2005). Some restoration responses, such as bare ground reduction, may be more resilient in the face of drought than others, such as natural tree recruitment. The capacity for natural recruitment in response to reduced grazing pressure is likely to markedly decline with reduced water availability for some riparian species (e.g. Stromberg *et al.* 2010; Horner *et al.* 2012).

A significant challenge for riparian restoration is dealing with uncertainty in the capacity to forecast trends in both climate and associated disturbance events (Reich and Lake 2014). Nevertheless, awareness of prevailing climatic conditions and predictable disturbances should guide the timing of efforts for maximal returns and could potentially be exploited to aid restoration. Dry conditions may deplete vegetation cover, with the decline of all but the most drought-tolerant species. In riparian environments with a grassy understorey, these conditions may favour drought-tolerant native perennials over annual exotics. If the timing of wetter conditions can be predicted, planting or seeding of native species could be timed to occur quickly after the first rains to take advantage of favourable moisture conditions and give native species a competitive advantage over exotics. For example, the ability to predict wetter years triggered by the El Niño Southern Oscillation climatic system has been used with some success in dryland systems to promote tree establishment and trigger phase-shifts in vegetation structure (Holmgren and Scheffer 2001; Holmgren *et al.* 2001; Holmgren 2009). In cases where prolonged droughts or major flooding occurs practitioners may need to revisit their initial expectations, adjust the frequency of monitoring and even alter their implementation techniques (e.g. Hardegree *et al.* 2012).

In response to changed environmental conditions, one pragmatic solution may be to focus on restoring the functions and dynamics of a system while altering the spatial scale of the restoration

(Rood *et al.* 2005). For example, if environmental change has reduced stream flow, instead of attempting to restore a degraded river and its floodplain vegetation to its extent before degradation, restoration efforts may attempt to restore the hydrological regime and associated community dynamics on a smaller scale. This has been done successfully in south-western Canada, where the restoration of seasonal flow regime bounded by a reduced mean discharge resulted in recruitment and establishment of cottonwoods along a newly defined, narrower river channel (Hall *et al.* 2011).

## Landscape context

Landscape context including the shape, location, connectivity and surrounding land use of riparian zones influences the capacity to restore a native, diverse and autogenic community as well as the benefits that a restored riparian zone may confer to adjacent waterways and landscapes. Riparian zones are typically linear systems within the landscape, a shape often accentuated by cleared land on the adjacent floodplain or hinterland. Riparian zones are also topographic sinks in the landscape, and thus water, sediment, nutrients and propagules tend to move towards them. These properties of riparian zones within a landscape heighten their importance as corridors for dispersal and as buffers for maintaining river health, but also present significant challenges for restoration.

Linear shapes along streams provide less core habitat and their high ratio of perimeter to area means they are more exposed and influenced by their surrounds (Forman 1995). For example, a recent study of 18 lowland riparian zones in south-eastern Australia found that soil properties in riparian zones and adjacent lands were often strongly correlated, but contingent on the condition of riparian vegetation (Burger *et al.* 2010). In a highly modified landscape, where riparian zones are particularly narrow, the riparian zone may receive substantial inputs of nutrients and propagules from the surrounding land. In an agricultural landscape, inputs may include high levels of nutrients transported via sediment or water from surrounding fertilised paddocks and large numbers of exotic propagules from naturalised species in crops or pasture: conditions that favour the proliferation of invasive species (Theoharides and Dukes 2007).

The location and shape of riparian zones means they behave as important corridors for plant and animal dispersal (Forman 1995; Naiman and Décamps 1997). This is particularly important in landscapes where remnant riparian trees may be the last vestige of native vegetation. The connectivity of riparian vegetation is essential for effective dispersal and the formation of metapopulations (Hanski 1998). Development of metapopulations and an autogenic community will be inhibited if patches of restored and/or remnant riparian vegetation are highly fragmented.

The capacity of riparian zones to effectively intercept and process nutrients before they reach a stream (i.e. the water quality benefits of riparian zones) depends strongly on characteristics of surrounding land and inputs as well as characteristics of the riparian zone itself (Gregory *et al.* 1991). For instance, Lowrance *et al.* (1997) estimated the amount of sediment, nitrogen and phosphorus retained by forested riparian buffers in each of the nine physiographic provinces that drain into Chesapeake Bay, USA. Removal efficiencies varied between the pollutants and provinces and these differences were attributed to changes in soil, slope, and hydrologic connectivity. Export of nutrients from different agricultural enterprises can differ markedly. For example, nitrate is highly mobile in the soil solution and therefore readily leached through the soil profile (Tinker and Nye 2000). In contrast, ammonium and phosphorus are much less mobile and tend to move with the soil particles to which they are bound. Furthermore, it is important to consider the rate at which different nutrient pools are turned over. For example, nitrogen is cycled rapidly through different pools, which differ in their mobility and availability to plants and soil microbes (Jackson *et al.* 2008). Gaps in the riparian zone may also be problematic (Weller *et al.* 1998); for example, subsurface farm and storm drains that bypass the riparian zone diminish the effectiveness of riparian buffers (Osborne and Kovacic 1993; Barton 1996). The capacity for a given riparian zone to buffer nutrient inputs from the surrounding landscape will also be positively correlated with its width (Hansen *et al.* 2015), connectivity (Osborne and Kovacic 1993; Barton 1996; Weller *et al.* 1998) and condition (Burger *et al.* 2010).

Landscape context should also help in planning and prioritising restoration efforts. For instance, if a riparian zone is adjacent to another waterbody (e.g. billabong or wetland), it may be advantageous to restore the riparian zone to connect the two. Prioritisation of sites for riparian restoration should be influenced by their position relative to areas of conservation. For example, degraded sites that are upstream of a high conservation area might be considered a high priority for restoration. Where waterways are adjacent to high-input agricultural systems, restored riparian zones will need to be wider if they are to effectively intercept and process nutrients (Allan 2004). However, regardless of their condition, continuity and lateral extent, riparian zones are not inexhaustible sinks for nutrients. The reduction of non-point source pollutants, especially nitrates, are most effectively achieved in headwater riparian systems (Lowe and Likens 2005). This means that restoration planning must also consider smaller waterways (both ephemeral and perennial) that may not necessarily be represented on maps used by waterway management (Baker *et al.* 2007).

The ecosystem functions performed by riparian zones will vary with landform and location and these, in turn, are influenced by the human activities performed on adjoining land. Thus, the objectives of riparian management should always be guided by knowledge of these settings and the key functions or uses of the riparian zone that are considered of greatest value.

## Initial conditions

Starting points matter. The condition of sites before restoration efforts – such as the presence of remnant trees, distance to remnant vegetation, diversity of extant native species, presence of highly invasive species, and the structural and chemical composition of the soil – influences the direction and rate of response to restoration efforts. As restoration sites begin at different points, it is unrealistic to expect sites to respond in the same way. Some sites respond and progress rapidly along a desired trajectory, while others may become choked by weeds.

Although intuitive, differences in initial condition often receive little attention in restoration efforts, which typically conform to a series of similar actions (e.g. destocking and replanting to a standard width), and have similar objectives and expectations irrespective of site condition. Even sites that appear superficially similar in structure and contemporary land use may be underlain by differing disturbance and land use histories (e.g. the intensity and duration of grazing). Although the management of the drivers of degradation (e.g. grazing, water supply, exotic species) is integral to restoration, past degradation may leave lasting legacies including depleted native propagule sources, disturbed or compacted soil and/or elevated nutrients. Site idiosyncrasies may mirror the detailed spatial mosaic of differing disturbance histories, ranging from those that vary over small (e.g. grazing intensities) to large spatial scales (e.g. regional climate patterns). These differences can influence responses to restoration efforts (Foster *et al.* 2003). For example, several lowland riparian sites that share characteristics including dominance by river red gum, similar channel size, livestock access and adjacent land use have exhibited different short-term responses to destocking and replanting (e.g. in the proliferation of exotic species; P. Reich, unpubl. data). Taken together, propagule sources, landscape context and historical legacies of past disturbances create detailed patchworks of variation within and among sites which determine trajectories and rates of response to restoration efforts and illustrate the importance of starting points to restoration success.

Initial conditions should be used to inform restoration actions, manage expectations and set achievable targets. In some cases, initial conditions may prevent certain restoration targets from being achieved simultaneously. For example, along degraded lowland streams in south-eastern Australia, two common targets of riparian restoration are to restore vegetation structure and to improve water quality in the adjacent stream by reducing nitrogen and sediment inputs. Degraded riparian sites in this region often have simplified vegetation such that little or no shrub cover remains and soils have elevated nitrogen levels due to past land use and inputs from adjacent agriculture. Restoration efforts at these sites typically focus on planting woody species (i.e. trees and shrubs), often planting large numbers of indigenous nitrogen-fixing *Acacia* species (Burger *et al.* 2010). Although consistent with the aim of reconstructing a 'pre-European' plant community at these sites, large increases in nitrogen-fixing species might be counteractive to reducing nitrogen loads in waterways (see Drake 2011). In this case, restoration may need to proceed in stages, where efforts are initially taken to reduce nitrogen loads in the soil while concurrently minimising potential sources of future inputs – the latter might pose a significant challenge for sites embedded in heavily nutrient-polluted agricultural landscapes. Alternatively, a trade-off may be required between targets related to function and to the re-establishment of a historical plant community. Such a case highlights the importance of defining clear and achievable restoration objectives and illustrates the need to consider initial condition and landscape context in restoration planning.

## Scale

Restoration success depends on appropriate spatial and temporal scales of restoration activities. Unfortunately, the scales of most restoration activities tend to be driven more by logistic, economic and social constraints than ecological understanding (Lake 2001; Bond and Lake 2003; Lake *et al.* 2007). Most restoration efforts occur at relatively small spatial extents on an ad hoc basis, leading to fragmented patches of restored land embedded in a landscape where degrading processes continue to operate. In many, if not most localities requiring restoration, the spatial extent of the disturbances generating degradation is invariably much larger than the spatial extent at which the restoration efforts are implemented. As a result, external and large-scale processes may dominate and compromise the recovery of restored land (e.g. Gregory *et al.* 2007; Lake *et al.* 2007).

Drivers of riparian degradation operate at different spatial scales. Some drivers that operate at large spatial extents such as drought, regulation of streamflow and inputs of nutrients from adjacent agriculture may override the functional and compositional effects of restoration activities that are undertaken at smaller spatial extents (see also 'Landscape context'). Riparian zones may be restored in patches to reduce nutrient inputs into streams, but without consideration of the runoff pathways at a landscape scale such small-scale interventions may fail to reduce nutrient inputs from catchment runoff (Dosskey *et al.* 2005). Catchment-scale disturbances such as nutrient inputs and altered hydrology can be major drivers of ecosystem dynamics, but are rarely incorporated into riparian restoration planning or explicitly built into expectations. In Australia, riparian restoration efforts are seldom integrated with catchment-scale management targets, such as reductions in catchment nutrient or sediment loadings (but see, for example, Bunn *et al.* 2010). Landscape change at the scale of entire catchments may have impacts too great for a riparian zone to moderate. Some have argued that the greater the impact, or more intensive the land use, the wider the riparian zone or 'buffer' necessary (Allan 2004; Hansen *et al.* 2015). For effective catchment management, a holistic approach is required, whereby the catchment is examined as an integration of land and water.

To circumvent issues of restoration at inadequate spatial extents, restoration efforts would ideally encompass whole catchments. However, this is rarely feasible. Instead, vexatious yet pragmatic questions arise from natural resource managers who need to know where, how much and how wide riparian restoration projects should be to maximise efficiency. To progress these issues we need much better information on the strengths and localities of the disturbances affecting waterways that could be reduced by riparian zone restoration at the appropriate spatial extents. With adequate monitoring of projects that occur at small spatial extents, it may be possible to detect those responses to restoration that might be 'scaled up' to much larger spatial extents (albeit predominantly linear ones, e.g. tree cover). However, this task is far from trivial and fraught with uncertainty, as many responses are likely to be complex and nonlinear (e.g. nutrient capture and processing, animal recruitment) or take many decades to eventuate. These issues of spatial scaling need to be addressed in the initial planning of restoration.

Riparian restoration is inherently a slow process. While some responses to riparian restoration activities might occur rapidly – for example, reduction in bare ground was observed within a year of livestock exclusion on lowland riparian sites (P. Reich, unpubl. data) and recovery of bird habitat was observed in some rainforest communities in south-eastern Australia within three years of restoration efforts (Peel 2010) – many responses and the re-establishment of community dynamics will require years of waiting and may require some level of ongoing intervention. For example, trees may take a long time to grow and mature, invasive species may require management, and if species are unable to self-recruit, supplementary planning or seeding or artificial triggers to promote recruitment may be required. Many of the aquatic benefits may take even longer; the development of shade, the accrual of carbon and coarse wood in stream channels may take many decades. Furthermore, changes in climate and environmental conditions are likely to occur within the time period of tree and plant community maturation following restoration efforts. Therefore, the challenge is for the restored community and its functions to be resilient to these changes.

## A WAY FORWARD

A dominant thread throughout this discussion of constraints and challenges facing successful riparian restoration is incomplete ecological knowledge. Most management activities that are termed riparian restoration are in fact exercises in replanting and livestock removal that adhere to few, if any, of the principles of stream restoration espoused by others (e.g. Jansson *et al.* 2005; Palmer *et al.* 2005) and are poorly monitored so little is learned from the process. Fortunately, notable exceptions exist and these provide valuable insights into improving restoration practices (e.g. Peel 2010; Gollan *et al.* 2011a, 2011b). To progress towards improving the effectiveness of riparian restoration in Australia, we echo many scientists and practitioners worldwide by emphasising that riparian restoration requires a clear vision and specific objectives. However, like all natural resource management issues, the success of riparian restoration also hinges on social and political will; solely focusing on technical, ecological constraints and solutions imperils progress. Restoration efforts are unlikely to instantly gratify and long-term commitments to maintaining restoration projects are essential to long-lasting success. We contend that a major reason for slow progress in achieving many of the objectives of riparian restoration in Australia is the failure to support, implement and coordinate sufficient monitoring of restoration efforts – we simply are not learning enough from the current investment.

Past restoration efforts from Australia and overseas may provide important lessons for improving efforts to restore Australian riparian zones. For example, through years of observation and experimentation, an effective approach has been developed and implemented to restore riparian sites dominated by exotic vegetation in East Gippsland (Peel 2010). This approach exploits principles of vegetation dynamics to restore rainforest communities. It has been successful in re-establishing vegetation and providing biodiversity benefits, with restored sites similar to un-impacted remnant locations in terms of native plant species richness and the presence of specialist bird species. The success of restoration at these sites emphasises the important role of monitoring and assessment as well as ecological issues such as the proximity of restoration sites to remnant vegetation (Peel 2010). In two examples from North America – restoration of the Chesapeake Bay watershed and rivers of the Pacific north-west – restoration efforts began as small, disconnected projects undertaken at inadequate spatial extents and with inadequate temporal continuity. In Chesapeake Bay, a large estuarine system in north-eastern USA, eutrophication led to algal blooms, fish kills and virtual collapse of oyster and blue crab populations, which once supported a lucrative commercial and recreational fishing industry. Together, these environmental and economic triggers instigated efforts towards restoring the bay. Degradation of rivers in the Pacific north-west, mostly from agriculture, land clearance and flow impoundments, coupled with overfishing, led to the decline of salmon, an iconic and economically valuable species in the region. Due to the link between salmon decline and the declining condition of rivers, the desire to increase salmon populations became the impetus of efforts towards the broader target of restoring river condition. Efforts to restore Chesapeake Bay were initially small and disconnected and achieved only minimal success because most of the nutrients and sediment responsible for poor water quality are generated far upstream in smaller tributaries (Hassett *et al.* 2005). Likewise, initial efforts towards restoring rivers in the Pacific north-west were typically small in scale, spatially fragmented and largely ineffective (Montgomery *et al.* 2003). In both Chesapeake Bay and the Pacific north-west, as well as several other notable examples (see Roni 2005), increased traction and effectiveness in achieving desired outcomes has been achieved through recognising the catchment scale of the problem (i.e. approach at system-wide level), integrating fragmented restoration efforts and implementing an adaptive management framework to monitor, revise and improve the approach of restoration efforts. These case studies reveal some key factors leading to the success and failure of restoration projects. These include a clear vision and objectives, appropriate spatial and temporal extents of restoration efforts, motivation for social and political will, and informative monitoring programs.

Restoration efforts need to be underpinned by a clear vision and specific targets (Palmer *et al.* 2005). Constant reminders about the imperative for such goal-setting (Hobbs and Norton 1996; Ehrenfeld 2000; SER 2004; Palmer *et al.* 2005) have become somewhat wearisome, yet goals often remain vague and poorly communicated in many riparian restoration projects. The complexity of riparian ecosystems themselves and multiple implicit objectives of projects, often spanning issues of water quality, aquatic and terrestrial plants and animals across various spatial scales, do not help. Restoration efforts often focus simultaneously on individual species (e.g. endangered and/or keystone species), vegetation structure, vegetation composition, functional aspects of vegetation (e.g. minimising stream inputs of sediment, providing terrestrial subsidies to support instream food webs), and more. Riparian management often has the intangible goal of restoring river or ecosystem health, presumably the ultimate objective of restoring (fully or partially) critical parts of the composition, structure, functions and dynamics of the original system. This is promoted by indices of stream conditions, which average across several metrics that are considered universally desirable or indicative of desirable functions. Such a diffuse target has advantages and disadvantages when it comes to designing and implementing restoration efforts. Implicit is the notion that restoring one ecosystem component will advantage others. However, restoring different components of riparian systems may require different actions, or several actions, that need to be prioritised and undertaken in appropriate order (Suding *et al.* 2004; Stewart-Koster *et al.* 2010). Objectives therefore need to be clear and explicitly linked to our scientific understanding and the assumptions that underpin any putative responses.

Riparian restoration is embedded in a social, economic and political environment and this will ultimately dictate restoration success. Understandably, landholders are reluctant to change their practices; lowland riparian zones are often the area of greatest fertility and moisture in a landscape and thus offer prime locations for grazing and cropping.

Yet, the success of restoration efforts will be underpinned by agreement among landholders and the community on a vision to restore riparian zones. Riparian restoration is inherently a long-term prospect and therefore requires a long-term commitment by surrounding landholders and management agencies for appropriate management of riparian land. Furthermore, restoration efforts will need to be coordinated across large spatial scales to have catchment-wide results and there will need to be agreement across communities for the need to tackle riparian degradation.

A catalyst for action is almost always required to galvanise sufficient public and political support. In restoring both Chesapeake Bay and the rivers of the Pacific north-west, public engagement and support for natural resource management was garnered by perceived urgency of environmental problems and conspicuous economic consequences. In Australia, some restoration efforts may be driven by similarly stark economic consequences and perceived crisis points (e.g. management of poor water quality generated in catchments draining into Moreton Bay or the Great Barrier Reef). However, many riverine systems in Australia, such as the Murray–Darling Basin, are in a chronically degraded state and more creative ways may be necessary to overcome inertia, motivate restoration actions and garner community support. Emphasising the minimal economic cost to the prevalent land use and/or agriculture in the region, as well as the significant advantages for biodiversity and river condition, could potentially help to stimulate support. Solutions may come from integrating and scaling-up grass roots efforts through economic incentives.

Market-based instruments, such as permit or reverse auction systems are one way of formally incorporating both the costs of restoration and environmental benefits. For instance, Roberts *et al.* (2009) estimated an annualised cost of US$1.3 million for establishing and maintaining a 45.7-m riparian buffer strip on all agricultural land adjoining a waterway in Tennessee's Harpeth River watershed. In this case, the primary motivation for creating vegetated buffer strips was to meet stringent nutrient reduction targets set by the United States Environmental Protection Agency. Roberts *et al.* (2009) estimated that for around 98% of agricultural land within the watershed, a riparian buffer strip would cost between $210 and $370 annually per hectare. These cost estimates, in conjunction with an analysis of the environmental benefits associated with the potential buffer strips, might be used as a starting point for a reverse auction process targeting riparian areas for restoration. Market-based incentives that target riparian management are currently being used successfully in some parts of Australia (e.g. Department of Sustainability and Environment Victoria 2008).

## CONCLUSIONS

Given the current state of Australia's riparian zones, restoration efforts need to occur now despite incomplete knowledge. As a consequence, mechanisms are required to deal with the uncertainty of outcomes and improve the accuracy of our assumptions into the future (see Reich *et al.* 2011). Adaptive management, characterised as 'learning by doing', has long been considered an appropriate framework in which to address uncertainty in management (Holling 1978). By explicitly incorporating learning and updating assumptions, adaptive management balances the urgency of action with the imperative to learn. We advocate focus on improving knowledge through implementing an iterative adaptive monitoring framework (Ralph and Poole 2003). The value of this monitoring framework comes from strong links between monitoring efforts (i.e. data collection), improving knowledge (i.e. data management and analysis) and revising management methods (i.e. communication and implementation). It is axiomatic that these restoration efforts be undertaken using the best available knowledge and thus links should be strengthened between ecological knowledge, policy and implementation. Monitoring should be proactive rather than reactive and designed concurrently with on-ground works or restoration activities before on-ground works are undertaken.

This provides an opportunity for restoration efforts to act as large-scale manipulative experiments that address crucial ecological questions using spatial and temporal scales that are realistic and informative for future restoration efforts. Monitoring must also be long term in order to reflect expected rates of ecological response.

A necessary compromise in the development of such a program is recognising and balancing competing expenditure on learning (i.e. monitoring) and doing (i.e. on-ground works), recognising the need to achieve outputs while ensuring these outputs are not misguided. Like most other scientific disciplines, restoration ecology grows as much from its mistakes as it does from it successes. The outcomes of large-scale restoration of riparian zones will not be apparent for many decades and thus learning must go hand in hand with implementation to ensure that mistakes can be identified early and solutions can be developed.

## ACKNOWLEDGEMENTS

The authors acknowledge the Murray–Darling Basin Authority for supporting this work. We also thank Fiona Coates for providing valuable input into an early version of the manuscript. TRC gratefully acknowledges the ARC for supporting his research via the award of a Future Fellowship (FT120100463).

## REFERENCES

Allan JD (2004) Landscapes and riverscapes: the influence of land use on stream ecosystems. *Annual Review of Ecology Evolution and Systematics* **35**, 257–284. doi:10.1146/annurev.ecolsys.35.120202.110122

Auld TD (1987) Population dynamics of the shrub *Acacia suaveolens* (Sm.) Willd.: survivorship throughout the life cycle, a synthesis. *Australian Journal of Ecology* **12**, 139–151. doi:10.1111/j.1442-9993.1987.tb00935.x

Baker M, Weller D, Jordan T (2007) Effects of stream map resolution on measures of riparian buffer distribution and nutrient retention potential. *Landscape Ecology* **22**, 973–992. doi:10.1007/s10980-007-9080-z

Barton D (1996) The use of percent model affinity to assess the effects of agriculture on benthic invertebrate communities in headwater streams of southern Ontario, Canada. *Freshwater Biology* **36**, 397–410. doi:10.1046/j.1365-2427.1996.00053.x

Bash JS, Ryan CM (2002) Stream restoration and enhancement projects: is anyone monitoring? *Environmental Management* **29**, 877–885. doi:10.1007/s00267-001-0066-3

Bernhardt ES, Palmer MA, Allan JD, Alexander G, Barnas K, Brooks SS, Carr J, Clayton S, Dahm C, Follstad Shah J, Galat D, Gloss S, Goodwin P, Hart D, Hassett B, Jenkinson R, Katz S, Kondolf GM, Lake PS, Lave R, Meyer JL, O'Donnell TK, Pagano L, Powell B, Sudduth E (2005) Synthesizing US river restoration efforts. *Science* **308**, 636–637. doi:10.1126/science.1109769

Bernhardt ES, Sudduth EB, Palmer MA, Allan JD, Meyer JL, Alexander G, Follastad-Shah J, Hassett B, Jenkinson R, Lave R, Rumps J, Pagano L (2007) Restoring rivers one reach at a time: results from a survey of US river restoration practitioners. *Restoration Ecology* **15**, 482–493. doi:10.1111/j.1526-100X.2007.00244.x

Bidwell S, Attiwill PM, Adams MA (2006) Nitrogen availability and weed invasion in a remnant native woodland in urban Melbourne. *Austral Ecology* **31**, 262–270. doi:10.1111/j.1442-9993.2006.01575.x

Blanch SJ, Ganf GG, Walker KF (1999) Tolerance of riverine plants to flooding and exposure indicated by water regime. *Regulated Rivers: Research and Management* **15**, 43–62. doi:10.1002/(SICI)1099-1646(199901/06)15:1/3<43::AID-RRR535>3.0.CO;2-Q

Blanton P, Marcus WA (2013) Transportation infrastructure, river confinement, and impacts on floodplain and channel habitat, Yakina and Chehalis rivers, Washington, USA. *Geomorphology* **189**, 55–65. doi:10.1016/j.geomorph.2013.01.016

Bohn BA, Kershner JL (2002) Establishing aquatic restoration priorities using a watershed approach.

*Journal of Environmental Management* **64**, 355–363. doi:10.1006/jema.2001.0496

Bond NR, Lake PS (2003) Local habitat restoration in streams: constraints on the effectiveness of restoration for stream biota. *Ecological Management & Restoration* **4**, 193–198. doi:10.1046/j.1442-8903.2003.00156.x

Bond NR, Lake PS (2005) Ecological restoration and large-scale ecological disturbance: the effects of drought on the response by fish to a habitat restoration experiment. *Restoration Ecology* **13**, 39–48. doi:10.1111/j.1526-100X.2005.00006.x

Bren LJ (1988) Effects of river regulation on flooding of a riparian red gum forest on the River Murray, Australia. *Regulated Rivers: Research and Management* **2**, 65–77. doi:10.1002/rrr.3450020202

Bren LJ (1992) Tree invasion of an intermittent wetland in relation to changes in the flooding frequency of the River Murray, Australia. *Austral Ecology* **17**, 395–408. doi:10.1111/j.1442-9993.1992.tb00822.x

Brooks SS, Lake PS (2007) River restoration in Victoria, Australia: change is in the wind, and none too soon. *Restoration Ecology* **15**, 584–591. doi:10.1111/j.1526-100X.2007.00253.x

Bunn S, Abal EG, Smith MJ, Choy SC, Fellows CS, Harch BD, Kennard MJ, Sheldon F (2010) Integration of science and monitoring of river ecosystem health to guide investments in catchment protection and rehabilitation *Freshwater Biology* **55** (Suppl. 1) 223–240. doi:10.1111/j.1365-2427.2009.02375.x

Burger B, Reich P, Cavagnaro TR (2010) Trajectories of change: riparian vegetation and soil conditions following livestock removal and replanting. *Austral Ecology* **35**, 980–987. doi:10.1111/j.1442-9993.2010.02112.x

Capon SJ (2005) Flood variability and spatial variation in plant community composition and structure on a large arid floodplain. *Journal of Arid Environments* **60**, 283–302. doi:10.1016/j.jaridenv.2004.04.004

Capon SJ, Brock MA (2006) Flooding, soil seed bank dynamics and vegetation resilience of a hydrologically variable desert floodplain. *Freshwater Biology* **51**, 206–223. doi:10.1111/j.1365-2427.2005.01484.x

Clarke A, Lake PS, O'Dowd DJ (2004) Ecological impacts on aquatic macroinvertebrates following upland stream invasion by a ponded pasture grass (*Glyceria maxima*) in southern Australia. *Marine and Freshwater Research* **55**, 709–713. doi:10.1071/MF04043

Cramer VA, Hobbs RJ, Standish RJ (2008) What's new about old fields? Land abandonment and ecosystem assembly. *Trends in Ecology & Evolution* **23**, 104–112. doi:10.1016/j.tree.2007.10.005

Crook DA, Robertson AI (1999) Relationships between riverine fish and woody debris: implications for lowland rivers. *Marine and Freshwater Research* **50**, 941–953. doi:10.1071/MF99072

Cunningham SC, Thomson JR, Read J, Baker PJ, Mac Nally R (2010) Does stand structure influence susceptibility of eucalypt floodplain forests to dieback? *Austral Ecology* **35**, 348–356. doi:10.1111/j.1442-9993.2009.02043.x

D'Antonio CM, Vitousek PM (1992) Biological invasions by exotic grasses, the grass/fire cycle, and global change. *Annual Review of Ecology and Systematics* **23**, 63–87. doi:10.1146/annurev.es.23.110192.000431

Department of Sustainability and Environment Victoria (2008) *EcoMarkets: Valuing our Environment*. Victorian Government Department of Sustainability and Environment, Melbourne.

Department of Sustainability and Environment Victoria (2010) 'Securing our rivers for future generations.' Victorian river health program report card 2002–2009. Department of Sustainability and Environment, Melbourne.

Dosskey MG, Eisenhauer DE, Helmers MJ (2005) Establishing conservation buffers using precision information. *Journal of Soil and Water Conservation* **60**, 349–354.

Douglas MM, O'Connor RA (2004) Weed invasion changes fuel characteristics: para grass (*Urochloa mutica* (Forssk.) TQ Nguyen) on a tropical floodplain. *Ecological Management & Restoration* **5**, 143–145. doi:10.1111/j.1442-8903.2004.201-7.x

Douglas MM, Bunn SE, Pidgeon RJW, Davies PM, Barrow P, O'Connor RA, Winning M (2001) 'Weed management and the biodiversity and ecological processes of tropical wetlands.' Report to Environment Australia and Land and Water Australia.

Drake DC (2011) Invasive legumes fix $N_2$ at high rates in riparian areas of an N-saturated, agricultural catchment. *Journal of Ecology* **99**, 515–523.

Earl J (2003) 'The distribution and impacts of lippia (*Phyla canescens*) in the Murray–Darling system.' Report to Murray–Darling Basin Lippia Working Group, Guyra, NSW.

Ehrenfeld JG (2000) Defining the limits of restoration: the need for realistic goals. *Restoration Ecology* **8**, 2–9. doi:10.1046/j.1526-100x.2000.80002.x

Forman RTT (1995) Some general principles of landscape and regional ecology. *Landscape Ecology* **10**, 133–142. doi:10.1007/BF00133027

Foster D, Swanson F, Aber J, Burke I, Brokaw N, Tilman D, Knapp A (2003) The importance of land-use legacies to ecology and conservation. *Bioscience* **53**, 77–88. doi:10.1641/0006-3568(2003)053[0077:TIOLUL]2.0.CO;2

George AK, Walker KF, Lewis MM (2005) Population status of eucalypt trees on the River Murray floodplain, South Australia. *River Research and Applications* **21**, 271–282. doi:10.1002/rra.846

Gilvear DJ, Casas-Mulet R, Spray CJ (2012) Trends and issues in delivery of integrated catchment scale river restoration: lessons learned from a national river restoration survey within Scotland. *River Research and Applications* **28**, 234–246. doi:10.1002/rra.1437

Gollan JR, de Bruyn LL, Reid N, Smith D, Wilkie L (2011a) Can ants be used as ecological indicators of restoration progress in dynamic environments? A case study in a revegetated riparian zone. *Ecological Indicators* **11**, 1517–1525. doi:10.1016/j.ecolind.2009.09.007

Gollan JR, Reid CAM, Barnes PB, Wilkie L (2011b) The ratio of exotic-to-native dung beetles can indicate habitat quality in riparian restoration. *Insect Conservation and Diversity* **4**, 123–131. doi:10.1111/j.1752-4598.2010.00115.x

Greenwood H, O'Dowd DJ, Lake PS (2004) Willow (*Salix* x *rubens*) invasion of the riparian zone in south-eastern Australia: reduced abundance and altered composition of terrestrial arthropods. *Diversity & Distributions* **10**, 485–492. doi:10.1111/j.1366-9516.2004.00104.x

Gregory SV, Swanson FJ, McKee WA, Cummins KW (1991) An ecosystem perspective of riparian zones: focus on links between land and water. *Bioscience* **41**, 540–551. doi:10.2307/1311607

Gregory S, Allen AW, Baker M, Boyer K, Dillaha T, Elliott J (2007) Realistic expectations of timing between conservation and restoration actions and ecological responses. In *Managing Agricultural Landscapes for Environmental Quality: Strengthening the Science Base*. (Eds M Schnepf and C Cox) pp. 115–144. Soil and Water Conservation Society, Ankeny, IA.

Hall AA, Rood SB, Higgins PS (2011) Resizing a river: a downscaled, seasonal flow regime promotes riparian restoration. *Restoration Ecology* **19**, 351–359. doi:10.1111/j.1526-100X.2009.00581.x

Hansen BD, Reich P, Cavagnaro TR, Lake PS (2015) Challenges in applying scientific evidence to width recommendations for riparian management in agricultural Australia. *Ecological Management and Restoration* **16**(1), 50–57.

Hanski I (1998) Metapopulation dynamics. *Nature* **396**, 41–49. doi:10.1038/23876

Hardegree SP, Cho J, Schneider J, Monaco T, Sheley R (2012) Weather variability, ecological processes, and optimization of soil microenvironment for rangeland restoration. In *Invasive Plant Ecology and Management: Linking Processes to Practice*. (Eds TA Monaco and RL Sheley) pp. 107–121. CAB International Wallingford, UK.

Hassett B, Palmer M, Bernhardt E, Smith S, Carr J, Hart D (2005) Restoring watersheds project by project: trends in Chesapeake Bay tributary restoration. *Frontiers in Ecology and the Environment* **3**, 259–267. doi:10.1890/1540-9295(2005)003[0259:RWPBPT]2.0.CO;2

Hobbs RJ, Cramer VA (2008) Restoration ecology: interventionist approaches for restoring and maintaining ecosystem function in the face of rapid environmental change. *Annual Review of Environment and Resources* **33**, 39–61. doi:10.1146/annurev.environ.33.020107.113631

Hobbs RJ, Norton DA (1996) Towards a conceptual framework for restoration ecology. *Restoration Ecology* **4**, 93–110. doi:10.1111/j.1526-100X.1996.tb00112.x

Hobbs RJ, Norton DA (2004) Ecological filters, thresholds and gradients in resistance to ecosystem reassembly. In *Assembly Rules And Restoration Ecology – Bridging The Gap Between Theory And Practice*. (Eds VM Temperton, RJ Hobbs, TJ Nuttle and S Halle) pp. 72–95. Island Press, Washington, DC.

Holland-Clift S, Davies J (2007) *Willows National Management Guide: Current Management and Control Options for Willows (*Salix *spp.) in Australia*. Victorian Department of Primary Industries, Melbourne.

Holling CS (Ed.) (1978) *Adaptive Environmental Assessment and Management*. Wiley, Chichester, UK.

Holmgren M (2009) Dryland dynamics and restoration: perspectives for the use of climatic swings. In *New Models for Ecosystem Dynamics and Restoration*. (Eds RJ Hobbs and KN Suding) pp. 112–123. Island Press, Washington, DC.

Holmgren M, Scheffer M (2001) El Niño as a window of opportunity for the restoration of degraded arid ecosystems. *Ecosystems* **4**, 151–159. doi:10.1007/s100210000065

Holmgren M, Scheffer M, Ezcurra E, Gutiérrez JR, Mohren GMJ (2001) El Niño effects on the dynamics of terrestrial ecosystems. *Trends in Ecology & Evolution* **16**, 89–94. doi:10.1016/S0169-5347(00)02052-8

Horner GJ, Baker PJ, Mac Nally R, Cunningham SC, Thomson JR, Hamilton F (2009) Mortality of developing floodplain forests subjected to a drying climate and water extraction. *Global Change Biology* **15**, 2176–2186. doi:10.1111/j.1365-2486.2009.01915.x

Horner GJ, Cunningham SC, Thomson JR, Baker PJ, Mac Nally R (2012) Forest structure, flooding and grazing predict understorey composition of floodplain forests in southeastern Australia. *Forest Ecology and Management* **286**, 148–158. doi:10.1016/j.foreco.2012.08.023

Jackson LE, Burger M, Cavagnaro TR (2008) Roots, nitrogen transformations, and ecosystem services. *Annual Review of Plant Biology* **59**, 341–363. doi:10.1146/annurev.arplant.59.032607.092932

Jansen A, Robertson AI (2001) Relationships between livestock management and the ecological condition of riparian habitats along an Australian floodplain river. *Journal of Applied Ecology* **38**, 63–75. doi:10.1046/j.1365-2664.2001.00557.x

Jansson R, Backx H, Boulton AJ, Dixon M, Dudgeon D, Hughes FMR, Nakamura K, Stanley EH, Tockner K (2005) Stating mechanisms and refining criteria for ecologically successful river restoration: a comment on Palmer *et al.* (2005). *Journal of Applied Ecology* **42**, 218–222. doi:10.1111/j.1365-2664.2005.01022.x

Jensen AE, Walker KF, Paton DC (2008) The role of seedbanks in restoration of floodplain woodlands. *River Research and Applications* **24**, 632–649. doi:10.1002/rra.1161

Johnson RK, Battarbee RW, Bennion H, Hering D, Soons MB, Verhoeven JTA (2010) Climate change: defining reference conditions and restoring freshwater ecosystems. In *Climate Change Impacts on Freshwater Ecosystems*. (Eds M Kernan, RW Battarbee and B Moss) pp. 203–235. Wiley-Blackwell, Chichester, UK.

Kauffman JB, Beschta RL, Otting N, Lytjen D (1997) An ecological perspective of riparian and stream restoration in the western United States. *Fisheries* **22**, 12–24. doi:10.1577/1548-8446(1997)022<0012:AEPORA>2.0.CO;2

Kingsford RT (2000) Ecological impacts of dams, water diversions and river management on floodplain wetlands in Australia. *Austral Ecology* **25**, 109–127. doi:10.1046/j.1442-9993.2000.01036.x

Kramer A, Reich P, Lake PS (2007) Preliminary insights into the status of ground-dwelling terrestrial arthropods at sites representing three riparian conditions. *Ecological Management & Restoration* **8**, 147–150. doi:10.1111/j.1442-8903.2007.00354.x

Kyle G, Schmidt T, Cooke J, Leishman MR (2008) Growth and survival of riparian plantings in relation to weeping willow canopy in the upper Hunter River. *Ecological Management & Restoration* **9**, 154–156. doi:10.1111/j.1442-8903.2008.00409.x

Lake PS (2000) Disturbance, patchiness, and diversity in streams. *Journal of the North American Benthological Society* **19**, 573–592. doi:10.2307/1468118

Lake PS (2001) On the maturing of restoration: linking ecological research and restoration. *Ecological Management & Restoration* **2**, 110–115. doi:10.1046/j.1442-8903.2001.00074.x

Lake PS, Bond N, Reich P (2007) Linking ecological theory with stream restoration. *Freshwater Biology* **52**, 597–615. doi:10.1111/j.1365-2427.2006.01709.x

Lindenmayer DB, Likens GE (2010) The science and application of monitoring. *Biological Conservation* **143**, 1317–1328. doi:10.1016/j.biocon.2010.02.013

Lovett S, Price P (1999) *Riparian Land Management Technical Guidelines, Volume One: Principles of Sound Management*. Land and Water Rural Research Development Corporation, Canberra.

Lowe WH, Likens GE (2005) Moving headwater streams to the head of the class. *Bioscience* **55**, 196–197. doi:10.1641/0006-3568(2005)055[0196:MHSTTH]2.0.CO;2

Lowrance R, Todd R, Fail J, Henrickson O, Leonard R, Asmussen L (1984) *Riparian Forests as Nutrient Filters in Agricultural Watersheds*. American Institute of Biological Sciences, New York, NY.

Lowrance R, Altier LS, Newbold JD, Schnabel RR, Groffman P, Denver JM, Correll DL, Gilliam JW, Robinson JL, Brinsfield RB, Staver KW, Lucas W, Todd AH (1997) Water quality functions of riparian forest buffers in Chesapeake Bay watersheds. *Environmental Management* **21**, 687–712. doi:10.1007/s002679900060

Lunt ID, Jansen A, Binns D, Kenny SA (2007) Long-term effects of exclusion of grazing stock on degraded herbaceous plant communities in riparian *Eucalyptus camaldulensis* forest in south-eastern Australia. *Austral Ecology* **32**, 937–949. doi:10.1111/j.1442-9993.2007.01782.x

Lymbery AJ, Doup RG, Pettit NE (2003) Effects of salinisation on riparian plant communities in experimental catchments on the Collie River, Western Australia. *Australian Journal of Botany* **51**, 667–672. doi:10.1071/BT02119

Mac Nally R, Horrocks G (2002) Habitat change and restoration: responses of a forest-floor mammal species to manipulations of fallen timber in floodplain forests. *Animal Biodiversity and Conservation* **25**, 41–52.

Macdonald MJ, Whalley W, Julien MH, Sindel BM, Duggin JA (2012) Flood-induced recruitment of the invasive perennial herb *Phyla canescens* lippia. *The Rangeland Journal* **34**, 269–276. doi:10.1071/RJ11094

McKergow LA, Weaver DM, Prosser IP, Grayson RB, Reed AEG (2003) Before and after riparian management: sediment and nutrient exports from a small agricultural catchment, Western Australia. *Journal of Hydrology* **270**, 253–272. doi:10.1016/S0022-1694(02)00286-X

McKergow LA, Prosser IP, Weaver DM, Grayson RB, Reed AEG (2006) Performance of grass and *Eucalyptus* riparian buffers in a pasture catchment, Western Australia, part 2: water quality. *Hydrological Processes* **20**, 2327–2346. doi:10.1002/hyp.6054

McLendon T, Redente EF (1991) Nitrogen and phosphorus effects on secondary succession dynamics on a semi-arid sagebrush site. *Ecology* **72**, 2016–2024. doi:10.2307/1941556

Meeson N, Robertson AI, Jansen A (2002) The effects of flooding and livestock on post-dispersal seed predation in river red gum habitats. *Journal of Applied Ecology* **39**, 247–258. doi:10.1046/j.1365-2664.2002.00706.x

Montgomery DR, Bolton SM, Booth DB, Wall L (2003) *Restoration of Puget Sound Rivers*. University of Washington Press, Seattle, WA.

Morghan KJR, Seastedt TR (1999) Effects of soil nitrogen reduction on nonnative plants in restored grasslands. *Restoration Ecology* **7**, 51–55. doi:10.1046/j.1526-100X.1999.07106.x

Morley SA, Karr JR (2002) Assessing and restoring the health of urban streams in the Puget Sound

Basin. *Conservation Biology* **16**, 1498–1509. doi:10.1046/j.1523-1739.2002.01067.x

Naiman R, Décamps H (1997) The ecology of interfaces: riparian zones. *Annual Review of Ecology and Systematics* **28**, 621–658. doi:10.1146/annurev.ecolsys.28.1.621

Naiman RJ, Décamps H, McClain ME (2005) *Riparia: Ecology, Conservation and Management of Streamside Communities.* Elsevier/Academic Press, San Diego, CA.

Natural Resources Conservation League of Victoria (1977) *Victoria's Resources.* Natural Resources Conservation League of Victoria, Springvale, Australia.

Norris RH, Prosser I, Young B, Liston P, Bauer N, Davies N, Dyer F, Linke S, Thoms M (2001) 'The assessment of river condition (ARC): an audit of the ecological condition of Australian rivers.' Report to the National Land and Water Resources Audit Office, Canberra.

Osborne LL, Kovacic DA (1993) Riparian vegetated buffer strips in water-quality restoration and stream management. *Freshwater Biology* **29**, 243–258. doi:10.1111/j.1365-2427.1993.tb00761.x

Palmer MA, Ambrose RF, Poff NL (1997) Ecological theory and community restoration ecology. *Restoration Ecology* **5**, 291–300. doi:10.1046/j.1526-100X.1997.00543.x

Palmer MA, Bernhardt ES, Allan JD, Lake PS, Alexander G, Brooks SS, Carr J, Clayton S, Dahm CN, Follstad Shah J, Galat DL, Loss GS, Goodwin P, Hart DD, Hassett B, Jenkinson R, Kondolf GM, Lave R, Meyer JL, O'Donnell TK, Pagano L, Sudduth E (2005) Standards for ecologically successful river restoration. *Journal of Applied Ecology* **42**, 208–217. doi:10.1111/j.1365-2664.2005.01004.x

Patten DT (2006) Restoration of wetland and riparian systems: the role of science, adaptive management, history, and values. *Journal of Contemporary Water Research & Education* **134**, 9–18. doi:10.1111/j.1936-704X.2006.mp134001003.x

Peel B (2010) *Rainforest Restoration Manual for South-eastern Australia.* CSIRO Publishing, Melbourne.

Pettit NE, Froend RH (2001) Availability of seed for recruitment of riparian vegetation: a comparison of a tropical and a temperate river ecosystem in Australia. *Australian Journal of Botany* **49**, 515–528. doi:10.1071/BT00059

Pettit N, Naiman R (2007) Fire in the riparian zone: characteristics and ecological consequences. *Ecosystems* **10**, 673–687. doi:10.1007/s10021-007-9048-5

Pickett STA, White PS (1985) *The Ecology of Natural Disturbance and Patch Dynamics.* Academic Press, New York, NY.

Planty-Tabacchi AM, Tabacchi E, Naiman RJ (1996) Invasibility of species-rich communities in riparian zones. *Conservation Biology* **10**, 598–607. doi:10.1046/j.1523-1739.1996.10020598.x

Prober SM, Thiele KR, Lunt ID, Koen TB (2005) Restoring ecological function in temperate grassy woodlands: manipulating soil nutrients, exotic annuals and native perennial grasses through carbon supplements and spring burns. *Journal of Applied Ecology* **42**, 1073–1085. doi:10.1111/j.1365-2664.2005.01095.x

Prosser IP, Rutherfurd ID, Olley JM, Young WJ, Wallbrink PJ, Moran CJ (2001) Large scale patterns of erosion and sediment transport in river networks, with examples from Australia. *Marine and Freshwater Research* **52**, 81–99. doi:10.1071/MF00033

Ralph S, Poole G (2003) Putting monitoring first: designing accountable ecosystem restoration and management plans. In *Restoration of Puget Sound Rivers.* (Eds DR Montgomery, S Bolton, DB Booth and L Wall) pp. 222–242. University of Washington Press, Seattle, WA.

Read MG, Barmuta LA (1999) Comparisons of benthic communities adjacent to riparian native eucalypt and introduced willow vegetation. *Freshwater Biology* **42**, 359–374. doi:10.1046/j.1365-2427.1999.444474.x

Reich P, Lake PS (2014) Extreme hydrological events and the ecological restoration of flowing waters. *Freshwater Biology*doi:10.1111/fwb.12508

Reich P, Lake PS, Williams L, Hale R (2011) On improving the science and practice of riparian restoration. *Ecological Management & Restoration* **12**, 4–5. doi:10.1111/j.1442-8903.2011.00561.x

Reid DJ, Lake PS, Quinn GP, Reich P (2008aAssociation of reduced riparian vegetation cover in agricultural landscapes with coarse detritus dynamics in lowland streams. *Marine and Freshwater Research* **59**, 998–1014. doi:10.1071/MF08012

Reid DJ, Quinn GP, Lake PS, Reich P (2008bTerrestrial detritus supports the food webs in lowland intermittent streams of south-eastern Australia: a stable isotope study. *Freshwater Biology* **53**, 2036–2050. doi:10.1111/j.1365-2427.2008.02025.x

Richardson DM, Holmes PM, Esler KJ, Galatowitsch S, Stromberg JC, Kirkman SP, Pysek P, Hobbs RJ (2007) Riparian vegetation: degradation, alien plant invasions, and restoration prospects. *Diversity & Distributions* **13**, 126–139. doi:10.1111/j.1366-9516.2006.00314.x

Roberts DC, Clark CD, English BC, Park WM, Roberts RK (2009) Estimating annualized riparian buffer costs for the Harpeth River watershed. *Applied Economic Perspectives and Policy* **31**, 894–913.

Robertson A, Rowling R (2000) Effects of livestock on riparian zone vegetation in an Australian dryland river. *Regulated Rivers: Research and Management* **16**, 527–541. doi:10.1002/1099-1646(200009/10)16:5<527::AID-RRR602>3.0.CO;2-W

Robinson D, Mann S (1998) 'Effects of grazing, fencing and licencing on the natural values of crown land frontages in the Goulburn-Broken catchment.' Goulburn Valley Environment Group, report to Goulburn-Broken Catchment Management Authority.

Roni P (2005) *Monitoring Stream and Watershed Restoration*. American Fisheries Society, Bethesda, MD.

Rood SB, Samuelson GM, Braatne JH, Gourley CR, Hughes FMR, Mahoney JM (2005) Managing river flows to restore floodplain forests. *Frontiers in Ecology and the Environment* **3**, 193–201. doi:10.1890/1540-9295(2005)003[0193:MRFTRF]2.0.CO;2

Ross L (1994) *The Effects of Instream Trees on Channel Morphology*. University of Melbourne, Melbourne.

Roth NE, Allan JD, Erickson DL (1996) Landscape influences on stream biotic integrity assessed at multiple spatial scales. *Landscape Ecology* **11**, 141–156. doi:10.1007/BF02447513

Rutherford JC, Marsh NA, Davies PM, Bunn SE (2004) Effects of patchy shade on stream water temperature: how quickly do small streams heat and cool? *Marine and Freshwater Research* **55**, 737–748. doi:10.1071/MF04120

Sabo J, Sponseller R, Dixon M, Gade K, Harms T, Heffernan J, Jani A, Katz G, Soykan C, Watts J, Welter J (2005) Riparian zones increase regional species richness by harboring different, not more, species. *Ecology* **86**, 56–62. doi:10.1890/04-0668

Sanjari G, Ghadiri H, Ciesiolka CAA, Yu B (2008) Comparing the effects of continuous and time-controlled grazing systems on soil characteristics in southeast Queensland. *Australian Journal of Soil Research* **46**, 348–358. doi:10.1071/SR07220

Sattler P, Creighton C (2002) *Australian Terrestrial Biodiversity Assessment 2002*. National Land and Water Resources Audit, Canberra.

Seavy NE, Gardali T, Golet GH, Griggs FT, Howell CA, Kelsey R, Small SL, Viers JH, Weigand JF (2009) Why climate change makes riparian restoration more important than ever: recommendations for practice and research. *Ecological Research* **27**, 330–338. doi:10.3368/er.27.3.330

SER (2004) *Society for Ecological Restoration International Science and Policy Working Group: the SER International Primer on Ecological Restoration*. Society for Ecological Restoration International, Tucson, AZ, <http://www.ser.org>.

Stewart-Koster B, Bunn SE, Mackay SJ, Poff NL, Naiman RJ, Lake PS (2010) The use of Bayesian networks to guide investments in flow and catchment restoration for impaired river ecosystems. *Freshwater Biology* **55**, 243–260. doi:10.1111/j.1365-2427.2009.02219.x

Stohlgren TJ, Chase TN, Pielke RA, Kittel TGF, Baron JS (1998) Evidence that local land use practices influence regional climate, vegetation, and stream flow patterns in adjacent natural areas. *Global Change Biology* **4**, 495–504. doi:10.1046/j.1365-2486.1998.t01-1-00182.x

Stohlgren TJ, Barnett DT, Kartesz JT (2003) The rich get richer: plant invasions in the United States.

*Frontiers in Ecology and the Environment* **1**, 11–14. doi:10.1890/1540-9295(2003)001[0011:TRGRPO]2.0.CO;2

Stromberg JC, Lite SJ, Dixon MD (2010) Effects of stream flow patterns on riparian vegetation of a semiarid river: implications for a changing climate. *River Research and Applications* **26**, 712–729.

Suding KN, Gross KL, Houseman GR (2004) Alternative states and positive feedbacks in restoration ecology. *Trends in Ecology & Evolution* **19**, 46–53. doi:10.1016/j.tree.2003.10.005

Tabacchi E, Correll DL, Hauer R, Pinay G, Planty-Tabacchi AM, Wissmar RC (1998) Development, maintenance and role of riparian vegetation in the river landscape. *Freshwater Biology* **40**, 497–516. doi:10.1046/j.1365-2427.1998.00381.x

Taylor B, Ganf GG (2005) Comparative ecology of two co-occurring floodplain plants: the native *Sporobolus mitchellii* and the exotic *Phyla canescens*. *Marine and Freshwater Research* **56**, 431–440. doi:10.1071/MF04196

Theoharides KA, Dukes JS (2007) Plant invasion across space and time: factors affecting non-indigenous species success during four stages of invasion. *New Phytologist* **176**, 256–273. doi:10.1111/j.1469-8137.2007.02207.x

Tinker PB, Nye PH (2000) *Solute Movement in the Rhizosphere*. Oxford University Press, Oxford, UK.

van Coller AL, Rogers KH, Heritage GL (2000) Riparian vegetation-environment relationships: complimentarity of gradients versus patch hierarchy approaches. *Journal of Vegetation Science* **11**, 337–350. doi:10.2307/3236626

Weller DE, Jordan TE, Correll DL (1998) Heuristic models for material discharge from landscapes with riparian buffers. *Ecological Applications* **8**, 1156–1169. doi:10.1890/1051-0761(1998)008[1156:HMFMDF]2.0.CO;2

Welsch DJ (1991) *Riparian Forest Buffers – Functional and Design Protection and Enhancement of Water Resources*. USDA Forest Service, Radnor, PA.

Williams L, Reich P, Capon SJ, Raulings E (2008) Soil seed banks of degraded riparian zones in south-eastern Australia and their potential contribution to the restoration of understorey vegetation. *River Research and Applications* **24**, 1002–1017. doi:10.1002/rra.1123

Williams LJ, Cavagnaro TR, Reich P, Lake PS (2009) '*Phalaris* in Australian ecosystems: understanding its biology and ecology as the first step towards effective management.' Report to Goulburn-Broken Catchment Management Authority, Australian Centre for Biodiversity, Monash University, Clayton, Victoria.

# 20

# Monitoring and assessment of vegetation in Australian riverine landscapes

*Nicholas Souter, Kasper Johansen and Michael Reid*

## INTRODUCTION

Australia's general aridity and its variable and unpredictable rainfall mean that rivers are critical ecological and economic resources. Dramatic modification of Australia's riverine landscapes began soon after Europeans settled and expanded across the continent. Since then a range of forces have contributed to the degradation of Australia's riverine systems including: catchment deterioration and erosion, point and non-point pollution, salinisation, reductions in riparian vegetation, flow reduction, river impoundment and the introduction of exotic biota (Lake 2006). The impact of humans on riverine systems alone justifies monitoring and assessment of riverine vegetation. However, this need is made even greater by the ecological importance of these biological communities. Riverine vegetation exists in the ecotone between terrestrial and aquatic systems and plays an important role in regulating and contributing to fluxes of energy and materials between these systems. As a result, riparian systems are diverse, productive and important providers of ecosystem goods and services (Capon *et al.* 2013).

In this chapter, we briefly describe the nature and purpose of monitoring and assessment of riverine vegetation in Australia. We then discuss the general issues and challenges associated with developing monitoring and assessment programs in riverine landscapes and follow this with more detailed summaries of the major techniques used for riverine vegetation monitoring and assessment including both on-ground and remote sensing approaches. We end the chapter with some concluding comments on the state and value of monitoring and assessment of riverine vegetation and future directions.

## THE NATURE AND PURPOSE OF MONITORING AND ASSESSMENT

Brydges (2004) classified ecological monitoring into four broad categories based on their nature and purpose: (1) simple monitoring; (2) survey monitoring; (3) surrogate or proxy monitoring; and (4) integrated monitoring. Simple monitoring refers to the recording of a single parameter, such as temperature or rainfall, at a single geographical location. While simple monitoring can provide background data that are valuable for interpreting results of riverine vegetation monitoring, it is unlikely to provide useful information on the condition of riverine vegetation itself.

Survey monitoring records current ecological conditions over a geographical area in locations subject to the effects of a stressor, as well as in locations not subject to that stressor. The approach is essentially a 'space for time' substitution that, in the absence of historical records of conditions before the introduction of the stressor, can be used to infer its impact. Although not identified as part of Brydges' (2004) schema, a variant on survey monitoring is more general surveillance monitoring which is aimed at providing a broader picture of environmental condition not necessarily in relation to the impact of a specific stressor, but rather in relation to the combined impacts of a suite of anthropogenic stressors (e.g. Davies *et al.* 2012). Surveillance monitoring typically makes use of real or conceptual reference sites or expected conditions to provide benchmarks to assess the degree of impact (Chessman and Royal 2004; Davies *et al.* 2012). Given the breadth and long history of stressors that may adversely impact riverine vegetation, broad-scale environmental auditing via survey or surveillance monitoring has a valuable role in ascertaining current condition and prioritising areas for management action. A range of methods have been developed to survey large areas of riverine vegetation, several of which are detailed later in this chapter.

Surrogate or proxy monitoring is defined by Brydges (2004) as referring to the use of information stored in environmental archives, such as sediments, to infer past, pre-impact conditions. Such approaches are widely used to infer past climates. Oxygen isotopes in ice cores, for example, are often examined to infer past global ice volumes and hence temperatures. Such approaches are less widely used in general environmental monitoring, although proxies for lake water pH, such as diatoms and chrysophytes in lake sediments, proved very valuable in establishing that lake acidification had occurred in North America and Europe (Charles and Whitehead 1986; Battarbee 1990; Dixit *et al.* 1992). In the context of Australia's riverine vegetation, surrogate or proxy approaches have not been widely used, although submerged plant-associated cladocera and diatoms preserved in sediments have provided strong evidence of submerged plant loss in billabongs in the southern Murray–Darling Basin (Ogden 2000; Reid *et al.* 2007; Reid 2008).

Information from simple, survey and surrogate monitoring is used largely to determine whether or not change has occurred. These approaches, however, do not usually provide sufficient or appropriate data to determine the cause of observed changes. Integrated monitoring includes a range of variables or indicators of ecological condition, as well as a range of explanatory variables, to explore not only whether changes have occurred but also why (Brydges 2004). An integrated monitoring program may measure all variables directly, or may draw together information from a range of sources across disciplines, potentially including the results of simple, survey or surrogate monitoring. Riverine vegetation monitoring conducted in Australia generally falls into the categories of either survey, surveillance or integrated monitoring. Examples of survey and surveillance monitoring include the Murray–Darling Basin Authority's sustainable rivers audit (Davies *et al.* 2012), the Victorian 'index of stream condition' monitoring (Ladson *et al.* 1999) and the condition monitoring of iconic floodplain trees such as *Eucalyptus camaldulensis* (river red gum) and *E. largiflorens* (black box) (Cunningham *et al.* 2011). Integrated monitoring is mostly carried out as part of research programs investigating the effects of altered watering regimes, for example, the integrated monitoring of environmental flows program established by the NSW Department of Land and Water Conservation (Chessman and Jones 2001) and the monitoring associated with environmental watering and other management actions under the Murray–Darling Basin Authority's The Living Murray program (e.g. Nicol *et al.* 2010; Ward 2011a, 2011b). These programs have often begun by utilising and/or undertaking simple and survey monitoring, with the stated (though not necessarily realised) aim of leading on to integrated monitoring over time.

## DEVELOPING MONITORING AND ASSESSMENT PROGRAMS: GENERAL CONSIDERATIONS AND CHALLENGES

High levels of spatial heterogeneity and temporal variability present major challenges for monitoring and assessment of riverine vegetation. A high degree of spatial heterogeneity is associated with the presence of strong physical gradients in these ecotone environments and the creation of a range of land forms by fluvial processes. Consequently, great sampling intensity is often needed to detect vegetation patterns against this background of variability. Similarly, high levels of temporal variability, a feature of fluvial systems generally but particularly important in the Australian context of climatic variability (Erskine and Warner 1998; Chiew and Prosser 2011), increase the time required to detect trends (Smith *et al.* 1996). A further associated issue is where monitoring programs are targeted, with assessing the potential for systems to respond when moisture limitation ceases a logistical challenge in an environment with long and unpredictable dry periods (Chiew and Prosser 2011).

Riverine systems are highly complex and several explanatory variables usually need to be collected in integrated monitoring programs to explain observed changes and meet monitoring objectives. In order to identify potential explanatory variables and begin to understand their relationship to riverine vegetation, a suitable conceptual model should be developed. Such models can take many forms – they may be pictorial, flow diagrams or simple statements regarding cause and effect. Pictorial models are ideal to communicate ideas to non-specialist audiences, while flow diagrams provide the basis for an integrated monitoring program. A conceptual model developed for floodplain tree condition on the Murray River's Chowilla floodplain provides an example (Fig. 20.1). This model depicts the diversity of environmental drivers, stressors and processes and their interactions that

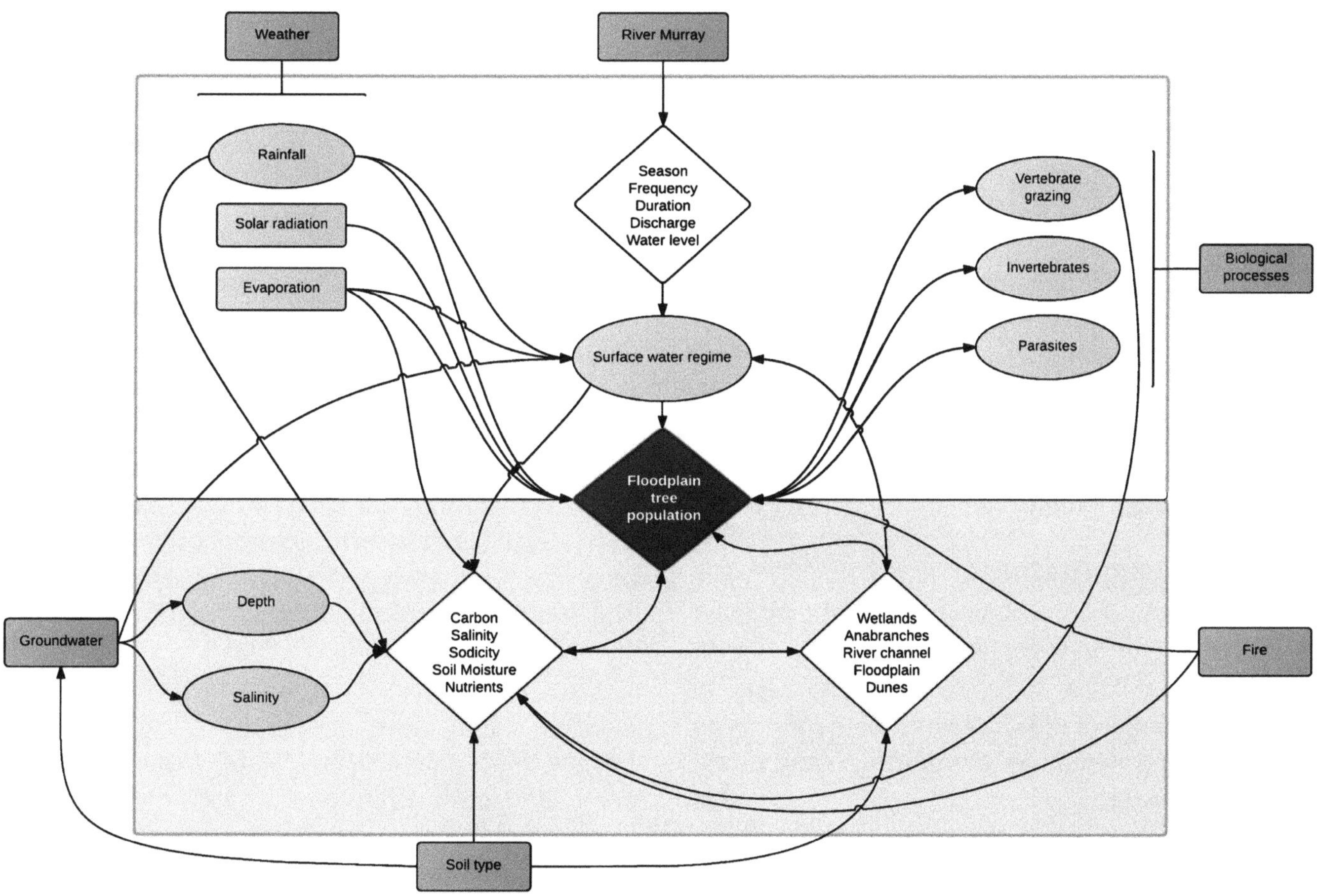


**Figure 20.1:** An example of a conceptual model developed for floodplain tree condition on the Murray River's Chowilla floodplain, Australia.

affect floodplain tree condition. By consolidating this information in a conceptual model the variables that explain aspects of a target variable (e.g. tree condition) or more general features of system structure and function can be more readily identified, monitored and, importantly, conveyed effectively to a broader audience.

With relationships between drivers and system components set out in a conceptual model, the next critical step is to select appropriate indicators to measure. The environmental monitoring and assessment program of the US EPA, established in the 1990s, developed a set of guidelines for selection of indicators (Jackson *et al.* 2000), which have subsequently formed the basis for selecting indicators in other large-scale monitoring programs worldwide, including the sustainable rivers audit in Australia (Davies *et al.* 2012). These guidelines cover conceptual relevance, feasibility of implementation, response variability and interpretation, and utility and provide a comprehensive checklist against which any potential indicator can be assessed (Jackson *et al.* 2000).

## TECHNIQUES

A wide variety of methods can be used to undertake monitoring and assessment of riverine vegetation. Here, we examine suitable approaches that can be used to assess vegetation at the scale of individual plants, populations and communities, and associated processes using both on-ground and remote sensing methods.

## On-ground approaches

Many on-ground approaches have been employed to monitor riverine vegetation in Australia (for an overview see Baldwin *et al.* 2005). This section describes a range of indicators that have either been used or offer potential for monitoring riverine vegetation in Australia.

### Detailed floristic approaches

Detailed floristic monitoring is widely used and involves identifying plants, usually to species level, in replicated quadrats or transects and measuring their frequency or cover. These data are then used to generate metrics such as taxon richness and species diversity and are well suited to analysis using multivariate statistics. Such analyses allow for quantitative comparisons between treatments, sites and over time. Detailed floristic approaches can be used for vegetation communities occurring within the wetted perimeter of streams (e.g. Mackay *et al.* 2003, 2010; Quinn *et al.* 2011) as well those associated with riparian and floodplain habitats (e.g. Robertson and Rowling 2000; Reid and Quinn 2004; Nicol *et al.* 2010). Floodplain understorey species are suited to this approach as they are diverse and understorey communities tend to be highly responsive to changes in water availability. Detailed floristic monitoring is also suited to integrated monitoring, particularly examining the effects of environmental flows, and can be used for long-term condition monitoring where there are multiple potentially interacting threats such as reduced flood frequency, grazing and salinisation.

Spatial and temporal variability in species distributions in all riverine contexts can be very high due to the importance of specific microhabitats, the presence of strong environmental gradients (Webb *et al.* 2006; James *et al.* 2007; Chessman *et al.* 2008) and the tendency, particularly among floodplain plants, for recruitment to occur episodically in response to flooding (Roberts and Marston 2011). Careful sampling designs and strategies are therefore needed to account for this variability. At the most general level, species-area curves can be used in a pilot stage to determine appropriate sampling intensity (Stohlgren 2007). Quadrat sizes, shapes and positioning can also be varied to control for spatial variability (Stohlgren 2007).

One approach frequently used in riverine habitats is to locate quadrats along belt transects. For example, heterogeneity in the vegetation associated with differences in water availability or depth may be assessed by sampling the vegetation of quadrats along transects perpendicular to the water's edge. When deploying belt transects their start and end points need to be determined. In small streams the low water mark and bankfull offer suitable points. On broad floodplains, however, deciding where to finish transects can be problematic, particularly if floodplains are kilometres wide. Where the riparian and terrestrial vegetation communities are distinct this can determine the end of a transect, aided potentially by statistical tests (Kröger *et al.* 2009). However, in subtropical and tropical streams where a distinct vegetation boundary may not occur, fluvial geomorphic definitions may be used. However, determining this may require the time-consuming collection of cross-sectional information. At its most intensive, the belt transect approach may utilise contiguous quadrats, the surveying of which can be very time-consuming. An alternative, more rapid approach is the use of line intercept methods, where plant species intercepting a discrete line (e.g. a tape measure) at predefined points along a transect are recorded and cover estimates made by dividing the number of plant intercepts by the total number of points. Line intercept methods tend not to record as many species as quadrat-based methods but are more precise and less prone to observer bias than visual estimates of cover in quadrats (Godínez-Alvarez *et al.* 2009).

Because riverine environments are ecotones, it is often useful to stratify sampling according to geomorphology, existing vegetation communities or hydrology (Nicol 2004; Barrett *et al.* 2010). Stratifying sampling can reduce within 'treatment' variability and hence increase the power to detect changes; however, delineation of strata does require prior knowledge of the system in relation to the variables being used to stratify. Another consideration is that delineation of boundaries is often subjective, although numerical approaches such as constrained classifications and moving split windows analysis can reduce the subjectivity of the process (Kröger *et al.* 2009). To reduce the extent of the environmental gradient covered by individual sampling units the width of quadrats with respect to the gradient is generally kept to a minimum (e.g. 1 m or less). Care also needs to be taken in analysing data to handle autocorrelation if applying statistical analyses because contiguous quadrats (and even proximate quadrats) cannot be considered independent (Legendre 1993).

Hypothesis testing can provide a template for arranging transects and quadrats. For example, White *et al.* (2009) examined the effect pumping water to a deflation basin on the lower Murray River floodplain on understorey vegetation. Three inundated treatment transects were placed across the basin while three dry control transects were located further away. Each transect was at least 100 m long. A 1 × 1 m quadrat was then surveyed every ten metres along each transect.

Sites used for detailed monitoring may be fixed (e.g. Reid and Quinn 2004; Alexander *et al.* 2008) or move between surveys in response to prevailing conditions. For example, when monitoring germination on receding lake shores, Nicol (2004) established transects 50 m apart but placed quadrats where the vegetation was germinating when the site was revisited during the lake drawdown.

### Individual species assessments

Individual plant species may be monitored for their intrinsic value (e.g. river red gum) or because of their sensitivity to specific stressors (e.g. grazing, flow regulation). Riverine vegetation such as submerged and emergent macrophytes, shrubs and trees are important structural elements in riverine ecosystems and are frequently monitored as indicators of overall ecosystem condition. While monitoring individual species requires less resourcing, particularly time and taxonomic expertise, than detailed floristic monitoring, individual species can present challenges and may require specific approaches due to variations in size, longevity and peculiarities of life form and habitat. The size of large adult trees means that approaches such as monitoring changes in the characteristics of individual plants are made over time, rather than surveying cover and abundance, as with smaller species or seedlings. For longer-lived species, such as trees and shrubs, distribution patterns tend to change slowly so assessing features other than simple presence or absence is more useful, such as condition, population age structure, growth or recruitment.

Floating aquatic plant species can be difficult to monitor because wind (cf Spencer *et al.* 1998) and flow can move rafts of these plants around. This reduces the effectiveness of quadrat-based methods and floating-leaved species are best monitored at site or reach scales, either by eye or remote sensing methods (e.g. aerial photography; Nohara 1991; Rogers and Owens 1995). This is particularly important for the management of several significant invasive floating aquatic plants, e.g. *Alternanthera philoxeroides* (alligator weed), *Salvinia molesta* (salvinia) and *Eichhornia crassipes* (water hyacinth).

Monitoring submerged plant species in Australia's frequently turbid rivers and streams also presents difficulties. When surveying submerged vegetation on the highly turbid Hawkesbury-Nepean River, Thiebaud and Williams (2008) only recorded macrophytes at or just below the surface of the water. In turbid waters, grapnels, rakes and grabs can be used to sample submerged plants. These methods can yield semiquantitative results (Johnson and Newman 2011), but their effectiveness depends on the plant species being surveyed (i.e. some material must break off and be collected), substrate type and the presence of obstructions. Where turbidity is not a problem and vegetation is visible, transects crossing the river or wetland can be surveyed (Henry and Amoros 1996; Schwarz and Snelder 1999; Mackay *et al.* 2003, 2010; Quinn *et al.* 2011).

A range of growth characteristics exhibited by common and widespread emergent vegetation species such as *Typha* spp. (cumbungi), *Schoenoplectus validus* (river club-rush), *Phragmites australis* (common reed) and *Bolboschoenus* spp. sedges have been used for monitoring as they respond to water availability. For example, three characteristics of these species were measured to assess the effect of weir pool surcharge on the lower Murray River. Measured in 1 × 1 m quadrats these were: cover, the proportion of live versus dead biomass and the proportion of all ramets with flowers (N.J. Souter, unpubl. data). Alternatively, Baldwin *et al.* (2005) suggest measuring the proportions of plants in a stand using eight categories: dead partial shoot, dead intact shoot, emerging shoot tip, live shoot tip and live reproductive, flowering, fruiting and dehiscence.

A range of riparian trees and shrubs are commonly monitored to determine their response to environmental changes. Assessments are most frequently made using visual assessment methods, although physiological methods are also sometimes used. Some parameters may be suitable for both survey and integrated monitoring (i.e. cover, proportion of live to dead biomass) while other measures (e.g. reproduction) are better suited to integrated monitoring.

The floodplain shrub *Duma florulenta* (tangled lignum) was also monitored to assess the effects of lower Murray River weir pool surcharge (Souter and Walter 2014). Monitored parameters included the proportion of either whole or partial branches with leaves and with flowers, and as tangled lignum can photosynthesise from its stems alone (Chong and Walker 2005), the proportion of whole or partial branches that are green. More generally, Baldwin *et al.* (2005) suggest eight demographic categories that can be used for monitoring shrubs: live seedlings, juvenile and mature shrubs; senescing; dead fallen seedlings and mature shrubs; and dead standing seedlings and mature shrubs.

The visual assessment of trees is used to monitor forest health throughout the world (UN/ECE 2006). In Australia, a standard visual assessment method is used for the two eucalypt species that dominate the floodplains of the Murray River, river red gum and black box (Souter *et al.* 2010). This method is targeted at using integrated monitoring to determine the outcomes of local scale management interventions taken to improve tree condition. Ten variables are measured from each tree (Souter *et al.* 2010), of which crown extent, crown density, diameter at breast height, mistletoe load and bark condition are used to assess long-term changes in tree condition, while new tip growth, epicormic growth, reproductive status and leaf die-off measure shorter-term response. The age structure of trees is also suitable for monitoring by assessing parameters such as the proportions of: live seedlings, saplings and mature trees; senescing and dead trees; fallen saplings, fallen mature, standing saplings and standing mature trees (Baldwin *et al.* 2005).

### Functional and life form groups

Plants are often classified into functional and life form groups. Functional groups are based on responses to disturbance or environmental change, which are then used to interpret and predict change in community function and dynamics (Noble and Gitay 1996). Life form groups categorise plants according to attributes such as size, class (monocots and dicots), structural formation (e.g. tree, shrub, sedge, forb) and habitat information.

Australian riverine plants are commonly assigned to functional groups according to their germination, growth and reproductive response to the water regime following the three-level hierarchical classification developed by Brock and Casanova (1997) and later amended by Casanova (2007). The first level in this classification broadly groups plants according to the amount of surface water in which they grow: terrestrial, amphibious and submerged. Terrestrial and submerged species are always either, respectively, fully terrestrial or aquatic. The amphibious group contains the great diversity of species found throughout the wet/dry ecotone and divides into two groups: fluctuation tolerators and fluctuation responders. These groups divide again with fluctuation-tolerators comprising emergent and low-growing species and fluctuation-responders comprising plastic and floating species (Brock and Casanova 1997). The terrestrial group divides at this level into dry and damp species. These functional group classifications, or variations on them, have been used in several monitoring and assessment studies of riverine vegetation (e.g. Reid and Quinn 2004; Alexander *et al.* 2008; Reid and Capon 2011).

Several alternative approaches to classifying riverine plants into groups have also been used. Victoria's 'index of wetland condition' monitoring program, for example, collects life form group data for wetland plants (DSE 2009) and groups them according to size and structural form: medium to large aquatic herbs; medium monocots; small to medium semi-aquatic herbs; and trees. Capon (2003) used both life history (annual versus perennial) and life form groups (shrubs, hydrophytes, annual grasses, perennial grasses, legumes and non-leguminous forbs) when analysing a floodplain plant community. At its simplest, assigning plants to functional and life form groups requires the ability to discern between groups and distinguish species within them, avoiding the need for high-level plant identification skills. Classifying plants into functional and life form groups can be used in both survey and integrated monitoring programs.

### Structural and rapid assessment approaches

Structural monitoring systems take an ecosystem approach to assessing riverine vegetation. While assessments of riverine vegetation are components of these systems, a variety of other aspects, including geomorphology, hydrology, water quality and physical form are also measured. These diverse indicators are often combined into an overall index of site condition or quality. Such systems are used to provide a rapid assessment of stream and riparian condition and are well suited to survey monitoring. In this section we provide an overview of the riverine vegetation components of some major structural and rapid assessment systems used in Australia.

Structural and rapid assessment approaches are typically assessed relative to some benchmark or reference condition. Where pre-European condition information is available, the benchmark condition may be a site relatively undisturbed by human activities. For tropical regions, where human disturbance is still relatively low, appropriate non-disturbed sites may be found. However, for most other regions of Australia undisturbed sites are unlikely to exist. In this situation there are a range of other options for setting a benchmark such as using previous studies and expert opinion to determine the condition with no or only minimal disturbance (i.e. a synthetic reference condition).

*Index of stream condition and index of wetland condition*

The index of stream condition (Ladson *et al.* 1999) and index of wetland condition (DSE 2005) are used for state-wide river health and wetland monitoring in Victoria. Index of stream condition assessments are conducted every five years and data are used to benchmark stream condition and set priorities for waterway restoration works. The index of stream condition comprises five sub-indices: hydrology, water quality, aquatic life, physical form and stream side zone. Of these, only the last two include measures of riverine vegetation: longitudinal continuity, tree canopy cover, understorey life forms, weed cover, debris, recruitment and regeneration. Data collected by the index of stream condition are compared with mapped riparian native vegetation classes in Victoria (i.e. ecological vegetation classes), which provide a reference condition describing the type and structure of the vegetation postulated before European settlement (i.e. 1750).

The index of wetland condition is designed for static water inland wetlands and is similar in many respects to the index of stream condition as many of its variables are compared against ecological vegetation classes (DSE 2009). The index of wetland condition measures four vegetation attributes: critical life form groupings (shrubs, herbs and graminoids in a range of size classes), indicators of altered processes (e.g. encroachment by terrestrial species or mortality of wetland plants as a result of hydrological change), weeds and vegetation structure and health; as well as physical form, hydrology, soil, water quality and catchment variables (DSE 2005).

*Quantifying the health of ephemeral rivers*

The 'quantifying the health of ephemeral rivers' assessment method incorporates hydrological, geomorphic and ecological indicators (including vegetation, water quality, ecosystem processes, stable isotope analysis and macroinvertebrates) (Costelloe and Sheldon 2006). While the streamside vegetation indicators are derived from those outlined in the 'rapid appraisal of riparian condition' protocol (Jansen *et al.* 2004 and below), they are assessed against a synthetic reference condition and have been adjusted to account for the unique characteristics of ephemeral streams. The synthetic reference condition was developed from conceptual models, expert opinion and long-term datasets. This approach allows critical bands for each indicator to be set for streams of different levels of ephemerality. Indicator bands can also be temporally divided to better reflect drying and flow conditions. Measured against the synthetic reference condition, the riparian or streamside vegetation indicators provide one measure of stream health.

Seven major streamside vegetation categories are assessed under the quantifying the health of ephemeral rivers assessment: habitat (canopy width and continuity), cover, native component of cover, debris, emergent macrophytes, native vegetation regeneration and native canopy health. In addition to these variables, a range of vegetation attributes are used in site descriptions: longitudinal continuity of woody vegetation; important plant taxa across various strata (including weeds); exposed tree roots; vegetation between bank top and toe; and, where present, an index of native canopy health, comprising estimates of crown size and density, number of dead branches and amount of epicormic growth.

*Rapid appraisal of riparian condition*

The 'rapid appraisal of riparian condition' (Jansen *et al.* 2004) assesses the ecological condition of riparian habitats using a suite of indicators that reflect functional aspects of the riparian zone. The rapid appraisal of riparian condition comprises five sub-indices: habitat continuity and extent; vegetation cover and structural complexity; standing dead trees and fallen logs and leaf litter; dominance of natives versus exotics and indicative features (regeneration and reeds). The condition of a site is assessed according to the departure from the maximum potential sub-index score (Jansen *et al.* 2004). Of the structural rapid systems examined here, this is simplest method as it does not rely on detailed assessments of pre-European condition (such as is required for the index of stream condition) or the generation of a synthetic reference condition (as used in quantifying the health of ephemeral rivers).

*Tropical river riparian assessments*

While the majority of riverine vegetation monitoring has occurred in the southern half of Australia, much of the north and west of the continent lacks defined and tested techniques. To address this, the 'tropical rapid appraisal of riparian condition' has been developed for tropical savanna streams with a well-defined channel and a distinct riparian zone (Dixon *et al.* 2006). This technique is an important development as tropical savannas cover ~25% of Australia (Mott *et al.* 1985). The tropical rapid appraisal of riparian condition is a rapid visual assessment technique made up of multiple indicators grouped into sub-indices. It is based on the rapid appraisal of riparian condition and includes sub-indices from two other tools: a rapid assessment protocol for riparian vegetation (Werren and Arthington 2002) developed for Queensland's rivers, and the waterway foreshore assessment tool developed for the Pilbara and Kimberley (Department of Water 2006). It has also drawn on the index of stream condition and quantifying the health of ephemeral rivers assessments. The tropical rapid appraisal of riparian condition assesses features that affect ecological function and enable the identification of suitable management actions (Dixon *et al.* 2006). It has a greater focus on vegetation than the other assessment approaches reviewed here, with three of the four sub-indices focussed on vegetation: plant cover, regeneration and weeds. Erosion is the other sub-index, and a pressure index is also calculated to identify the causes of any change in condition. The tropical rapid appraisal of riparian condition is currently being field tested in northern Australia.

## Process-based approaches

Riverine vegetation can be used to monitor a range of ecological processes. This may include direct measurements of recruitment and vegetation condition, which, while being vital for the persistence of the vegetation itself, often provide important habitat for other species of flora and fauna. Detailed monitoring can also be carried out to assess processes such as water flux, nutrient cycling and primary productivity.

*Recruitment*

Recruitment is a valuable process-based variable as it indicates the long-term viability of riverine vegetation. The index of stream condition assesses the recruitment of woody native vegetation and defines recruitment as the establishment of individual plants beyond the initial seedling (or early germinant) stage. Immature woody plants that contain no evidence of flowering or fruiting material, and seedlings that have had at least the full annual cycle of seasons since germination or first establishment are considered recruits. As a functional measure the index of stream condition also considers older suppressed individuals that would likely function as recruits if competition were reduced. Recruitment is assessed against the relevant ecological vegetation class and the recruitment strategy (either episodic disturbance driven or continuous) is considered.

Recruitment can also be monitored for non-woody plants. For example, when monitoring the effects of weir pool manipulation on the lower Murray River, the response of *Centipeda cunninghamii* (sneeze weed) was assessed as it was considered representative of many smaller herbs with seed banks that enable them to rapidly colonise new habitat (N.J. Souter, unpubl. data). While the cover and condition of sneeze weed was assessed in this case, recruitment could also have been measured easily by counting flowering individuals to show that the species had reached reproductive maturity and thus recruited successfully. The difference between these two approaches to measuring recruitment highlights the need to carefully consider both the monitoring question and choice of indicator. While the index of stream condition measures only large perennial woody species recruitment, as they are generally less susceptible to short-term spatial and temporal changes, monitoring a small perennial herb can be used to assess short-term changes. Measures of recruitment can be used for both survey and integrated monitoring, although they are better suited to the latter.

*Ecological process indicators*

Physiological changes can provide early warning of deteriorating condition before it becomes evident in morphological or abundance data. The physiology of a particular plant species, range of species or a vegetation community may have important implications for primary productivity and nutrient cycling. Ecophysiological techniques that measure the amount of water in plants, fluxes of water through the soil–plant–atmosphere continuum or plant functioning (Eamus *et al.* 2006) can be used to monitor these ecological processes.

Plant water stress can be inferred from leaf or stem xylem water potential measured using a pressure chamber or Scholander pressure bomb (Eamus *et al.* 2006). The rate of tree water use can be estimated by measuring its sap flow – the movement of water through the plant (Eamus *et al.* 2006). Sap flow can be assessed via one of three approaches: heat pulse, thermal dissipation and stem heat balance (Eamus *et al.* 2006). Measurements of sap flow provide an indication of transpiration and the fluxes of water through the soil–plant–atmosphere continuum. Open-top chambers are used to measure the rate of evapotranspiration from low-lying vegetation, while micro-meterological techniques can measure evapotranspiration from larger areas of vegetation. However, such large area techniques are limited as they cannot separate the origin of water vapour from bare earth, overstorey or understorey vegetation (Eamus *et al.* 2006).

Methods that measure aspects of plant functioning may also be suitable as monitoring indicators. Leaf area index is a canopy scale measurement that assesses the ratio of total canopy leaf area to the ground area covered by the canopy. The leaf area index can be used to measure changes in vegetation health across the landscape (Eamus *et al.* 2006). While a range of methods are available to measure leaf area index (canopy harvesting, leaf area index meter and hemispherical photography) the use of hemispherical photography is the most user and environmentally friendly, despite the need to take photos at dawn and dusk. The use of hemispherical photos is an important component of The Living Murray stand condition monitoring method (Cunningham *et al.* 2009).

The rate of stem growth can be measured using one of two similar techniques. Dendrometers are narrow (~1 cm), thin (1 mm) metal bands that encircle the stem at 1.3 m above the ground and can measure growth rates of less than 1 mm per month. Diameter at breast height can be measured using a tape measure. As variation among diameter at breast height measurements is far greater than a dendrometer (which requires six trees per species) greater replication is required when measuring diameter at breast height (Eamus *et al.* 2006).

These methods are best employed in answering very specific questions as part of an integrated monitoring program. However, the need to deploy specialist and often expensive equipment tends to limit the scale at which these methods can be deployed.

## Summary of on-ground monitoring techniques

A variety of techniques have been used in on-ground monitoring of riverine and riparian vegetation in Australia. Techniques range from the detailed and fine scale, such as floristic monitoring and measures of plant physiology and function, to broad and rapid assessment techniques. The choice of method will depend on the specific aims of the monitoring and assessment program. Detailed methods are able to detect subtle changes, but may be time-consuming or use expensive equipment. Rapid assessment methods can be used to monitor a wide area at relatively low cost, yet their coarse nature can fail to detect small, potentially significant changes over time. On-ground methods for monitoring riverine vegetation continue to develop as monitoring effort increases, e.g. the tropical rivers assessment protocol. While the tropical rapid appraisal of riparian condition fulfils the need for a rapid approach to monitoring riverine and riparian vegetation in this vast area of Australia, the adoption and modification of more detailed approaches awaits.

## Remote sensing approaches

Remote sensing typically refers to the extraction of information from airborne or satellite imagery. Remote sensing can be used to monitor changes in streams and riparian zones if two or more datasets of the same area have been captured at different times (Johansen *et al.* 2008a). For multi-temporal and large spatial extent (>100 km of stream length) assessments of riparian vegetation structure, remote sensing techniques provide the only feasible and reliable means for obtaining quantitative measurements because of their spatially extensive, non-invasive, repetitive nature and ability to be utilised in inaccessible areas (Johansen *et al.* 2007). Remote sensing data are suitable for both survey and integrated monitoring. Through assessment of trends in riparian biophysical parameters and land cover classes in image datasets, useful information can be derived on changes in the environmental condition of these areas (e.g. to assess the success of management activities) (Goetz 2006).

### Datasets suitable for riparian zone monitoring

High spatial resolution image data (pixel size <5 × 5 m) are required to map and monitor riparian zones because of their narrow width and spatial heterogeneity (Congalton *et al.* 2002). Suitable remotely sensed datasets can be derived from high spatial resolution multi-spectral satellite sensors such as Ikonos, QuickBird, GeoEye-1 and WorldView-2, which include spectral bands located in the blue, green, red and near infrared parts of the spectrum. Airborne high spatial resolution sensors are increasingly being used as alternatives to satellite sensors, because of their higher spatial resolution, rapid deployment and reduced costs for mapping large areas (Johansen *et al.* 2008b). Aerial photography dating back in time may also provide important datasets for monitoring of riparian zones.

Airborne light detection and ranging (LiDAR) sensors derive information on the elevation and reflectance of terrain and vegetation from a pulse or continuous wave laser fired from an airborne transmitter, for which its position is precisely and accurately measured (Mallet and Bretar 2009). Processing of the reflected LiDAR signal provides an accurate measure of distance between the transmitter and reflecting surfaces based on the time of travel of the laser and position of the sensor (Lefsky *et al.* 2002). LiDAR data are useful for riparian zone mapping because airborne laser scanning systems provide precise three-dimensional canopy, understorey and geomorphic information (Johansen *et al.* 2010a, 2010c). No matter which type of remote sensor is used for riparian zone monitoring, it is imperative that for multi-temporal analysis, multiple datasets of the same area are accurately georeferenced and spectrally normalised to ensure accurate overlay and comparison allowing for change detection.

### Remotely sensed monitoring approaches

Whereas on-ground approaches allow detailed and accurate measurements of most riparian zone attributes to be made over small areas, remote sensing of riparian zones can provide maps of several variables with complete spatial coverage for large areas (Ladson *et al.* 2006; Johansen *et al.* 2007). Remote sensing of riparian zones has been most successfully used in assessing riparian zone attributes related to vegetation structure and physical form through mapping of land cover classes and biophysical parameters. Thematic mapping on land cover classes can be used for assessment of the tree canopy, canopy gaps, grass, bare ground and waterbodies. Biophysical parameters assessed from high spatial resolution datasets include stream bed width, bankfull width, riparian zone width, plant/foliage projective cover, longitudinal continuity of the riparian zone, vegetation height classes, large trees, large instream wood, vegetation overhanging the stream, and bank stability (Johansen *et al.* 2010a). However, the majority of research has focused on mapping riparian zones (survey monitoring), with very limited efforts directed towards monitoring these environments over time or explaining the reason behind observed changes (integrated monitoring). Past attempts have mainly used manual image interpretation (Ferreira *et al.* 2005). Unfortunately, manual delineation of riparian zones and interpretation of canopy characteristics are time-consuming and labour-intensive for monitoring large areas (>100 km of stream length) and biased because of inter-operator variability and lack of metadata describing the mapping approach used (Lillesand *et al.* 2008). Therefore, results of multi-temporal image analysis are often not reliable and cannot be consistently repeated in the same manner for future studies. Hence, remotely sensed monitoring of riparian vegetation requires field and/or high spatial resolution ancillary data to be used for calibration, validation and modelling of output thematic maps of land cover classes and quantitative maps of riparian zone biophysical parameters (Johansen *et al.* 2008a). This approach has been adopted for monitoring floodplain tree condition under The Living Murray program (Cunningham *et al.* 2009). Here a modelled relationship between stand condition (which is based on ground surveys of percentage live basal area, plant area index and crown vigour) and reflectance values, as well as a derived 'normalised difference vegetation index' from remotely sensed Landsat imagery are used to predict the condition of river red gum and black box across the floodplain. Note that although the normalised difference vegetation index is widely used in remote sensing studies of vegetation, it is a spectral band ratio rather than a true biophysical parameter, and hence it is less effective for monitoring and quantifying change over time (see Box 20.1).

### Box 20.1: Limitations of normalised difference vegetation index (NDVI) for monitoring change over time

Image-based spectral vegetation indices are commonly used as indicators of canopy chlorophyll content and vertical structure to estimate forest canopy condition (Levesque and King 2003). The normalised difference vegetation index (NDVI), which is derived from the reflectance in the red and near-infrared portions of the spectrum, has been used widely as it compensates for changing illumination conditions, surface slope, aspect and other extraneous factors (Gao 1996; Lillesand and Kiefer 2000). Most research has tried to relate spectral vegetation indices to leaf area index and photosynthetically active radiation (Li and Moreau 1996; Huemmrich and Goward 1997; Coops *et al.* 1998; Wulder *et al.* 1998; Gao *et al.* 2000; Ishiyama *et al.* 2001; Myneni *et al.* 2002; Wylie *et al.* 2002), but the relationship between these parameters has proved to be affected by factors such as background reflectance, tree species and tree structure. Dawson *et al.* (2003) demonstrated that NDVI is highly sensitive to both canopy foliage and understory chlorophyll content, which can account for significant errors in remotely sensed estimates of canopy foliage parameters.

While NDVI is commonly used in remote sensing of vegetation, it is important to highlight that this is a spectral band ratio and not a biophysical parameter. Whereas remotely sensed quantitative measurements of biophysical parameters such as leaf area index and foliage projective cover are useful for monitoring and change-detection studies of riparian vegetation, NDVI and other spectral vegetation indices collected for the same area over time by the same or different sensors cannot be directly compared and used for change studies, as it is a unitless band ratio with multiple limitations for vegetation time-series assessment. For example, areas with a leaf area index of 3 or greater result in NDVI saturation (Gamon *et al.* 1995; Gao 1996; Huete *et al.* 1999). This is due to the greater amount of horizontal stratification of leaves, which is difficult to identify with remote sensing instruments that operate from a vertical perspective (Wulder *et al.* 1998; Sims and Gamon 2002). Gitelson *et al.* (1996) also found that NDVI saturates due to the red chlorophyll absorption band causing NDVI to become insensitive at chlorophyll-a concentrations above 3–5 $\mu g/cm^2$. The main disadvantage of ratio-based indices tends to be their non-linearities resulting in asymptotic behaviours of spectral vegetation indices in relation to the vegetation properties being examined, which lead to insensitivities to vegetation variations over certain land cover conditions. Ratios also fail to account for the spectral dependencies of additive atmospheric effects (path radiance), canopy background interactions, and canopy bi-directional reflectance anisotropies, particularly those associated with canopy shadowing (Huete *et al.* 1999; Sims and Gamon 2002; Asner and Warner 2003; Levesque and King 2003).

Automated approaches (i.e. the use of rules and algorithms) for mapping and monitoring riparian zone attributes are less biased and can be used for assessing large areas. Only limited case studies have been conducted focusing on remotely sensed change detection in riparian zones. For example, Bryant and Gilvear (1999) used airborne image data and pixel-based image classification for detecting changes in riparian geomorphic and land cover characteristics in Scotland. Johansen *et al.* (2008a) assessed the ability to detect change in several indicators of riparian condition in northern Australia from multi-temporal high spatial resolution QuickBird image data, including plant projective cover, organic litter, canopy continuity, bank stability, flood damage, riparian zone width, river width and tree clearing. It was found that these riparian condition indicators could be monitored with accuracies feasible for multi-temporal analysis within the study area and that reporting reliability and accuracy measures are imperative for map interpretation and before initiating any management actions based on the mapping results.

## Riparian land cover classification

One major limitation of using optical image datasets for riparian zone monitoring is the requirement of a very accurate image classification of land cover classes such as waterbodies, bare ground, grass cover and canopy cover. For example, waterbodies are difficult to map accurately because of obstruction and shadows from tree crowns along the water's edge (Arroyo *et al.* 2010). Also, pixel-based algorithms used to map plant projective cover and the longitudinal continuity of riparian zones cannot distinguish between woody (e.g. tree cover) and non-woody (e.g. grasses) vegetation based on spectral information only. Therefore, accurate image classification of woody and non-woody vegetation is first required to exclude non-woody vegetation from the analysis and mapping of plant projective cover and longitudinal continuity. High spatial resolution image data are generally difficult to accurately classify into land cover classes because of the large reflectance variation and heterogeneous nature of pixels within individual features at these spatial resolutions (Blaschke 2010). As a result, the fundamental assumption of per-pixel image classification (i.e. homogenous features) is not met (Desclée *et al.* 2006). Object-based image analysis is a suitable means to achieve accurate image classification results, and promising results for multi-temporal and multi-site assessment have been reported (Blaschke 2010). Object-based image analysis is based on the assumption that image objects provide a more appropriate scale to map environmental features, and allows features with significant variations in their spectral reflectance signature to be mapped at specific spatial scales (Blaschke and Hay 2001; Burnett and Blaschke 2003; Flanders *et al.* 2003; Benz *et al.* 2004). Johansen *et al.* (2010b) showed that object-based classification accuracies of riparian vegetation were significantly higher than those of the pixel-based classifications, because of the ability of the object-based approaches to: reduce effects of slight misregistration between the two images; reduce high spatial frequency noise; include context relationships and object shape information; reduce effects of shadows from trees; and reduce effects of differences in sensor viewing geometry and illumination angle. Hence, specialised image-processing techniques, such as object-based image analysis, are required for accurate, automated and unbiased monitoring of riparian land cover classes.

## Riparian zone assessment from LiDAR data

The use of LiDAR data for mapping and monitoring can overcome some of the issues related to the use of optical image datasets, as image classification of individual land cover classes is not required. For example, woody and non-woody vegetation can simply be separated based on height differences between the two vegetation types. LiDAR data can also be used to directly derive information on quantitative riparian zone biophysical parameters such as bank slope, stream profile, fractional canopy cover, canopy height, number of vegetation height classes, longitudinal continuity and large trees. Other riparian zone attributes that require more advanced processing to be mapped from LiDAR data include waterbodies, stream bed width, bankfull width, riparian zone width, vegetation overhanging the stream and bank stability (Johansen *et al.* 2010a, 2011; Michez *et al.* 2013). To maximise results, the capture of LiDAR data with suitable acquisition specifications is important. For monitoring, anniversary image data (i.e. data collected at the same time of the year) are ideal for detection of changes (Jensen 2005). As LiDAR data are not affected by clouds above the plane and can be captured both day and night, it provides more flexibility in the timing of capture than the use of optical sensors, which are restricted to midday hours. Large spatial extents can also be covered by LiDAR over shorter periods of time than optical data. Based on current research, LiDAR data allow more riparian zone biophysical parameters to be mapped accurately than optical image datasets using automated image processing routines (Johansen *et al.* 2010c). To obtain the best possible results of multi-temporal analysis, it is important that all processing steps are clearly documented and that acquisition specifications are similar for the datasets used to repeat automated processing routines and to reduce bias. Development and testing of monitoring approaches with LiDAR data still require more research before operational monitoring of riparian zones can be performed. It is likely that LiDAR data will also prove more feasible than optical image data for monitoring purposes, especially with the latest invention of full waveform LiDAR, which allows more detailed assessment the riparian canopy structure and terrain (Mallet and Bretar 2009).

### Summary of remote sensing monitoring techniques

Remote sensing is a cost- and time-effective technique for monitoring riparian zones. With advances in image data, processing techniques and processing power, the application of remote sensing for monitoring riparian zones is likely to increase in the future. This is substantiated by the increasing availability of full waveform small-footprint LiDAR sensors and software packages currently being developed for multi-temporal three-dimensional assessments of LiDAR data and for partial automation of mapping and monitoring routines. The capacity to monitor riparian zones in the future is also likely to improve because of the use of increasing data dimensionality through integration and combined processing of different image datasets such as high spatial resolution satellite or airborne imagery and airborne LiDAR data (Arroyo *et al.* 2010). While remote sensing of riparian vegetation has significantly advanced in recent years, there is still a lack of capabilities to consistently map and monitor tree and understorey species composition, recruitment and weed cover in an operational manner. Further research, including high spatial resolution image data combined with LiDAR data, and potentially other ancillary data, is needed to address these shortfalls in the future.

## CONCLUSIONS AND FUTURE DIRECTIONS

Australian riverine vegetation provides a large and diverse set of monitoring indicators (Table 20.1). Yet, vegetation is not used to monitor Australian riverine landscapes to the same extent as water quality variables, aquatic macroinvertebrates and fish. This should change as interest in the ephemeral systems that cover the majority of Australia grows. Riverine vegetation indicators are well suited to these systems, as large woody riverine vegetation persists when water is absent and it can be monitored remotely, a major advantage in these difficult to access areas. It is obvious from this chapter that the majority of monitoring and vegetation studies have been conducted in the temperate south-east of the country. While methods to monitor Australia's northern riverine systems are being developed (e.g. Dixon *et al.* 2006), more will be needed as the call to develop northern Australia's water resources grows louder.

Future monitoring of riverine vegetation is likely also to involve more integration of on-ground and remote techniques. This is already seen in The Living Murray program (Cunningham *et al.* 2009)

**Table 20.1.** Summary of the range of approaches for monitoring the vegetation of riverine landscapes

Abbreviations: TLM, The Living Murray; ISC, index of stream condition; IWC, index of wetland condition; QHER, quantifying the health of ephemeral rivers; RARC, rapid appraisal of riparian condition; TRARC, tropical rapid appraisal of riparian condition.

| Approach | Examples | General purpose | Advantages | Disadvantages |
|---|---|---|---|---|
| Detailed floristic survey | TLM | Variable, but most useful for integrated monitoring | Data rich; sensitive; relatively objective and insensitive to inter-operator bias | Time-consuming; requires expert knowledge and/or extensive training; cannot be surveyed remotely |
| Individual species assessments | TLM | Survey and integrated monitoring | Sensitive (in relation to specific questions); requires less training or expertise than detailed floristic surveys; generally less time-consuming than floristic surveys; in some cases can be surveyed remotely (see below) | Potential to miss important changes among other taxa not monitored |
| Functional and life form groups | ISC, IWC | Survey and integrated monitoring | Focussed on functional and process relationships with drivers or stressors; requires less training or expertise than detailed floristic surveys; generally less time-consuming than floristic surveys | Potential to miss important changes associated with shifts in species distribution and abundance |
| Structural and rapid assessment | ISC, IWC, QHER, RARC, TRARC | Survey monitoring | Requires less training or expertise than detailed floristic surveys; generally less time-consuming than floristic surveys | Potential to miss important changes associated with shifts in species distribution and abundance; ecological significance of observed changes and relationships to drivers or stressors may not be clear |
| Process-based approaches | ISC, TLM | Survey and integrated monitoring | Focussed on functional and process relationships with drivers or stressors; ecophysiological measures may provide 'early warning' | Can be time-consuming; may require extensive training |
| Multi-spectral (satellite and airborne) | TLM, ISC | Survey and integrated monitoring | Large spatial coverage, quantitative; data readily archived and suitable for re-analysis as techniques are refined; no accessibility restrictions | Classification can be time-consuming; interpretation dependent on detailed ground-truthing and calibration; air-borne data collection costly |
| LiDAR | ISC | Survey and integrated monitoring | Large spatial coverage, quantitative; data readily archived and suitable for re-analysis as techniques are refined; no accessibility restrictions | Classification can be time-consuming; interpretation dependent on detailed ground-truthing and calibration; collection costly |

and is set to be part of the long-term intervention monitoring across the Murray–Darling Basin which seeks to determine the environmental effects of water sourced from Commonwealth Water buy-backs that is returned to the environment. Such integrations should drive increases in coverage (addressing spatial variability issues associated with on-ground methods) and accuracy and interpretive scope (in relation to remote sensing methods).

New vegetation-based monitoring methods and systems are likely to be developed within an evidence-based management framework. Here monitoring programs provide information that is used to optimise management objectives, and this approach is being used in a range of ecological systems (for a marine example see Gerber *et al.* 2005). Given that the response of riparian and floodplain vegetation is frequently used to assess management actions (e.g. the application of environmental water), such an approach should be seriously considered when designing monitoring programs to assess the impact of management interventions. The outcome of such a monitoring program is to make the best decision for management, rather than simply assessing the outcome of a management action. One way that this can be achieved is by integrating monitoring data into a 'decision support system', which aids management by integrating complex models, knowledge bases, expert systems and database technologies into a user-friendly software interface (Coulson *et al.* 1987; Jones 1989). A decision support system improves the ability of decision-makers to use often highly technical information and models. By integrating vegetation monitoring into a decision support system, monitoring data are directly tied to management, their usefulness made explicit in the management process and their utility constantly tested.

Australia's riverine plants are diverse in species richness, life form and in their response to environmental change. Aspects of their biology can be measured, from the short-term physiological response of individual plants to large-scale, long-term landscape change. These characteristics make riverine plants ideal for environmental monitoring. However, we believe that riverine plants have been underused as monitoring indicators in Australia. We hope that this review helps to increase their use, leading to improved management of Australia's inland waterways.

## REFERENCES


Alexander P, Nielsen DL, Nias D (2008) Response of wetland plant communities to inundation within floodplain wetlands. *Ecological Management & Restoration* **9**, 187–195. doi:10.1111/j.1442-8903.2008.00417.x

Arroyo LA, Johansen K, Phinn S (2010) Integration of LiDAR and QuickBird imagery for mapping riparian biophysical parameters and land cover types in Australian tropical savannas. *Forest Ecology and Management* **259**, 598–606. doi:10.1016/j.foreco.2009.11.018

Asner GP, Warner AS (2003) Canopy shadow in IKONOS satellite observations of tropical forests and savannas. *Remote Sensing of Environment* **87**, 521–533.

Baldwin DS, Nielsen DL, Bowen T, Williams J (2005) *Recommended Methods for Monitoring Floodplains and Wetlands*. MDBC publication no. 72/04, Murray–Darling Basin Commission, Canberra.

Barrett R, Nielsen DL, Croome R (2010) Associations between the plant communities of floodplain wetlands, water regime and wetland type. *River Research and Applications* **26**(7), 866–876. doi:10.1002/rra.1299

Battarbee RW (1990) The causes of lake acidification, with special reference to the role of acid deposition. *Philosophical Transactions of the Royal Society of London. Series B, Biological Sciences* **327**, 339–347. doi:10.1098/rstb.1990.0071

Benz UC, Hofmann P, Willhauck G, Lingenfelder I, Heynen M (2004) Multi-resolution, object-oriented fuzzy analysis of remote sensing data for GIS-ready information. *ISPRS Journal of Photogrammetry and Remote Sensing* **58**, 239–258. doi:10.1016/j.isprsjprs.2003.10.002

Blaschke T (2010) Object based image analysis for remote sensing. *ISPRS Journal of Photogrammetry and Remote Sensing* **65**, 2–16. doi:10.1016/j.isprsjprs.2009.06.004

Blaschke T, Hay GJ (2001) Object-oriented image analysis and scale-space: theory and methods for modeling and evaluating multiscale landscape structure. *International Archives of Photogrammetry and Remote Sensing* **34**, 22–29.

Brock M, Casanova M (1997) Plant life at the edge of wetlands: ecological responses to wetting and drying patterns. In *Frontiers in Ecology: Building the Links*. (Eds N Klomp N and I Lunt) pp. 181–192. Elsevier Science, Oxford, UK.

Bryant RG, Gilvear DJ (1999) Quantifying geomorphic and riparian land cover changes either side of a large flood event using airborne remote sensing: River Tay, Scotland. *Geomorphology* **29**, 307–321. doi:10.1016/S0169-555X(99)00023-9

Brydges T (2004) Basic concepts and applications of environmental monitoring. In *Environmental Monitoring*. (Ed. BG Wiersma) pp. 83–110. CRC Press, Boca Raton, FL.

Burnett C, Blaschke T (2003) A multi-scale segmentation/object relationship modelling methodology for landscape analysis. *Ecological Modelling* **168**, 233–249. doi:10.1016/S0304-3800(03)00139-X

Capon S (2003) Plant community responses to wetting and drying in a large arid floodplain. *River Research and Applications* **19**, 509–520. doi:10.1002/rra.730

Capon SJ, Chambers LE, Mac Nally RA, Naiman RJ, Davies P, Marshall N, Pittock J, Reid MA, Capon T, Douglas M, Catford J, Baldwin DS, Stewardson M, Roberts J, Parsons M, Williams SE (2013) Riparian ecosystems in the 21st century: hotspots for climate change adaptation? *Ecosystems* **16**, 359–381. doi:10.1007/s10021-013-9656-1

Casanova M (2007) *Functional Groups: Where to From Here?* Joint meeting New Zealand Freshwater Sciences Society and the Australian Society for Limnology, Queenstown, NZ.

Charles DF, Whitehead DR (1986) The PIRLA project: paleolimnological investigation of recent lake acidification. *Hydrobiologia* **143**, 13–20. doi:10.1007/BF00026639

Chessman BC, Jones H (2001) 'Integrated monitoring of environmental flows (IMEF): design report.' NSW Department of Land and Water Conservation, Sydney.

Chessman BC, Royal MJ (2004) Bioassessment without reference sites: use of environmental filters to predict natural assemblages of river macroinvertebrates. *Journal of the North American Benthological Society* **23**, 599–615.

Chessman BC, Royal MJ, Muschal M (2008) Does water abstraction from unregulated streams affect aquatic macrophyte assemblages? An evaluation based on comparisons with reference sites. *Ecohydrology* **1**, 67–75. doi:10.1002/eco.8

Chiew F, Prosser I (2011) Water and climate. In *Water: Science and Solutions for Australia*. (Ed. IP Prosser) pp. 29–46. CSIRO Publishing, Melbourne.

Chong C, Walker KF (2005) Does lignum rely on a soil seed bank? Germination and reproductive phenology of *Muehlenbeckia florulenta* (Polygonaceae). *Australian Journal of Botany* **53**(5), 407–415. doi:10.1071/BT04130

Congalton RG, Birch K, Jones R, Schriever J (2002) Evaluating remotely sensed techniques for mapping riparian vegetation. *Computers and Electronics in Agriculture* **37**, 113–126. doi:10.1016/S0168-1699(02)00108-4

Coops NC, Waring RH, Landsberg JJ (1998) Assessing forest productivity in Australia and New Zealand using a physiologically-based model driven with averaged monthly weather data and satellite-derived estimates of canopy photosynthetic capacity. *Forest Ecology and Management* **104**, 113–127.

Costelloe J, Sheldon F (2006) 'Quantifying the health of ephemeral rivers: protocols for assessing the health of ephemeral rivers.' Final report to Land and Water Australia, Project UOC24.

Coulson RN, Folse LJ, Loh DK (1987) Artificial intelligence and natural resource management. *Science* **237**, 262–267. doi:10.1126/science.237.4812.262

Cunningham SC, MacNally R, Read J, Baker PJ, White M, Thomson JR, Griffionen P (2009) A robust technique for mapping vegetation condition across a major river system. *Ecosystems* **12**, 207–219. doi:10.1007/s10021-008-9218-0

Cunningham SC, Griffioen P, White M, Mac Nally R (2011) 'Mapping the condition of river red gum (*Eucalyptus camaldulensis* Dehnh.) and black box (*Eucalyptus largiflorens* F. Muell.) stands in The Living Murray icon sites.' Stand condition report 2010. Murray–Darling Basin Authority, Canberra.

Davies PE, Stewardson MJ, Hillman TJ, Roberts JR, Thoms MC (2012) *Sustainable Rivers Audit 2: the Ecological Health of Rivers in the Murray–Darling Basin at the End of the Millennium Drought (2008–2010). Volume 1.* Prepared by the Independent Sustainable Rivers Audit Group for the Murray–Darling Basin Ministerial Council, Canberra.

Dawson TP, North PRJ, Plummer SE, Curran PJ (2003) Forest ecosystem chlorophyll content: implications for remotely sensed estimates of net primary productivity. *International Journal of Remote Sensing* **24**(3), 611–618.

Department of Water (2006) *DRAFT: Waterway Foreshore Assessment Tool for Pilbara and Kimberley.* Department of Water, Western Australia Government, Perth.

Desclée B, Bogaert P, Defourny P (2006) Forest change detection by statistical object-based method. *Remote Sensing of Environment* **102**, 1–11. doi:10.1016/j.rse.2006.01.013

Dixit SS, Smol JP, Kingston JC, Charles DF (1992) Diatoms: powerful indicators of environmental change. *Environmental Science & Technology* **26**(1), 22–33. doi:10.1021/es00025a002

Dixon I, Douglas M, Dowe J, Burrows D (2006) *Tropical Rapid Appraisal of Riparian Condition Version 1 (For Use in Tropical Savannas).* River and Riparian Land Management Technical Guideline Number 7, Land and Water Australia, Canberra.

DSE (2005) *Index of Wetland Condition.* Conceptual framework and selection of measures, Department of Sustainability and Environment, Melbourne.

DSE (2009) *Index of Wetland Condition, Assessment of Wetland Vegetation Update – September 2009.* Department of Sustainability and Environment, Melbourne.

Eamus D, Hatton T, Cook P, Colvin C (2006) *Ecohydrology: Vegetation Function, Water and Resource Management.* CSIRO Publishing, Melbourne.

Erskine WD, Warner RF (1998) Further assessment of flood- and drought-dominated regimes in southeastern Australia. *The Australian Geographer* **29**(2), 257–261. doi:10.1080/00049189808703218

Ferreira MT, Aguiar FC, Nogueira C (2005) Changes in riparian woods over space and time: influence of environment and land use. *Forest Ecology and Management* **212**, 145–159. doi:10.1016/j.foreco.2005.03.010

Flanders D, Hall-Beyer M, Pereverzoff J (2003) Preliminary evaluation of eCognition object-oriented software for cut block delineation and feature extraction. *Canadian Journal of Remote Sensing* **29**, 441–452. doi:10.5589/m03-006

Gamon JA, Field CB, Goulden ML, Griffin KL, Hartley AE, Joel G, Penuelas J, Valentini R (1995) Relationships between NDVI, canopy structure, and photosynthesis in three Californian vegetation types. *Ecological Applications* **5**(1), 28–41.

Gao B (1996) NDWI – a normalised difference water index for remote sensing of vegetation liquid water from space. *Remote Sensing of Environment* **58**, 257–266.

Gao X, Huete AR, Ni W, Miura T (2000) Optical-biophysical relationships of vegetation spectra without background contamination. *Remote Sensing of Environment* **74**, 609–620.

Gerber LR, Beger M, McCarthy MA, Possingham HP (2005) A theory for optimal monitoring of marine reserves. *Ecology Letters* **8**, 829–837. doi:10.1111/j.1461-0248.2005.00784.x

Gitelson AA, Kaufman YJ, Merzlyak MN (1996) Use of a green channel in remote sensing of global vegetation from EOS-MODIS. *Remote Sensing of Environment* **58**, 289–298.

Godínez-Alvarez H, Herrick J, Mattocks M, Toledo D, Van Zee J (2009) Comparison of three vegetation monitoring methods: their relative utility for ecological assessment and monitoring. *Ecological Indicators* **9**(5), 1001–1008. doi:10.1016/j.ecolind.2008.11.011

Goetz SJ (2006) Remote sensing of riparian buffers: past progress and future prospects. *Journal of the American Water Resources Association* **42**(1), 133–143. doi:10.1111/j.1752-1688.2006.tb03829.x

Henry CP, Amoros C (1996) Restoration ecology of riverine wetlands. III. Vegetation survey and monitoring optimization. *Ecological Engineering* **7**, 35–58. doi:10.1016/0925-8574(95)00064-X

Huemmrich KF, Goward SN (1997) Vegetation canopy PAR absorptance and NDVI: an assessment for ten tree species with the sail model. *Remote Sensing of Environment* **61**, 254–269.

Huete A, Justice C, van Leeuwen W (1999) *MODIS Vegetation Index (MOD 13). Algorithm Theoretical Basis Document. Version 3.* University of Arizona, University of Virginia, Charlotteville, Tucson, AZ.

Ishiyama T, Tanaka S, Uchida K, Fujikawa S, Yamashita Y, Kato M (2001) Relationship among vegetation variables and vegetation features of arid lands derived from satellite data. *Advances in Space Research* **28**(1), 183–188.

Jackson LE, Kurtz JC, Fisher WS (Eds) (2000) *Evaluation Guidelines for Ecological Indicators*. US Environmental Protection Agency, Office of Research and Development, Triangle Park, NC.

James CS, Capon SJ, White MG, Rayburg SC, Thoms MC (2007) Spatial variability of the soil seed bank in a heterogeneous ephemeral wetland system in semi-arid Australia. *Plant Ecology* **190**(2), 205–217. doi:10.1007/s11258-006-9201-9

Jansen A, Robertson A, Thompson L, Wilson A (2004) 'Development and application of a method for the rapid appraisal of riparian condition.' River and Riparian Land Management Technical Guideline.

Jensen JR (2005) *Introductory Digital Image Processing: a Remote Sensing Perspective*. Prentice Hall, Upper Saddle River, NJ.

Johansen K, Dixon D, Douglas M, Phinn S, Lowry J (2007) Comparison of image and rapid field assessments of riparian zone condition in Australian tropical savannas. *Forest Ecology and Management* **240**, 42–60. doi:10.1016/j.foreco.2006.12.015

Johansen K, Phinn S, Lowry J, Douglas D (2008a) Quantifying indicators of riparian condition in Australian tropical savannas: integrating high spatial resolution imagery and field survey data. *International Journal of Remote Sensing* **29**(23), 7003–7028. doi:10.1080/01431160802220201

Johansen K, Roelfsema C, Phinn S (2008b) Special feature – high spatial resolution remote sensing for environmental monitoring and management. *Journal of Spatial Science* **53**(1), 43–47. doi:10.1080/14498596.2008.9635134

Johansen K, Arroyo LA, Armston J, Phinn S, Witte C (2010a) Mapping riparian condition indicators in a sub-tropical savanna environment from discrete return LiDAR data using object-based image analysis. *Ecological Indicators* **10**(4), 796–807. doi:10.1016/j.ecolind.2010.01.001

Johansen K, Arroyo LA, Phinn S (2010b) Comparison of geo-object based and pixel-based change detection of riparian environments using high spatial resolution multi-spectral imagery. *Photogrammetric Engineering and Remote Sensing* **76**(2), 123–136. doi:10.14358/PERS.76.2.123

Johansen K, Phinn S, Witte C (2010c) Mapping of riparian zone attributes using discrete return LiDAR, QuickBird, and SPOT-5 imagery: assessing accuracy and costs. *Remote Sensing of Environment* **114**(11), 2679–2691. doi:10.1016/j.rse.2010.06.004

Johansen K, Tiede D, Blaschke T, Arroyo L, Phinn S (2011) Automatic geographic object based mapping streambed and riparian zone extent from LiDAR data in a temperate rural urban environment, Australia. *Remote Sensing* **3**, 1139–1156. doi:10.3390/rs3061139

Johnson JA, Newman RM (2011) A comparison of two methods for sampling biomass of aquatic plants. *Journal of Aquatic Plant Management* **49**, 1–8.

Jones JW (1989) Integrating models with expert systems and data bases for decision making. In *Climate & Agriculture: Systems Approaches to Decision Making*. (Ed. A Weiss) pp. 194–211. University of Nebraska, Lincoln, NE.

Kröger R, Khomo LM, Levick S, Rogers KH (2009) Moving window analysis and riparian boundary delineation on the northern plains of Kruger National Park, South Africa. *Acta Oecologica* **35**(5), 573–580. doi:10.1016/j.actao.2009.05.007

Ladson AR, White LJ, Doolan JA, Finlayson BL, Hart BT, Lake PS, Tilleard JW (1999) Development and testing of an index of stream condition for waterway management in Australia. *Freshwater Biology* **41**, 453–468. doi:10.1046/j.1365-2427.1999.00442.x

Ladson AR, Grayson RB, Jawecki B, White LJ (2006) Effect of sampling density on the measurement of stream condition indicators in two lowland Australian streams. *River Research and Applications* **22**, 853–869. doi:10.1002/rra.940

Lake PS (2006) Of floods and droughts: river and stream ecosystems in Australia. In *River and*

*Stream Ecosystems of the World*. (Eds CE Cushing, KW Cummins and GW Minshall) pp. 659–694. University of California Press Berkeley and Los Angeles, CA.

Lefsky MA, Cohen WB, Parker GG, Harding DJ (2002) Lidar remote sensing for ecosystem studies. *Bioscience* **52**, 19–30. doi:10.1641/0006-3568(2002)052[0019:LRSFES]2.0.CO;2

Legendre P (1993) Spatial autocorrelation: trouble or new paradigm? *Ecology* **74**(6), 1659–1673. doi:10.2307/1939924

Levesque J, King DJ (2003) Spatial analysis of radiometric fraction from high-resolution multispectral imagery from modelling individual tree crown and forest canopy structure and health. *Remote Sensing of Environment* **85**, 589–602.

Li Z, Moreau L (1996) A new approach for remote sensing of canopy-absorbed photosynthetically active radiation. I: total surface absorption. *Remote Sensing of Environment* **55**, 175–191.

Lillesand TM, Kiefer RW (2000) *Remote Sensing and Image Interpretation*. John Wiley & Sons, New York.

Lillesand TM, Kiefer RW, Chipman JW (2008) *Remote Sensing and Image Interpretation*. 6th edn. John Wiley & Sons, Hoboken, NJ.

Mackay SJ, Arthington AH, Kennard MJ, Pusey BJ (2003) Spatial variation in the distribution and abundance of submersed macrophytes in an Australian subtropical river. *Aquatic Botany* **77**(3), 169–186. doi:10.1016/S0304-3770(03)00103-7

Mackay SJ, James CS, Arthington AH (2010) Macrophytes as indicators of stream condition in the wet tropics region, northern Queensland, Australia. *Ecological Indicators* **10**(2), 330–340. doi:10.1016/j.ecolind.2009.06.017

Mallet C, Bretar F (2009) Full-waveform topographic lidar: state-of-the-art. *ISPRS Journal of Photogrammetry and Remote Sensing* **64**, 1–16. doi:10.1016/j.isprsjprs.2008.09.007

Michez A, Piegay H, Toromanoff F, Brogna D, Bonnet S, Lejeune P, Claessens H (2013) LiDAR derived ecological integrity indicators for riparian zones: application to the Houille River in southern Belgium/northern France. *Ecological Indicators* **34**, 627–640. doi:10.1016/j.ecolind.2013.06.024

Mott JJ, Williams J, Andrew MA, Gillison AN (1985) Australian savanna ecosystems. In *Ecology and Management of the World's Savannas*. (Eds JC Tothill and JJ Mott) pp. 56–82. Australian Academy of Science, Canberra.

Myneni RB, Hoffman S, Knyazikhin Y, Privette JL, Glassy J, Tian Y, Wang Y, Song X, Zhang Y, Smith GR, Lotsch A, Friedl M, Morisette JT, Votava P, Nemani RR, Running SW (2002) Global products of vegetation leaf area and fraction absorbed PAR from year one of MODIS data. *Remote Sensing of Environment* **83**, 214–231.

Nicol JM (2004) Vegetation dynamics of the Menindee Lakes with reference to the seed bank. PhD thesis. School of Earth and Environmental Sciences, The University of Adelaide.

Nicol JM, Weedon JT, Marsland K (2010) 'Understorey vegetation monitoring of Chowilla environmental watering sites 2004–08.' SARDI publication number RD F2010/000632–1. SARDI research report series no. 499. South Australian Research and Development Institute (Aquatic Sciences), Adelaide.

Noble IR, Gitay H (1996) A functional classification for predicting the dynamics of landscapes. *Journal of Vegetation Science* **7**, 329–336. doi:10.2307/3236276

Nohara S (1991) A study on annual changes in surface cover of floating-leaved plants in a lake using aerial photography. *Plant Ecology* **97**, 125–136. doi:10.1007/BF00035386

Ogden RW (2000) Modern and historical variation in aquatic macrophyte cover of billabongs associated with catchment development. *Regulated Rivers: Research and Management* **16**, 497–512. doi:10.1002/1099-1646(200009/10)16:5<497::AID-RRR600>3.0.CO;2-Y

Quinn LD, Schooler SS, Van Klinken RD (2011) Effects of land use and environment on alien and native macrophytes: lessons from a large-scale survey of Australian rivers. *Diversity & Distributions* **17**(1), 132–143. doi:10.1111/j.1472-4642.2010.00726.x

Reid MA (2008) Evidence for catastrophic shifts in the trophic structure of floodplain lakes associated with soil erosion. In *Sediment Dynamics in Changing Environments*. (Eds J Schmidt, T Cochrane, C Phillips, S Elliot, T Davies and L Basher) pp. 584–590. IAHS Press, Wallingford, UK.

Reid MA, Capon S (2011) The role of the soil seed bank in vegetation response to environmental flows on a drought-affected floodplain. *River Systems* **19**(3), 249–259. doi:10.1127/1868-5749/2011/019-0022

Reid MA, Quinn GP (2004) Hydrologic regime and macrophyte assemblages in temporary floodplain wetlands: implications for detecting responses to environmental water allocations. *Wetlands* **24**(3), 586–599. doi:10.1672/0277-5212(2004)024[0586:HRAMAI]2.0.CO;2

Reid MA, Sayer CD, Kershaw AP, Heijnis H (2007) Palaeolimnological evidence for submerged plant loss in a floodplain lake associated with accelerated catchment soil erosion (Murray River, Australia). *Journal of Paleolimnology* **38**, 191–208. doi:10.1007/s10933-006-9067-9

Roberts J, Marston F (2011) *Water Regime for Wetland and Floodplain Plants: a Source Book for the Murray–Darling Basin*. National Water Commission, Canberra.

Robertson AI, Rowling RW (2000) Effects of livestock on riparian zone vegetation in an Australian dryland river. *Regulated Rivers: Research and Management* **16**, 527–541. doi:10.1002/1099-1646(200009/10)16:5<527::AID-RRR602>3.0.CO;2-W

Rogers SJ, Owens TW (1995) *Long Term Resource Monitoring Program Procedures: Vegetation Monitoring*. National Biological Service, Environmental Management Technical Center, Onalaska, WI.

Schwarz A-M, Snelder T (1999) Integrated submerged aquatic vegetation management in an urban New Zealand river. *Hydrobiologia* **415**, 235–241. doi:10.1023/A:1003839402075

Sims DA, Gamon JA (2002) Estimation of vegetation water content and photosynthetic tissue area from spectral reflectance: a comparison of indices based on liquid water and chlorophyll absorption features. *Remote Sensing of Environment* **84**, 526–537.

Smith DG, McBride GB, Bryers GG, Wisse J, Mink DFJ (1996) Trends in New Zealand's national river water quality network. *New Zealand Journal of Marine and Freshwater Research* **30**, 485–500. doi:10.1080/00288330.1996.9516737

Souter NJ, Walter M (2014) An unexpected lack of response in lignum (*Muehlenbeckia florulenta* Meissn.) to a managed rise in the level of the lower River Murray, South Australia. *Transactions of the Royal Society of South Australia* **138**, 124–134. doi:10.1080/03721426.2014.10887196

Souter NJ, Cunningham S, Little S, Wallace T, McCarthy B, Henderson M (2010) Evaluation of a visual assessment method for tree condition of eucalypt floodplain forests. *Ecological Management & Restoration* **11**, 210–214. doi:10.1111/j.1442-8903.2010.00551.x

Spencer C, Robertson AL, Curtis A (1998) Development and testing of a rapid appraisal wetland condition index in south-eastern Australia. *Journal of Environmental Management* **54**, 143–159. doi:10.1006/jema.1998.0212

Stohlgren TJ (2007) *Measuring Plant Diversity: Lessons From the Field*. Oxford University Press, Oxford, UK.

Thiebaud I, Williams RJ (2008) *Distribution of Freshwater Macrophytes in the Hawkesbury Nepean River From Warragamba Dam to Wisemans Ferry, 2007*. NSW Department of Primary Industries, Port Stephens Fisheries Centre, Nelson Bay, NSW.

UN/ECE (2006) Manual on methods and criteria for harmonized sampling, assessment, monitoring and analysis of the effects of air pollution on forests, part II visual assessment of crown condition. <http://www.icp-forests.org/pdf/Chapt2_compl06.pdf>.

Ward PA (2011a) 'Monitoring understorey vegetation response to flooding in Barmah–Millewa Forest, 2011/12: Progress report – Spring 2011.' Consultant progress report prepared as part of The Living Murray Condition Monitoring Program for the Barmah–Millewa Icon Site, managed by the NSW Department of Environment, Climate Change and Water, Griffith, on behalf of the Murray–Darling Basin Authority, Canberra.

Ward PA (2011b) 'Monitoring understorey vegetation response to flooding in Barmah–Millewa Forest, 2011/12: Progress report – Summer 2011/12.' Consultant progress report prepared as part of The Living Murray Condition Monitoring Program for the Barmah–Millewa Icon Site, managed by the NSW Department of Environment, Climate Change and Water, Griffith, on behalf of the Murray–Darling Basin Authority, Canberra.

Webb M, Reid M, Capon S, Thoms M, Rayburg S, James C (2006) Are floodplain-wetland plant communities determined by seed bank composition or inundation periods? *IAHS Publication* **306**, 241–248.

Werren G, Arthington A (2002) The assessment of riparian vegetation as an indicator of stream condition, with particular emphasis on the rapid assessment of flow-related impacts. In *Landscape Health of Queensland*. (Eds A Shapcott, J Playford and AJ Franks) pp. 194–222. The Royal Society of Queensland, St Lucia, QLD.

White M, Berens V, Souter N (2009) 'Bookpurnong Living Murray pilot project: artificial inundation of *Eucalyptus camaldulensis* on a floodplain to improve vegetation condition.' DWLBC report 2009/19. Department of Water Land and Biodiversity Conservation, Adelaide.

Wulder MA, LeDrew EF, Franklin SE, Lavigne MB (1998) Aerial image texture information in the estimation of northern deciduous and mixed wood forest leaf area index (LAI). *Remote Sensing of Environment* **64**, 64–76.

Wylie BK, Meyer DJ, Tieszen LL, Mannel S (2002) Satellite mapping of surface biophysical parameters at the biome scale over the North American grasslands. A case study. *Remote Sensing of Environment* **79**, 266–278.

# 21

# Climate change and the future of Australian riverine vegetation

*Cassandra James, Michael Reid and Samantha Capon*

## INTRODUCTION

Freshwater and riparian ecosystems are likely to be highly vulnerable to the effects of the current phase of rapid, anthropogenic climate change because of their high levels of exposure and sensitivity to changes in climatic stimuli. Furthermore, the adaptive capacity of these ecosystems has been limited due to human activities (Capon *et al.* 2013; Capon and Bunn 2015). Riverine vegetation is exposed to all the direct influences of rising $CO_2$ and the climatic changes driven by this described for terrestrial vegetation, including elevated temperatures, altered precipitation and evapotranspiration patterns. Additionally, climate change has many indirect effects on riverine vegetation as a result of changes in soil moisture, hydrology and, in coastal situations, sea level rise and storm surges. Arguably, no other vegetation type is subject to the same levels of exposure and yet relatively little is known about how plants in Australian riverine habitats may respond.

While climate has similar influences on riparian vegetation across the globe, the Australian context differs in the highly weathered and low-nutrient soils that characterise much of the continent. Furthermore, Australian inland rivers are among the most hydrologically variable in the world (Puckridge *et al.* 1998) due to highly variable temperatures, precipitation and corresponding hydrological regimes. Australia's uniqueness has important implications for how the vegetation of Australian riverine systems may respond to a changing climate and their capacity to adapt. Riverine vegetation is found throughout many Australian landscapes, including desert, alpine, temperate, subtropical, tropical and Mediterranean. Riverine vegetation often differs in structure, composition and distinctiveness from the surrounding terrestrial vegetation. Under rising $CO_2$ and other climate change effects, the structure and function of riverine ecosystems in most landscapes are likely to change although the direction and magnitude of this will differ between regions, landscape position and ecosystem type (Catford *et al.* 2013).

Predicting climate change impacts on riverine vegetation is difficult due to the high potential for complex interactions and feedbacks between drivers and responses further compounded by the considerable uncertainties associated with projecting changes to precipitation and its amplification for runoff. Many Australian riverine plants and vegetation communities may be able to adapt to changing conditions, having evolved in the face of the inherent climatic variability that characterises their habitats. Indeed, there is evidence to suggest that key species of riverine plants in Australia survived past climatic changes by contracting their ranges and expanding when favourable conditions returned (e.g. Butcher *et al.* 2009). The capacity of riverine plants and vegetation communities to adapt to the current period of climate change may be compromised, however, by both the rapid pace of change and constraints on their adaptive capacity due to high levels of habitat degradation and anthropogenic pressures. Consequently, there is a high potential for loss of biodiversity and dramatic shifts in the composition and structure of vegetation across riverine landscapes. Such impacts are likely to have significant ramifications for the critical functions for which riparian zones are valued, as well as for their own intrinsic worth. Additionally, the critical role riverine vegetation plays in the broader landscape means that such impacts will likely cascade through other systems and the biota they support. Planned, managed adaptation will be crucial if such impacts are to be minimised.

In this chapter, we review recent observed and projected changes to the Australian climate. We then examine how major elements of the climate system drive riverine vegetation dynamics and how changes in these are likely to affect riverine plants and vegetation. We briefly discuss past climate change and vegetation responses in Australia as well as major factors influencing how riverine plants and vegetation will respond to the current phase of climate change. Finally, we survey relevant adaptation approaches for the conservation and management of riverine plants and vegetation.

## AUSTRALIA'S CHANGING CLIMATE

Prior to the industrial revolution, the world's atmospheric $CO_2$ levels were relatively stable for several millennia (Raupach and Fraser 2011). Since 1800, however, $CO_2$ levels have risen sharply, from 280 ppm in 1800 to 400 ppm in 2013, with increasing rates since then of around 2 ppm annually (Tans and Keeling 2015). The dramatic rise in $CO_2$ concentrations is widely attributed to emissions associated with human activities, including the burning of fossil fuels and deforestation.

Climate projections for Australia over the 21st century vary regionally and depend on the scenario for global greenhouse emissions and global climate model considered. Temperatures are projected to rise by up to 4.8°C by 2081–2100 in addition to the temperature increase of around 1°C that has already occurred recently (Figs 21.1 and 21.2; CSIRO and Bureau of Meteorology 2015). While more warming is projected for inland than coastal

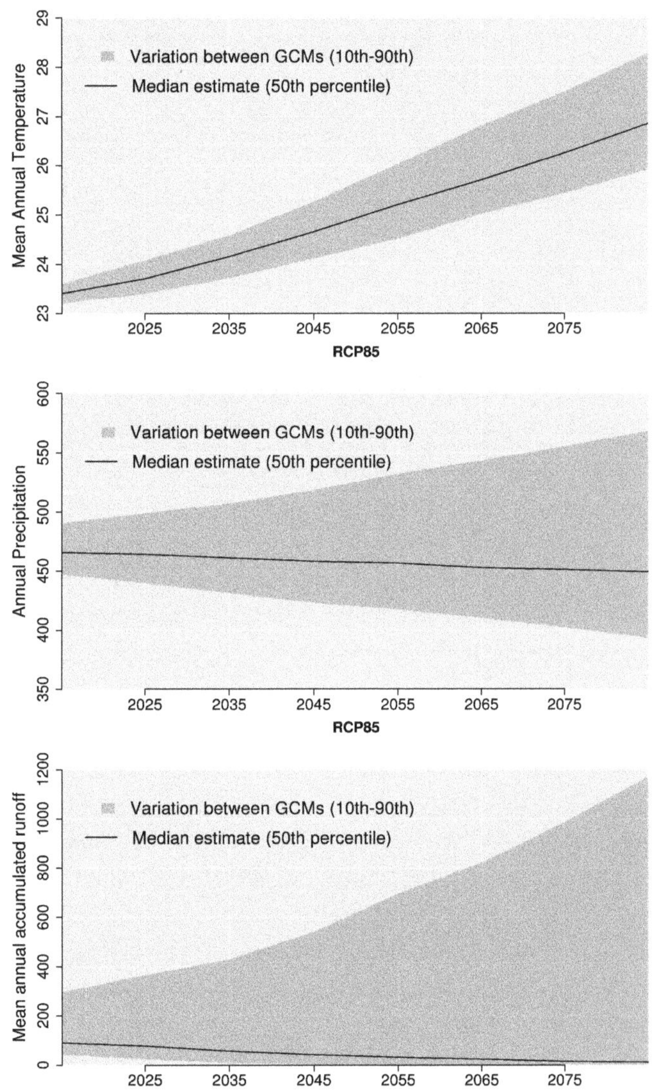


**Figure 21.1:** Projected mean daily temperature (°C), mean annual rainfall (mm) and mean annual accumulated runoff (ML) for Australia. Values represent the median estimate across the country assuming anthropogenic greenhouse gas emissions continue to rise throughout the 21st century (RCP, representative concentration pathway 85). Shaded areas show variation across 18 global climate models (GCMs) (see James *et al.* 2013 for further details).

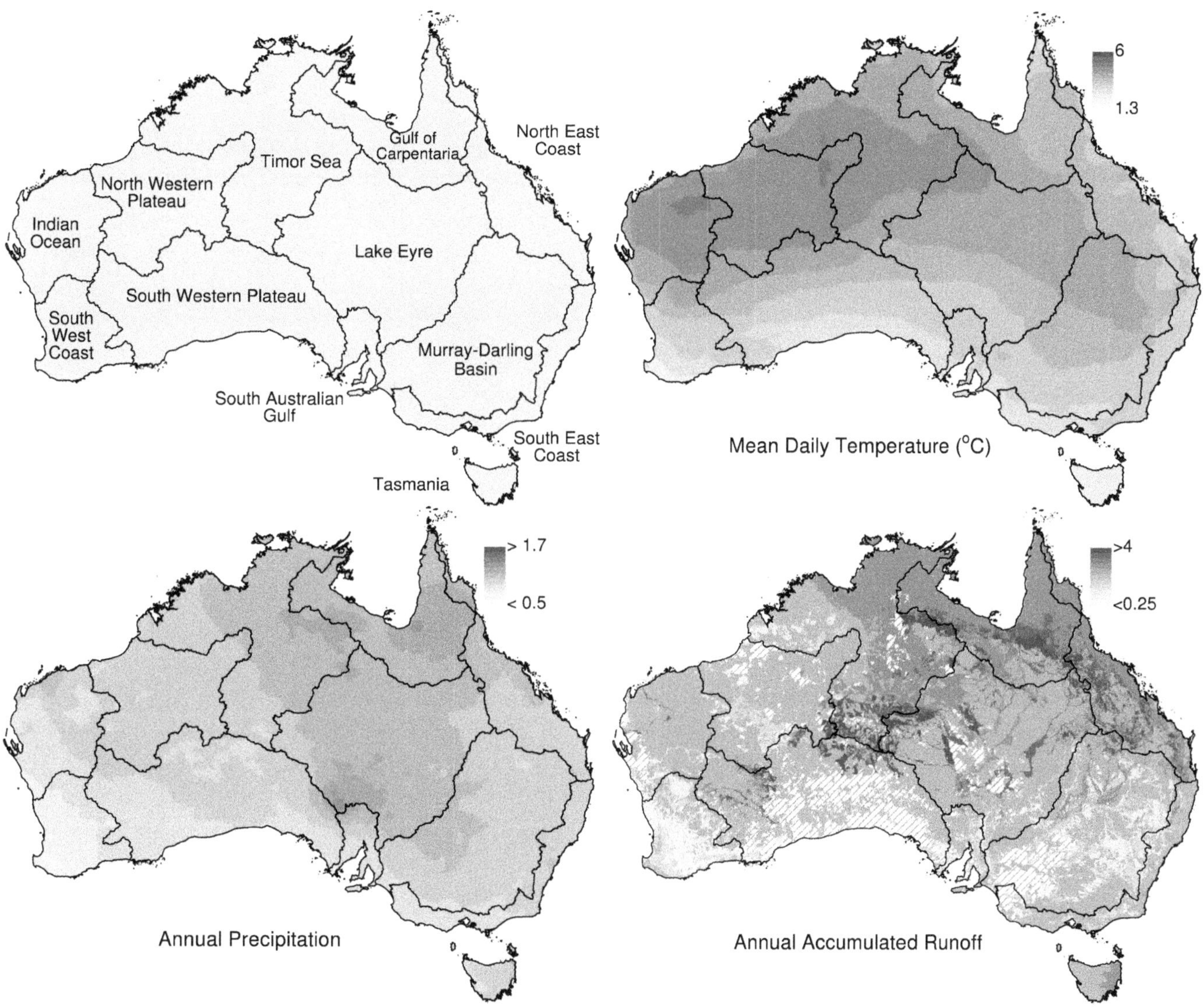


**Figure 21.2:** Changes (future minus current) in mean daily temperature (°C) and proportional changes (future divided by current) in annual rainfall and annual mean accumulated runoff for Australia. Values represent the median estimate across 18 global climate models assuming continued anthropogenic greenhouse gas emissions anthropogenic greenhouse gas emissions continue to rise throughout the 21st century. Figures adapted from James *et al.* (2013). Hatched areas indicate no data.

regions (Whetton 2011), coastal regions are actually more likely to experience temperatures that are outside current ranges because of their narrower thermal window (James *et al.* 2013). Warming is predicted across all seasons, with declines in the frequency of extreme cold events, particularly in the south, and increases in the frequency of warm extremes and heatwaves, the latter especially in the north-east (Braganza and Church 2011). In Melbourne, for example, around twice the number of days over 35°C currently experienced are predicted by 2070 (Whetton 2011).

Changes to future rainfall patterns are highly uncertain and there is low consensus across global climate models in terms of either the direction or magnitude of change. Average rainfall across the continent has been increasing largely due to increases in wet season rainfall in northern Australia (CSIRO and Bureau of Meteorology 2015). Models are broadly consistent in predicting a decline in rainfall in south-west Western Australia and a general trend for drying in southern and eastern Australia. In the northern gulf region, predictions are for either a slight increase (Figs 21.1 and 21.2; James *et al.* 2013) or no change in annual rainfall as the consensus across a number of the global climate models (Petheram *et al.* 2012). There is general agreement that rainfall will decrease during winter and spring for most of the continent, a continuing trend for the south-west and south-east of the continent (CSIRO and Bureau of Meteorology 2015). An intensification of rainfall events is predicted due to warming increasing the moisture-holding capacity of the atmosphere (Whetton 2011), while droughts in intervening periods are expected to become more prolonged, especially in southern Australia. Snowfall is projected to decline with areas averaging at least one day of snow cover for Victoria reducing 35–75% by 2050 under a high-emission scenario (Bhend *et al.* 2012). Decreased snowfall will accompany increased snowmelt and ultimately a reduction in snow cover (Whetton 2011). Shifts in precipitation patterns, at least until 2030, are likely to be modest relative to current levels of inter-annual variation, which are very high (James *et al.* 2013; CSIRO and Bureau of Meteorology 2015).

Declines in soil moisture are most confidently predicted for areas in which increased evaporation is accompanied by decreases in rainfall. Hence, decreases are most likely in southern Australia, particularly in the winter and spring. Elsewhere, soil moisture trends are extremely hard to predict because of the uncertainty surrounding rainfall projections, although evaporation demands are likely to increase due to rising temperatures. Uncertainty in rainfall is amplified when translated into runoff (Chiew and McMahon 2002), with an estimated 1% change in precipitation equating to a 2–3.5% change in mean annual streamflow (Chiew 2006). The greatest consistency across models is shown for south-west Western Australia, where a projected decrease in rainfall is accompanied by a decrease in annual average runoff (Figs 21.1 and 21.2). For other areas, runoff predictions are highly variable in both direction and magnitude although broadly suggest a decrease in runoff for south-eastern Australia for the median of the models considered (e.g. James *et al.* 2013; CSIRO and Bureau of Meteorology 2015). For northern Australia, predictions are for a slight increase in annual runoff (James *et al.* 2013), although Petheram *et al.* (2012) found no consistent trends for tropical savannas and semi-arid rangelands.

Changes to hydrological regimes could be far greater than those predicted from changes in annual average runoff. Alterations to seasonality and increased frequency of extreme rainfall events, for instance, are likely to translate into altered flow regimes although exactly how these will manifest into the future is very difficult to predict. Broadly speaking, predictions of flow regime changes include decreases in average flows, lower base flows, higher peak flows and longer periods of intervening low or no flow. For arid areas, the implications are for increasing aridity in central Australia and reduced hydrological persistence of arid zone aquatic habitats, particularly those linked directly to surface climate regimes (Davis *et al.* 2013).

Riverine vegetation in coastal regions is also vulnerable to sea level rises, storm surges and salt-water intrusion. Sea levels have risen by around 20 cm since 1900 (CSIRO and Bureau of Meteorology 2015) and are projected to rise throughout the 21st century. By 2090, the average sea level around Australia is projected to rise equivalently or slightly more than the global mean projected rise of up to 82 cm under a high-emission scenario. This could, however, be several tenths of a metre higher if marine-based sectors of the Antarctic ice sheet collapse (CSIRO and Bureau of Meteorology 2015). Marine incursion further inland may be exacerbated by storm surges and potentially lower fresh-water runoff for some regions.

Other indirect effects of climate change include changes to fire regimes, which are highly uncertain, depending on variations in ignition rates, fuel loads and climate. An increase in fire risk is predicted where weather gets hotter and drier, particularly in southern Australia, although fuel loads may decrease for areas where drought reduces plant productivity (Williams *et al.* 2009).

## CLIMATE CHANGE EFFECTS ON RIVERINE PLANTS AND VEGETATION

Climate plays a fundamental role in shaping the composition, structure and condition of vegetation across the Earth's surface. Key elements of the climate system including atmospheric carbon dioxide levels, temperature and precipitation regimes exert direct controls on plants as well as affecting non-climatic drivers of plant growth and vegetation dynamics, such as competition and herbivory. Changes in climate will also drive alterations to the landscape water balance, altering patterns of soil moisture and runoff, which are key drivers of the distribution, abundance and diversity of river plants (Merritt *et al.* 2010). Rising $CO_2$ and associated climatic changes have fundamental implications for global biogeochemical cycles altering the rate, intensity and quantities of matter being transferred from one component to another. Riverine vegetation will itself be affected by changes to biogeochemical cycles and will, in turn, impact on nutrient and water cycles, although our understanding of these interactions and potential feedbacks is poor (Wassen *et al.* 2013). Other indirect effects of climate change (e.g. altered fire regimes) will also have significant implications for riverine vegetation of some regions.

### Rising carbon dioxide concentrations

Photosynthetic rates typically rise with increasing $CO_2$ concentrations until other factors (e.g. light, temperature, water or nutrients) become limiting. This process is often referred to as $CO_2$ 'fertilisation' or 'enrichment' (Steffen *et al.* 2009). Plant species respond differently to elevated $CO_2$ and this variation in itself may result in changes to vegetation. Plants with a $C_3$ photosynthetic pathway are likely to be particularly responsive to rises in $CO_2$ because, unlike $C_4$ photosynthesis, $C_3$ photosynthesis is not considered saturated at current $CO_2$ concentrations (Perry *et al.* 2012; Drake 2014). Effects of $CO_2$ enrichment also change with plant life history stage, with older plants generally less responsive compared with younger plants (Steffen and Canadell 2005).

Among the different plant growth forms, trees are probably the most responsive to elevated $CO_2$ (Ainsworth and Long 2005). Submerged aquatic macrophytes may also respond to elevated $CO_2$, particularly in lentic waters. Slow rates of gas diffusion usually limit $CO_2$ availability underwater, constraining photosynthesis and the growth of submerged macrophytes (see Chapter 6). Elevated $CO_2$ concentrations may benefit some macrophytes, reducing their reliance on alternative forms of carbon, such as bicarbonate, which incur metabolic costs (Raven and Lucas 1985).

### Warming

Rising temperatures are concomitant with $CO_2$ increases under climate change. While the importance of riparian vegetation in mitigating warming for instream environments under climate change has been widely recognised (Davies 2010; Bowler *et al.* 2012), warming impacts on riverine vegetation itself are rarely acknowledged. Temperature has wide-ranging influences on plants, shaping their metabolism, growth, reproduction and phenology. Plant responses to climate change relate to rising average temperatures and altered temperature variability and extremes. How these translate to leaf temperatures, which can be considerably higher than surrounding air temperatures, is particularly significant (Berry and Bjorkman 1980; Larcher 2000). Water temperatures are also likely to rise and, although unlikely to mirror air temperature rises exactly due to myriad other influencing factors (Hannah *et al.* 2008; van Vliet *et al.* 2013), may nevertheless influence instream vegetation photosynthesis and growth (Santamaría and vanVierssen 1997).

Plant species differ in their sensitivity to warming. $C_4$ grasses, for example, tend to have greater high temperature tolerance than $C_3$ plants. Furthermore, under enriched $CO_2$ concentrations temperature optimums may also rise (Long 1991; Cernusak *et al.* 2013). Photosynthesis is a temperature-dependent process and enhanced photosynthesis is generally predicted with elevated temperatures (Saxe *et al.* 2001) until some physiological temperature threshold is reached above which photosynthesis declines. Temperatures exceeding between 50 and 55°C, however, will typically damage or kill leaf tissue in most plants.

Information on the thermal responses of Australian riverine species is biased heavily towards a few widespread and economically important species (e.g. river red gum, *Eucalyptus camaldulensis*, and forest red gum, *E. tereticornis*). For many other riverine species, responses to temperature are almost completely unknown and geographic distributions may be used to infer temperature thresholds, although these can be misleading where plant distributions are not controlled primarily by temperature. In many northern regions of Australia, riverine species are conceivably closer to physiological temperature thresholds as temperatures already regularly exceed 45°C. Such species may therefore be more susceptible to further increases in temperature which can lead to heat-induced inhibition of photosynthesis, tissue damage and mortality. For riverine species of northern savannas, the extremely dry and arid regions of central Australia will probably limit expansion southwards, effectively reducing suitable habitat into the future. Tropical riverine species of northern Australia could also be particularly susceptible to rising temperatures as tropical species tend to have narrower optimal temperature tolerances and less acclimation potential (Cunningham and Read 2002, 2003). Even modest rises in temperature may therefore shift the new climate outside of species optimum ranges.

For temperate species with more southerly distributions, bioclimatic modelling (e.g. Hughes *et al.* 1996; Butt *et al.* 2013) predicts a contraction in suitable climate space as plant distributions are bound by the southern limit of the continent and plant distributions are essentially squeezed up against the south coast. For example, the mainland distribution of *Eucalyptus ovata*, a species found in south-east Australia and Tasmania, is predicted to contract considerably by 2050 (Harrison *et al.* 2013). Predictions for Australia's alpine flora under climate change are likewise pessimistic due to increasing temperatures and decreasing snow cover (Pickering and Armstrong 2003), with altitudinal shifts limited by available habitat and suitability of that habitat in other respects (e.g. slope and substrate).

Plants of arid and semi-arid regions may be less sensitive to rises in temperature as many of these species already experience widely fluctuating diurnal and seasonal temperature regimes and have morphologies, physiologies and/or life history traits that facilitate their survival under extreme heat and water stress. Features common to many riverine species of inland habitats, such as small thick leaves with low specific leaf areas and drooping foliage, minimise canopy exposure to direct sun, prevent overheating and increase water use efficiency. Australian arid-zone plants show considerable capacity to tolerate high temperature (Curtis *et al.* 2014), and this is probably also the case for riverine species of these regions. *Eucalyptus camaldulensis*, for example, has a capacity to tolerate heat in excess of many other *Eucalyptus* species (Sa ardavut *et al.* 1984), and heat damage in river red gums was not observed at temperatures of 50°C (Karschon and Pinchas 1971; Curtis *et al.* 2014). A broad temperature optimum for photosynthesis is typical of many trees (Berry and Bjorkman 1980), including at least some riverine species (e.g. *E. camaldulensis*). This is consistent with species from thermally variable environments where such a response allows plants to maintain photosynthesis under changing temperatures, with only minimal adjustment to photosynthetic rates.

Early plant life history stages are likely to be more vulnerable to rising temperatures and, in particular, heatwaves. Seed banks, for example, which are critically important for many annual and short lived perennial species of Australian riverine desert habitats (see Chapter 13), are susceptible to increasing temperatures (Ooi *et al.* 2012). Soil temperature rises may be proportionally greater than rises in air temperatures, particularly in sparsely vegetated desert areas which are exposed to direct sunlight and lack sufficient vegetation and litter cover to effectively shade soils. Warmer soils also have implications for low-growing species and early life history stages such as tree seedlings (Helgerson 1990). Microclimate amelioration through plant–plant interactions, which recent studies have shown contribute significantly to Australian riverine vegetation diversity (Good *et al.* 2014; James *et al.* in press), may have increasing importance under future climate conditions.

## Hydrologic changes

Changes to precipitation, and their interaction with warming and other climate change effects, have significant implications for the hydrology of riverine habitats including changes to soil moisture and groundwater levels, runoff, stream flow, flooding and water quality (Capon *et al.* 2013). Changes in runoff and river flows will affect floodplain and wetland connectivity, modifying wetting and drying cycles and changing the fundamental character of riverine habitats. The resultant hydrological changes will, in many areas, reduce the frequency of inundation and prolong the periods between wetting. Riverine vegetation dynamics are tightly coupled to flow regimes (Nilsson and Svedmark 2002) and flow has a major influence on the composition, structure and condition of Australian riverine vegetation across a broad spectrum of temperate, arid, subtropical and tropical settings (e.g. Blanch *et al.* 1999; Pettit *et al.* 2001; Mackay *et al.* 2003; Capon 2005; Greet *et al.* 2011; Chalmers *et al.* 2012). Consequently, altered hydrology is likely to be one of the most significant effects of climate change on riverine vegetation.

The future distributions of many riverine species are dependent on the regeneration responses to declining rainfall and runoff, as the regeneration stages are often the most vulnerable to the effects of abiotic and biotic stresses. Flooding is considered important for reproduction and recruitment of several common woody riparian species, with mass episodic recruitment of woody vegetation often observed following flooding (e.g. Good *et al.* 2014). Subsequent seedling survival may depend on water table depths and groundwater replenishment from surface waters although later floods could be detrimental to seedling survival. Likewise, for shorter-lived annual and perennial species floodplain wetting and subsequent drying presents a critical opportunity to grow, reproduce and replenish the seed bank. Reduced flooding will have both short-term effects on extant vegetation composition, growth and regeneration, and longer-term implications for vegetation persistence, although the latter effects are much less understood.

Changes in high flows and flooding intensity also have consequences for riverine vegetation. Increases in high flow frequency or intensity of flows is likely to shift composition towards younger age cohorts and pioneer species able to withstand or rapidly regenerate following flood disturbance (Stromberg *et al.* 2010). Frequently flooded areas of inland regions are typically dominated by annuals (Capon 2003, 2005) and, hence, annuals may become more prevalent where flood frequency increases (Stromberg *et al.* 2010). In tropical regions, riparian vegetation will have to contend with cyclonic winds where cyclone intensity increases, resulting in shifts in species composition to disturbance-tolerant species at the expense of less tolerant species (Turton 2013).

Geomorphological shifts in fluvial landforms will occur as a result of, among other phenomena, changes in flow regimes and shifts in vegetation composition and structure in response to rising $CO_2$ and climate change (Goudie 2006). These are likely to result in long-term changes to physical river landscapes and the development and modification of in-channel and floodplain landforms such as levees and billabongs. Such landforms are crucial to the maintenance of floodplain heterogeneity and, hence, a diversity of habitats for riverine vegetation.

Changes to flow regimes and hydrological connectivity between aquatic habitats have implications for the dispersal of riverine plants. Plants are passive dispersers and depend on a vector to carry them to suitable habitat. Dispersal constraints may restrict the capacity of plants to migrate and track suitable climate (Corlett and Westcott 2013; Mokany *et al.* 2014), and rarer longer-distance dispersal may play an important role in allowing vegetation to track suitable climate. Riverine species can capitalise on longer-distance dispersal mechanisms such as hydrochory (water-borne dispersal) and zoochory (specifically waterbirds). They may therefore have the capacity to migrate and track suitable climate great distances and at rapid rates. These movements, however, are reliant on the direction and timing of vector movement and the absence of restrictions to dispersal (i.e. both innate life history constraints and external barriers to movement).

Impacts of drought may be offset to some extent by raised $CO_2$ levels which reduce plant transpiration water loss and increase water use efficiency. Increased water use efficiency under elevated $CO_2$ may allow encroachment of more mesic vegetation into drier regions at vegetation boundaries (Hughes 2003). In general, enriched $CO_2$ stimulates plant growth to a greater extent during drought (Polley *et al.* 2002) and this interactive effect has been noted for at least one *Eucalyptus* species found in riverine environments (*E. tereticornis*; Atwell *et al.* 2007). There is evidence from other eucalypts, however, that factors such as tree size and rooting depth may alter expected responses considerably (Duursma *et al.* 2011). Differential responses to enriched $CO_2$ and drought have also been noted for several species invasive to Australian northern riverine habitats. Polley *et al.* (2002) note that recruitment of invasive drought-tolerant species such as *Parkinsonia aculeata* may be favoured over more mesic species such as *Acacia nilotica*, as the latter is generally restricted to watercourses in arid regions. In water-limited regions, elevated $CO_2$ may also reinforce interspecific differences in drought tolerance among indigenous Australian species, benefitting those species that already show considerable drought tolerance. Increases in $CO_2$ and plant water use efficiency, however, are unlikely to compensate fully in a drier, hotter climate (Perry *et al.* 2013).

Many regions of inland and southern Australia are predicted to get both hotter and drier, which could reduce streamflows considerably in some areas. Reduced streamflows are likely, at the most basic level, to result in altered distributions of species across elevation gradients and encroachment of xeric species into more mesic habitats, as has been observed as a result of flow regime changes in the past (e.g. Bren 1992). Where this results in vegetation establishing in the channels themselves, there are likely to be adjustments in fluvial dynamics, channel morphology and aquatic habitats which, in turn, influence vegetation (Miller *et al.* 2013). At its extreme, complete loss of wetland habitats may occur under reduced water inputs (i.e. both flow and precipitation), higher temperatures and greater evaporative losses. A decline in streamflows and reduced recharge of soil and groundwater stores, which are critical to riparian vegetation health, are likely to impact on riverine vegetation condition, structure and composition (Horner *et al.* 2009; Cunningham *et al.* 2011; Kath *et al.* 2014). The 'millennium drought' between 1997 and 2009 saw a decline in health and mortality of *E. camaldulensis* along the Murray River (Horner *et al.* 2009; Doody *et al.* 2014; Catelotti *et al.* 2015). In other areas, such as south-west Western Australia, lowering of groundwater has impacted on riparian vegetation distributions and condition (Groom *et al.* 2001). These observed responses are indicative of potential impacts of a drier climate in the future.

## Sea level rise and storm surge

Riverine vegetation in coastal areas is vulnerable to rising sea levels as a result of climate change. Salt marsh and mangrove communities will be affected by changes in salinity gradients and salinity regimes associated with the upstream migration of the marine–freshwater boundary. Mangroves in particular may be able to cope with rising sea levels through natural sediment trapping and aggradation, and studies suggest some communities are keeping pace with recent sea level rise (Kirwan and Megonigal 2013; Silva *et al.* 2013; Di Nitto *et al.* 2014). The capacity of mangrove communities to vertically accumulate sediment at a sufficient rate, however, varies with supply of mineral sediments, natural subsidence, dominant species and overall productivity (Kumara *et al.* 2010; Lovelock *et al.* 2011; Schile *et al.* 2014) and other studies indicate that sediment accretion is not keeping pace with sea level rise (Sanders *et al.* 2012).

For freshwater coastal environments, saltwater intrusion will alter floodplain hydrology and water quality, probably resulting in shifts in vegetation to more salt-tolerant species (Catford *et al.* 2013). Depending on the geomorphic context, inland migration of wetland habitats and the creation of new habitats may be possible with sea level rise although migration may be constrained in many situations, either topographically or by human settlements and infrastructure.

### Other effects

In some instances the indirect effects of climate change on riverine vegetation may outweigh the direct effects of $CO_2$ rises and climate change. New species assemblages will result from the proliferation of non-indigenous species and the movement of indigenous species into areas that were previous climatically unsuitable. Altered hydrological conditions have facilitated invasion of exotic species into Australian riverine landscapes (Catford *et al.* 2011; Catford *et al.* 2014) and future hydrological alteration as a result of climate change is likely to further advantage invasive species in some riverine landscapes (Catford *et al.* 2013). Changes in fire regimes will also impact fire-sensitive riparian vegetation (Andersen *et al.* 2005). Increases in the incidence of fire are likely to modify species composition, promote fire-resistant species including invasive species and alter the structure of vegetation through differential mortality of different size classes (Radford *et al.* 2008). In addition to myriad direct and indirect effects from $CO_2$ rises and climate change are the effects wrought on riverine systems by humans as they attempt to respond to the changing climate. Management of water to compensate for drier lowland environments, for example, will have implications for upstream riverine communities including alpine riparian flora (see Chapter 11).

## ADAPTIVE CAPACITY OF AUSTRALIAN RIVERINE VEGETATION

Riverine vegetation is likely to have high capacity to adapt to climate change compared with that of many other ecosystems, because it has developed and evolved under conditions characterised by high levels of environmental variability (Capon *et al.* 2013). Plant traits that enable riverine species to respond to current disturbances, for instance, such as the capacity to re-sprout, may also allow them to cope with current and future climate change to some degree. Further adaptive adjustments to climate change may occur in situ, involving genetic change among plants or shifts in composition or structure at a community level. As natural corridors linking a range of topographic elevations, and therefore different local climates, riparian zones may also provide routes for plant species to migrate along in response to shifting climatic conditions.

### Historical responses to climate change

Riverine environments are naturally dynamic, and the suite of climate and associated changes outlined above are not without precedent in past millennia. These changes, particularly those of the past 30 000 years are described in Chapter 4, but are addressed here briefly to provide context for ongoing anthropogenic climate change. The glacial–interglacial cycles that have characterised the Quaternary period (~past 2 million years) resulted in climate fluctuations of similar magnitude to the changes that are predicted to occur as a result of anthropogenic climate change. Most dramatically and recently, the last glacial maximum (LGM) at around 22 ka before present (BP) saw average global temperatures more than 4°C cooler than pre-industrial temperatures (Shakun and Carlson 2010). Importantly, this global average is reduced due to the relatively muted changes in temperature at low latitudes and over oceans, meaning that the changes in temperature from the LGM to today over land and at higher latitudes were substantially greater. East Antarctica, Greenland and northern Europe, for example, are estimated to have been 10 to 15°C cooler at the LGM (Clark *et al.* 2012). Estimates for LGM to modern temperature differences in Australia are as great as 9°C for south-eastern Australia and the interior (Reeves *et al.* 2013).

Combined with these reduced temperatures, a more complex and uncertain pattern of differences in rainfall and water balance is likely to have existed across the continent at the LGM (Petherick *et al.* 2013; Reeves *et al.* 2013). Perhaps most dramatically, sea levels were some 120 m lower than today at the LGM, meaning that vast areas of coastal plain that existed then are now beneath the ocean.

Periods also occurred in the geologically recent past when temperatures were warmer than today. During the last interglacial period, at around 130 ka BP, global averages were around 1.5°C warmer than today (Turney and Jones 2010), with estimates for high latitudes up to 5°C warmer (Kopp *et al.* 2013). More recently, global average temperatures during the early Holocene period (around 8 ka BP) are estimated to have been 0.5°C warmer than the 1961–90 average. Partly due to higher temperatures during the early Holocene, but also due to regional isostatic adjustments, sea levels were 1–2 m higher than today during much of the Holocene (Sloss *et al.* 2007). These Holocene sea level changes, as well as sediment accretion and progradation, resulted in highly dynamic deltaic environments during the Holocene, particularly in northern Australia where mangrove communities are extensive and coastal gradients are low (Mulrennan and Woodroffe 1998; Grindrod *et al.* 1999).

Although the climate fluctuations of the Quaternary did not result in the continental-scale advance and retreat of glaciers in Australia, the environmental results of temperature changes were nonetheless substantial for riverine vegetation. In addition to the effects of fluctuating sea levels on communities associated with estuaries, the changes in temperature at higher altitudes drove variation in treelines of several hundred metres, while in the middle reaches of rivers changes in temperature and altered rainfall patterns drove substantial changes in water balance and river character (Bowler 1978; Page *et al.* 2009). Despite this, there is little evidence that the substantial climatic and environmental transformation that occurred from the LGM to the Holocene resulted in widespread extinctions or increases in extinction rates (De Deckker 1986; Hampe *et al.* 2013). This is perhaps not surprising given that multiple glacial–interglacial cycles of the Quaternary, on top of the gradual cooling and drying of the Tertiary period, are likely to have had a winnowing effect on the species pools of riverine vegetation. Indeed, the fact that Australian riverine vegetation is characterised by relatively few, yet widely distributed woody species (see Chapter 8) is consistent with such an effect. Similarly, the phylogeography of Australia's most widespread species of eucalypt, *E. camaldulensis*, is consistent with a pattern of contraction to and expansion from refugia in more humid areas in response to successive deteriorations and ameliorations in climate (Butcher *et al.* 2009).

## Constraints on adaptive capacity

Although the vegetation of Australia's riverine landscapes coped with the climate fluctuations of the Quaternary, including those of the (relatively) recent past, its capacity to adapt to current and future anthropogenic climate change is uncertain. This uncertainty arises from several factors. First, climate change projections are increasingly suggesting that global climates may be warming to a degree that the earth has not experienced for at least 200 ka and perhaps even 2 million years (Overpeck *et al.* 2005). Attempts to understand possible future climates have often drawn on palaeo-climate reconstructions to provide analogues for future climate conditions (Lamb 1987; Overpeck *et al.* 2005; Turney and Jones 2010). Current projections of future climates, however, suggest 'analogues' will need to be found in the increasingly distant past. Recent predictions of warming have global climates warmer than the Holocene altithermal (+0.5°C) by 2100 under the lowest greenhouse gas concentration (Representative Concentration Pathway 2.6, the greenhouse gas concentration trajectory representing stringent mitigation of emissions) (0.3–1.7; mean = 1.0) (IPCC 2014). Indeed, given that the global climate has warmed by 0.65–1.06°C since 1880, it would seem that the global temperatures of the Holocene altithermal may already have been exceeded by current Anthropocene conditions.

In view of this, the last interglacial period has been presented as a possible global climate analogue for future climate conditions (Overpeck *et al.* 2005; Turney and Jones 2010; Kopp *et al.* 2013). Although estimates of global temperatures during the last interglacial period are more in line with mid-range scenarios for future climate, there remains substantial uncertainty as to how comparable conditions might be. The data on which climate reconstructions of the last interglacial period are based are far more sparse than those that reconstructions of the early Holocene altithermal or the medieval warm period are based. Moreover, comparison of conditions in relation to orbital forcing during the last interglacial period and a future warmer earth today suggests that the last interglacial period was characterised by stronger seasonality due to a more eccentric orbit (Kopp *et al.* 2013). More pessimistic predictions that take global temperatures well beyond last interglacial period conditions (+4°C by 2100) must also be considered.

The potential for 'runaway' global climate warming of +4°C, notwithstanding, perhaps, the greatest uncertainty regarding the capacity of vegetation to adapt to future warmer climates, stems from interactions with other anthropogenic stressors (Mackey *et al.* 2008; Palmer *et al.* 2008, 2009; Capon *et al.* 2013). Riverine vegetation is subject to a range of pressures, as outlined elsewhere in this volume, that are likely to both amplify the effects of anthropogenic climate change and aggravate existing stressors. Additionally, many riparian zones are highly fragmented as a result of agriculture and human settlements and infrastructure, limiting the ability of species to move in response to climate change. Indeed, riverine landscapes are often considered to be among the most modified and degraded in the world (Tockner and Stanford 2002) and, consequently, their inherent adaptive capacity is likely to be significantly hampered by the many anthropogenic pressures to which they are subjected (Capon *et al.* 2013; Capon and Bunn 2015).

In particular, water resource development has already resulted in reduced water availability and altered seasonality of flooding for floodplain vegetation. The effects of these hydrological changes and floodplain modifications on floodplain vegetation have been detailed elsewhere in this volume (see Chapter 14). In essence, however, these changes have made large areas of floodplain unsuitable as habitat for the former dominant aquatic and amphibious species and species associations, and facilitated the expansion of terrestrial taxa, including invasive species, into floodplains and wetlands. In the Murray–Darling Basin, for example, it has been estimated that 75% of the wetlands of the Gwydir River floodplain have been lost (Wettin *et al.* 1994). Similarly, 76% of the floodplain vegetation of the lower Murrumbidgee is estimated to have been lost or severely degraded as a result of water resource development (Kingsford and Thomas 2004). Compounding these effects, physical modifications to floodplains associated with water resource development, along with other agricultural developments, have also led to substantial loss and modification of habitat through vegetation clearance, secondary salinisation, acidification, grazing and trampling by stock, and erosion and deposition of sediment. Riverine vegetation communities are further affected by invasive species introduced inadvertently or deliberately by people, the establishment and spread of which is facilitated by the disturbance and habitat fragmentation associated with the aforementioned stressors as well as reduced flood frequency from water resource development (see Chapter 18). The vegetation of riverine landscapes is thus both heavily fragmented and subject to a wide range of interacting stressors. In these circumstances, the capacity of riverine plants and vegetation communities to autonomously adapt, either in situ or by shifting or contracting distributions, is likely to be severely compromised (Capon *et al.* 2013).

## MANAGED ADAPTATION

Given the constraints on the adaptive capacity of natural riverine ecosystems outlined above, protecting and maintaining the functions and values of riverine vegetation in the face of climate change are likely to depend on planned and management adaptation by people. Aside from 'do nothing' or 'wait and see' approaches (e.g. Reuveny 2007; Glenk and Fischer 2010), the three main strategies for responding to climate change in the conservation and natural resources management sectors can be broadly categorised as (1) 'hold the line', (2) 'directed change' and (3) 'resilience-building'. These strategies are not necessarily mutually exclusive and may be applied simultaneously with respect to different goals or locations, or as part of a broader strategy over time.

Adaptation actions associated with the first category, a 'hold the line' approach to climate change, broadly include those that attempt to control climate change impacts, such as sea level rise, or mitigate their effects (e.g. drought or the movement of invasive species) in an attempt to maintain current conditions. Hard engineering options such as dams, weirs, barrages or sea walls typically fall into this category, for example, the ongoing environmental works and measures program in the Murray–Darling Basin (Capon 2014). Changes to the operation of existing infrastructure (e.g. environmental water releases from storages to flood drought-affected wetlands) might also be considered in this category where the goals are to protect existing local populations (e.g. stressed river red gums). Other actions that might 'hold the line' with respect to riverine vegetation could include planting using only local seed or seedlings to maintain current vegetation patterns in the face of changed climatic conditions.

The second main strategy for adaptation of conservation and natural resources management involves directing climate in line with particular goals such as the preservation of species or the maintenance of specific ecosystem services (Capon and Bunn 2015). Many ecological engineering strategies, for example, aim to maintain certain ecosystem functions at the expense of vegetation composition or structure (e.g. planting of non-native species to control bank erosion). Other actions may even seek to enhance current ecological functions to compensate for predicted impacts. Dense planting of fast-growing, high-shade species in riparian zones (i.e. 'over-restoration'), for instance, has been proposed as a means of protecting instream biota from rising temperatures (Davies 2010). Species translocation is another example of directed change aimed at conserving biodiversity rather than function.

Finally, resilience-building actions encompass those that aim to increase or restore the natural adaptive capacity of species, ecosystems or landscapes such that they are better able to cope with climate change either through migration or in situ adjustments. Management of existing pressures and restoration activities come under this category, especially when these are planned and implemented with respect to the risks associated with projected climate change. Riparian restoration is likely to be particularly important in relation to managing climate change impacts across catchments more broadly (Seavy *et al.* 2009; Capon *et al.* 2013). Improving networks of protected areas and corridors between them is also an important strategy for building ecological resilience (Steffen *et al.* 2009), especially where reserves are prioritised in areas that provide climate refuges for species. Programs that maintain or promote genetic diversity may allow some species to evolve and adapt to future environmental conditions in situ. A more dramatic approach to resilience-building involves retreat, such as the removal of infrastructure to provide space for ecosystems to respond to climate change free from the constraints these impose (Capon *et al.* 2013). With respect to riverine ecosystems, retreat options therefore include the decommissioning and removal of barriers, for example, weirs or artificial levees (Stanley and Doyle 2003) or the rezoning of floodplains and coastal flats to limit development and allow rivers and estuarine ecosystems to also move in response to changing conditions.

As previously stated, these strategies are not mutually exclusive. Resilience-building, for instance, may be employed initially with hold the line type actions introduced at a later date when very high value assets become threatened. Alternatively, holding the line may be a temporary (i.e. band-aid) approach while other actions are executed to enable a shift to directed change or resilience-building at a later stage (Capon and Bunn 2015).

All adaptation options, as with any intervention, may be associated with unexpected and perverse outcomes. Resilience-building actions, however, are often viewed as being low-regret options with multiple and far-reaching benefits, many of which may occur regardless of the climate change trajectory that eventuates. In contrast, hold the line options may entail a greater risk of maladaptation due to having high opportunity costs and likelihood of mass failure during extreme events (Capon *et al.* 2013). Substantial safety margins, regular reviews and short lifespans are therefore recommended (Hallegatte 2009; Pittock and Hartmann 2011).

While on-ground adaptation actions necessarily occur at local scales, the scale of climate change (as well as that of many other pressures with which it interacts) necessitates catchment- and landscape-scale planning as well as cross-sectoral integration. Principles of systematic conservation planning may be useful in informing such complex planning at these scales across a range of possible scenarios and with the involvement of multiple stakeholders (Hermoso *et al.* 2012; Adame *et al.* 2015).

## CONCLUSIONS

The future of riverine vegetation in Australia is likely to be shaped by the manifestations of climate change. Impacts from changes in soil moisture, surface water supply, and groundwater levels are likely to be critically important due to the aridity of much of the continent. A high level of uncertainty, however, is associated with these effects as a result of the amplification of uncertainty in projections when translated into river flows, interactions and feedbacks, and outcomes of novel species combinations. The exact interplay of these factors in any one context will determine the direction and magnitude of change and may result in quite different outcomes for different areas. Although many of the projected changes to rainfall and runoff are relatively small compared with current variability and changes due to land use, water abstraction and river regulation (McMahon and Finlayson 2003; Jenkins *et al.* 2011), the risks posed by climate change should not be ignored. Indeed, it is the potential for strong interactions between the direct and indirect effects of climate change with non-climatic stressors that may pose the greatest threat to the conservation of riverine vegetation and the maintenance of the many critical functions it supports.

In general, Australian riverine landscapes are hydrologically variable and dynamic. Plants and vegetation communities persisting under these conditions may therefore be better able to cope with climate change than those from many other environments. At the same time, however, it may be difficult to detect climate change impacts that exceed the limits of the natural spatial and temporal variability that typifies vegetation of these habitats. There is a great risk that key vegetation processes such as recruitment and regeneration might be fundamentally altered by the time any impact is detected. 'Do nothing' and 'wait and see' approaches to managed adaptation therefore probably entail unacceptable levels of risk in this respect. Additionally, present-day Australian riverine vegetation is unlikely to be in equilibrium (even a dynamic equilibrium) with its contemporary environment and is probably still responding to the past 200 or so years of intensive land use and habitat modification. Despite these complexities, however, there is an urgent need to pre-empt climate change impacts on riverine vegetation and consider managed adaptation approaches in earnest now, especially for critical communities and landscapes, given the importance of riparian vegetation, both intrinsically and for the critical functions it provides.

## REFERENCES

Adame MF, Hermoso V, Perhans K, Lovelock CE, Herrera-Silveira JA (2015) Selecting cost-effective areas for restoration of ecosystem services. *Conservation Biology* **29**, 493–502. doi:10.1111/cobi.12391

Ainsworth EA, Long SP (2005) What have we learned from 15 years of free-air $CO_2$ enrichment (FACE)? A meta-analytic review of the responses of photosynthesis, canopy. *New Phytologist* **165**, 351–372. doi:10.1111/j.1469-8137.2004.01224.x

Andersen AN, Cook GD, Corbett LK, Douglas MM, Eager RW, Russell-Smith J, Setterfield SA, Williams RJ, Woinarski JCZ (2005) Fire frequency and biodiversity conservation in Australian tropical savannas: implications from the Kapalga fire experiment. *Austral Ecology* **30**, 155–167. doi:10.1111/j.1442-9993.2005.01441.x

Atwell BJ, Henery ML, Rogers GS, Seneweera SP, Treadwell M, Conroy JP (2007) Canopy development and hydraulic function in *Eucalyptus tereticornis* grown in drought in $CO_2$-enriched atmospheres. *Functional Plant Biology* **34**, 1137–1149. doi:10.1071/FP06338

Berry J, Bjorkman O (1980) Photosynthetic response and adaptation to temperature in higher-plants. *Annual Review of Plant Physiology and Plant Molecular Biology* **31**, 491–543. doi:10.1146/annurev.pp.31.060180.002423

Bhend J, Bathols J, Hennessy K (2012) *Climate Change Impacts on Snow in Victoria*. CSIRO Publishing, Melbourne.

Blanch SJ, Ganf GG, Walker KF (1999) Tolerance of riverine plants to flooding and exposure indicated by water regime. *Regulated Rivers: Research and Management* **15**, 43–62. doi:10.1002/(SICI)1099-1646(199901/06)15:1/3<43::AID-RRR535>3.0.CO;2-Q

Bowler JM (1978) Quaternary climate and tectonics in the evolution of the Riverine Plain, south-eastern Australia. In *Landform Evolution in Australasia*. (Eds JL Davies and MAJ Williams) pp. 70–112. Australian National University Press, Canberra.

Bowler DE, Mant R, Orr H, Hannah DM, Pullin AS (2012) What are the effects of wooded riparian zones on stream temperature? *Environmental Evidence* **1**, 3. doi:10.1186/2047-2382-1-3

Braganza K, Church JA (2011) Observations of global and Australian climate. In *Climate Change: Science and Solutions for Australia*. (Eds H Cleugh, M Stafford Smith, M Battaglis and P Graham) pp. 1–14. CSIRO Publishing, Collingwood.

Bren LJ (1992) Tree invasion of an intermittent wetland in relation to changes in the flooding frequency of the River Murray, Australia. *Australian Journal of Ecology* **17**, 395–408. doi:10.1111/j.1442-9993.1992.tb00822.x

Butcher PA, McDonald MW, Bell JC (2009) Congruence between environmental parameters, morphology and genetic structure in Australia's most widely distributed eucalypt, *Eucalyptus camaldulensis*. *Tree Genetics & Genomes* **5**, 189–210. doi:10.1007/s11295-008-0169-6

Butt N, Pollock LJ, McAlpine CA (2013) Eucalypts face increasing climate stress. *Ecology and Evolution* **3**, 5011–5022. doi:10.1002/ece3.873

Capon SJ (2003) Plant community responses to wetting and drying in a large arid floodplain. *River Research and Applications* **19**, 509–520. doi:10.1002/rra.730

Capon SJ (2005) Flood variability and spatial variation in plant community composition and structure on a large and floodplain. *Journal of Arid Environments* **60**, 283–302. doi:10.1016/j.jaridenv.2004.04.004

Capon S (2014) Adapting water resources management to drought and water scarcity in Australia's Murray–Darling Basin: impacts and legacies of 'The Big Dry'. In *Adaptation to Climate Change Through Water Resources Management: Capacity, Equity and Sustainability*. (Eds D Stucker and E Lopez-Gunn) pp. 110–133. Routledge, Milton Park, UK.

Capon SJ, Bunn SE (2015) Assessing climate change risks and prioritising adaptation options using a water ecosystem services approach. In *Water Ecosystem Services: a Global Perspective*. (Eds J Martin-Ortega, RC Ferrier, IJ Gordon and S Khan) pp. 17–25. Cambridge University Press, Cambridge, UK.

Capon SJ, Chambers LE, Mac Nally R, Naiman RJ, Davies P, Marshall N, Pittock J, Reid M, Capon T, Douglas M, Catford J, Baldwin DS, Stewardson M, Roberts J, Parsons M, Williams SE (2013) Riparian ecosystems in the 21st century:

hotspots for climate change adaptation? *Ecosystems* **16**, 359–381. doi:10.1007/s10021-013-9656-1

Catelotti K, Kingsford RT, Bino G, Bacon P (2015) Inundation requirements for persistence and recovery of river red gums (*Eucalyptus camaldulensis*) in semi-arid Australia. *Biological Conservation* **184**, 346–356. doi:10.1016/j.biocon.2015.02.014

Catford JA, Downes BJ, Gippel CJ, Vesk PA (2011) Flow regulation reduces native plant cover and facilitates exotic invasion in riparian wetlands. *Journal of Applied Ecology* **48**, 432–442. doi:10.1111/j.1365-2664.2010.01945.x

Catford JA, Naiman RJ, Chambers LE, Roberts J, Douglas M, Davies P (2013) Predicting novel riparian ecosystems in a changing climate. *Ecosystems* **16**, 382–400. doi:10.1007/s10021-012-9566-7

Catford JA, Morris WK, Vesk PA, Gippel CJ, Downes BJ (2014) Species and environmental characteristics point to flow regulation and drought as drivers of riparian plant invasion. *Diversity & Distributions* **20**, 1084–1096. doi:10.1111/ddi.12225

Cernusak LA, Ubierna N, Winter K, Holtum JAM, Marshall JD, Farquhar GD (2013) Environmental and physiological determinants of carbon isotope discrimination in terrestrial plants. *New Phytologist* **200**, 950–965. doi:10.1111/nph.12423

Chalmers AC, Erskine WD, Keene AF, Bush RT (2012) Relationship between vegetation, hydrology and fluvial landforms on an unregulated sand-bed stream in the Hunter Valley, Australia. *Austral Ecology* **37**, 193–203. doi:10.1111/j.1442-9993.2011.02262.x

Chiew FHS (2006) Estimation of rainfall elasticity of streamflow in Australia. *Hydrological Sciences Journal* **51**, 613–625. doi:10.1623/hysj.51.4.613

Chiew FHS, McMahon TA (2002) Modelling the impacts of climate change on Australian streamflow. *Hydrological Processes* **16**, 1235–1245. doi:10.1002/hyp.1059

Clark PU, Shakun JD, Baker PA, Bartlein PJ, Brewer S, Brook E, Carlson AE, Cheng H, Kaufman DS, Liu ZY, Marchitto TM, Mix AC, Morrill C, Otto-Bliesner BL, Pahnke K, Russell JM, Whitlock C, Adkins JF, Blois JL, Clark J, Colman SM, Curry WB, Flower BP, He F, Johnson TC, Lynch-Stieglitz J, Markgraf V, McManus J, Mitrovica JX, Moreno PI, Williams JW (2012) Global climate evolution during the last deglaciation. *Proceedings of the National Academy of Sciences of the United States of America* **109**, E1134–E1142. doi:10.1073/pnas.1116619109

Corlett RT, Westcott DA (2013) Will plant movements keep up with climate change? *Trends in Ecology & Evolution* **28**, 482–488. doi:10.1016/j.tree.2013.04.003

CSIRO and Bureau of Meteorology (2015) 'Climate change in Australia information for Australia's natural resource management regions: technical report.' CSIRO and Bureau of Meteorology, Canberra.

Cunningham SC, Read J (2002) Comparison of temperate and tropical rainforest tree species: photosynthetic responses to growth temperature. *Oecologia* **133**, 112–119. doi:10.1007/s00442-002-1034-1

Cunningham SC, Read J (2003) Do temperate rainforest trees have a greater ability to acclimate to changing temperatures than tropical rainforest trees? *New Phytologist* **157**, 55–64. doi:10.1046/j.1469-8137.2003.00652.x

Cunningham SC, Thomson JR, Mac Nally R, Read J, Baker PJ (2011) Groundwater change forecasts widespread forest dieback across an extensive floodplain system. *Freshwater Biology* **56**, 1494–1508. doi:10.1111/j.1365-2427.2011.02585.x

Curtis EM, Knight CA, Petrou K, Leigh A (2014) A comparative analysis of photosynthetic recovery from thermal stress: a desert plant case study. *Oecologia* **175**, 1051–1061. doi:10.1007/s00442-014-2988-5

Davies PM (2010) Climate change implications for river restoration in global biodiversity hotspots. *Restoration Ecology* **18**, 261–268. doi:10.1111/j.1526-100X.2009.00648.x

Davis J, Pavlova A, Thompson R, Sunnucks P (2013) Evolutionary refugia and ecological refuges: key concepts for conserving Australian arid zone freshwater biodiversity under climate change. *Global Change Biology* **19**, 1970–1984. doi:10.1111/gcb.12203

De Deckker P (1986) What happened to the Australian aquatic biota 18 000 years ago? In *Limnology in Australia*. (Eds P de Deckker and WD Williams) pp. 487–496. CSIRO Publishing, Melbourne.

Di Nitto D, Neukermans G, Koedam N, Defever H, Pattyn F, Kairo JG, Dahdouh-Guebas F (2014) Mangroves facing climate change: landward migration potential in response to projected scenarios of sea level rise. *Biogeosciences* **11**, 857–871. doi:10.5194/bg-11-857-2014

Doody TM, Benger SN, Pritchard JL, Overton IC (2014) Ecological response of *Eucalyptus camaldulensis* (river red gum) to extended drought and flooding along the River Murray, South Australia (1997–2011) and implications for environmental flow management. *Marine and Freshwater Research* **65**, 1082–1093. doi:10.1071/MF13247

Drake BG (2014) Rising sea level, temperature, and precipitation impact plant and ecosystem responses to elevated $CO_2$ on a Chesapeake Bay wetland: review of a 28-year study. *Global Change Biology* **20**, 3329–3343. doi:10.1111/gcb.12631

Duursma RA, Barton CVM, Eamus D, Medlyn BE, Ellsworth DS, Forster MA, Tissue DT, Linder S, McMurtrie RE (2011) Rooting depth explains $[CO_2] \times$ drought interaction in *Eucalyptus saligna*. *Tree Physiology* **31**, 922–931. doi:10.1093/treephys/tpr030

Glenk K, Fischer A (2010) Insurance, prevention or just wait and see? Public preferences for water management strategies in the context of climate change. *Ecological Economics* **69**, 2279–2291. doi:10.1016/j.ecolecon.2010.06.022

Good MK, Clarke PJ, Price JN, Reid N (2014) Seasonality and facilitation drive tree establishment in a semi-arid floodplain savanna. *Oecologia* **175**, 261–271. doi:10.1007/s00442-014-2886-x

Goudie AS (2006) Global warming and fluvial geomorphology. *Geomorphology* **79**, 384–394. doi:10.1016/j.geomorph.2006.06.023

Greet J, Webb JA, Downes BJ (2011) Flow variability maintains the structure and composition of in-channel riparian vegetation. *Freshwater Biology* **56**, 2514–2528. doi:10.1111/j.1365-2427.2011.02676.x

Grindrod J, Moss P, Van der Kaars S (1999) Late Quaternary cycles of mangrove development and decline on the north Australian continental shelf. *Journal of Quaternary Science* **14**, 465–470. doi:10.1002/(SICI)1099-1417(199908)14:5<465::AID-JQS473>3.0.CO;2-E

Groom PK, Froend RH, Mattiske EM, Gurner RP (2001) Long-term changes in vigour and distribution of *Banksia* and *Melaleuca* overstorey species on the Swan Coastal Plain. *Journal of the Royal Society of Western Australia* **84**, 63–69.

Hallegatte S (2009) Strategies to adapt to an uncertain climate change. *Global Environmental Change – Human and Policy Dimensions* **19**, 240–247. doi:10.1016/j.gloenvcha.2008.12.003

Hampe A, Rodríguez-Sánchez F, Dobrowski S, Hu FS, Gavin DG (2013) Climate refugia: from the last glacial maximum to the twenty-first century. *New Phytologist* **197**, 16–18. doi:10.1111/nph.12059

Hannah DM, Webb BW, Nobilis F (2008) River and stream temperature: dynamics, processes, models and implications. *Hydrological Processes* **22**, 899–901. doi:10.1002/hyp.6997

Harrison PA, Worth JRP, Vaillancourt RE, Potts BM (2013) *Predicting the Distribution of* Eucalyptus ovata *Under Past, Current, and Future Climates.* Graduate Research Conference Sharing Excellence in Research, Hobart.

Helgerson OT (1990) Heat damage in tree seedlings and its prevention. *New Forests* **3**, 333–358. doi:10.1007/BF00030044

Hermoso V, Pantus F, Olley J, Linke S, Mugodo J, Lea P (2012) Systematic planning for river rehabilitation: integrating multiple ecological and economic objectives in complex decisions. *Freshwater Biology* **57**, 1–9. doi:10.1111/j.1365-2427.2011.02693.x

Horner GJ, Baker PJ, Mac Nally R, Cunningham SC, Thomson JR, Hamilton F (2009) Mortality of developing floodplain forests subjected to a drying climate and water extraction. *Global*

*Change Biology* **15**, 2176–2186. doi:10.1111/j.1365-2486.2009.01915.x

Hughes L (2003) Climate change and Australia: trends, projections and impacts. *Austral Ecology* **28**, 423–443. doi:10.1046/j.1442-9993.2003.01300.x

Hughes L, Cawsey EM, Westoby M (1996) Climatic range sizes of Eucalyptus species in relation to future climate change. *Global Ecology and Biogeography Letters* **5**, 23–29. doi:10.2307/2997467

IPCC (2014) 'Climatic change 2014: synthesis report.' Contribution of Working Groups I, II and III to the Fifth Assessment Report of the Intergovernmental Panel on Climate Change, IPCC, Geneva, Switzerland.

James C, vanDerWal J, Capon S, Hodgson L, Waltham N, Ward D, Anderson B, Pearson R (2013) *Identifying Climate Refuges for Freshwater Biodiversity Across Australia*. National Climate Change Adaptation Research Facility, Gold Coast, Queensland.

James CS, Capon SJ, Quinn GP (2015) Nurse plant effects of a dominant shrub (*Duma florulenta*) on understorey vegetation in a large, semi-arid wetland in relation to flood frequency and drying. *Journal of Vegetation Science* **26,** 985–994. doi:10.1111/jvs.12297

Jenkins KM, Kingsford RT, Closs GP, Wolfenden BJ, Matthaei CD, Hay SE (2011) Climate change and freshwater ecosystems in Oceania: an assessment of vulnerability and adaptation opportunities. Pacific Conservation Biology 17, 201–219.

Karschon R, Pinchas L (1971) Variations in heat resistance of ecotypes of *Eucalyptus camaldulensis* Dehn and their significance. *Australian Journal of Botany* **19**, 261–272. doi:10.1071/BT9710261

Kath J, Reardon-Smith K, Le Brocque AF, Dyer FJ, Dafny E, Fritz L, Batterham M (2014) Groundwater decline and tree change in floodplain landscapes: identifying non-linear threshold responses in canopy condition. *Global Ecology and Conservation* **2**, 148–160. doi:10.1016/j.gecco.2014.09.002

Kingsford RT, Thomas RF (2004) Destruction of wetlands and waterbird populations by dams and irrigation on the Murrumbidgee River in arid Australia. *Environmental Management* **34**, 383–396. doi:10.1007/s00267-004-0250-3

Kirwan ML, Megonigal JP (2013) Tidal wetland stability in the face of human impacts and sea-level rise. *Nature* **504**, 53–60. doi:10.1038/nature12856

Kopp RE, Simons FJ, Mitrovica JX, Maloof AC, Oppenheimer M (2013) A probabilistic assessment of sea level variations within the last interglacial stage. *Geophysical Journal International* **193**, 711–716. doi:10.1093/gji/ggt029

Kumara MP, Jayatissa LP, Krauss KW, Phillips DH, Huxham M (2010) High mangrove density enhances surface accretion, surface elevation change, and tree survival in coastal areas susceptible to sea-level rise. *Oecologia* **164**, 545–553. doi:10.1007/s00442-010-1705-2

Lamb PJ (1987) On the development of regional climatic scenarios for policy-oriented climatic-impact assessment. *Bulletin of the American Meteorological Society* **68**, 1116–1123. doi:10.1175/1520-0477(1987)068<1116:OTDORC>2.0.CO;2

Larcher W (2000) Temperature stress and survival ability of Mediterranean sclerophyllous plants. *Plant Biosystems* **134**, 279–295. doi:10.1080/11263500012331350455

Long SP (1991) Modification of the response of photosynthetic productivity to rising temperature by atmospheric $CO_2$ concentrations – has its importance been underestimated. *Plant, Cell & Environment* **14**, 729–739. doi:10.1111/j.1365-3040.1991.tb01439.x

Lovelock CE, Bennion V, Grinham A, Cahoon DR (2011) The role of surface and subsurface processes in keeping pace with sea level rise in intertidal wetlands of Moreton Bay, Queensland, Australia. *Ecosystems* **14**, 745–757. doi:10.1007/s10021-011-9443-9

Mackay SJ, Arthington AH, Kennard MJ, Pusey BJ (2003) Spatial variation in the distribution and abundance of submersed macrophytes in an Australian subtropical river. *Aquatic Botany* **77**, 169–186. doi:10.1016/S0304-3770(03)00103-7

Mackey BG, Watson JEM, Hope G, Gilmore S (2008) Climate change, biodiversity conservation, and the role of protected areas: an Australian perspective. *Biodiversity* **9**, 11–18. doi:10.1080/14888386.2008.9712902

McMahon TA, Finlayson BL (2003) Droughts and anti-droughts: the low flow hydrology of Australian rivers. *Freshwater Biology* **48**, 1147–1160. doi:10.1046/j.1365-2427.2003.01098.x

Merritt DM, Scott ML, Poff NL, Auble GT, Lytle DA (2010) Theory, methods and tools for determining environmental flows for riparian vegetation: riparian vegetation-flow response guilds. *Freshwater Biology* **55**, 206–225. doi:10.1111/j.1365-2427.2009.02206.x

Miller KA, Webb JA, de Little SC, Stewardson MJ (2013) Environmental flows can reduce the encroachment of terrestrial vegetation into river channels: a systematic literature review. *Environmental Management* **52**, 1202–1212. doi:10.1007/s00267-013-0147-0

Mokany K, Prasad S, Westcott DA (2014) Loss of frugivore seed dispersal services under climate change. *Nature Communications* **5**, 3971. doi:10.1038/ncomms4971

Mulrennan ME, Woodroffe CD (1998) Saltwater intrusion into the coastal plains of the lower Mary River, Northern Territory, Australia. *Journal of Environmental Management* **54**, 169–188. doi:10.1006/jema.1998.0229

Nilsson C, Svedmark M (2002) Basic principles and ecological consequences of changing water regimes: riparian plant communities. *Environmental Management* **30**, 468–480. doi:10.1007/s00267-002-2735-2

Ooi MKJ, Auld TD, Denham AJ (2012) Projected soil temperature increase and seed dormancy response along an altitudinal gradient: implications for seed bank persistence under climate change. *Plant and Soil* **353**, 289–303. doi:10.1007/s11104-011-1032-3

Overpeck J, Cole J, Bartlein P (2005) A 'paleoperspective' on climate variability and change. In *Climate Change and Biodiversity*. (Eds TE Lovejoy and L Hannah) pp. 91–108. Yale University Press, New Haven, CT.

Page KJ, Kemp J, Nanson GC (2009) Late Quaternary evolution of riverine plain paleochannels, southeastern Australia. *Australian Journal of Earth Sciences* **56**, S19–S33. doi:10.1080/08120090902870772

Palmer MA, Liermann CAR, Nilsson C, Florke M, Alcamo J, Lake PS, Bond N (2008) Climate change and the world's river basins: anticipating management options. *Frontiers in Ecology and the Environment* **6**, 81–89. doi:10.1890/060148

Palmer MA, Lettenmaier DP, Poff NL, Postel SL, Richter B, Warner R (2009) Climate change and river ecosystems: protection and adaptation options. *Environmental Management* **44**, 1053–1068. doi:10.1007/s00267-009-9329-1

Perry LG, Andersen DC, Reynolds LV, Nelson SM, Shafroth PB (2012) Vulnerability of riparian ecosystems to elevated $CO_2$ and climate change in arid and semiarid western North America. *Global Change Biology* **18**, 821–842. doi:10.1111/j.1365-2486.2011.02588.x

Perry LG, Shafroth PB, Blumenthal DM, Morgan JA, LeCain DR (2013) Elevated $CO_2$ does not offset greater water stress predicted under climate change for native and exotic riparian plants. *New Phytologist* **197**, 532–543. doi:10.1111/nph.12030

Petheram C, Rustomji P, McVicar TR, Cai WJ, Chiew FHS, Vleeshouwer J, Van Niel TG, Li LT, Cresswell RG, Donohue RJ, Teng J, Perraud JM (2012) Estimating the impact of projected climate change on runoff across the tropical savannas and semiarid rangelands of northern Australia. *Journal of Hydrometeorology* **13**, 483–503. doi:10.1175/JHM-D-11-062.1

Petherick L, Bostock H, Cohen TJ, Fitzsimmons K, Tibby J, Fletcher MS, Moss P, Reeves J, Mooney S, Barrows T, Kemp J, Jansen J, Nanson G, Dosseto A (2013) Climatic records over the past 30 ka from temperate Australia – a synthesis from the Oz-INTIMATE workgroup. *Quaternary Science Reviews* **74**, 58–77. doi:10.1016/j.quascirev.2012.12.012

Pettit NE, Froend RH, Davies PM (2001) Identifying the natural flow regime and the relationship with riparian vegetation for two contrasting western Australian rivers. *Regulated Rivers: Research and Management* **17**, 201–215. doi:10.1002/rrr.624

Pickering CM, Armstrong T (2003) The potential impact of climatic change on plant communities in the Kosciuszko alpine zone. *The Victorian Naturalist* **120**, 15–24.

Pittock J, Hartmann J (2011) Taking a second look: climate change, periodic relicensing and improved management of dams. *Marine and Freshwater Research* **62**, 312–320. doi:10.1071/MF09302

Polley HW, Tischler CR, Johnson HB, Derner JD (2002) Growth rate and survivorship of drought: $CO_2$ effects on the presumed tradeoff in seedlings of five woody legumes. *Tree Physiology* **22**, 383–391. doi:10.1093/treephys/22.6.383

Puckridge JT, Sheldon F, Walker KF, Boulton AJ (1998) Flow variability and the ecology of large rivers. *Marine and Freshwater Research* **49**, 55–72. doi:10.1071/MF94161

Radford IJ, Grice AC, Abbott BN, Nicholas DM, Whiteman L (2008) Impacts of changed fire regimes on tropical riparian vegetation invaded by an exotic vine. *Austral Ecology* **33**, 151–167. doi:10.1111/j.1442-9993.2007.01803.x

Raupach M, Fraser P (2011) Climate and greenhouse gases. In *Climate Change: Science and Solutions for Australia.* (Eds H Cleugh, M Stafford Smith, M Battaglis and P Graham) pp. 15–34. CSIRO Publishing, Melbourne.

Raven JA, Lucas WJ (1985) Energy costs of carbon acquisition. In *Inorganic Carbon Uptake by Aquatic Photosynthetic Organisms. Proceedings of an International Workshop on Bicarbonate Use in Photosynthesis.* August 18–22. (Eds WJ Lucas and JA Berry) pp. 305–324. The American Society of Plant Physiologists, Rockville, MD.

Reeves JM, Barrows TT, Cohen TJ, Kiem AS, Bostock HC, Fitzsimmons KE, Jansen JD, Kemp J, Krause C, Petherick L, Phipps SJ, OZ-INTIMATE Members (2013) Climate variability over the last 35 000 years recorded in marine and terrestrial archives in the Australian region: an OZ-INTIMATE compilation. *Quaternary Science Reviews* **74**, 21–34. doi:10.1016/j.quascirev.2013.01.001

Reuveny R (2007) Climate change-induced migration and violent conflict. *Political Geography* **26**, 656–673. doi:10.1016/j.polgeo.2007.05.001

Sa ardavut P, Shepherd KR, Awe JO (1984) Effects of air and soil temperatures on growth of provenances of *Eucalyptus camaldulensis* Dehnh. seedlings. *Australian Forest Research* **14**, 57–66.

Sanders CJ, Smoak JM, Waters MN, Sanders LM, Brandini N, Patchineelam SR (2012) Organic matter content and particle size modifications in mangrove sediments as responses to sea level rise. *Marine Environmental Research* **77**, 150–155. doi:10.1016/j.marenvres.2012.02.004

Santamaría L, vanVierssen W (1997) Photosynthetic temperature responses of fresh- and brackish-water macrophytes: a review. *Aquatic Botany* **58**, 135–150. doi:10.1016/S0304-3770(97)00015-6

Saxe H, Cannell MGR, Johnsen B, Ryan MG, Vourlitis G (2001) Tree and forest functioning in response to global warming. *New Phytologist* **149**, 369–399. doi:10.1046/j.1469-8137.2001.00057.x

Schile LM, Callaway JC, Morris JT, Stralberg D, Parker VT, Kelly M (2014) Modeling tidal marsh distribution with sea-level rise: evaluating the role of vegetation, sediment, and upland habitat in marsh resiliency. *PLoS ONE* **9**, e88760. doi:10.1371/journal.pone.0088760

Seavy NE, Gardali T, Golet GH, Griggs TF, Howell C, Kelsey R, Small S, Viers JH, Weigand J (2009) Why climate change makes riparian restoration more important than ever: recommendations for practice and research. *Ecological Research* **27**, 330–338. doi:10.3368/er.27.3.330

Shakun JD, Carlson AE (2010) A global perspective on last glacial maximum to Holocene climate change. *Quaternary Science Reviews* **29**, 1801–1816. doi:10.1016/j.quascirev.2010.03.016

Silva TA, Freitas MC, Andrade C, Taborda R, Freire P, Schmidt S, Antunes C (2013) Geomorphological response of the salt-marshes in the Tagus estuary to sea level rise. *Journal of Coastal Research* **65**, 582–587. doi:10.2112/SI65-099.1

Sloss CR, Murray-Wallace CV, Jones BG (2007) Holocene sea-level change on the southeast coast of Australia: a review. *The Holocene* **17**, 999–1014. doi:10.1177/0959683607082415

Stanley EH, Doyle MW (2003) Trading off: the ecological removal effects of dam removal. *Frontiers in Ecology and the Environment* **1**, 15–22. doi:10.1890/1540-9295(2003)001[0015:TOTEEO]2.0.CO;2

Steffen W, Canadell JG (2005) 'Carbon dioxide fertilisation and climate change policy.' Consultancy report prepared for the Australian Greenhouse Office. Department of the Environment and Heritage, Commonwealth of Australia, Canberra.

Steffen W, Burbidge AA, Hughes L, Kitching R, Lindenmayer D, Musgrave W, Stafford Smith M, Werner PA (2009) *Australia's Biodiversity and Climate Change*. CSIRO Publishing, Melbourne.

Stromberg JC, Lite SJ, Dixon MD (2010) Effects of stream flow patterns on riparian vegetation of a semiarid river: implications for a changing climate. *River Research and Applications* **26**, 712–729.

Tans P, Keeling R (2015) *Trends in Atmospheric Carbon Dioxide*. ESRL Global Monitoring Division – Global Greenhouse Gas Reference Network, National Oceanic and Atmospheric Administration, Scripps Institution of Oceanography, <http://www.esrl.noaa.gov/gmd/ccgg/trends/>.

Tockner K, Stanford JA (2002) Riverine flood plains: present state and future trends. *Environmental Conservation* **29**, 308–330. doi:10.1017/S037689290200022X

Turney CS, Jones RT (2010) Does the Agulhas current amplify global temperatures during super-interglacials? *Journal of Quaternary Science* **25**, 839–843. doi:10.1002/jqs.1423

Turton SM (2013) Tropical cyclones and forest dynamics under a changing climate: what are the long-term implications for tropical forest canopies in the cyclone belt? In *Treetops at Risk: Challenges of Global Canopy Ecology and Conservation*. (Eds M Lowman, D Soubadra and T Ganesh) pp. 105–111. Springer, New York, NY.

van Vliet MTH, Franssen WHP, Yearsley JR, Ludwig F, Haddeland I, Lettenmaier DP, Kabat P (2013) Global river discharge and water temperature under climate change. *Global Environmental Change* **23**, 450–464. doi:10.1016/j.gloenvcha.2012.11.002

Wassen MJ, de Boer HJ, Fleischer K, Rebel KT, Dekker SC (2013) Vegetation-mediated feedback in water, carbon, nitrogen and phosphorus cycles. *Landscape Ecology* **28**, 599–614. doi:10.1007/s10980-012-9843-z

Wettin P, Jacobs T, Johnston P, Ryan N (1994) Water sharing and floodplain wetlands in the Murray–Darling Basin. In *Murray–Darling Basin Floodplain Wetlands Management: Proceedings of the Floodplain Wetlands Management Workshop*. (Eds A Sharley and C Huggan) pp. 21–41. Murray–Darling Basin Commission, Canberra.

Whetton P (2011) Future Australian climate scenarios. In *Climate Change: Science and Solutions for Australia*. (Eds H Cleugh, M Stafford Smith, M Battaglis and P Graham) pp. 35–44. CSIRO Publishing, Melbourne.

Williams RJ, Bradstock RA, Cary GJ, Gill AM, Liedloff AC, Lucas C, Whelan RJ, Andersen AA, Bowman DJMS, Clarke P, Cook GJ, Hennessy K, York A (2009) 'Interactions between climate, fire regimes and biodiversity in Australia: a preliminary assessment.' Report to Australian Government, Department of Climate change and Department of Environment, Water, Heritage and the Arts, Canberra.

# INDEX

www.ingramcontent.com/pod-product-compliance
Lightning Source LLC
LaVergne TN
LVHW070846210926
848482LV00016B/879